Student Solutions Manual To Accompany

Chemical Principles

SEVENTH EDITION

Steven S. Zumdahl
Illinois State University

Prepared by

Thomas Hummel
University of Illinois

BROOKS/COLE
CENGAGE Learning·

Australia · Brazil · Japan · Korea · Mexico · Singapore · Spain · United Kingdom · United States

For product information and technology assistance, contact us at **Cengage Learning Customer & Sales Support, 1-800-354-9706**

For permission to use material from this text or product, submit all requests online at **www.cengage.com/permissions** Further permissions questions can be emailed to **permissionrequest@cengage.com**

ISBN-13: 978-1-133-10923-5
ISBN-10: 1-133-10923-3

Brooks/Cole
20 Davis Drive
Belmont, CA 94002-3098
USA

Cengage Learning is a leading provider of customized learning solutions with office locations around the globe, including Singapore, the United Kingdom, Australia, Mexico, Brazil, and Japan. Locate your local office at: **www.cengage.com/global**

Cengage Learning products are represented in Canada by Nelson Education, Ltd.

To learn more about Brooks/Cole, visit **www.cengage.com/brookscole**

Purchase any of our products at your local college store or at our preferred online store **www.cengagebrain.com**

Printed in the United States of America
1 2 3 4 5 6 7 15 14 13 12 11

TABLE OF CONTENTS

TO THE STUDENT: HOW TO USE THIS GUIDE

Solutions to odd-numbered chapter exercises are in this manual. This "Solutions Guide" can be a valuable resource if you use it properly. The way <u>NOT</u> to use it is to look at an exercise in the book and then immediately check the solution, often saying to yourself, "That's easy, I can do it." Developing problem solving skills takes practice. Don't look up a solution to a problem until you have tried to work it on your own. If you are completely stuck, see if you can find a similar problem in the Sample Exercises in the chapter. Only look up the solution as a last resort. If you do this for a problem, look for a similar problem in the end of chapter exercises and try working it. The more problems you do, the easier chemistry becomes. It is also in your self interest to try to work as many problems as possible. Most exams that you will take in chemistry will involve a lot of problem solving. If you have worked several problems similar to the ones on an exam, you will do much better than if the exam is the first time you try to solve a particular type of problem. No matter how much you read and study the text, or how well you think you understand the material, you don't really understand it until you have taken the information in the text and applied the principles to problem solving. You will make mistakes, but the good students learn from their mistakes.

In this manual we have worked problems as in the textbook. We have shown intermediate answers to the correct number of significant figures and used the rounded answer in later calculations. Thus, some of your answers may differ slightly from ours. When we have not followed this convention, we have usually noted this in the solution. The most common exception is when working with the natural logarithm (ln) function, where we usually carried extra significant figures in order to reduce round-off error. In addition, we tried to use constants and conversion factors reported to at least one more significant figure as compared to numbers given in the problem. For some problems, this required the use of more precise atomic masses for H, C, N, and O as given in Chapter 3. This practice of carrying one extra significant figure in constants helps minimize round-off error.

TJH
SSZ

v

CHAPTER 2

ATOMS, MOLECULES, AND IONS

Development of the Atomic Theory

19. From Avogadro's hypothesis (law), volume ratios are equal to molecule ratios at constant temperature and pressure. Therefore, we can write a balanced equation using the volume data, $Cl_2 + 5 F_2 \rightarrow 2 X$. Two molecules of X contain 10 atoms of F and two atoms of Cl. The formula of X is ClF_5 for a balanced equation.

21. Avogadro's hypothesis (law) implies that volume ratios are equal to molecule ratios at constant temperature and pressure. Here, 1 volume of N_2 reacts with 3 volumes of H_2 to produce 2 volumes of the gaseous product or in terms of molecule ratios:

    $$1 N_2 + 3 H_2 \rightarrow 2 \text{ product}$$

    In order for the equation to be balanced, the product must be NH_3.

23. Hydrazine: 1.44×10^{-1} g H/g N; ammonia: 2.16×10^{-1} g H/g N; hydrogen azide: 2.40×10^{-2} g H/g N. Let's try all of the ratios:

    $$\frac{0.144}{0.0240} = 6.00; \quad \frac{0.216}{0.0240} = 9.00; \quad \frac{0.0240}{0.0240} = 1.00; \quad \frac{0.216}{0.144} = 1.50 = \frac{3}{2}$$

    All the masses of hydrogen in these three compounds can be expressed as simple whole-number ratios. The g H/g N in hydrazine, ammonia, and hydrogen azide are in the ratios $6 : 9 : 1$.

25. To get the atomic mass of H to be 1.00, we divide the mass that reacts with 1.00 g of oxygen by 0.126, that is, $0.126/0.126 = 1.00$. To get Na, Mg, and O on the same scale, we do the same division.

    $$Na: \frac{2.875}{0.126} = 22.8; \quad Mg: \frac{1.500}{0.126} = 11.9; \quad O: \frac{1.00}{0.126} = 7.94$$

	H	O	Na	Mg
Relative value	1.00	7.94	22.8	11.9
Accepted value	1.0079	15.999	22.99	24.31

1

The atomic masses of O and Mg are incorrect. The atomic masses of H and Na are close. Something must be wrong about the assumed formulas of the compounds. It turns out that the correct formulas are H_2O, Na_2O, and MgO. The smaller discrepancies result from the error in the assumed atomic mass of H.

The Nature of the Atom

27.   From section 2.6, the nucleus has "a diameter of about 10^{-13} cm" and the electrons "move about the nucleus at an average distance of about 10^{-8} cm from it." We will use these statements to help determine the densities. Density of hydrogen nucleus (contains one proton only):

$$V_{nucleus} = \frac{4}{3}\pi r^3 = \frac{4}{3}(3.14)(5 \times 10^{-14} \text{ cm})^3 = 5 \times 10^{-40} \text{ cm}^3$$

$$d = \text{density} = \frac{1.67 \times 10^{-24} \text{ g}}{5 \times 10^{-40} \text{ cm}^3} = 3 \times 10^{15} \text{ g/cm}^3$$

Density of H atom (contains one proton and one electron):

$$V_{atom} = \frac{4}{3}(3.14)(1 \times 10^{-8} \text{ cm})^3 = 4 \times 10^{-24} \text{ cm}^3$$

$$d = \frac{1.67 \times 10^{-24} \text{ g} + 9 \times 10^{-28} \text{ g}}{4 \times 10^{-24} \text{ cm}^3} = 0.4 \text{ g/cm}^3$$

29.   First, divide all charges by the smallest quantity, 6.40×10^{-13}.

$$\frac{2.56 \times 10^{-12}}{6.40 \times 10^{-13}} = 4.00; \quad \frac{7.68}{0.640} = 12.00; \quad \frac{3.84}{0.640} = 6.00$$

Because all charges are whole-number multiples of 6.40×10^{-13} zirkombs, the charge on one electron could be 6.40×10^{-13} zirkombs. However, 6.40×10^{-13} zirkombs could be the charge of two electrons (or three electrons, etc.). All one can conclude is that the charge of an electron is 6.40×10^{-13} zirkombs or an integer fraction of 6.40×10^{-13}.

31.   If the plum pudding model were correct (a diffuse positive charge with electrons scattered throughout), then α particles should have traveled through the thin foil with very minor deflections in their path. This was not the case because a few of the α particles were deflected at very large angles. Rutherford reasoned that the large deflections of these α particles could be caused only by a center of concentrated positive charge that contains most of the atom's mass (the nuclear model of the atom).

Elements, Ions, and the Periodic Table

33.   The atomic number of an element is equal to the number of protons in the nucleus of an atom of that element. The mass number is the sum of the number of protons plus neutrons in the nucleus. The atomic mass is the actual mass of a particular isotope (including electrons). As is discussed in Chapter 3, the average mass of an atom is taken from a measurement made on a large number of atoms. The average atomic mass value is listed in the periodic table.

35.   a.   The noble gases are He, Ne, Ar, Kr, Xe, and Rn (helium, neon, argon, krypton, xenon, and radon). Radon has only radioactive isotopes. In the periodic table, the whole number enclosed in parentheses is the mass number of the longest-lived isotope of the element.

       b.   promethium (Pm) and technetium (Tc)

37.   Use the periodic table to identify the elements.

       a.   Cl; halogen   b.   Be; alkaline earth metal

       c.   Eu; lanthanide metal   d.   Hf; transition metal

       e.   He; noble gas   f.   U; actinide metal

       g.   Cs; alkali metal

39.   For lighter, stable isotopes, the number of protons in the nucleus is about equal to the number of neutrons. When the number of protons and neutrons is equal to each other, the mass number (protons + neutrons) will be twice the atomic number (protons). Therefore, for lighter isotopes, the ratio of the mass number to the atomic number is close to 2. For example, consider ^{28}Si, which has 14 protons and $(28 - 14 =)$ 14 neutrons. Here, the mass number to atomic number ratio is $28/14 = 2.0$. For heavier isotopes, there are more neutrons than protons in the nucleus. Therefore, the ratio of the mass number to the atomic number increases steadily upward from 2 as the isotopes get heavier and heavier. For example, ^{238}U has 92 protons and $(238 - 92 =)$ 146 neutrons. The ratio of the mass number to the atomic number for ^{238}U is $238/92 = 2.6$.

41.   a. $^{24}_{12}Mg$: 12 protons, 12 neutrons, 12 electrons

       b. $^{24}_{12}Mg^{2+}$: 12 p, 12 n, 10 e   c. $^{59}_{27}Co^{2+}$: 27 p, 32 n, 25 e

       d. $^{59}_{27}Co^{3+}$: 27 p, 32 n, 24 e   e. $^{59}_{27}Co$: 27 p, 32 n, 27 e

       f. $^{79}_{34}Se$: 34 p, 45 n, 34 e   g. $^{79}_{34}Se^{2-}$: 34 p, 45 n, 36 e

       h. $^{63}_{28}Ni$: 28 p, 35 n, 28 e   i. $^{59}_{28}Ni^{2+}$: 28 p, 31 n, 26 e

43.   Atomic number = 63 (Eu); net charge = $+63 - 60 = 3+$; mass number = $63 + 88 = 151$; symbol: $^{151}_{63}Eu^{3+}$

Atomic number = 50 (Sn);   mass number = 50 + 68 = 118;   net charge = +50 − 48 = 2+;
symbol: $^{118}_{50}Sn^{2+}$.

45.   In ionic compounds, metals lose electrons to form cations, and nonmetals gain electrons to form anions. Group 1A, 2A, and 3A metals form stable 1+, 2+, and 3+ charged cations, respectively. Group 5A, 6A, and 7A nonmetals form 3−, 2−, and 1− charged anions, respectively.

   a.   Lose 2 e$^−$ to form Ra^{2+}.   b.   Lose 3 e$^−$ to form In^{3+}.   c.   Gain 3 e$^−$ to form P$^{3−}$.

   d.   Gain 2 e$^−$ to form Te$^{2−}$.   e.   Gain 1 e$^−$ to form Br$^−$.   f.   Lose 1 e$^−$ to form Rb$^+$.

Nomenclature

47.   AlCl$_3$, aluminum chloride; CrCl$_3$, chromium(III) chloride; ICl$_3$, iodine trichloride; AlCl$_3$ and CrCl$_3$ are ionic compounds following the rules for naming ionic compounds. The major difference is that CrCl$_3$ contains a transition metal (Cr) that generally exhibits two or more stable charges when in ionic compounds. We need to indicate which charged ion we have in the compound. This is generally true whenever the metal in the ionic compound is a transition metal. ICl$_3$ is made from only nonmetals and is a covalent compound. Predicting formulas for covalent compounds is extremely difficult. Because of this, we need to indicate the number of each nonmetal in the binary covalent compound. The exception is when there is only one of the first species present in the formula; when this is the case, mono- is not used (it is assumed).

49.   a.   sulfur difluoride   b.   dinitrogen tetroxide

      c.   iodine trichloride   d.   tetraphosphorus hexoxide

51.   a.   copper(I) iodide   b.   copper(II) iodide   c.   cobalt(II) iodide

      d.   sodium carbonate   e.   sodium hydrogen carbonate or sodium bicarbonate

      f.   tetrasulfur tetranitride   g.   selenium tetrabromide   h.   sodium hypochlorite

      i.   barium chromate   j.   ammonium nitrate

53.   a.   SO$_2$ b.   SO$_3$ c.   Na$_2$SO$_3$ d.   KHSO$_3$

      e.   Li$_3$N   f.   Cr$_2$(CO$_3$)$_3$ g.   Cr(C$_2$H$_3$O$_2$)$_2$ h.   SnF$_4$

      i.   NH$_4$HSO$_4$: composed of NH$_4^+$ and HSO$_4^−$ ions

      j.   (NH$_4$)$_2$HPO$_4$ k.   KClO$_4$ l.   NaH

      m.   HBrO   n.   HBr

55.   a.   Pb(C$_2$H$_3$O$_2$)$_2$; lead(II) acetate   b.   CuSO$_4$; copper(II) sulfate

      c.   CaO; calcium oxide   d.   MgSO$_4$; magnesium sulfate

      e.   Mg(OH)$_2$; magnesium hydroxide   f.   CaSO$_4$; calcium sulfate

      g.   N$_2$O; dinitrogen monoxide or nitrous oxide (common name)

57.   a. nitric acid, HNO_3 b. perchloric acid, $HClO_4$ c. acetic acid, $HC_2H_3O_2$

       d. sulfuric acid, H_2SO_4 e. phosphoric acid, H_3PO_4

Additional Exercises

59.   The equation for the reaction between the elements of sodium and chlorine is $2\ Na(s) + Cl_2(g)$ $\rightarrow$ $2\ NaCl(s)$. The sodium reactant exists as singular sodium atoms packed together very tightly and in a very organized fashion. This type of packing of atoms represents the solid phase. The chlorine reactant exists as Cl_2 molecules. In the picture of chlorine, there is a lot of empty space present. This only occurs in the gaseous phase. When sodium and chlorine react, the ionic compound NaCl is the product. NaCl exists as separate Na^+ and Cl^- ions. Because the ions are packed very closely together and are packed in a very organized fashion, NaCl is depicted in the solid phase.

61.   From the law of definite proportions, a given compound always contains exactly the same proportion of elements by mass. The first sample of chloroform has a total mass of 12.0 g C + 106.4 g Cl + 1.01 g H = 119.41 g (carrying extra significant figures). The mass percent of carbon in this sample of chloroform is:

$$\frac{12.0\ g\ C}{119.41\ g\ total} \times 100 = 10.05\%\ C\ by\ mass$$

       From the law of definite proportions, the second sample of chloroform must also contain 10.05% C by mass. Let x = mass of chloroform in the second sample:

$$\frac{30.0\ g\ C}{x} \times 100 = 10.05, \quad x = 299\ g\ chloroform$$

63.   From the Na_2X formula, X has a 2– charge. Because 36 electrons are present, X has 34 protons, $79 - 34 = 45$ neutrons, and is selenium.

       a. True. Nonmetals bond together using covalent bonds and are called covalent compounds.

       b. False. The isotope has 34 protons.

       c. False. The isotope has 45 neutrons.

       d. False. The identity is selenium, Se.

65.   From the XBr_2 formula, the charge on element X is 2+. Therefore, the element has 88 protons, which identifies it as radium, Ra. $230 - 88 = 142$ neutrons.

67.   In the case of sulfur, SO_4^{2-} is sulfate, and SO_3^{2-} is sulfite. By analogy:

       SeO_4^{2-}: selenate; SeO_3^{2-}: selenite; TeO_4^{2-}: tellurate; TeO_3^{2-}: tellurite

69.   If the formula is InO, then one atomic mass of In would combine with one atomic mass of O, or:

$$\frac{A}{16.00} = \frac{4.784 \text{ g In}}{1.000 \text{ g O}}, \quad A = \text{atomic mass of In} = 76.54$$

If the formula is In_2O_3, then two times the atomic mass of In will combine with three times the atomic mass of O, or:

$$\frac{2A}{(3)16.00} = \frac{4.784 \text{ g In}}{1.000 \text{ g O}}, \quad A = \text{atomic mass of In} = 114.8$$

The latter number is the atomic mass of In used in the modern periodic table.

71.   The cation has 51 protons and 48 electrons. The number of protons corresponds to the atomic number. Thus this is element 51, antimony. There are 3 fewer electrons than protons. Therefore, the charge on the cation is 3+. The anion has one-third the number of protons of the cation which corresponds to 17 protons; this is element 17, chlorine. The number of electrons in this anion of chlorine is 17 + 1 = 18 electrons. The anion must have a charge of 1−.

The formula of the compound formed between Sb^{3+} and Cl^- is $SbCl_3$. The name of the compound is antimony(III) chloride. The Roman numeral is used to indicate the charge of Sb because the predicted charge is not obvious from the periodic table.

73.   Because this is a relatively small number of neutrons, the number of protons will be very close to the number of neutrons present. The heavier elements have significantly more neutrons than protons in their nuclei. Because this element forms anions, it is a nonmetal and will be a halogen because halogens form stable 1− charged ions in ionic compounds. From the halogens listed, chlorine, with an average atomic mass of 35.45, fits the data. The two isotopes are ^{35}Cl and ^{37}Cl, and the number of electrons in the 1− ion is 18. Note that because the atomic mass of chlorine listed in the periodic table is closer to 35 than 37, we can assume that ^{35}Cl is the more abundant isotope. This is discussed in Chapter 3.

Challenge Problems

75.   a. Both compounds have C_2H_6O as the formula. Because they have the same formula, their mass percent composition will be identical. However, these are different compounds with different properties because the atoms are bonded together differently. These compounds are called isomers of each other.

b. When wood burns, most of the solid material in wood is converted to gases, which escape. The gases produced are most likely CO_2 and H_2O.

c. The atom is not an indivisible particle but is instead composed of other smaller particles, for example, electrons, neutrons, and protons.

d. The two hydride samples contain different isotopes of either hydrogen and/or lithium. Although the compounds are composed of different isotopes, their properties are similar because different isotopes of the same element have similar properties (except, of course, their mass).

77.   Compound I: $\dfrac{14.0\ \text{g R}}{3.00\ \text{g Q}} = \dfrac{4.67\ \text{g R}}{1.00\ \text{g Q}}$;   Compound II: $\dfrac{7.00\ \text{g R}}{4.50\ \text{g Q}} = \dfrac{1.56\ \text{g R}}{1.00\ \text{g Q}}$

The ratio of the masses of R that combines with 1.00 g Q is $\dfrac{4.67}{1.56} = 2.99 \approx 3$.

As expected from the law of multiple proportions, this ratio is a small whole number.

Because compound I contains three times the mass of R per gram of Q as compared with compound II (RQ), the formula of compound I should be R_3Q.

79.   Avogadro proposed that equal volumes of gases (at constant temperature and pressure) contain equal numbers of molecules. In terms of balanced equations, Avogadro's hypothesis (law) implies that volume ratios will be identical to molecule ratios. Assuming one molecule of octane reacts, then 1 molecule of C_xH_y produces 8 molecules of CO_2 and 9 molecules of H_2O. $C_xH_y + n\ O_2 \rightarrow 8\ CO_2 + 9\ H_2O$. Because all the carbon in octane ends up as carbon in CO_2, octane must contain 8 atoms of C. Similarly, all hydrogen in octane ends up as hydrogen in H_2O, so one molecule of octane must contain $9 \times 2 = 18$ atoms of H. Octane formula = C_8H_{18} and the ratio of C:H = 8:18 or 4:9.

CHAPTER 3

STOICHIOMETRY

Atomic Masses and the Mass Spectrometer

23.   Average atomic mass = A = 0.0800(45.95269) + 0.0730(46.951764) + 0.7380(47.947947)

$$+ 0.0550(48.947841) + 0.0540(49.944792) = 47.88 \text{ amu}$$

This is element Ti (titanium).

25.   If silver is 51.82% ^{107}Ag, then the remainder is ^{109}Ag (48.18%). Determining the atomic mass (A) of ^{109}Ag:

$$107.868 = \frac{51.82(106.905) + 48.18(A)}{100}$$

$$10786.8 = 5540. + (48.18)A, \quad A = 108.9 \text{ amu} = \text{atomic mass of } ^{109}Ag$$

27. $186.207 = 0.6260(186.956) + 0.3740(A), \quad 186.207 - 117.0 = 0.3740(A)$

$$A = \frac{69.2}{0.3740} = 185 \text{ amu} \ (A = 184.95 \text{ amu without rounding to proper significant figures})$$

29.   There are three peaks in the mass spectrum, each 2 mass units apart. This is consistent with two isotopes, differing in mass by two mass units. The peak at 157.84 corresponds to a Br_2 molecule composed of two atoms of the lighter isotope. This isotope has mass equal to 157.84/2, or 78.92. This corresponds to ^{79}Br. The second isotope is ^{81}Br with mass equal to 161.84/2 = 80.92. The peaks in the mass spectrum correspond to $^{79}Br_2$, $^{79}Br^{81}Br$, and $^{81}Br_2$ in order of increasing mass. The intensities of the highest and lowest masses tell us the two isotopes are present at about equal abundance. The actual abundance is 50.68% ^{79}Br and 49.32% ^{81}Br.

31.   GaAs can be either $^{69}GaAs$ or $^{71}GaAs$. The mass spectrum for GaAs will have two peaks at 144 (= 69 + 75) and 146 (= 71 + 75) with intensities in the ratio of 60 : 40 or 3 : 2.

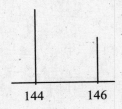

144   146

8

Ga_2As_2 can be $^{69}Ga_2As_2$, $^{69}Ga^{71}GaAs_2$, or $^{71}Ga_2As_2$. The mass spectrum will have three peaks at 288, 290, and 292 with intensities in the ratio of 36 : 48 : 16 or 9 : 12 : 4. We get this ratio from the following probability table:

	^{69}Ga (0.60)	^{71}Ga (0.40)
^{69}Ga (0.60)	0.36	0.24
^{71}Ga (0.40)	0.24	0.16

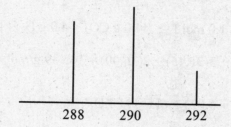

Moles and Molar Masses

33.   a. $9(12.011) + 8(1.0079) + 4(15.999) = 180.158 \text{ g/mol}$

   b. $500. \text{ mg} \times \dfrac{1 \text{ g}}{1000 \text{ mg}} \times \dfrac{1 \text{ mol}}{180.16 \text{ g}} = 2.78 \times 10^{-3} \text{ mol}$

   $2.78 \times 10^{-3} \text{ mol} \times \dfrac{6.022 \times 10^{23} \text{ molecules}}{\text{mol}} = 1.67 \times 10^{21} \text{ molecules}$

35.   a. $20.0 \text{ mg } C_8H_{10}N_4O_2 \times \dfrac{1 \text{ g}}{1000 \text{ mg}} \times \dfrac{1 \text{ mol}}{194.20 \text{ g}} = 1.03 \times 10^{-4} \text{ mol } C_8H_{10}N_4O_2$

   b. $2.72 \times 10^{21} \text{ molecules } C_2H_5OH \times \dfrac{1 \text{ mol}}{6.022 \times 10^{23} \text{ molecules}} = 4.52 \times 10^{-3} \text{ mol } C_2H_5OH$

   c. $1.50 \text{ g } CO_2 \times \dfrac{1 \text{ mol}}{44.01 \text{ g}} = 3.41 \times 10^{-2} \text{ mol } CO_2$

37. $4.0 \text{ g } H_2 \times \dfrac{1 \text{ mol } H_2}{2.016 \text{ g } H_2} \times \dfrac{2 \text{ mol } H}{1 \text{ mol } H_2} \times \dfrac{6.022 \times 10^{23} \text{ atoms } H}{1 \text{ mol } H} = 2.4 \times 10^{24} \text{ atoms}$

   $4.0 \text{ g } He \times \dfrac{1 \text{ mol } He}{4.003 \text{ g } He} \times \dfrac{6.022 \times 10^{23} \text{ atoms } He}{1 \text{ mol } He} = 6.0 \times 10^{23} \text{ atoms}$

$$1.0 \text{ mol } F_2 \times \frac{2 \text{ mol F}}{1 \text{ mol } F_2} \times \frac{6.022 \times 10^{23} \text{ atoms F}}{1 \text{ mol F}} = 1.2 \times 10^{24} \text{ atoms}$$

$$44.0 \text{ g } CO_2 \times \frac{1 \text{ mol } CO_2}{44.01 \text{ g } CO_2} \times \frac{3 \text{ mol atoms}(1 \text{ C} + 2 \text{ O})}{1 \text{ mol } CO_2} \times \frac{6.022 \times 10^{23} \text{ atoms}}{1 \text{ mol atoms}}$$
$$= 1.81 \times 10^{24} \text{ atoms}$$

$$146 \text{ g } SF_6 \times \frac{1 \text{ mol } SF_6}{146.07 \text{ g } SF_6} \times \frac{7 \text{ mol atoms}(1 \text{ S} + 6 \text{ F})}{1 \text{ mol } SF_6} \times \frac{6.022 \times 10^{23} \text{ atoms}}{1 \text{ mol atoms}} = 4.21 \times 10^{24} \text{ atoms}$$

The order is: $4.0 \text{ g He} < 1.0 \text{ mol } F_2 < 44.0 \text{ g } CO_2 < 4.0 \text{ g } H_2 < 146 \text{ g } SF_6$

39.   a. $2(12.01) + 3(1.008) + 3(35.45) + 2(16.00) = 165.39 \text{ g/mol}$

     b. $500.0 \text{ g} \times \dfrac{1 \text{ mol}}{165.39 \text{ g}} = 3.023 \text{ mol } C_2H_3Cl_3O_2$

     c. $2.0 \times 10^{-2} \text{ mol} \times \dfrac{165.39 \text{ g}}{\text{mol}} = 3.3 \text{ g } C_2H_3Cl_3O_2$

     d. $5.0 \text{ g } C_2H_3Cl_3O_2 \times \dfrac{1 \text{ mol}}{165.39 \text{ g}} \times \dfrac{6.02 \times 10^{23} \text{ molecules}}{\text{mol}} \times \dfrac{3 \text{ atoms Cl}}{\text{molecule}}$
$$= 5.5 \times 10^{22} \text{ atoms of chlorine}$$

     e. $1.0 \text{ g Cl} \times \dfrac{1 \text{ mol Cl}}{35.45 \text{ g}} \times \dfrac{1 \text{ mol } C_2H_3Cl_3O_2}{3 \text{ mol Cl}} \times \dfrac{165.39 \text{ g } C_2H_3Cl_3O_2}{\text{mol } C_2H_3Cl_3O_2} = 1.6 \text{ g chloral hydrate}$

     f. $500 \text{ molecules} \times \dfrac{1 \text{ mol}}{6.022 \times 10^{23} \text{ molecules}} \times \dfrac{165.39 \text{ g}}{\text{mol}} = 1.373 \times 10^{-19} \text{ g}$

Percent Composition

41.   Molar mass $= 20(12.01) + 29(1.008) + 19.00 + 3(16.00) = 336.43 \text{ g/mol}$

    Mass % C $= \dfrac{20(12.01) \text{ g C}}{336.43 \text{ g compound}} \times 100 = 71.40\% \text{ C}$

    Mass % H $= \dfrac{29(1.008) \text{ g H}}{336.43 \text{ g compound}} \times 100 = 8.689\% \text{ H}$

$$\text{Mass \% F} = \frac{19.00 \text{ g F}}{336.43 \text{ g compound}} \times 100 = 5.648\% \text{ F}$$

$$\text{Mass \% O} = 100.00 - (71.40 + 8.689 + 5.648) = 14.26\% \text{ O or:}$$

$$\text{Mass \% O} = \frac{3(16.00) \text{ g O}}{336.43 \text{ g compound}} \times 100 = 14.27\% \text{ O}$$

43.   In 1 mole of $YBa_2Cu_3O_7$, there are 1 mole of Y, 2 moles of Ba, 3 moles of Cu, and 7 moles of O.

$$\text{Molar mass} = 1 \text{ mol Y} \left(\frac{88.91 \text{ g Y}}{\text{mol Y}} \right) + 2 \text{ mol Ba} \left(\frac{137.3 \text{ g Ba}}{\text{mol Ba}} \right)$$

$$+ 3 \text{ mol Cu} \left(\frac{63.55 \text{ g Cu}}{\text{mol Cu}} \right) + 7 \text{ mol O} \left(\frac{16.00 \text{ g O}}{\text{mol O}} \right)$$

$$\text{Molar mass} = 88.91 + 274.6 + 190.65 + 112.00 = 666.2 \text{ g/mol}$$

$$\text{Mass \% Y} = \frac{88.91 \text{ g}}{666.2 \text{ g}} \times 100 = 13.35\% \text{ Y}; \quad \text{mass \% Ba} = \frac{274.6 \text{ g}}{666.2 \text{ g}} \times 100 = 41.22\% \text{ Ba}$$

$$\text{Mass \% Cu} = \frac{190.65 \text{ g}}{666.2 \text{ g}} \times 100 = 28.62\% \text{ Cu}; \quad \text{mass \% O} = \frac{112.0 \text{ g}}{666.2 \text{ g}} \times 100 = 16.81\% \text{ O}$$

45.   NO: $\text{Mass \% N} = \dfrac{14.01 \text{ g N}}{30.01 \text{ g NO}} \times 100 = 46.68\% \text{ N}$

NO_2: $\text{Mass \% N} = \dfrac{14.01 \text{ g N}}{46.01 \text{ g NO}_2} \times 100 = 30.45\% \text{ N}$

N_2O: $\text{Mass \% N} = \dfrac{2(14.01) \text{ g N}}{44.02 \text{ g N}_2\text{O}} \times 100 = 63.65\% \text{ N}$

From the calculated mass percents, only NO is 46.7% N by mass, so NO could be this species. Any other compound having NO as an empirical formula could also be the compound.

47.   There are 0.390 g Cu for every 100.000 g of fungal laccase. Let's assume 100.000 g fungal laccase.

$$\text{Mol fungal laccase} = 0.390 \text{ g Cu} \times \frac{1 \text{ mol Cu}}{63.55 \text{ g Cu}} \times \frac{1 \text{ mol fungal laccase}}{4 \text{ mol Cu}} = 1.53 \times 10^{-3} \text{ mol}$$

$$\frac{x \text{ g fungal laccase}}{1 \text{ mol fungal laccase}} = \frac{100.000 \text{ g}}{1.53 \times 10^{-3} \text{ mol}}, \quad x = \text{molar mass} = 6.54 \times 10^4 \text{ g/mol}$$

Empirical and Molecular Formulas

49.   a.   Molar mass of $CH_2O = 1 \text{ mol C}\left(\dfrac{12.011 \text{ g}}{\text{mol C}}\right) + 2 \text{ mol H}\left(\dfrac{1.0079 \text{ g}}{\text{mol H}}\right)$

$$+ 1 \text{ mol O}\left(\dfrac{15.999 \text{ g}}{\text{mol O}}\right) = 30.026 \text{ g/mol}$$

$$\% \text{ C} = \dfrac{12.011 \text{ g C}}{30.026 \text{ g } CH_2O} \times 100 = 40.002\% \text{ C}; \quad \% \text{ H} = \dfrac{2.0158 \text{ g H}}{30.026 \text{ g } CH_2O} \times 100 = 6.7135\% \text{ H}$$

$$\% \text{ O} = \dfrac{15.999 \text{ g O}}{30.026 \text{ g } CH_2O} \times 100 = 53.284\% \text{ O} \text{ or } \% \text{ O} = 100.000 - (40.002 + 6.7135)$$

$$= 53.285\%$$

b.   Molar mass of $C_6H_{12}O_6 = 6(12.011) + 12(1.0079) + 6(15.999) = 180.155 \text{ g/mol}$

$$\% \text{ C} = \dfrac{72.066 \text{ g C}}{180.155 \text{ g } C_6H_{12}O_6} \times 100 = 40.002\%; \quad \% \text{ H} = \dfrac{12(1.0079) \text{ g}}{180.155 \text{ g}} \times 100 = 6.7136\%$$

$$\% \text{ O} = 100.00 - (40.002 + 6.7136) = 53.284\%$$

c.   Molar Mass of $HC_2H_3O_2 = 2(12.011) + 4(1.0079) + 2(15.999) = 60.052 \text{ g/mol}$

$$\% \text{ C} = \dfrac{24.022 \text{ g}}{60.052 \text{ g}} \times 100 = 40.002\%; \quad \% \text{ H} = \dfrac{4.0316 \text{ g}}{60.052 \text{ g}} \times 100 = 6.7135\%$$

$$\% \text{ O} = 100.000 - (40.002 + 6.7135) = 53.285\%$$

All three compounds have the same empirical formula, CH_2O, and different molecular formulas. The composition of all three in mass percent is also the same (within rounding differences). Therefore, elemental analysis will give us only the empirical formula.

51.   a.   SNH: Empirical formula mass = 32.07 + 14.01 + 1.008 = 47.09 g/mol

$$\dfrac{188.35 \text{ g}}{47.09 \text{ g}} = 4.000; \text{ so the molecular formula is } (SNH)_4 \text{ or } S_4N_4H_4.$$

b. $NPCl_2$: Empirical formula mass = 14.01 + 30.97 + 2(35.45) = 115.88 g/mol

$$\dfrac{347.64 \text{ g}}{115.88 \text{ g}} = 3.0000; \text{ molecular formula is } (NPCl_2)_3 \text{ or } N_3P_3Cl_6.$$

c. CoC_4O_4: $58.93 + 4(12.01) + 4(16.00) = 170.97$ g/mol

$$\frac{341.94 \text{ g}}{170.97 \text{ g}} = 2.0000; \text{ molecular formula: } Co_2C_8O_8$$

d.   SN: $32.07 + 14.01 = 46.08$ g/mol; $\dfrac{184.32 \text{ g}}{46.08 \text{ g}} = 4.000$; molecular formula: S_4N_4

53.   Compound I: mass O = 0.6498 g Hg_xO_y − 0.6018 g Hg = 0.0480 g O

$$0.6018 \text{ g Hg} \times \frac{1 \text{ mol Hg}}{200.6 \text{ g Hg}} = 3.000 \times 10^{-3} \text{ mol Hg}$$

$$0.0480 \text{ g O} \times \frac{1 \text{ mol O}}{16.00 \text{ g O}} = 3.00 \times 10^{-3} \text{ mol O}$$

The mole ratio between Hg and O is 1 : 1, so the empirical formula of compound I is HgO.

Compound II: mass Hg = 0.4172 g Hg_xO_y − 0.016 g O = 0.401 g Hg

$$0.401 \text{ g Hg} \times \frac{1 \text{ mol Hg}}{200.6 \text{ g Hg}} = 2.00 \times 10^{-3} \text{ mol Hg}; \ \ 0.016 \text{ g O} \times \frac{1 \text{ mol O}}{16.00 \text{ g O}} = 1.0 \times 10^{-3} \text{ mol O}$$

The mole ratio between Hg and O is 2 : 1, so the empirical formula is Hg_2O.

55.   First, we will determine composition in mass percent. We assume that all the carbon in the 0.213 g CO_2 came from the 0.157 g of the compound and that all the hydrogen in the 0.0310 g H_2O came from the 0.157 g of the compound.

$$0.213 \text{ g CO}_2 \times \frac{12.01 \text{ g C}}{44.01 \text{ g CO}_2} = 0.0581 \text{ g C}; \ \ \% \text{ C} = \frac{0.0581 \text{ g C}}{0.157 \text{ g compound}} \times 100 = 37.0\% \text{ C}$$

$$0.0310 \text{ g H}_2O \times \frac{2.016 \text{ g H}}{18.02 \text{ g H}_2O} = 3.47 \times 10^{-3} \text{ g H}; \ \ \% \text{ H} = \frac{3.47 \times 10^{-3} \text{ g}}{0.157 \text{ g}} \times 100 = 2.21\% \text{ H}$$

We get the mass percent of N from the second experiment:

$$0.0230 \text{ g NH}_3 \times \frac{14.01 \text{ g N}}{17.03 \text{ g NH}_3} = 1.89 \times 10^{-2} \text{ g N}$$

$$\% \text{ N} = \frac{1.89 \times 10^{-2} \text{ g}}{0.103 \text{ g}} \times 100 = 18.3\% \text{ N}$$

The mass percent of oxygen is obtained by difference:

$$\% \text{ O} = 100.00 - (37.0 + 2.21 + 18.3) = 42.5\% \text{ O}$$

So, out of 100.00 g of compound, there are:

$$37.0 \text{ g C} \times \frac{1 \text{ mol C}}{12.01 \text{ g C}} = 3.08 \text{ mol C}; \quad 2.21 \text{ g H} \times \frac{1 \text{ mol H}}{1.008 \text{ g H}} = 2.19 \text{ mol H}$$

$$18.3 \text{ g N} \times \frac{1 \text{ mol N}}{14.01 \text{ g N}} = 1.31 \text{ mol N}; \quad 42.5 \text{ g O} \times \frac{1 \text{ mol O}}{16.00 \text{ g O}} = 2.66 \text{ mol O}$$

Lastly, and often the hardest part, we need to find simple whole number ratios. Divide all mole values by the smallest number:

$$\frac{3.08}{1.31} = 2.35; \quad \frac{2.19}{1.31} = 1.67; \quad \frac{1.31}{1.31} = 1.00; \quad \frac{2.66}{1.31} = 2.03$$

Multiplying all these ratios by 3 gives an empirical formula of $C_7H_5N_3O_6$.

57.   Assuming 100.0 g of compound:

$$26.7 \text{ g P} \times \frac{1 \text{ mol P}}{30.97 \text{ g P}} = 0.862 \text{ mol P}; \quad 12.1 \text{ g N} \times \frac{1 \text{ mol N}}{14.01 \text{ g N}} = 0.864 \text{ mol N}$$

$$61.2 \text{ g Cl} \times \frac{1 \text{ mol Cl}}{35.45 \text{ g Cl}} = 1.73 \text{ mol Cl}$$

$$\frac{1.73}{0.862} = 2.01; \text{ the empirical formula is } PNCl_2.$$

The empirical formula mass is $\approx 31.0 + 14.0 + 2(35.5) = 116$ g/mol.

$$\frac{\text{Molar mass}}{\text{Empirical formula mass}} = \frac{580}{116} = 5.0; \text{ the molecular formula is } (PNCl_2)_5 = P_5N_5Cl_{10}.$$

59.   First, we will determine composition by mass percent:

$$16.01 \text{ mg CO}_2 \times \frac{1 \text{ g}}{1000 \text{ mg}} \times \frac{12.011 \text{ g C}}{44.009 \text{ g CO}_2} \times \frac{1000 \text{ mg}}{\text{g}} = 4.369 \text{ mg C}$$

$$\% \text{ C} = \frac{4.369 \text{ mg C}}{10.68 \text{ mg compound}} \times 100 = 40.91\% \text{ C}$$

$$4.37 \text{ mg H}_2\text{O} \times \frac{1 \text{ g}}{1000 \text{ mg}} \times \frac{2.016 \text{ g H}}{18.02 \text{ g H}_2\text{O}} \times \frac{1000 \text{ mg}}{\text{g}} = 0.489 \text{ mg H}$$

$$\% \text{ H} = \frac{0.489 \text{ mg}}{10.68 \text{ mg}} \times 100 = 4.58\% \text{ H}; \quad \% \text{ O} = 100.00 - (40.91 + 4.58) = 54.51\% \text{ O}$$

So, in 100.00 g of the compound, we have:

$$40.91 \text{ g C} \times \frac{1 \text{ mol C}}{12.011 \text{ g C}} = 3.406 \text{ mol C}; \quad 4.58 \text{ g H} \times \frac{1 \text{ mol H}}{1.008 \text{ g H}} = 4.54 \text{ mol H}$$

$$54.51 \text{ g O} \times \frac{1 \text{ mol O}}{15.999 \text{ g O}} = 3.407 \text{ mol O}$$

Dividing by the smallest number: $\dfrac{4.54}{3.406} = 1.33 \cdot \dfrac{4}{3}$; the empirical formula is $C_3H_4O_3$.

The empirical formula mass of $C_3H_4O_3$ is $\approx 3(12) + 4(1) + 3(16) = 88$ g.

Because $\dfrac{176.1}{88} = 2.0$, the molecular formula is $C_6H_8O_6$.

Balancing Chemical Equations

61.   Only one product is formed in this representation. This product has two Ys bonded to an X. The other substance present in the product mixture is just the excess of one of the reactants (Y). The best equation has smallest whole numbers. Here, answer c would be this smallest whole number equation (X + 2 Y → XY$_2$). Answers a and b have incorrect products listed, and for answer d, an equation only includes the reactants that go to produce the product; excess reactants are not shown in an equation.

63.   When balancing reactions, start with elements that appear in only one of the reactants and one of the products, and then go on to balance the remaining elements.

a. $C_6H_{12}O_6(s) + O_2(g) \rightarrow CO_2(g) + H_2O(g)$

Balance C atoms: $C_6H_{12}O_6 + O_2 \rightarrow 6\ CO_2 + H_2O$

Balance H atoms: $C_6H_{12}O_6 + O_2 \rightarrow 6\ CO_2 + 6\ H_2O$

Lastly, balance O atoms: $C_6H_{12}O_6(s) + 6\ O_2(g) \rightarrow 6\ CO_2(g) + 6\ H_2O(g)$

b. $Fe_2S_3(s) + HCl(g) \rightarrow FeCl_3(s) + H_2S(g)$

Balance Fe atoms: $Fe_2S_3 + HCl \rightarrow 2\ FeCl_3 + H_2S$

Balance S atoms: $Fe_2S_3 + HCl \rightarrow 2\ FeCl_3 + 3\ H_2S$

There are 6 H and 6 Cl on right, so balance with 6 HCl on left:

$Fe_2S_3(s) + 6\ HCl(g) \rightarrow 2\ FeCl_3(s) + 3\ H_2S(g)$

c. $CS_2(l) + NH_3(g) \rightarrow H_2S(g) + NH_4SCN(s)$

C and S are balanced; balance N:

$CS_2 + 2 NH_3 \rightarrow H_2S + NH_4SCN$

H is also balanced. $CS_2(l) + 2 NH_3(g) \rightarrow H_2S(g) + NH_4SCN(s)$.

65.   a. $16 Cr(s) + 3 S_8(s) \rightarrow 8 Cr_2S_3(s)$

b. $2 NaHCO_3(s) \rightarrow Na_2CO_3(s) + CO_2(g) + H_2O(g)$

c. $2 KClO_3(s) \rightarrow 2 KCl(s) + 3 O_2(g)$

d. $2 Eu(s) + 6 HF(g) \rightarrow 2 EuF_3(s) + 3 H_2(g)$

e. $2 C_6H_6(l) + 15 O_2(g) \rightarrow 12 CO_2(g) + 6 H_2O(g)$

Reaction Stoichiometry

67. $1.000 \text{ kg Al} \times \dfrac{1000 \text{ g Al}}{\text{kg Al}} \times \dfrac{1 \text{ mol Al}}{26.98 \text{ g Al}} \times \dfrac{3 \text{ mol NH}_4\text{ClO}_4}{3 \text{ mol Al}} \times \dfrac{117.49 \text{ g NH}_4\text{ClO}_4}{\text{mol NH}_4\text{ClO}_4}$

$$= 4355 \text{ g NH}_4\text{ClO}_4$$

69. $Fe_2O_3(s) + 2 Al(s) \rightarrow 2 Fe(l) + Al_2O_3(s)$

$15.0 \text{ g Fe} \times \dfrac{1 \text{ mol Fe}}{55.85 \text{ g Fe}} = 0.269 \text{ mol Fe};\ \ 0.269 \text{ mol Fe} \times \dfrac{2 \text{ mol Al}}{2 \text{ mol Fe}} \times \dfrac{26.98 \text{ g Al}}{\text{mol Al}} = 7.26 \text{ g Al}$

$0.269 \text{ mol Fe} \times \dfrac{1 \text{ mol Fe}_2O_3}{2 \text{ mol Fe}} \times \dfrac{159.70 \text{ g Fe}_2O_3}{\text{mol Fe}_2O_3} = 21.5 \text{ g Fe}_2O_3$

$0.269 \text{ mol Fe} \times \dfrac{1 \text{ mol Al}_2O_3}{2 \text{ mol Fe}} \times \dfrac{101.96 \text{ g Al}_2O_3}{\text{mol Al}_2O_3} = 13.7 \text{ g Al}_2O_3$

71. $2 LiOH(s) + CO_2(g) \rightarrow Li_2CO_3(aq) + H_2O(l)$

The total volume of air exhaled each minute for the 7 astronauts is $7 \times 20. = 140 \text{ L/min}$.

$25,000 \text{ g LiOH} \times \dfrac{1 \text{ mol LiOH}}{23.95 \text{ g LiOH}} \times \dfrac{1 \text{ mol CO}_2}{2 \text{ mol LiOH}} \times \dfrac{44.01 \text{ g CO}_2}{\text{mol CO}_2} \times \dfrac{100 \text{ g air}}{4.0 \text{ g CO}_2} \times$

$\dfrac{1 \text{ mL air}}{0.0010 \text{ g air}} \times \dfrac{1 \text{ L}}{1000 \text{ mL}} \times \dfrac{1 \text{ min}}{140 \text{ L air}} \times \dfrac{1 \text{ h}}{60 \text{ min}} = 68 \text{ h} = 2.8 \text{ days}$

73. 1.0×10^3 g phosphorite $\times \dfrac{75 \text{ g Ca}_3(\text{PO}_4)_2}{100 \text{ g phosphorite}} \times \dfrac{1 \text{ mol Ca}_3(\text{PO}_4)_2}{310.18 \text{ g Ca}_3(\text{PO}_4)_2} \times$

$$\dfrac{1 \text{ mol P}_4}{2 \text{ mol Ca}_3(\text{PO}_4)_2} \times \dfrac{123.88 \text{ g P}_4}{\text{mol P}_4} = 150 \text{ g P}_4$$

Limiting Reactants and Percent Yield

75.   The product formed in the reaction is NO_2; the other species present in the product picture is excess O_2. Therefore, NO is the limiting reactant. In the pictures, 6 NO molecules react with 3 O_2 molecules to form 6 NO_2 molecules.

$$6 \text{ NO(g)} + 3 \text{ O}_2(g) \rightarrow 6 \text{ NO}_2(g)$$

For smallest whole numbers, the balanced reaction is:

$$2 \text{ NO(g)} + \text{O}_2(g) \rightarrow 2 \text{ NO}_2(g)$$

77. $1.50 \text{ g BaO}_2 \times \dfrac{1 \text{ mol BaO}_2}{169.3 \text{ g BaO}_2} = 8.86 \times 10^{-3} \text{ mol BaO}_2$

$25.0 \text{ mL} \times \dfrac{0.0272 \text{ g HCl}}{\text{mL}} \times \dfrac{1 \text{ mol HCl}}{36.46 \text{ g HCl}} = 1.87 \times 10^{-2} \text{ mol HCl}$

The required mole ratio from the balanced reaction is 2 mol HCl to 1 mol BaO_2. The actual mole ratio is:

$$\dfrac{1.87 \times 10^{-2} \text{ mol HCl}}{8.86 \times 10^{-3} \text{ mol BaO}_2} = 2.11$$

Because the actual mole ratio is larger than the required mole ratio, the denominator (BaO_2) is the limiting reagent.

$8.86 \times 10^{-3} \text{ mol BaO}_2 \times \dfrac{1 \text{ mol H}_2\text{O}_2}{\text{mol BaO}_2} \times \dfrac{34.02 \text{ g H}_2\text{O}_2}{\text{mol H}_2\text{O}_2} = 0.301 \text{ g H}_2\text{O}_2$

The amount of HCl reacted is:

$8.86 \times 10^{-3} \text{ mol BaO}_2 \times \dfrac{2 \text{ mol HCl}}{\text{mol BaO}_2} = 1.77 \times 10^{-2} \text{ mol HCl}$

Excess mol HCl = 1.87×10^{-2} mol $-\ 1.77 \times 10^{-2}$ mol $= 1.0 \times 10^{-3}$ mol HCl

Mass of excess HCl = $1.0 \times 10^{-3} \text{ mol HCl} \times \dfrac{36.46 \text{ g HCl}}{\text{mol HCl}} = 3.6 \times 10^{-2} \text{ g HCl}$

79. $2.50 \text{ metric tons Cu}_3\text{FeS}_3 \times \dfrac{1000 \text{ kg}}{\text{metric ton}} \times \dfrac{1000 \text{ g}}{\text{kg}} \times \dfrac{1 \text{ mol Cu}_3\text{FeS}_3}{342.71 \text{ g}} \times \dfrac{3 \text{ mol Cu}}{1 \text{ mol Cu}_3\text{FeS}_3} \times$

$$\dfrac{63.55 \text{ g}}{\text{mol Cu}} = 1.39 \times 10^6 \text{ g Cu (theoretical)}$$

$$1.39 \times 10^6 \text{ g Cu (theoretical)} \times \dfrac{86.3 \text{ g Cu (actual)}}{100. \text{ g Cu (theoretical)}} = 1.20 \times 10^6 \text{ g Cu} = 1.20 \times 10^3 \text{ kg Cu}$$

$$= 1.20 \text{ metric tons Cu (actual)}$$

81.   An alternative method to solve limiting-reagent problems is to assume that each reactant is limiting and then calculate how much product could be produced from each reactant. The reactant that produces the smallest amount of product will run out first and is the limiting reagent.

$$5.00 \times 10^6 \text{ g NH}_3 \times \dfrac{1 \text{ mol NH}_3}{17.03 \text{ g NH}_3} \times \dfrac{2 \text{ mol HCN}}{2 \text{ mol NH}_3} = 2.94 \times 10^5 \text{ mol HCN}$$

$$5.00 \times 10^6 \text{ g O}_2 \times \dfrac{1 \text{ mol O}_2}{32.00 \text{ g O}_2} \times \dfrac{2 \text{ mol HCN}}{3 \text{ mol O}_2} = 1.04 \times 10^5 \text{ mol HCN}$$

$$5.00 \times 10^6 \text{ g CH}_4 \times \dfrac{1 \text{ mol CH}_4}{16.04 \text{ g CH}_4} \times \dfrac{2 \text{ mol HCN}}{2 \text{ mol CH}_4} = 3.12 \times 10^5 \text{ mol HCN}$$

O_2 is limiting because it produces the smallest amount of HCN. Although more product could be produced from NH_3 and CH_4, only enough O_2 is present to produce 1.04×10^5 mol HCN. The mass of HCN that can be produced is:

$$1.04 \times 10^5 \text{ mol HCN} \times \dfrac{27.03 \text{ g HCN}}{\text{mol HCN}} = 2.81 \times 10^6 \text{ g HCN}$$

$$5.00 \times 10^6 \text{ g O}_2 \times \dfrac{1 \text{ mol O}_2}{32.00 \text{ g O}_2} \times \dfrac{6 \text{ mol H}_2\text{O}}{3 \text{ mol O}_2} \times \dfrac{18.02 \text{ g H}_2\text{O}}{1 \text{ mol H}_2\text{O}} = 5.63 \times 10^6 \text{ g H}_2\text{O}$$

83. $P_4(s) + 6 \, F_2(g) \rightarrow 4 \, PF_3(g)$; the theoretical yield of PF_3 is:

$$120. \text{ g PF}_3 \text{ (actual)} \times \dfrac{100.0 \text{ g PF}_3 \text{ (theoretical)}}{78.1 \text{ g PF}_3 \text{ (actual)}} = 154 \text{ g PF}_3 \text{ (theoretical)}$$

$$154 \text{ g PF}_3 \times \dfrac{1 \text{ mol PF}_3}{87.97 \text{ g PF}_3} \times \dfrac{6 \text{ mol F}_2}{4 \text{ mol PF}_3} \times \dfrac{38.00 \text{ g F}_2}{\text{mol F}_2} = 99.8 \text{ g F}_2$$

99.8 g F_2 are needed to actually produce 120. g of PF_3 if the percent yield is 78.1%.

Additional Exercises

85. $17.3 \text{ g H} \times \dfrac{1 \text{ mol H}}{1.008 \text{ g H}} = 17.2 \text{ mol H}; \quad 82.7 \text{ g C} \times \dfrac{1 \text{ mol C}}{12.01 \text{ g C}} = 6.89 \text{ mol C}$

$\dfrac{17.2}{6.89} = 2.50;$ the empirical formula is C_2H_5.

The empirical formula mass is ~29 g, so two times the empirical formula would put the compound in the correct range of the molar mass. Molecular formula = $(C_2H_5)_2 = C_4H_{10}$

$2.59 \times 10^{23} \text{ atoms H} \times \dfrac{1 \text{ molecule } C_4H_{10}}{10 \text{ atoms H}} \times \dfrac{1 \text{ mol } C_4H_{10}}{6.022 \times 10^{23} \text{ molecules}}$

$= 4.30 \times 10^{-2} \text{ mol } C_4H_{10}$

$4.30 \times 10^{-2} \text{ mol } C_4H_{10} \times \dfrac{58.12 \text{ g}}{\text{mol } C_4H_{10}} = 2.50 \text{ g } C_4H_{10}$

87.   Molar mass $X_2 = \dfrac{0.105 \text{ g}}{8.92 \times 10^{20} \text{ molecules} \times \dfrac{1 \text{ mol}}{6.022 \times 10^{23} \text{ molecules}}} = 70.9 \text{ g/mol}$

The mass of X = 1/2(70.9 g/mol) = 35.5 g/mol. This is the element chlorine.

Assuming 100.00 g of MX_3 compound:

$54.47 \text{ g Cl} \times \dfrac{1 \text{ mol}}{35.45 \text{ g}} = 1.537 \text{ mol Cl}$

$1.537 \text{ mol Cl} \times \dfrac{1 \text{ mol M}}{3 \text{ mol Cl}} = 0.5123 \text{ mol M}$

Molar mass of M = $\dfrac{45.53 \text{ g M}}{0.5123 \text{ mol M}} = 88.87 \text{ g/mol M}$

M is the element yttrium (Y), and the name of YCl_3 is yttrium(III) chloride.

The balanced equation is $2 \text{ Y} + 3 \text{ Cl}_2 \rightarrow 2 \text{ YCl}_3$.

Assuming Cl_2 is limiting:

$1.00 \text{ g Cl}_2 \times \dfrac{1 \text{ mol Cl}_2}{70.90 \text{ g Cl}_2} \times \dfrac{2 \text{ mol YCl}_3}{3 \text{ mol Cl}_2} \times \dfrac{195.26 \text{ g YCl}_3}{1 \text{ mol YCl}_3} = 1.84 \text{ g YCl}_3$

Assuming Y is limiting:

$$1.00 \text{ g Y} \times \frac{1 \text{ mol Y}}{88.91 \text{ g Y}} \times \frac{2 \text{ mol YCl}_3}{2 \text{ mol Y}} \times \frac{195.26 \text{ g YCl}_3}{1 \text{ mol YCl}_3} = 2.20 \text{ g YCl}_3$$

Because Cl_2, when it all reacts, produces the smaller amount of product, Cl_2 is the limiting reagent, and the theoretical yield is 1.84 g YCl_3.

89.   Mass of H_2O = 0.755 g $CuSO_4 \bullet xH_2O$ − 0.483 g $CuSO_4$ = 0.272 g H_2O

$$0.483 \text{ g CuSO}_4 \times \frac{1 \text{ mol CuSO}_4}{159.62 \text{ g CuSO}_4} = 0.00303 \text{ mol CuSO}_4$$

$$0.272 \text{ g H}_2O \times \frac{1 \text{ mol H}_2O}{18.02 \text{ g H}_2O} = 0.0151 \text{ mol H}_2O$$

$$\frac{0.0151 \text{ mol H}_2O}{0.00303 \text{ mol CuSO}_4} = \frac{4.98 \text{ mol H}_2O}{1 \text{ mol CuSO}_4} ; \text{ compound formula} = CuSO_4 \bullet 5H_2O, \ x = 5$$

91.   Consider the case of aluminum plus oxygen. Aluminum forms Al^{3+} ions; oxygen forms O^{2-} anions. The simplest compound of the two elements is Al_2O_3. Similarly, we would expect the formula of a Group 6A element with Al to be Al_2X_3. Assuming this, out of 100.00 g of compound, there are 18.56 g Al and 81.44 g of the unknown element, X. Let's use this information to determine the molar mass of X, which will allow us to identify X from the periodic table.

$$18.56 \text{ g Al} \times \frac{1 \text{ mol Al}}{26.98 \text{ g Al}} \times \frac{3 \text{ mol X}}{2 \text{ mol Al}} = 1.032 \text{ mol X}$$

81.44 g of X must contain 1.032 mol of X.

$$\text{Molar mass of X} = \frac{81.44 \text{ g X}}{1.032 \text{ mol X}} = 78.91 \text{ g/mol}$$

From the periodic table, the unknown element is selenium, and the formula is Al_2Se_3.

93. $$1.20 \text{ g CO}_2 \times \frac{1 \text{ mol CO}_2}{44.01 \text{ g}} \times \frac{1 \text{ mol C}}{\text{mol CO}_2} \times \frac{1 \text{ mol C}_{24}H_{30}N_3O}{24 \text{ mol C}} \times \frac{376.51 \text{ g}}{\text{mol C}_{24}H_{30}N_3O}$$
$$= 0.428 \text{ g C}_{24}H_{30}N_3O$$

$$\frac{0.428 \text{ g C}_{24}H_{30}N_3O}{1.00 \text{ g sample}} \times 100 = 42.8\% \text{ C}_{24}H_{30}N_3O \text{ (LSD)}$$

95. $2 NaNO_3(s) \rightarrow 2 NaNO_2(s) + O_2(g)$;   the amount of $NaNO_3$ in the impure sample is:

$$0.2864 \text{ g NaNO}_2 \times \frac{1 \text{ mol NaNO}_2}{69.00 \text{ g NaNO}_2} \times \frac{2 \text{ mol NaNO}_3}{2 \text{ mol NaNO}_2} \times \frac{85.00 \text{ g NaNO}_3}{\text{mol NaNO}_3}$$
$$= 0.3528 \text{ g NaNO}_3$$

Mass percent $NaNO_3$ = $\dfrac{0.3528 \text{ g } NaNO_3}{0.4230 \text{ g sample}}$ × 100 = 83.40%

97.   453 g Fe × $\dfrac{1 \text{ mol Fe}}{55.85 \text{ g Fe}}$ × $\dfrac{1 \text{ mol } Fe_2O_3}{2 \text{ mol Fe}}$ × $\dfrac{159.70 \text{ g } Fe_2O_3}{\text{mol } Fe_2O_3}$ = 648 g Fe_2O_3

Mass % Fe_2O_3 = $\dfrac{648 \text{ g } Fe_2O_3}{752 \text{ g ore}}$ × 100 = 86.2%

99. $\dfrac{^{85}Rb \text{ atoms}}{^{87}Rb \text{ atoms}}$ = 2.591;   If we had exactly 100 atoms, x = number of ^{85}Rb atoms and $100 - x$ = number of ^{87}Rb atoms.

$\dfrac{x}{100 - x}$ = 2.591, x = 259.1 − (2.591)x, x = $\dfrac{259.1}{3.591}$ = 72.15; 72.15% ^{85}Rb

0.7215(84.9117) + 0.2785(A) = 85.4678,   A = $\dfrac{85.4678 - 61.26}{0.2785}$ = 86.92 amu

101.   The volume of a gas is proportional to the number of molecules of gas. Thus the formulas are:

I: NH_3 II: N_2H_4 III: HN_3

The mass ratios are:

I: $\dfrac{4.634 \text{ g N}}{\text{g H}}$ II: $\dfrac{6.949 \text{ g N}}{\text{g H}}$ III: $\dfrac{41.7 \text{ g N}}{\text{g H}}$

If we set the atomic mass of H equal to 1.008, then the atomic mass, A, for nitrogen is:

I: 14.01   II: 14.01   III. 14.0

For example, for compound I: $\dfrac{A}{3(1.008)}$ = $\dfrac{4.634}{1}$, A = 14.01

103.   1.375 g AgI × $\dfrac{1 \text{ mol AgI}}{234.8 \text{ g AgI}}$ = 5.856 × 10^{-3} mol AgI = 5.856 × 10^{-3} mol I

1.375 g AgI × $\dfrac{126.9 \text{ g I}}{234.8 \text{ g AgI}}$ = 0.7431 g I; XI_2 contains 0.7431 g I and 0.257 g X.

5.856 × 10^{-3} mol I × $\dfrac{1 \text{ mol X}}{2 \text{ mol I}}$ = 2.928 × 10^{-3} mol X

Molar mass = $\dfrac{0.257 \text{ g X}}{2.928 \times 10^{-3} \text{ mol X}}$ = $\dfrac{87.8 \text{ g}}{\text{mol}}$; atomic mass = 87.8 amu (X is Sr.)

105.   Assuming 1 mole of vitamin A (286.4 g vitamin A):

$$\text{mol C} = 286.4 \text{ g vitamin A} \times \frac{0.8396 \text{ g C}}{\text{g vitamin A}} \times \frac{1 \text{ mol C}}{12.011 \text{ g C}} = 20.00 \text{ mol C}$$

$$\text{mol H} = 286.4 \text{ g vitamin A} \times \frac{0.1056 \text{ g H}}{\text{g vitamin A}} \times \frac{1 \text{ mol H}}{1.0079 \text{ g H}} = 30.01 \text{ mol H}$$

Because 1 mole of vitamin A contains 20 mol C and 30 mol H, the molecular formula of vitamin A is $C_{20}H_{30}E$. To determine E, lets calculate the molar mass of E:

$$286.4 \text{ g} = 20(12.01) + 30(1.008) + \text{molar mass E},\ \text{molar mass E} = 16.0 \text{ g/mol}$$

From the periodic table, E = oxygen, and the molecular formula of vitamin A is $C_{20}H_{30}O$.

Challenge Problems

107.   When the discharge voltage is low, the ions present are in the form of molecules. When the discharge voltage is increased, the bonds in the molecules are broken, and the ions present are in the form of individual atoms. Therefore, the high discharge data indicate that the ions $^{16}O^{+}$, $^{18}O^{+}$, and $^{40}Ar^{+}$ are present. The only combination of these individual ions that can explain the mass data at low discharge is $^{16}O^{16}O^{+}$ (mass = 32), $^{16}O^{18}O^{+}$ (mass = 34), and $^{40}Ar^{+}$ (mass = 40). Therefore, the gas mixture contains $^{16}O^{16}O$, $^{16}O^{18}O$, and ^{40}Ar. To determine the percent composition of each isotope, we use the relative intensity data from the high discharge data to determine the percentage that each isotope contributes to the total relative intensity. For ^{40}Ar:

$$\frac{1.0000}{0.7500 + 0.0015 + 1.0000} \times 100 = \frac{1.0000}{1.7515} \times 100 = 57.094\% \ ^{40}Ar$$

For ^{16}O: $\dfrac{0.7500}{1.7515} \times 100 = 42.82\% \ ^{16}O$;   for ^{18}O: $\dfrac{0.0015}{1.7515} \times 100 = 8.6 \times 10^{-2}\% \ ^{18}O$

Note: ^{18}F instead of ^{18}O could also explain the data. However, OF(g) is not a stable compound. This is why ^{18}O is the best choice because O_2(g) does form.

109.   10.00 g XCl_2 + excess $Cl_2 \rightarrow$ 12.55 g XCl_4; 2.55 g Cl reacted with XCl_2 to form XCl_4. XCl_4 contains 2.55 g Cl and 10.00 g XCl_2. From mole ratios, 10.00 g XCl_2 must also contain 2.55 g Cl; mass X in XCl_2 = 10.00 − 2.55 = 7.45 g X:

$$2.55 \text{ g Cl} \times \frac{1 \text{ mol Cl}}{35.45 \text{ g Cl}} \times \frac{1 \text{ mol } XCl_2}{2 \text{ mol Cl}} \times \frac{1 \text{ mol X}}{\text{mol } XCl_2} = 3.60 \times 10^{-2} \text{ mol X}$$

So, 3.60×10^{-2} mol X has a mass equal to 7.45 g X. The molar mass of X is:

$$\frac{7.45 \text{ g X}}{3.60 \times 10^{-2} \text{ mol X}} = 207 \text{ g/mol X}; \quad \text{atomic mass} = 207 \text{ amu, so X is Pb.}$$

111.   For a gas, density and molar mass are directly proportional to each other.

$$\text{Molar mass XH}_n = 2.393(32.00) = \frac{76.58 \text{ g}}{\text{mol}}$$

$$0.803 \text{ g H}_2\text{O} \times \frac{2 \text{ mol H}}{18.02 \text{ g H}_2\text{O}} = 8.91 \times 10^{-2} \text{ mol H}$$

$$\frac{8.91 \times 10^{-2} \text{ mol H}}{2.23 \times 10^{-2} \text{ mol XH}_n} = \frac{4 \text{ mol H}}{\text{mol XH}_n}$$

Molar mass X = $76.58 - 4(1.008 \text{ g}) = 72.55$ g/mol;   the element is Ge.

113. $4.000 \text{ g M}_2\text{S}_3 \rightarrow 3.723 \text{ g MO}_2$

There must be twice as many moles of MO_2 as moles of M_2S_3 in order to balance M in the reaction. Setting up an equation for $2(\text{mol M}_2\text{S}_3) = \text{mol MO}_2$ where A = molar mass M:

$$2\left(\frac{4.000 \text{ g}}{2A + 3(32.07)}\right) = \frac{3.723 \text{ g}}{A + 2(16.00)}, \quad \frac{8.000}{2A + 96.21} = \frac{3.723}{A + 32.00}$$

$(8.000)A + 256.0 = (7.446)A + 358.2, \quad (0.554)A = 102.2, \quad A = 184$ g/mol; atomic mass
$= 184$ amu

115.   The balanced equations are:

$$C(s) + 1/2 \text{ O}_2(g) \rightarrow CO(g) \text{ and } C(s) + O_2(g) \rightarrow CO_2(g)$$

If we have 100.0 mol of products, then we have 72.0 mol CO_2, 16.0 mol CO, and 12.0 mol O_2. The initial moles of C equals 72.0 (from CO_2) + 16.00 (from CO) = 88.0 mol C and the initial moles of O_2 equals 72.0 (from CO_2) + 16.0/2 (from CO) + 12.0 (unreacted O_2) = 92.0 mol O_2. The initial reaction mixture contained:

$$\frac{92.0 \text{ mol O}_2}{88.0 \text{ mol C}} = 1.05 \text{ mol O}_2/\text{mol C}$$

117. $LaH_{2.90}$ is the formula. If only La^{3+} is present, LaH_3 would be the formula. If only La^{2+} is present, LaH_2 would be the formula. Let x = mol La^{2+} and y = mol La^{3+}:

$$(La^{2+})_x(La^{3+})_y\text{H}_{(2x + 3y)} \text{ where } x + y = 1.00 \text{ and } 2x + 3y = 2.90$$

Solving by simultaneous equations:

$$2x + 3y = 2.90$$
$$\underline{-2x - 2y = -2.00}$$
$$y = 0.90 \text{ and } x = 0.10$$

$LaH_{2.90}$ contains $\dfrac{1}{10}$ La^{2+}, or 10.% La^{2+}, and $\dfrac{9}{10}$ La^{3+}, or 90.% La^{3+}.

119.   Let x = mass KCl and y = mass KNO_3. Assuming 100.0 g of mixture, $x + y = 100.0$ g.

Molar mass KCl = 74.55 g/mol; molar mass KNO_3 = 101.11 g/mol

Mol KCl = $\dfrac{x}{74.55}$; mol KNO_3 = $\dfrac{y}{101.11}$

Knowing that the mixture is 43.2% K, then in the 100.0 g mixture:

$$39.10\left(\dfrac{x}{74.55} + \dfrac{y}{101.11}\right) = 43.2$$

We have two equations and two unknowns:

$$(0.5245)x + (0.3867)y = 43.2$$
$$x + y = 100.0$$

Solving, x = 32.9 g KCl; $\dfrac{32.9\text{ g}}{100.0\text{ g}} \times 100$ = 32.9% KCl

121.   The balanced equations are:

$$4\ NH_3(g) + 5\ O_2(g) \rightarrow 4\ NO(g) + 6\ H_2O(g) \text{ and } 4\ NH_3(g) + 7\ O_2(g) \rightarrow 4\ NO_2(g)$$
$$+ 6\ H_2O(g)$$

Let $4x$ = number of moles of NO formed, and let $4y$ = number of moles of NO_2 formed. Then:

$$4x\ NH_3 + 5x\ O_2 \rightarrow 4x\ NO + 6x\ H_2O \text{ and } 4y\ NH_3 + 7y\ O_2 \rightarrow 4y\ NO_2 + 6y\ H_2O$$

All the NH_3 reacted, so $4x + 4y = 2.00$.

$10.00 - 6.75 = 3.25$ mol O_2 reacted, so $5x + 7y = 3.25$.

Solving by the method of simultaneous equations:

$$20x + 28y = 13.0$$
$$\underline{-20x - 20y = -10.0}$$
$$8y = 3.0, \ y = 0.38; \ 4x + 4 \times 0.38 = 2.00, \ x = 0.12$$

Mol NO = $4x = 4 \times 0.12 = 0.48$ mol NO formed

CHAPTER 4

TYPES OF CHEMICAL REACTIONS AND SOLUTION STOICHIOMETRY

Aqueous Solutions: Strong and Weak Electrolytes

11. Solution A: $\dfrac{4 \text{ molecules}}{1.0 \text{ L}}$; solution B: $\dfrac{6 \text{ molecules}}{4.0 \text{ L}} = \dfrac{1.5 \text{ molecules}}{1.0 \text{ L}}$

Solution C: $\dfrac{4 \text{ molecules}}{2.0 \text{ L}} = \dfrac{2 \text{ molecules}}{1.0 \text{ L}}$; solution D: $\dfrac{6 \text{ molecules}}{2.0 \text{ L}} = \dfrac{3 \text{ molecules}}{1.0 \text{ L}}$

Solution A has the most molecules per unit volume so solution A is most concentrated. This is followed by solution D, then solution C. Solution B has the fewest molecules per unit volume, so solution B is least concentrated.

13. a. Polarity is a term applied to covalent compounds. Polar covalent compounds have an unequal sharing of electrons in bonds that results in unequal charge distribution in the overall molecule. Polar molecules have a partial negative end and a partial positive end. These are not full charges as in ionic compounds but are charges much smaller in magnitude. Water is a polar molecule and dissolves other polar solutes readily. The oxygen end of water (the partial negative end of the polar water molecule) aligns with the partial positive end of the polar solute, whereas the hydrogens of water (the partial positive end of the polar water molecule) align with the partial negative end of the solute. These opposite charge attractions stabilize polar solutes in water. This process is called hydration. Nonpolar solutes do not have permanent partial negative and partial positive ends; nonpolar solutes are not stabilized in water and do not dissolve.

b. KF is a soluble ionic compound, so it is a strong electrolyte. KF(aq) actually exists as separate hydrated K^+ ions and hydrated F^- ions in solution. $C_6H_{12}O_6$ is a polar covalent molecule that is a nonelectrolyte. $C_6H_{12}O_6$ is hydrated as described in part a.

c. RbCl is a soluble ionic compound, so it exists as separate hydrated Rb^+ ions and hydrated Cl^- ions in solution. AgCl is an insoluble ionic compound so the ions stay together in solution and fall to the bottom of the container as a precipitate.

d. HNO_3 is a strong acid and exists as separate hydrated H^+ ions and hydrated NO_3^- ions in solution. CO is a polar covalent molecule and is hydrated as explained in part a.

15. a. $Ba(NO_3)_2(aq) \rightarrow Ba^{2+}(aq) + 2 NO_3^-(aq)$; picture iv represents the Ba^{2+} and NO_3^- ions present in $Ba(NO_3)_2(aq)$.

25

b. $NaCl(aq) \rightarrow Na^+(aq) + Cl^-(aq)$; picture ii represents $NaCl(aq)$.

c. $K_2CO_3(aq) \rightarrow 2\ K^+(aq) + CO_3^{2-}(aq)$; picture iii represents $K_2CO_3(aq)$.

d. $MgSO_4(aq) \rightarrow Mg^{2+}(aq) + SO_4^{2-}(aq)$; picture i represents $MgSO_4(aq)$.

Solution Concentration: Molarity

17. a. $2.00\ L \times \dfrac{0.250\ mol\ NaOH}{L} \times \dfrac{40.00\ g\ NaOH}{mol} = 20.0\ g\ NaOH$

Place 20.0 g NaOH in a 2-L volumetric flask; add water to dissolve the NaOH, and fill to the mark with water, mixing several times along the way.

b. $2.00\ L \times \dfrac{0.250\ mol\ NaOH}{L} \times \dfrac{1\ L\ stock}{1.00\ mol\ NaOH} = 0.500\ L$

Add 500. mL of 1.00 M NaOH stock solution to a 2-L volumetric flask; fill to the mark with water, mixing several times along the way.

c. $2.00\ L \times \dfrac{0.100\ mol\ K_2CrO_4}{L} \times \dfrac{194.20\ g\ K_2CrO_4}{mol\ K_2CrO_4} = 38.8\ g\ K_2CrO_4$

Similar to the solution made in part a, instead using 38.8 g K_2CrO_4.

d. $2.00\ L \times \dfrac{0.100\ mol\ K_2CrO_4}{L} \times \dfrac{1\ L\ stock}{1.75\ mol\ K_2CrO_4} = 0.114\ L$

Similar to the solution made in part b, instead using 114 mL of the 1.75 M K_2CrO_4 stock solution.

19. Molar mass of NaOH = 22.99 + 16.00 + 1.008 = 40.00 g/mol

Mass NaOH = $0.2500\ L \times \dfrac{0.400\ mol\ NaOH}{L} \times \dfrac{40.00\ g\ NaOH}{mol\ NaOH} = 4.00\ g\ NaOH$

21. Mol solute = volume (L) $\times$ molarity$\left(\dfrac{mol}{L}\right)$; $AlCl_3(s) \rightarrow Al^{3+}(aq) + 3\ Cl^-(aq)$

Mol $Cl^- = 0.1000\ L \times \dfrac{0.30\ mol\ AlCl_3}{L} \times \dfrac{3\ mol\ Cl^-}{mol\ AlCl_3} = 9.0 \times 10^{-2}\ mol\ Cl^-$

$MgCl_2(s) \rightarrow Mg^{2+}(aq) + 2\ Cl^-(aq)$

Mol $Cl^- = 0.0500\ L \times \dfrac{0.60\ mol\ MgCl_2}{L} \times \dfrac{2\ mol\ Cl^-}{mol\ MgCl_2} = 6.0 \times 10^{-2}\ mol\ Cl^-$

$$NaCl(s) \rightarrow Na^+(aq) + Cl^-(aq)$$

$$\text{Mol } Cl^- = 0.2000 \text{ L} \times \frac{0.40 \text{ mol NaCl}}{L} \times \frac{1 \text{ mol } Cl^-}{\text{mol NaCl}} = 8.0 \times 10^{-2} \text{ mol } Cl^-$$

100.0 mL of 0.30 M $AlCl_3$ contains the most moles of Cl^- ions.

23.　$$\text{Mol } Na_2CO_3 = 0.0700 \text{ L} \times \frac{3.0 \text{ mol } Na_2CO_3}{L} = 0.21 \text{ mol } Na_2CO_3$$

$$Na_2CO_3(s) \rightarrow 2 \, Na^+(aq) + CO_3^{2-}(aq); \text{ mol } Na^+ = 2(0.21 \text{ mol}) = 0.42 \text{ mol}$$

$$\text{Mol } NaHCO_3 = 0.0300 \text{ L} \times \frac{1.0 \text{ mol } NaHCO_3}{L} = 0.030 \text{ mol } NaHCO_3$$

$$NaHCO_3(s) \rightarrow Na^+(aq) + HCO_3^-(aq); \text{ mol } Na^+ = 0.030 \text{ mol}$$

$$M_{Na^+} = \frac{\text{total mol } Na^+}{\text{total volume}} = \frac{0.42 \text{ mol} + 0.030 \text{ mol}}{0.0700 \text{ L} + 0.0300 \text{ L}} = \frac{0.45 \text{ mol}}{0.1000 \text{ L}} = 4.5 \, M \, Na^+$$

25.　$$\text{Stock solution} = \frac{10.0 \text{ mg}}{500.0 \text{ mL}} = \frac{10.0 \times 10^{-3} \text{ g}}{500.0 \text{ mL}} = \frac{2.00 \times 10^{-5} \text{ g steroid}}{\text{mL}}$$

$$100.0 \times 10^{-6} \text{ L stock} \times \frac{1000 \text{ mL}}{L} \times \frac{2.00 \times 10^{-5} \text{ g steroid}}{\text{mL}} = 2.00 \times 10^{-6} \text{ g steroid}$$

This is diluted to a final volume of 100.0 mL.

$$\frac{2.00 \times 10^{-6} \text{ g steroid}}{100.0 \text{ mL}} \times \frac{1000 \text{ mL}}{L} \times \frac{1 \text{ mol steroid}}{336.4 \text{ g steroid}} = 5.95 \times 10^{-8} \, M \text{ steroid}$$

27.　a.　$$5.0 \text{ ppb Hg in water} = \frac{5.0 \text{ ng Hg}}{\text{mL } H_2O} = \frac{5.0 \times 10^{-9} \text{ g Hg}}{\text{mL } H_2O}$$

$$\frac{5.0 \times 10^{-9} \text{ g Hg}}{\text{mL}} \times \frac{1 \text{ mol Hg}}{200.6 \text{ g Hg}} \times \frac{1000 \text{ mL}}{L} = 2.5 \times 10^{-8} \, M \text{ Hg}$$

　　b.　$$\frac{1.0 \times 10^{-9} \text{ g } CHCl_3}{\text{mL}} \times \frac{1 \text{ mol } CHCl_3}{119.4 \text{ g } CHCl_3} \times \frac{1000 \text{ mL}}{L} = 8.4 \times 10^{-9} \, M \, CHCl_3$$

　　c.　$$10.0 \text{ ppm As} = \frac{10.0 \text{ µg As}}{\text{mL}} = \frac{10.0 \times 10^{-6} \text{ g As}}{\text{mL}}$$

$$\frac{10.0 \times 10^{-6} \text{ g As}}{\text{mL}} \times \frac{1 \text{ mol As}}{74.92 \text{ g As}} \times \frac{1000 \text{ mL}}{L} = 1.33 \times 10^{-4} \, M \text{ As}$$

d. $\dfrac{0.10 \times 10^{-6}\ \text{g DDT}}{\text{mL}} \times \dfrac{1\ \text{mol DDT}}{354.5\ \text{g DDT}} \times \dfrac{1000\ \text{mL}}{\text{L}} = 2.8 \times 10^{-7}\ M\ \text{DDT}$

Precipitation Reactions

29.   Use the solubility rules in Table 4.1. Some soluble bromides by Rule 2 would be NaBr, KBr, and NH_4Br (there are others). The insoluble bromides by Rule 3 would be AgBr, $PbBr_2$, and Hg_2Br_2. Similar reasoning is used for the other parts to this problem.

Sulfates: Na_2SO_4, K_2SO_4, and $(NH_4)_2SO_4$ (and others) would be soluble, and $BaSO_4$, $CaSO_4$, and $PbSO_4$ (or Hg_2SO_4) would be insoluble.

Hydroxides: NaOH, KOH, $Ca(OH)_2$ (and others) would be soluble, and $Al(OH)_3$, $Fe(OH)_3$, and $Cu(OH)_2$ (and others) would be insoluble.

Phosphates: Na_3PO_4, K_3PO_4, $(NH_4)_3PO_4$ (and others) would be soluble, and Ag_3PO_4, $Ca_3(PO_4)_2$, and $FePO_4$ (and others) would be insoluble.

Lead: $PbCl_2$, $PbBr_2$, PbI_2, $Pb(OH)_2$, $PbSO_4$, and PbS (and others) would be insoluble. $Pb(NO_3)_2$ would be a soluble Pb^{2+} salt.

31.   Use Table 4.1 to predict the solubility of the possible products.

a. Possible products = Hg_2SO_4 and $Cu(NO_3)_2$; precipitate = Hg_2SO_4

b. Possible products = $NiCl_2$ and $Ca(NO_3)_2$; both salts are soluble so no precipitate forms.

c. Possible products = KI and $MgCO_3$; precipitate = $MgCO_3$

d. Possible products = NaBr and $Al_2(CrO_4)_3$; precipitate = $Al_2(CrO_4)_3$

33.   For the following answers, the balanced molecular equation is first, followed by the complete ionic equation, and then the net ionic equation.

a. $(NH_4)_2SO_4(aq) + Ba(NO_3)_2(aq) \rightarrow 2\ NH_4NO_3(aq) + BaSO_4(s)$

$2\ NH_4^+(aq) + SO_4^{2-}(aq) + Ba^{2+}(aq) + 2\ NO_3^-(aq) \rightarrow 2\ NH_4^+(aq) + 2\ NO_3^-(aq) + BaSO_4(s)$

$Ba^{2+}(aq) + SO_4^{2-}(aq) \rightarrow BaSO_4(s)$ is the net ionic equation (spectator ions omitted).

b. $Pb(NO_3)_2(aq) + 2\ NaCl(aq) \rightarrow PbCl_2(s) + 2\ NaNO_3(aq)$

$Pb^{2+}(aq) + 2\ NO_3^-(aq) + 2\ Na^+(aq) + 2\ Cl^-(aq) \rightarrow PbCl_2(s) + 2\ Na^+(aq) + 2\ NO_3^-(aq)$

$Pb^{2+}(aq) + 2\ Cl^-(aq) \rightarrow PbCl_2(s)$

c. The possible products, potassium phosphate and sodium nitrate, are both soluble in water. Therefore, no reaction occurs.

d. No reaction occurs because all possible products are soluble.

e. $CuCl_2(aq) + 2\ NaOH(aq) \rightarrow Cu(OH)_2(s) + 2\ NaCl(aq)$

$Cu^{2+}(aq) + 2\ Cl^-(aq) + 2\ Na^+(aq) + 2\ OH^-(aq) \rightarrow Cu(OH)_2(s) + 2\ Na^+(aq) + 2\ Cl^-(aq)$

$Cu^{2+}(aq) + 2\ OH^-(aq) \rightarrow Cu(OH)_2(s)$

35.   a.   When $CuSO_4(aq)$ is added to $Na_2S(aq)$, the precipitate that forms is $CuS(s)$. Therefore, Na^+ (the gray spheres) and SO_4^{2-} (the bluish green spheres) are the spectator ions.

$CuSO_4(aq) + Na_2S(aq) \rightarrow CuS(s) + Na_2SO_4(aq);\ \ Cu^{2+}(aq) + S^{2-}(aq) \rightarrow CuS(s)$

b.   When $CoCl_2(aq)$ is added to $NaOH(aq)$, the precipitate that forms is $Co(OH)_2(s)$. Therefore, Na^+ (the gray spheres) and Cl^- (the green spheres) are the spectator ions.

$CoCl_2(aq) + 2\ NaOH(aq) \rightarrow Co(OH)_2(s) + 2\ NaCl(aq)$

$Co^{2+}(aq) + 2\ OH^-(aq) \rightarrow Co(OH)_2(s)$

c.   When $AgNO_3(aq)$ is added to $KI(aq)$, the precipitate that forms is $AgI(s)$. Therefore, K^+ (the red spheres) and NO_3^- (the blue spheres) are the spectator ions.

$AgNO_3(aq) + KI(aq) \rightarrow AgI(s) + KNO_3(aq);\ \ Ag^+(aq) + I^-(aq) \rightarrow AgI(s)$

37.   Because a precipitate formed with Na_2SO_4, the possible cations are Ba^{2+}, Pb^{2+}, Hg_2^{2+}, and Ca^{2+} (from the solubility rules). Because no precipitate formed with KCl, Pb^{2+}, and Hg_2^{2+} cannot be present. Because both Ba^{2+} and Ca^{2+} form soluble chlorides and soluble hydroxides, both these cations could be present. Therefore, the cations could be Ba^{2+} and Ca^{2+} (by the solubility rules in Table 4.1). For students who do a more rigorous study of solubility, Sr^{2+} could also be a possible cation (it forms an insoluble sulfate salt, whereas the chloride and hydroxide salts of strontium are soluble).

39. $2\ AgNO_3(aq) + CaCl_2(aq) \rightarrow 2\ AgCl(s) + Ca(NO_3)_2(aq)$

$$\text{Mol } AgNO_3 = 0.1000\ L \times \frac{0.20\ \text{mol } AgNO_3}{L} = 0.020\ \text{mol } AgNO_3$$

$$\text{Mol } CaCl_2 = 0.1000\ L \times \frac{0.15\ \text{mol } CaCl_2}{L} = 0.015\ \text{mol } CaCl_2$$

The required mol $AgNO_3$ to mol $CaCl_2$ ratio is 2 : 1 (from the balanced equation). The actual mole ratio present is 0.020/0.015 = 1.3 (1.3 : 1). Therefore, $AgNO_3$ is the limiting reagent.

$$\text{Mass } AgCl = 0.020\ \text{mol } AgNO_3 \times \frac{1\ \text{mol } AgCl}{1\ \text{mol } AgNO_3} \times \frac{143.4\ \text{g } AgCl}{\text{mol } AgCl} = 2.9\ \text{g } AgCl$$

The net ionic equation is $Ag^+(aq) + Cl^-(aq) \rightarrow AgCl(s)$. The ions remaining in solution are the unreacted Cl^- ions and the spectator ions, NO_3^- and Ca^{2+} (all Ag^+ is used up in forming AgCl).   The moles of each ion present initially (before reaction) can be easily determined from the moles of each reactant. 0.020 mol $AgNO_3$ dissolves to form 0.020 mol Ag^+ and 0.020 mol NO_3^-. 0.015 mol $CaCl_2$ dissolves to form 0.015 mol Ca^{2+} and $2(0.015) = 0.030$ mol Cl^-.

Mol unreacted $Cl^- = 0.030$ mol Cl^- initially $-$ 0.020 mol Cl^- reacted

Mol unreacted $Cl^- = 0.010$ mol Cl^-

$$M_{Cl^-} = \frac{0.010 \text{ mol } Cl^-}{\text{total volume}} = \frac{0.010 \text{ mol } Cl^-}{0.1000 \text{ L} + 0.1000 \text{ L}} = 0.050 \ M \ Cl^-$$

The molarity of the spectator ions are:

$$M_{NO_3^-} = \frac{0.020 \text{ mol } NO_3^-}{0.2000 \text{ L}} = 0.10 \ M \ NO_3^-; \quad M_{Ca_2^+} = \frac{0.015 \text{ mol } Ca^{2+}}{0.2000 \text{ L}} = 0.075 \ M \ Ca^{2+}$$

41. $2 \ AgNO_3(aq) + Na_2CrO_4(aq) \rightarrow Ag_2CrO_4(s) + 2 \ NaNO_3(aq)$

$$0.0750 \text{ L} \times \frac{0.100 \text{ mol } AgNO_3}{\text{L}} \times \frac{1 \text{ mol } Na_2CrO_4}{2 \text{ mol } AgNO_3} \times \frac{161.98 \text{ g } Na_2CrO_4}{\text{mol } Na_2CrO_4} = 0.607 \text{ g } Na_2CrO_4$$

43.   Use aluminum in the formulas to convert from mass of $Al(OH)_3$ to mass of $Al_2(SO_4)_3$ in the mixture.

$$0.107 \text{ g } Al(OH)_3 \times \frac{1 \text{ mol } Al(OH)_3}{78.00 \text{ g}} \times \frac{1 \text{ mol } Al^{3+}}{\text{mol } Al(OH)_3} \times \frac{1 \text{ mol } Al_2(SO_4)_3}{2 \text{ mol } Al^{3+}} \times$$

$$\frac{342.17 \text{ g } Al_2(SO_4)_3}{\text{mol } Al_2(SO_4)_3} = 0.235 \text{ g } Al_2(SO_4)_3$$

$$\text{Mass } \% \ Al_2(SO_4)_3 = \frac{0.235 \text{ g}}{1.45 \text{ g}} \times 100 = 16.2\%$$

45.   All the sulfur in $BaSO_4$ came from the saccharin. The conversion from $BaSO_4$ to saccharin uses the molar masses and formulas of each compound.

$$0.5032 \text{ g } BaSO_4 \times \frac{32.07 \text{ g S}}{233.4 \text{ g } BaSO_4} \times \frac{183.9 \text{ g saccharin}}{32.07 \text{ g S}} = 0.3949 \text{ g saccharin}$$

$$\frac{\text{Average mass}}{\text{Tablet}} = \frac{0.3949 \text{ g}}{10 \text{ tablets}} = \frac{3.949 \times 10^{-2} \text{ g}}{\text{tablet}} = \frac{39.49 \text{ mg}}{\text{tablet}}$$

$$\text{Average mass } \% = \frac{0.3949 \text{ g saccharin}}{0.5894 \text{ g}} \times 100 = 67.00\% \text{ saccharin by mass}$$

47. $M_2SO_4(aq) + CaCl_2(aq) \rightarrow CaSO_4(s) + 2 \ MCl(aq)$

$$1.36 \text{ g } CaSO_4 \times \frac{1 \text{ mol } CaSO_4}{136.15 \text{ g } CaSO_4} \times \frac{1 \text{ mol } M_2SO_4}{\text{mol } CaSO_4} = 9.99 \times 10^{-3} \text{ mol } M_2SO_4$$

From the problem, 1.42 g M_2SO_4 was reacted, so:

$$\text{molar mass} = \frac{1.42 \text{ g } M_2SO_4}{9.99 \times 10^{-3} \text{ mol } M_2SO_4} = 142 \text{ g/mol}$$

142 amu = 2(atomic mass M) + 32.07 + 4(16.00), atomic mass M = 23 amu

From periodic table, M is Na (sodium).

Acid-Base Reactions

49.   a.   Perchloric acid reacted with potassium hydroxide is a possibility.

$$HClO_4(aq) + KOH(aq) \rightarrow H_2O(l) + KClO_4(aq)$$

b.   Nitric acid reacted with cesium hydroxide is a possibility.

$$HNO_3(aq) + CsOH(aq) \rightarrow H_2O(l) + CsNO_3(aq)$$

c.   Hydroiodic acid reacted with calcium hydroxide is a possibility.

$$2 \text{ HI}(aq) + Ca(OH)_2(aq) \rightarrow 2 \text{ H}_2O(l) + CaI_2(aq)$$

51.   If we begin with 50.00 mL of 0.100 M NaOH, then:

$$50.00 \times 10^{-3} \text{ L} \times \frac{0.100 \text{ mol}}{L} = 5.00 \times 10^{-3} \text{ mol NaOH to be neutralized.}$$

a. $NaOH(aq) + HCl(aq) \rightarrow NaCl(aq) + H_2O(l)$

$$5.00 \times 10^{-3} \text{ mol NaOH} \times \frac{1 \text{ mol HCl}}{\text{mol NaOH}} \times \frac{1 \text{ L soln}}{0.100 \text{ mol}} = 5.00 \times 10^{-2} \text{ L or 50.0 mL}$$

b. $2 \text{ NaOH}(aq) + H_2SO_3(aq) \rightarrow 2 \text{ H}_2O(l) + Na_2SO_3(aq)$

$$5.00 \times 10^{-3} \text{ mol NaOH} \times \frac{1 \text{ mol H}_2SO_3}{2 \text{ mol NaOH}} \times \frac{1 \text{ L soln}}{0.100 \text{ mol H}_2SO_3} = 2.50 \times 10^{-2} \text{ L or 25.0 mL}$$

c. $3 \text{ NaOH}(aq) + H_3PO_4(aq) \rightarrow Na_3PO_4(aq) + 3 \text{ H}_2O(l)$

$$5.00 \times 10^{-3} \text{ mol NaOH} \times \frac{1 \text{ mol H}_3PO_4}{3 \text{ mol NaOH}} \times \frac{1 \text{ L soln}}{0.200 \text{ mol H}_3PO_4} = 8.33 \times 10^{-3} \text{ L or 8.33 mL}$$

d. $HNO_3(aq) + NaOH(aq) \rightarrow H_2O(l) + NaNO_3(aq)$

$$5.00 \times 10^{-3} \text{ mol NaOH} \times \frac{1 \text{ mol HNO}_3}{\text{mol NaOH}} \times \frac{1 \text{ L soln}}{0.150 \text{ mol HNO}_3} = 3.33 \times 10^{-2} \text{ L or 33.3 mL}$$

e. $HC_2H_3O_2(aq) + NaOH(aq) \rightarrow H_2O(l) + NaC_2H_3O_2(aq)$

$$5.00 \times 10^{-3} \text{ mol NaOH} \times \frac{1 \text{ mol } HC_2H_3O_2}{\text{mol NaOH}} \times \frac{1 \text{ L soln}}{0.200 \text{ mol } HC_2H_3O_2} = 2.50 \times 10^{-2} \text{ L}$$

or 25.0 mL

f. $H_2SO_4(aq) + 2 \text{ NaOH}(aq) \rightarrow 2 H_2O(l) + Na_2SO_4(aq)$

$$5.00 \times 10^{-3} \text{ mol NaOH} \times \frac{1 \text{ mol } H_2SO_4}{2 \text{ mol NaOH}} \times \frac{1 \text{ L soln}}{0.300 \text{ mol } H_2SO_4} = 8.33 \times 10^{-3} \text{ L or } 8.33 \text{ mL}$$

53.   The acid is a diprotic acid (H_2A) meaning that it has two H^+ ions in the formula to donate to a base. The reaction is $H_2A(aq) + 2 \text{ NaOH}(aq) \rightarrow 2 H_2O(l) + Na_2A(aq)$, where A^{2-} is what is left over from the acid formula when the two protons (H^+ ions) are reacted.

For the HCl reaction, the base has the ability to accept two protons. The most common examples are $Ca(OH)_2$, $Sr(OH)_2$, and $Ba(OH)_2$. A possible reaction would be $2 \text{ HCl}(aq) + Ca(OH)_2(aq) \rightarrow 2 H_2O(l) + CaCl_2(aq)$.

55.   The pertinent reactions are:

$$2 \text{ NaOH}(aq) + H_2SO_4(aq) \rightarrow Na_2SO_4(aq) + 2 H_2O(l)$$

$$HCl(aq) + NaOH(aq) \rightarrow NaCl(aq) + H_2O(l)$$

Amount of NaOH added $= 0.0500 \text{ L} \times \dfrac{0.213 \text{ mol}}{L} = 1.07 \times 10^{-2} \text{ mol NaOH}$

Amount of NaOH neutralized by HCl:

$$0.01321 \text{ L HCl} \quad H \quad \frac{0.103 \text{ mol HCl}}{\text{L HCl}} \times \frac{1 \text{ mol NaOH}}{\text{mol HCl}} = 1.36 \times 10^{-3} \text{ mol NaOH}$$

The difference, 9.3×10^{-3} mol, is the amount of NaOH neutralized by the sulfuric acid.

$$9.3 \times 10^{-3} \text{ mol NaOH} \times \frac{1 \text{ mol } H_2SO_4}{2 \text{ mol NaOH}} = 4.7 \times 10^{-3} \text{ mol } H_2SO_4$$

Concentration of $H_2SO_4 = \dfrac{4.7 \times 10^{-3} \text{ mol}}{0.1000 \text{ L}} = 4.7 \times 10^{-2} \, M \, H_2SO_4$

57. $HC_2H_3O_2(aq) + NaOH(aq) \rightarrow H_2O(l) + NaC_2H_3O_2(aq)$

a. $16.58 \times 10^{-3} \text{ L soln} \quad H \quad \dfrac{0.5062 \text{ mol NaOH}}{\text{L soln}} \times \dfrac{1 \text{ mol acetic acid}}{\text{mol NaOH}}$

$$= 8.393 \times 10^{-3} \text{ mol acetic acid}$$

Concentration of acetic acid $= \dfrac{8.393 \times 10^{-3} \text{ mol}}{0.01000 \text{ L}} = 0.8393 \, M \, HC_2H_3O_2$

b. If we have 1.000 L of solution: total mass = 1000. mL $\times \dfrac{1.006\ g}{mL}$ = 1006 g solution

Mass of $HC_2H_3O_2$ = 0.8393 mol $\times \dfrac{60.052\ g}{mol}$ = 50.40 g $HC_2H_3O_2$

Mass % acetic acid = $\dfrac{50.40\ g}{1006\ g} \times 100$ = 5.010%

59. HCl and HNO_3 are strong acids; $Ca(OH)_2$ and RbOH are strong bases. The net ionic equation that occurs is $H^+(aq) + OH^-(aq) \rightarrow H_2O(l)$.

Mol H^+ = 0.0500 L $\times \dfrac{0.100\ mol\ HCl}{L} \times \dfrac{1\ mol\ H^+}{mol\ HCl}$ +

0.1000 L $\times \dfrac{0.200\ mol\ HNO_3}{L} \times \dfrac{1\ mol\ H^+}{mol\ HNO_3}$ = 0.00500 + 0.0200 = 0.0250 mol H^+

Mol OH^- = 0.5000 L $\times \dfrac{0.0100\ mol\ Ca(OH)_2}{L} \times \dfrac{2\ mol\ OH^-}{mol\ Ca(OH)_2}$ +

0.2000 L $\times \dfrac{0.100\ mol\ RbOH}{L} \times \dfrac{1\ mol\ OH^-}{mol\ RbOH}$ = 0.0100 + 0.0200 = 0.0300 mol OH^-

We have an excess of OH^-, so the solution is basic (not neutral). The moles of excess OH^- = 0.0300 mol OH^- initially – 0.0250 mol OH^- reacted (with H^+) = 0.0050 mol OH^- excess.

$$M_{OH^-} = \dfrac{0.0050\ mol\ OH^-}{(0.05000 + 0.1000 + 0.5000 + 0.2000)\ L} = \dfrac{0.0050\ mol}{0.8500\ L} = 5.9 \times 10^{-3}\ M$$

61. $Ba(OH)_2(aq) + 2\ HCl(aq) \rightarrow BaCl_2(aq) + 2\ H_2O(l)$; $H^+(aq) + OH^-(aq) \rightarrow H_2O(l)$

75.0 $\times 10^{-3}$ L $\times \dfrac{0.250\ mol\ HCl}{L}$ = 1.88×10^{-2} mol HCl = 1.88×10^{-2} mol H^+ +
1.88×10^{-2} mol Cl^-

225.0 $\times 10^{-3}$ L $\times \dfrac{0.0550\ mol\ Ba(OH)_2}{L}$ = 1.24×10^{-2} mol $Ba(OH)_2$ = 1.24×10^{-2} mol Ba^{2+} +
2.48×10^{-2} mol OH^-

The net ionic equation requires a 1 : 1 mol ratio between OH^- and H^+. The actual mol OH^- to mol H^+ ratio is greater than 1 : 1, so OH^- is in excess. Because 1.88×10^{-2} mol OH^- will be neutralized by the H^+, we have $(2.48 - 1.88) \times 10^{-2}$ = 0.60×10^{-2} mol OH^- in excess.

$$M_{OH^-} = \dfrac{mol\ OH^-\ excess}{total\ volume} = \dfrac{6.0 \times 10^{-3}\ mol\ OH^-}{0.0750\ L + 0.2250\ L} = 2.0 \times 10^{-2}\ M\ OH^-$$

Oxidation-Reduction Reactions

63.   a.   The species reduced is the element that gains electrons. The reducing agent causes reduction to occur by itself being oxidized. The reducing agent generally refers to the entire formula of the compound/ion that contains the element oxidized.

b.   The species oxidized is the element that loses electrons. The oxidizing agent causes oxidation to occur by itself being reduced. The oxidizing agent generally refers to the entire formula of the compound/ion that contains the element reduced.

c.   For simple binary ionic compounds, the actual charge on the ions are the same as the oxidation states. For covalent compounds and ions, nonzero oxidation states are imaginary charges the elements would have if they were held together by ionic bonds (assuming the bond is between two different nonmetals). Nonzero oxidation states for elements in covalent compounds are not actual charges. Oxidation states for covalent compounds are a bookkeeping method to keep track of electrons in a reaction.

65.   Apply rules in Table 4.3.

a. $KMnO_4$ is composed of K^+ and MnO_4^- ions. Assign oxygen an oxidation state value of -2, which gives manganese a $+7$ oxidation state because the sum of oxidation states for all atoms in MnO_4^- must equal the $1-$ charge on MnO_4^-. K, $+1$; O, -2; Mn, $+7$.

b.   Assign O a -2 oxidation state, which gives nickel a $+4$ oxidation state. Ni, $+4$; O, -2.

c. $K_4Fe(CN)_6$ is composed of K^+ cations and $Fe(CN)_6^{4-}$ anions. $Fe(CN)_6^{4-}$ is composed of iron and CN^- anions. For an overall anion charge of $4-$, iron must have a $+2$ oxidation state.

d. $(NH_4)_2HPO_4$ is made of NH_4^+ cations and HPO_4^{2-} anions. Assign $+1$ as oxidation state of H and -2 as the oxidation state of O. For N in NH_4^+: $x + 4(+1) = +1$, $x = -3 =$ oxidation state of N. For P in HPO_4^{2-}: $+1 + y + 4(-2) = -2$, $y = +5 =$ oxidation state of P.

e.   O, -2; P, $+3$ f.   O, -2; Fe, $+8/3$

g.   O, -2; F, -1; Xe, $+6$ h.   F, -1; S, $+4$

i.   O, -2; C, $+2$ j.   H, $+1$; O, -2; C, 0

67.   a. $SrCr_2O_7$: Composed of Sr^{2+} and $Cr_2O_7^{2-}$ ions. Sr, $+2$; O, -2; Cr, $2x + 7(-2) = -2$, $x = +6$

b.   Cu, $+2$; Cl, -1;   c.   O, 0;   d.   H, $+1$; O, -1

e. Mg^{2+} and CO_3^{2-} ions present. Mg, $+2$; O, -2; C, $+4$;   f.   Ag, 0

g. Pb^{2+} and SO_3^{2-} ions present. Pb, $+2$; O, -2; S, $+4$;   h.   O, -2; Pb, $+4$

i. Na^+ and $C_2O_4^{2-}$ ions present. Na, $+1$; O, -2; C, $2x + 4(-2) = -2$, $x = +3$

j.   O, −2; C, +4

k.   Ammonium ion has a 1+ charge (NH_4^+), and sulfate ion has a 2− charge (SO_4^{2-}). Therefore, the oxidation state of cerium must be +4 (Ce^{4+}). H, +1; N, −3; O, −2; S, +6

l.   O, −2; Cr, +3

69.   a. $Al(s) + 3 HCl(aq) \rightarrow AlCl_3(aq) + 3/2 H_2(g)$ or $2 Al(s) + 6 HCl(aq) \rightarrow 2 AlCl_3(aq) +$

$$3 H_2(g)$$

Hydrogen is reduced (goes from the +1 oxidation state to the 0 oxidation state), and aluminum Al is oxidized (0 → +3).

b.   Balancing S is most complicated because sulfur is in both products. Balance C and H first; then worry about S.

$$CH_4(g) + 4 S(s) \rightarrow CS_2(l) + 2 H_2S(g)$$

Sulfur is reduced (0 → −2), and carbon is oxidized (−4 → +4).

c.   Balance C and H first; then balance O.

$$C_3H_8(g) + 5 O_2(g) \rightarrow 3 CO_2(g) + 4 H_2O(l)$$

Oxygen is reduced (0 → −2), and carbon is oxidized (−8/3 → +4).

d.   Although this reaction is mass balanced, it is not charge balanced. We need 2 mol of silver on each side to balance the charge.

$$Cu(s) + 2 Ag^+(aq) \rightarrow 2 Ag(s) + Cu^{2+}(aq)$$

Silver is reduced (+1 → 0), and copper is oxidized (0 → +2).

71.   a.   Review Section 4.11 of the text for rules on balancing by the half-reaction method. The first step is to separate the reaction into two half-reactions, and then balance each half-reaction separately.

$(Cu \rightarrow Cu^{2+} + 2 e^-) \times 3$ $NO_3^- \rightarrow NO + 2 H_2O$
                                                  $(3 e^- + 4 H^+ + NO_3^- \rightarrow NO + 2 H_2O) \times 2$

Adding the two balanced half-reactions so electrons cancel:

$$3 Cu \rightarrow 3 Cu^{2+} + 6 e^-$$
$$6 e^- + 8 H^+ + 2 NO_3^- \rightarrow 2 NO + 4 H_2O$$

$$\overline{3 Cu(s) + 8 H^+(aq) + 2 NO_3^-(aq) \rightarrow 3 Cu^{2+}(aq) + 2 NO(g) + 4 H_2O(l)}$$

The final step is to simplify the equation by cancelling identical species on both sides of the equations. Other than the electrons, this equation has no identical species to cancel, so this is the balanced equation. Typically, H^+ and H_2O are the species which can be cancelled in this final step (other than the electrons).

b. $(2 \text{ Cl}^- \rightarrow \text{Cl}_2 + 2 \text{ e}^-) \times 3$

$$\text{Cr}_2\text{O}_7{}^{2-} \rightarrow 2 \text{ Cr}^{3+} + 7 \text{ H}_2\text{O}$$
$$6 \text{ e}^- + 14 \text{ H}^+ + \text{Cr}_2\text{O}_7{}^{2-} \rightarrow 2 \text{ Cr}^{3+} + 7 \text{ H}_2\text{O}$$

Add the two balanced half-reactions with six electrons transferred:

$$6 \text{ Cl}^- \rightarrow 3 \text{ Cl}_2 + 6 \text{ e}^-$$
$$6 \text{ e}^- + 14 \text{ H}^+ + \text{Cr}_2\text{O}_7{}^{2-} \rightarrow 2 \text{ Cr}^{3+} + 7 \text{ H}_2\text{O}$$

$$\overline{14 \text{ H}^+(aq) + \text{Cr}_2\text{O}_7{}^{2-}(aq) + 6 \text{ Cl}^-(aq) \rightarrow 3 \text{ Cl}_2(g) + 2 \text{ Cr}^{3+}(aq) + 7 \text{ H}_2\text{O}(l)}$$

c.

$$\text{Pb} \rightarrow \text{PbSO}_4 \qquad\qquad\qquad \text{PbO}_2 \rightarrow \text{PbSO}_4$$
$$\text{Pb} + \text{H}_2\text{SO}_4 \rightarrow \text{PbSO}_4 + 2 \text{ H}^+ \qquad\qquad \text{PbO}_2 + \text{H}_2\text{SO}_4 \rightarrow \text{PbSO}_4 + 2 \text{ H}_2\text{O}$$
$$\text{Pb} + \text{H}_2\text{SO}_4 \rightarrow \text{PbSO}_4 + 2 \text{ H}^+ + 2 \text{ e}^- \quad 2 \text{ e}^- + 2 \text{ H}^+ + \text{PbO}_2 + \text{H}_2\text{SO}_4 \rightarrow \text{PbSO}_4 + 2 \text{ H}_2\text{O}$$

Add the two half-reactions with two electrons transferred:

$$2 \text{ e}^- + 2 \text{ H}^+ + \text{PbO}_2 + \text{H}_2\text{SO}_4 \rightarrow \text{PbSO}_4 + 2 \text{ H}_2\text{O}$$
$$\text{Pb} + \text{H}_2\text{SO}_4 \rightarrow \text{PbSO}_4 + 2 \text{ H}^+ + 2 \text{ e}^-$$

$$\overline{\text{Pb}(s) + 2 \text{ H}_2\text{SO}_4(aq) + \text{PbO}_2(s) \rightarrow 2 \text{ PbSO}_4(s) + 2 \text{ H}_2\text{O}(l)}$$

This is the reaction that occurs in an automobile lead storage battery.

d.

$$\text{Mn}^{2+} \rightarrow \text{MnO}_4{}^-$$
$$(4 \text{ H}_2\text{O} + \text{Mn}^{2+} \rightarrow \text{MnO}_4{}^- + 8 \text{ H}^+ + 5 \text{ e}^-) \times 2$$

$$\text{NaBiO}_3 \rightarrow \text{Bi}^{3+} + \text{Na}^+$$
$$6 \text{ H}^+ + \text{NaBiO}_3 \rightarrow \text{Bi}^{3+} + \text{Na}^+ + 3 \text{ H}_2\text{O}$$
$$(2 \text{ e}^- + 6 \text{ H}^+ + \text{NaBiO}_3 \rightarrow \text{Bi}^{3+} + \text{Na}^+ + 3 \text{ H}_2\text{O}) \times 5$$

$$8 \text{ H}_2\text{O} + 2 \text{ Mn}^{2+} \rightarrow 2 \text{ MnO}_4{}^- + 16 \text{ H}^+ + 10 \text{ e}^-$$
$$10 \text{ e}^- + 30 \text{ H}^+ + 5 \text{ NaBiO}_3 \rightarrow 5 \text{ Bi}^{3+} + 5 \text{ Na}^+ + 15 \text{ H}_2\text{O}$$

$$\overline{8 \text{ H}_2\text{O} + 30 \text{ H}^+ + 2 \text{ Mn}^{2+} + 5 \text{ NaBiO}_3 \rightarrow 2 \text{ MnO}_4{}^- + 5 \text{ Bi}^{3+} + 5 \text{ Na}^+ + 15 \text{ H}_2\text{O} + 16 \text{ H}^+}$$

Simplifying:

$$14 \text{ H}^+(aq) + 2 \text{ Mn}^{2+}(aq) + 5 \text{ NaBiO}_3(s) \rightarrow 2 \text{ MnO}_4{}^-(aq) + 5 \text{ Bi}^{3+}(aq) + 5 \text{ Na}^+(aq) + 7 \text{ H}_2\text{O}(l)$$

e.

$$\text{H}_3\text{AsO}_4 \rightarrow \text{AsH}_3 \qquad\qquad (Zn \rightarrow \text{Zn}^{2+} + 2 \text{ e}^-) \times 4$$
$$\text{H}_3\text{AsO}_4 \rightarrow \text{AsH}_3 + 4 \text{ H}_2\text{O}$$
$$8 \text{ e}^- + 8 \text{ H}^+ + \text{H}_3\text{AsO}_4 \rightarrow \text{AsH}_3 + 4 \text{ H}_2\text{O}$$

$$8 \text{ e}^- + 8 \text{ H}^+ + \text{H}_3\text{AsO}_4 \rightarrow \text{AsH}_3 + 4 \text{ H}_2\text{O}$$
$$4 \text{ Zn} \rightarrow 4 \text{ Zn}^{2+} + 8 \text{ e}^-$$

$$\overline{8 \text{ H}^+(aq) + \text{H}_3\text{AsO}_4(aq) + 4 \text{ Zn}(s) \rightarrow 4 \text{ Zn}^{2+}(aq) + \text{AsH}_3(g) + 4 \text{ H}_2\text{O}(l)}$$

f.$\qquad$ $As_2O_3 \rightarrow H_3AsO_4$

$\qquad\qquad$ $As_2O_3 \rightarrow 2\ H_3AsO_4$

$\qquad$ $(5\ H_2O + As_2O_3 \rightarrow 2\ H_3AsO_4 + 4\ H^+ + 4\ e^-) \times 3$

$\qquad\qquad\qquad\qquad\qquad\qquad\qquad\qquad$ $NO_3^- \rightarrow NO + 2\ H_2O$

$\qquad\qquad\qquad\qquad\qquad\qquad\qquad$ $4\ H^+ + NO_3^- \rightarrow NO + 2\ H_2O$

$\qquad\qquad\qquad\qquad\qquad$ $(3\ e^- + 4\ H^+ + NO_3^- \rightarrow NO + 2\ H_2O) \times 4$

$\qquad\qquad\qquad\qquad$ $12\ e^- + 16\ H^+ + 4\ NO_3^- \rightarrow 4\ NO + 8\ H_2O$

$\qquad\qquad\qquad\qquad\qquad$ $15\ H_2O + 3\ As_2O_3 \rightarrow 6\ H_3AsO_4 + 12\ H^+ + 12\ e^-$

$\rule{12cm}{0.4pt}$

$7\ H_2O(l) + 4\ H^+(aq) + 3\ As_2O_3(s) + 4\ NO_3^-(aq) \rightarrow 4\ NO(g) + 6\ H_3AsO_4(aq)$

g.$\quad$ $(2\ Br^- \rightarrow Br_2 + 2\ e^-) \times 5$ $\qquad\qquad\qquad$ $MnO_4^- \rightarrow Mn^{2+} + 4\ H_2O$

$\qquad\qquad\qquad\qquad\qquad\qquad$ $(5\ e^- + 8\ H^+ + MnO_4^- \rightarrow Mn^{2+} + 4\ H_2O) \times 2$

$\qquad\qquad\qquad\qquad$ $10\ Br^- \rightarrow 5\ Br_2 + 10\ e^-$

$\qquad\qquad$ $10\ e^- + 16\ H^+ + 2\ MnO_4^- \rightarrow 2\ Mn^{2+} + 8\ H_2O$

$\rule{12cm}{0.4pt}$

$16\ H^+(aq) + 2\ MnO_4^-(aq) + 10\ Br^-(aq) \rightarrow 5\ Br_2(l) + 2\ Mn^{2+}(aq) + 8\ H_2O(l)$

h.$\quad$ $CH_3OH \rightarrow CH_2O$ $\qquad\qquad\qquad\qquad$ $Cr_2O_7^{2-} \rightarrow Cr^{3+}$

$\quad$ $(CH_3OH \rightarrow CH_2O + 2\ H^+ + 2\ e^-) \times 3$ $\qquad$ $14\ H^+ + Cr_2O_7^{2-} \rightarrow 2\ Cr^{3+} + 7\ H_2O$

$\qquad\qquad\qquad\qquad\qquad\qquad$ $6\ e^- + 14\ H^+ + Cr_2O_7^{2-} \rightarrow 2\ Cr^{3+} + 7\ H_2O$

$\qquad\qquad\qquad$ $3\ CH_3OH \rightarrow 3\ CH_2O + 6\ H^+ + 6\ e^-$

$\qquad\qquad\qquad$ $6\ e^- + 14\ H^+ + Cr_2O_7^{2-} \rightarrow 2\ Cr^{3+} + 7\ H_2O$

$\rule{12cm}{0.4pt}$

$8\ H^+(aq) + 3\ CH_3OH(aq) + Cr_2O_7^{2-}(aq) \rightarrow 2\ Cr^{3+}(aq) + 3\ CH_2O(aq) + 7\ H_2O(l)$

73.$\quad$ a.$\quad$ HCl(aq) dissociates to $H^+(aq) + Cl^-(aq)$. For simplicity, let's use H^+ and Cl^- separately.

$\qquad$ $H^+ \rightarrow H_2$ $\qquad\qquad\qquad\qquad\qquad$ $Fe \rightarrow HFeCl_4$

$\quad$ $(2\ H^+ + 2\ e^- \rightarrow H_2) \times 3$ $\qquad\qquad$ $(H^+ + 4\ Cl^- + Fe \rightarrow HFeCl_4 + 3\ e^-) \times 2$

$\qquad\qquad$ $6\ H^+ + 6\ e^- \rightarrow 3\ H_2$

$\qquad$ $2\ H^+ + 8\ Cl^- + 2\ Fe \rightarrow 2\ HFeCl_4 + 6\ e^-$

$\rule{9cm}{0.4pt}$

$\qquad$ $8\ H^+ + 8\ Cl^- + 2\ Fe \rightarrow 2\ HFeCl_4 + 3\ H_2$

or$\qquad$ $8\ HCl(aq) + 2\ Fe(s) \rightarrow 2\ HFeCl_4(aq) + 3\ H_2(g)$

b.
$$IO_3^- \rightarrow I_3^-$$
$$3\ IO_3^- \rightarrow I_3^-$$
$$3\ IO_3^- \rightarrow I_3^- + 9\ H_2O$$
$$16\ e^- + 18\ H^+ + 3\ IO_3^- \rightarrow I_3^- + 9\ H_2O$$

$$I^- \rightarrow I_3^-$$
$$(3\ I^- \rightarrow I_3^- + 2\ e^-) \times 8$$

$$16\ e^- + 18\ H^+ + 3\ IO_3^- \rightarrow I_3^- + 9\ H_2O$$
$$24\ I^- \rightarrow 8\ I_3^- + 16\ e^-$$

$$18\ H^+ + 24\ I^- + 3\ IO_3^- \rightarrow 9\ I_3^- + 9\ H_2O$$

Reducing: $6\ H^+(aq) + 8\ I^-(aq) + IO_3^-(aq) \rightarrow 3\ I_3^-(aq) + 3\ H_2O(l)$

c. $(Ce^{4+} + e^- \rightarrow Ce^{3+}) \times 97$
$$Cr(NCS)_6^{4-} \rightarrow Cr^{3+} + NO_3^- + CO_2 + SO_4^{2-}$$
$$54\ H_2O + Cr(NCS)_6^{4-} \rightarrow Cr^{3+} + 6\ NO_3^- + 6\ CO_2 + 6\ SO_4^{2-} + 108\ H^+$$

Charge on left $= -4$. Charge on right $= +3 + 6(-1) + 6(-2) + 108(+1) = +93$. Add 97 e^- to the product side, and then add the two balanced half-reactions with a common factor of 97 e^- transferred.

$$54\ H_2O + Cr(NCS)_6^{4-} \rightarrow Cr^{3+} + 6\ NO_3^- + 6\ CO_2 + 6\ SO_4^{2-} + 108\ H^+ + 97\ e^-$$
$$97\ e^- + 97\ Ce^{4+} \rightarrow 97\ Ce^{3+}$$

$$97\ Ce^{4+}(aq) + 54\ H_2O(l) + Cr(NCS)_6^{4-}(aq) \rightarrow 97\ Ce^{3+}(aq) + Cr^{3+}(aq) + 6\ NO_3^-(aq)$$
$$+ 6\ CO_2(g) + 6\ SO_4^{2-}(aq) + 108\ H^+(aq)$$

This is very complicated. A check of the net charge is a good check to see if the equation is balanced. Left: charge $= 97(+4) - 4 = +384$. Right: charge $= 97(+3) + 3 + 6(-1) + 6(-2) + 108(+1) = +384$.

d.
$$CrI_3 \rightarrow CrO_4^{2-} + IO_4^-$$
$$(16\ H_2O + CrI_3 \rightarrow CrO_4^{2-} + 3\ IO_4^- + 32\ H^+ + 27\ e^-) \times 2$$

$$Cl_2 \rightarrow Cl^-$$
$$(2\ e^- + Cl_2 \rightarrow 2\ Cl^-) \times 27$$

Common factor is a transfer of 54 e^-.

$$54\ e^- + 27\ Cl_2 \rightarrow 54\ Cl^-$$
$$32\ H_2O + 2\ CrI_3 \rightarrow 2\ CrO_4^{2-} + 6\ IO_4^- + 64\ H^+ + 54\ e^-$$

$$32\ H_2O + 2\ CrI_3 + 27\ Cl_2 \rightarrow 54\ Cl^- + 2\ CrO_4^{2-} + 6\ IO_4^- + 64\ H^+$$

Add 64 OH^- to both sides and convert 64 H^+ into 64 H_2O.

$$64\ OH^- + 32\ H_2O + 2\ CrI_3 + 27\ Cl_2 \rightarrow 54\ Cl^- + 2\ CrO_4^{2-} + 6\ IO_4^- + 64\ H_2O$$

Reducing gives:

$$64\ OH^-(aq) + 2\ CrI_3(s) + 27\ Cl_2(g) \rightarrow 54\ Cl^-(aq) + 2\ CrO_4^{2-}(aq) + 6\ IO_4^-(aq)$$
$$+ 32\ H_2O(l)$$

e.
$$Ce^{4+} \rightarrow Ce(OH)_3$$
$$(e^- + 3 H_2O + Ce^{4+} \rightarrow Ce(OH)_3 + 3 H^+) \times 61$$

$$Fe(CN)_6^{4-} \rightarrow Fe(OH)_3 + CO_3^{2-} + NO_3^-$$
$$Fe(CN)_6^{4-} \rightarrow Fe(OH)_3 + 6 CO_3^{2-} + 6 NO_3^-$$

There are 39 extra O atoms on right. Add 39 H_2O to left, then add 75 H^+ to right to balance H^+.

$$39 H_2O + Fe(CN)_6^{4-} \rightarrow Fe(OH)_3 + 6 CO_3^{2-} + 6 NO_3^- + 75 H^+$$
$$\text{net charge} = 4- \qquad\qquad \text{net charge} = 57+$$

Add 61 e^- to the product side, and then add the two balanced half-reactions with a common factor of 61 e^- transferred.

$$39 H_2O + Fe(CN)_6^{4-} \rightarrow Fe(OH)_3 + 6 CO_3^- + 6 NO_3^- + 75 H^+ + 61 e^-$$
$$61 e^- + 183 H_2O + 61 Ce^{4+} \rightarrow 61 Ce(OH)_3 + 183 H^+$$

$$\overline{222 H_2O + Fe(CN)_6^{4-} + 61 Ce^{4+} \rightarrow 61 Ce(OH)_3 + Fe(OH)_3 + 6 CO_3^{2-} + 6 NO_3^- + 258 H^+}$$

Adding 258 OH^- to each side, and then reducing gives:

$$258 OH^-(aq) + Fe(CN)_6^{4-}(aq) + 61 Ce^{4+}(aq) \rightarrow 61 Ce(OH)_3(s) + Fe(OH)_3(s)$$
$$+ 6 CO_3^{2-}(aq) + 6 NO_3^-(aq) + 36 H_2O(l)$$

75. $(H_2C_2O_4 \rightarrow 2 CO_2 + 2 H^+ + 2 e^-) \times 5$ $(5 e^- + 8 H^+ + MnO_4^- \rightarrow Mn^{2+} + 4 H_2O) \times 2$

$$5 H_2C_2O_4 \rightarrow 10 CO_2 + 10 H^+ + 10 e^-$$
$$10 e^- + 16 H^+ + 2 MnO_4^- \rightarrow 2 Mn^{2+} + 8 H_2O$$

$$\overline{6 H^+(aq) + 5 H_2C_2O_4(aq) + 2 MnO_4^-(aq) \rightarrow 10 CO_2(g) + 2 Mn^{2+}(aq) + 8 H_2O(l)}$$

$$0.1058 \text{ g } H_2C_2O_4 \times \frac{1 \text{ mol } H_2C_2O_4}{90.034 \text{ g}} \times \frac{2 \text{ mol } MnO_4^-}{5 \text{ mol } H_2C_2O_4} = 4.700 \times 10^{-4} \text{ mol } MnO_4^-$$

$$\text{Molarity} = \frac{4.700 \times 10^{-4} \text{ mol } MnO_4^-}{28.97 \text{ mL}} \times \frac{1000 \text{ mL}}{L} = 1.622 \times 10^{-2} \, M \, MnO_4^-$$

77.
$$(Fe^{2+} \rightarrow Fe^{3+} + e^-) \times 5$$
$$5 e^- + 8 H^+ + MnO_4^- \rightarrow Mn^{2+} + 4 H_2O$$

$$\overline{8 H^+(aq) + MnO_4^-(aq) + 5 Fe^{2+}(aq) \rightarrow 5 Fe^{3+}(aq) + Mn^{2+}(aq) + 4 H_2O(l)}$$

From the titration data we can get the number of moles of Fe^{2+}. We then convert this to a mass of iron and calculate the mass percent of iron in the sample.

$$38.37 \times 10^{-3} \text{ L MnO}_4^- \times \frac{0.0198 \text{ mol MnO}_4^-}{\text{L}} \times \frac{5 \text{ mol Fe}^{2+}}{\text{mol MnO}_4^-} = 3.80 \times 10^{-3} \text{ mol Fe}^{2+}$$

$$= 3.80 \times 10^{-3} \text{ mol Fe present}$$

$$3.80 \times 10^{-3} \text{ mol Fe} \times \frac{55.85 \text{ g Fe}}{\text{mol Fe}} = 0.212 \text{ g Fe}$$

$$\text{Mass \% Fe} = \frac{0.212 \text{ g}}{0.6128 \text{ g}} \times 100 = 34.6\% \text{ Fe}$$

79. $Mg(s) + 2 HCl(aq) \rightarrow MgCl_2(aq) + H_2(g)$

$$3.00 \text{ g Mg} \times \frac{1 \text{ mol Mg}}{24.31 \text{ g Mg}} \times \frac{2 \text{ mol HCl}}{\text{mol Mg}} \times \frac{1 \text{ L HCl}}{5.0 \text{ mol HCl}} = 0.0494 \text{ L} = 49.4 \text{ mL HCl}$$

Additional Exercises

81. $\text{Mol CaCl}_2 \text{ present} = 0.230 \text{ L CaCl}_2 \times \frac{0.275 \text{ mol CaCl}_2}{\text{L CaCl}_2} = 6.33 \times 10^{-2} \text{ mol CaCl}_2$

The volume of $CaCl_2$ solution after evaporation is:

$$6.33 \times 10^{-2} \text{ mol CaCl}_2 \times \frac{1 \text{ L CaCl}_2}{1.10 \text{ mol CaCl}_2} = 5.75 \times 10^{-2} \text{ L} = 57.5 \text{ mL CaCl}_2$$

Volume H_2O evaporated = 230. mL − 57.5 mL = 173 mL H_2O evaporated

83.   a. $MgCl_2(aq) + 2 AgNO_3(aq) \rightarrow 2 AgCl(s) + Mg(NO_3)_2(aq)$

$$0.641 \text{ g AgCl} \times \frac{1 \text{ mol AgCl}}{143.4 \text{ g AgCl}} \times \frac{1 \text{ mol MgCl}_2}{2 \text{ mol AgCl}} \times \frac{95.21 \text{ g}}{\text{mol MgCl}_2} = 0.213 \text{ g MgCl}_2$$

$$\frac{0.213 \text{ g MgCl}_2}{1.50 \text{ g mixture}} \times 100 = 14.2\% \text{ MgCl}_2$$

b. $0.213 \text{ g MgCl}_2 \times \dfrac{1 \text{ mol MgCl}_2}{95.21 \text{ g}} \times \dfrac{2 \text{ mol AgNO}_3}{\text{mol MgCl}_2} \times \dfrac{1 \text{ L}}{0.500 \text{ mol AgNO}_3} \times \dfrac{1000 \text{ mL}}{1 \text{ L}}$

$$= 8.95 \text{ mL AgNO}_3$$

85.   a. $0.308 \text{ g AgCl} \times \dfrac{35.45 \text{ g Cl}}{143.4 \text{ g AgCl}} = 0.0761 \text{ g Cl}; \quad \% \text{ Cl} = \dfrac{0.0761 \text{ g}}{0.256 \text{ g}} \times 100 = 29.7\% \text{ Cl}$

Cobalt(III) oxide, Co_2O_3: 2(58.93) + 3(16.00) = 165.86 g/mol

$$0.145 \text{ g Co}_2\text{O}_3 \times \frac{117.86 \text{ g Co}}{165.86 \text{ g Co}_2\text{O}_3} = 0.103 \text{ g Co}; \quad \% \text{ Co} = \frac{0.103 \text{ g}}{0.416 \text{ g}} \times 100 = 24.8\% \text{ Co}$$

The remainder, $100.0 - (29.7 + 24.8) = 45.5\%$, is water.

Assuming 100.0 g of compound:

$$45.5 \text{ g H}_2\text{O} \times \frac{2.016 \text{ g H}}{18.02 \text{ g H}_2\text{O}} = 5.09 \text{ g H}; \quad \% \text{ H} = \frac{5.09 \text{ g H}}{100.0 \text{ g compound}} \times 100 = 5.09\% \text{ H}$$

$$45.5 \text{ g H}_2\text{O} \times \frac{16.00 \text{ g O}}{18.02 \text{ g H}_2\text{O}} = 40.4 \text{ g O}; \quad \% \text{ O} = \frac{40.4 \text{ g O}}{100.0 \text{ g compound}} \times 100 = 40.4\% \text{ O}$$

The mass percent composition is 24.8% Co, 29.7% Cl, 5.09% H, and 40.4% O.

b. Out of 100.0 g of compound, there are:

$$24.8 \text{ g Co} \times \frac{1 \text{ mol}}{58.93 \text{ g Co}} = 0.421 \text{ mol Co}; \quad 29.7 \text{ g Cl} \times \frac{1 \text{ mol}}{35.45 \text{ g Cl}} = 0.838 \text{ mol Cl}$$

$$5.09 \text{ g H} \times \frac{1 \text{ mol}}{1.008 \text{ g H}} = 5.05 \text{ mol H}; \quad 40.4 \text{ g O} \times \frac{1 \text{ mol}}{16.00 \text{ g O}} = 2.53 \text{ mol O}$$

Dividing all results by 0.421, we get $CoCl_2 \cdot 6H_2O$ for the empirical formula, which is also the molecular formula.

c. $CoCl_2 \cdot 6H_2O(aq) + 2 \text{ AgNO}_3(aq) \rightarrow 2 \text{ AgCl}(s) + Co(NO_3)_2(aq) + 6 H_2O(l)$

$CoCl_2 \cdot 6H_2O(aq) + 2 \text{ NaOH}(aq) \rightarrow Co(OH)_2(s) + 2 \text{ NaCl}(aq) + 6 H_2O(l)$

$Co(OH)_2 \rightarrow Co_2O_3$ This is an oxidation-reduction reaction. Thus we also need to include an oxidizing agent. The obvious choice is O_2.

$4 \text{ Co(OH)}_2(s) + O_2(g) \rightarrow 2 \text{ Co}_2\text{O}_3(s) + 4 H_2O(l)$

87. $Ag^+(aq) + Cl^-(aq) \rightarrow AgCl(s)$; let x = mol NaCl and y = mol KCl.

$(22.90 \times 10^{-3} \text{ L}) \times 0.1000 \text{ mol/L} = 2.290 \times 10^{-3} \text{ mol Ag}^+ = 2.290 \times 10^{-3} \text{ mol Cl}^-$ total

$x + y = 2.290 \times 10^{-3} \text{ mol Cl}^-$, $x = 2.290 \times 10^{-3} - y$

Because the molar mass of NaCl is 58.44 g/mol and the molar mass of KCl is 74.55 g/mol:

$(58.44)x + (74.55)y = 0.1586 \text{ g}$

$58.44(2.290 \times 10^{-3} - y) + (74.55)y = 0.1586, \; (16.11)y = 0.0248, \; y = 1.54 \times 10^{-3}$ mol KCl

$$\text{Mass \% KCl} = \frac{1.54 \times 10^{-3} \text{ mol} \times 74.55 \text{ g/mol}}{0.1586 \text{ g}} \times 100 = 72.4\% \text{ KCl}$$

% NaCl = 100.0 − 72.4 = 27.6% NaCl

89. $Cr(NO_3)_3(aq) + 3 \, NaOH(aq) \rightarrow Cr(OH)_3(s) + 3 \, NaNO_3(aq)$

Mol NaOH used = $2.06 \text{ g Cr(OH)}_3 \times \dfrac{1 \text{ mol Cr(OH)}_3}{103.02 \text{ g}} \times \dfrac{3 \text{ mol NaOH}}{\text{mol Cr(OH)}_3} = 6.00 \times 10^{-2}$ mol
to form precipitate

$NaOH(aq) + HCl(aq) \rightarrow NaCl(aq) + H_2O(l)$

Mol NaOH used = $0.1000 \text{ L} \times \dfrac{0.400 \text{ mol HCl}}{\text{L}} \times \dfrac{1 \text{ mol NaOH}}{\text{mol HCl}} = 4.00 \times 10^{-2}$ mol
to react with HCl

$$M_{NaOH} = \frac{\text{total mol NaOH}}{\text{volume}} = \frac{6.00 \times 10^{-2} \text{ mol} + 4.00 \times 10^{-2} \text{ mol}}{0.0500 \text{ L}} = 2.00 \, M \text{ NaOH}$$

91. Mol KHP used = $0.4016 \text{ g} \times 1 \text{ mol}/204.22 \text{ g} = 1.967 \times 10^{-3}$ mol KHP

Because 1 mole of NaOH reacts completely with 1 mole of KHP, the NaOH solution contains 1.967×10^{-3} mol NaOH.

$$\text{Molarity of NaOH} = \frac{1.967 \times 10^{-3} \text{ mol}}{25.06 \times 10^{-3} \text{ L}} = \frac{7.849 \times 10^{-2} \text{ mol NaOH}}{\text{L}}$$

$$\text{Maximum molarity} = \frac{1.967 \times 10^{-3} \text{ mol}}{25.01 \times 10^{-3} \text{ L}} = \frac{7.865 \times 10^{-2} \text{ mol NaOH}}{\text{L}}$$

$$\text{Minimum molarity} = \frac{1.967 \times 10^{-3} \text{ mol}}{25.11 \times 10^{-3} \text{ L}} = \frac{7.834 \times 10^{-2} \text{ mol NaOH}}{\text{L}}$$

We can express this as 0.07849 ±0.00016 M. An alternate way is to express the molarity as 0.0785 ±0.0002 M. This second way shows the actual number of significant figures in the molarity.   The advantage of the first method is that it shows that we made all our individual measurements to four significant figures.

93. Mol $C_6H_8O_7 = 0.250 \text{ g } C_6H_8O_7 \times \dfrac{1 \text{ mol } C_6H_8O_7}{192.1 \text{ g } C_6H_8O_7} = 1.30 \times 10^{-3}$ mol $C_6H_8O_7$

Let H_xA represent citric acid, where x is the number of acidic hydrogens. The balanced neutralization reaction is:

$$H_xA(aq) + x\ OH^-(aq) \rightarrow x\ H_2O(l) + A^{x-}(aq)$$

$$\text{Mol } OH^- \text{ reacted} = 0.0372\ L \times \frac{0.105\ \text{mol } OH^-}{L} = 3.91 \times 10^{-3}\ \text{mol } OH^-$$

$$x = \frac{\text{mol } OH^-}{\text{mol citric acid}} = \frac{3.91 \times 10^{-3}\ \text{mol}}{1.30 \times 10^{-3}\ \text{mol}} = 3.01$$

Therefore, the general acid formula for citric acid is H_3A, meaning that citric acid has three acidic hydrogens per citric acid molecule (citric acid is a triprotic acid).

95. $H_2SO_4(aq) + 2\ NaOH(aq) \rightarrow Na_2SO_4(aq) + 2\ H_2O(l)$

$$0.02844\ L \times \frac{0.1000\ \text{mol NaOH}}{L} \times \frac{1\ \text{mol } H_2SO_4}{2\ \text{mol NaOH}} \times \frac{1\ \text{mol } SO_2}{\text{mol } H_2SO_4} \times \frac{32.07\ \text{g S}}{\text{mol } SO_2}$$

$$= 4.560 \times 10^{-2}\ \text{g S}$$

$$\text{Mass \% S} = \frac{0.04560\ g}{1.325\ g} \times 100 = 3.442\%$$

Challenge Problems

97.   a.   Let x = mass of Mg, so $10.00 - x$ = mass of Zn. $Ag^+(aq) + Cl^-(aq) \rightarrow AgCl(s)$.

From the given balanced equations, there is a 2 : 1 mole ratio between mol Mg and mol Cl^-. The same is true for Zn. Because mol Ag^+ = mol Cl^- present, one can setup an equation relating mol Cl^- present to mol Ag^+ added.

$$x\ \text{g Mg} \times \frac{1\ \text{mol Mg}}{24.31\ \text{g Mg}} \times \frac{2\ \text{mol } Cl^-}{\text{mol Mg}} + (10.00 - x)\ \text{g Zn} \times \frac{1\ \text{mol Zn}}{65.38\ \text{g Zn}} \times \frac{2\ \text{mol } Cl^-}{\text{mol Zn}}$$

$$= 0.156\ L \times \frac{3.00\ \text{mol } Ag^+}{L} \times \frac{1\ \text{mol } Cl^-}{\text{mol } Ag^+} = 0.468\ \text{mol } Cl^-$$

$$\frac{2x}{24.31} + \frac{2(10.00 - x)}{65.38} = 0.468, \quad 24.31 \times 65.38 \left(\frac{2x}{24.31} + \frac{20.00 - 2x}{65.38} = 0.468 \right)$$

$(130.8)x + 486.2 - (48.62)x = 743.8$ (carrying 1 extra significant figure)

$(82.2)x = 257.6, \quad x = 3.13\ \text{g Mg};$ % Mg $= \dfrac{3.13\ \text{g Mg}}{10.00\ \text{g mixture}} \times 100 = 31.3\%$ Mg

b. $0.156\ L \times \dfrac{3.00\ \text{mol } Ag^+}{L} \times \dfrac{1\ \text{mol } Cl^-}{\text{mol } Ag^+} = 0.468\ \text{mol } Cl^- = 0.468\ \text{mol HCl added}$

$$M_{HCl} = \frac{0.468\ \text{mol}}{0.0780\ L} = 6.00\ M\ \text{HCl}$$

99. $Zn(s) + 2\ AgNO_2(aq) \rightarrow 2\ Ag(s) + Zn(NO_2)_2(aq)$

Let x = mass of Ag and y = mass of Zn after the reaction has stopped. Then $x + y = 29.0$ g. Because the moles of Ag produced will equal two times the moles of Zn reacted:

$$(19.0 - y)\ \text{g Zn} \times \frac{1\ \text{mol Zn}}{65.38\ \text{g Zn}} \times \frac{2\ \text{mol Ag}}{1\ \text{mol Zn}} = x\ \text{g Ag} \times \frac{1\ \text{mol Ag}}{107.9\ \text{g Ag}}$$

Simplifying:

$$3.059 \times 10^{-2}(19.0 - y) = (9.268 \times 10^{-3})x$$

Substituting $x = 29.0 - y$ into the equation gives:

$$3.059 \times 10^{-2}(19.0 - y) = 9.268 \times 10^{-3}(29.0 - y)$$

Solving:

$$0.581 - (3.059 \times 10^{-2})y = 0.269 - (9.268 \times 10^{-3})y,\ (2.132 \times 10^{-2})y = 0.312,\ y = 14.6\ \text{g Zn}$$

14.6 g Zn are present, and $29.0 - 14.6 = 14.4$ g Ag are also present after the reaction is stopped.

101.   Molar masses: KCl, $39.10 + 35.45 = 74.55$ g/mol; KBr, $39.10 + 79.90 = 119.00$ g/mol, AgCl, $107.9 + 35.45 = 143.4$ g/mol; AgBr, $107.9 + 79.90 = 187.8$ g/mol

Let x = number of moles of KCl in mixture and y = number of moles of KBr in mixture. $Ag^+ + Cl^- \rightarrow AgCl$ and $Ag^+ + Br^- \rightarrow AgBr$; so, x = moles AgCl and y = moles AgBr.

Setting up two equations from the given information:

$$0.1024\ \text{g} = (74.55)x + (119.0)y\ \text{ and }\ 0.1889\ \text{g} = (143.4)x + (187.8)y$$

Multiply the first equation by $\dfrac{187.8}{119.0}$, and then subtract from the second.

$$
\begin{aligned}
0.1889 &=\ \ \ (143.4)x + (187.8)y \\
-0.1616 &= -(117.7)x - (187.8)y \\
\hline
0.0273 &=\ \ \ (25.7)x, \qquad\qquad x = 1.06 \times 10^{-3}\ \text{mol KCl}
\end{aligned}
$$

$$1.06 \times 10^{-3}\ \text{mol KCl} \times \frac{74.55\ \text{g KCl}}{\text{mol KCl}} = 0.0790\ \text{g KCl}$$

$$\text{Mass \% KCl} = \frac{0.0790\ \text{g}}{0.1024\ \text{g}} \times 100 = 77.1\%,\ \ \%\ \text{KBr} = 100.0 - 77.1 = 22.9\%$$

103.   a. $C_{12}H_{10-n}Cl_n + n\,Ag^+ \rightarrow n\,AgCl$; molar mass of AgCl = 143.4 g/mol

Molar mass of PCB = $12(12.01) + (10 - n)(1.008) + n(35.45) = 154.20 + (34.44)n$

Because n mol AgCl are produced for every 1 mol PCB reacted, $n(143.4)$ g of AgCl will be produced for every $[154.20 + (34.44)n]$ g of PCB reacted.

$$\frac{\text{Mass of AgCl}}{\text{Mass of PCB}} = \frac{(143.4)n}{154.20 + (34.44)n} \text{ or mass}_{AgCl}[154.20 + (34.44)n] = \text{mass}_{PCB}(143.4)n$$

b. $0.4971[154.20 + (34.44)n] = 0.1947(143.4)n, \quad 76.65 + (17.12)n = (27.92)n$

$76.65 = (10.80)n, \quad n = 7.097$

105.   a.   Flow rate = 5.00×10^4 L/s + 3.50×10^3 L/s = 5.35×10^4 L/s

b. $C_{HCl} = \dfrac{3.50 \times 10^3 (65.0)}{5.35 \times 10^4} = 4.25$ ppm HCl

c.   1 ppm = 1 mg/kg H_2O = 1 mg/L   (assuming density = 1.00 g/mL)

$$8.00 \text{ h} \times \frac{60 \text{ min}}{h} \times \frac{60 \text{ s}}{min} \times \frac{1.80 \times 10^4 \text{ L}}{s} \times \frac{4.25 \text{ mg HCl}}{L} \times \frac{1 \text{ g}}{1000 \text{ mg}} = 2.20 \times 10^6 \text{ g HCl}$$

$$2.20 \times 10^6 \text{ g HCl} \times \frac{1 \text{ mol HCl}}{36.46 \text{ g HCl}} \times \frac{1 \text{ mol CaO}}{2 \text{ mol HCl}} \times \frac{56.08 \text{ g Ca}}{\text{mol CaO}} = 1.69 \times 10^6 \text{ g CaO}$$

d.   The concentration of Ca^{2+} going into the second plant was:

$$\frac{5.00 \times 10^4 (10.2)}{5.35 \times 10^4} = 9.53 \text{ ppm}$$

The second plant used: 1.80×10^4 L/s $\times (8.00 \times 60 \times 60)$ s = 5.18×10^8 L of water.

$$1.69 \times 10^6 \text{ g CaO} \times \frac{40.08 \text{ g Ca}^{2+}}{56.08 \text{ g CaO}} = 1.21 \times 10^6 \text{ g Ca}^{2+} \text{ was added to this water.}$$

$$C_{Ca^{2+}} \text{ (plant water)} = 9.53 + \frac{1.21 \times 10^9 \text{ mg}}{5.18 \times 10^8 \text{ L}} = 9.53 + 2.34 = 11.87 \text{ ppm}$$

Because 90.0% of this water is returned, $(1.80 \times 10^4) \times 0.900 = 1.62 \times 10^4$ L/s of water with 11.87 ppm Ca^{2+} is mixed with $(5.35 - 1.80) \times 10^4 = 3.55 \times 10^4$ L/s of water containing 9.53 ppm Ca^{2+}.

$$C_{Ca^{2+}} \text{ (final)} = \frac{(1.62 \times 10^4 \text{ L/s})(11.87 \text{ ppm}) + (3.55 \times 10^4 \text{ L/s})(9.53 \text{ ppm})}{1.62 \times 10^4 \text{ L/s} + 3.55 \times 10^4 \text{ L/s}} = 10.3 \text{ ppm}$$

107.   a. $YBa_2Cu_3O_{6.5}$:

$$+3 + 2(+2) + 3x + 6.5(-2) = 0$$

$$7 + 3x - 13 = 0, \quad 3x = 6, \quad x = +2 \qquad \text{Only } Cu^{2+} \text{ present.}$$

$YBa_2Cu_3O_7$:

$$+3 + 2(+2) + 3x + 7(-2) = 0, \quad x = +2 \; 1/3 \text{ or } 2.33$$

This corresponds to two Cu^{2+} and one Cu^{3+} present.

$YBa_2Cu_3O_8$:

$$+3 + 2(+2) + 3x + 8(-2) = 0, \quad x = +3; \quad \text{Only } Cu^{3+} \text{ present.}$$

b.

$$\frac{\begin{array}{l}(e^- + Cu^{2+} + I^- \rightarrow CuI) \times 2\\ 3I^- \rightarrow I_3^- + 2\,e^-\end{array}}{2\;Cu^{2+}(aq) + 5\;I^-(aq) \rightarrow 2\;CuI(s) + I_3^-(aq)} \qquad \frac{\begin{array}{l}2\,e^- + Cu^{3+} + I^- \rightarrow CuI\\ 3I^- \rightarrow I_3^- + 2\,e^-\end{array}}{Cu^{3+}(aq) + 4\;I^-(aq) \rightarrow CuI(s) + I_3^-(aq)}$$

$$\frac{\begin{array}{l}2\;S_2O_3^{2-} \rightarrow S_4O_6^{2-} + 2\,e^-\\ 2\,e^- + I_3^- \rightarrow 3\;I^-\end{array}}{2\;S_2O_3^{2-}(aq) + I_3^-(aq) \rightarrow 3\;I^-(aq) + S_4O_6^{2-}(aq)}$$

c.   Step II data: All Cu is converted to Cu^{2+}. *Note*: Superconductor abbreviated as "123."

$$22.57 \times 10^{-3}\;L \times \frac{0.1000\;mol\;S_2O_3^{2-}}{L} \times \frac{1\;mol\;I_3^-}{2\;mol\;S_2O_3^{2-}} \times \frac{2\;mol\;Cu^{2+}}{mol\;I_3^-}$$
$$= 2.257 \times 10^{-3}\;mol\;Cu^{2+}$$

$$2.257 \times 10^{-3}\;mol\;Cu \times \frac{1\;mol\;"123"}{3\;mol\;Cu} = 7.523 \times 10^{-4}\;mol\;"123"$$

$$\text{Molar mass of } YBa_2Cu_3O_x = \frac{0.5402\;g}{7.523 \times 10^{-4}\;mol} = 670.2\;g/mol$$

$$670.2 = 88.91 + 2(137.3) + 3(63.55) + x(16.00), \quad 670.2 = 554.2 + x(16.00)$$

$$x = 7.250; \quad \text{formula is } YBa_2Cu_3O_{7.25}.$$

Check with Step I data: Both Cu^{2+} and Cu^{3+} present.

$$37.77 \times 10^{-3}\;L \times \frac{0.1000\;mol\;S_2O_3^{2-}}{L} \times \frac{1\;mol\;I_3^-}{2\;mol\;S_2O_3^{2-}} = 1.889 \times 10^{-3}\;mol\;I_3^-$$

We get 1 mol I_3^- per mol Cu^{3+} and 1 mol I_3^- per 2 mol Cu^{2+}. Let $n_{Cu^{3+}}$ = mol Cu^{3+} and $n_{Cu^{2+}}$ = mol Cu^{2+}, then:

$$n_{Cu^{3+}} + \frac{n_{Cu^{2+}}}{2} = 1.889 \times 10^{-3} \text{ mol}$$

In addition: $\dfrac{0.5625 \text{ g}}{670.2 \text{ g/mol}} = 8.393 \times 10^{-4}$ mol "123"; this amount of "123" contains:

$$3(8.393 \times 10^{-4}) = 2.518 \times 10^{-3} \text{ mol Cu total} = n_{Cu^{3+}} + n_{Cu^{2+}}$$

Solving by simultaneous equations:

$$n_{Cu^{3+}} + n_{Cu^{2+}} = 2.518 \times 10^{-3}$$
$$-n_{Cu^{3+}} - \frac{n_{Cu^{2+}}}{2} = -1.889 \times 10^{-3}$$
$$\overline{\qquad\qquad\qquad\qquad\qquad}$$
$$\frac{n_{Cu^{2+}}}{2} = 6.29 \times 10^{-4}$$

$n_{Cu^{2+}} = 1.26 \times 10^{-3}$ mol Cu^{2+}; $n_{Cu^{3+}} = 2.518 \times 10^{-3} - 1.26 \times 10^{-3} = 1.26 \times 10^{-3}$ mol Cu^{3+}

This sample of superconductor contains equal moles of Cu^{2+} and Cu^{3+}. Therefore, 1 mole of $YBa_2Cu_3O_x$ contains 1.50 mol Cu^{2+} and 1.50 mol Cu^{3+}. Solving for x using oxidation states:

$$+3 + 2(+2) + 1.50(+2) + 1.50(+3) + x(-2) = 0, \quad 14.50 = 2x, \quad x = 7.25$$

The two experiments give the same result, $x = 7.25$ with formula $YBa_2Cu_3O_{7.25}$.

Average oxidation state of Cu:

$$+3 + 2(+2) + 3(x) + 7.25(-2) = 0, \quad 3x = 7.50, \quad x = +2.50$$

As determined from Step I data, this superconductor sample contains equal moles of Cu^{2+} and Cu^{3+}, giving an average oxidation state of +2.50.

109.   There are three unknowns so we need three equations to solve for the unknowns. Let $x =$ mass $AgNO_3$, $y =$ mass $CuCl_2$, and $z =$ mass $FeCl_3$. Then $x + y + z = 1.0000$ g. The Cl^- in $CuCl_2$ and $FeCl_3$ will react with the excess $AgNO_3$ to form the precipitate $AgCl(s)$. Assuming silver has an atomic mass of 107.90:

$$\text{Mass of Cl in mixture} = 1.7809 \text{ g AgCl} \times \frac{35.45 \text{ g Cl}}{143.35 \text{ g AgCl}} = 0.4404 \text{ g Cl}$$

$$\text{Mass of Cl from } CuCl_2 = y \text{ g } CuCl_2 \times \frac{2(35.45) \text{ g Cl}}{134.45 \text{ g } CuCl_2} = (0.5273)y$$

$$\text{Mass of Cl from } FeCl_3 = z \text{ g } FeCl_3 \times \frac{3(35.45) \text{ g Cl}}{162.20 \text{ g } FeCl_3} = (0.6557)z$$

The second equation is: $0.4404 \text{ g Cl} = (0.5273)y + (0.6557)z$

Similarly, let's calculate the mass of metals in each salt.

Mass of Ag in $AgNO_3$ = x g $AgNO_3 \times \dfrac{107.9 \text{ g Ag}}{169.91 \text{ g } AgNO_3} = (0.6350)x$

For $CuCl_2$ and $FeCl_3$, we already calculated the amount of Cl in each initial amount of salt; the remainder must be the mass of metal in each salt.

Mass of Cu in $CuCl_2 = y - (0.5273)y = (0.4727)y$

Mass of Fe in $FeCl_3 = z - (0.6557)z = (0.3443)z$

The third equation is: $0.4684 \text{ g metals} = (0.6350)x + (0.4727)y + (0.3443)z$

We now have three equations with three unknowns. Solving:

$$-0.6350\,(1.0000 = \quad x \quad + \quad y \quad + \quad z)$$
$$0.4684 = (0.6350)x + (0.4727)y + (0.3443)z$$

$$\rule{7cm}{0.4pt}$$

$$-0.1666 = \qquad\qquad -(0.1623)y - (0.2907)z$$

$$\dfrac{0.5273}{0.1623}\,[-0.1666 = -(0.1623)y - (0.2907)z]$$

$$0.4404 = \quad (0.5273)y + (0.6557)z$$

$$\rule{7cm}{0.4pt}$$

$$-0.1009 = \qquad\qquad -(0.2888)z, \quad z = \dfrac{0.1009}{0.2888} = 0.3494 \text{ g } FeCl_3$$

$0.4404 = (0.5273)y + 0.6557(0.3494), \ y = 0.4007 \text{ g } CuCl_2$

$x = 1.0000 - y - z = 1.0000 - 0.4007 - 0.3494 = 0.2499 \text{ g } AgNO_3$

Mass % $AgNO_3 = \dfrac{0.2499 \text{ g}}{1.0000 \text{ g}} \times 100 = 24.99\% \ AgNO_3$

Mass % $CuCl_2 = \dfrac{0.4007 \text{ g}}{1.0000 \text{ g}} \times 100 = 40.07\% \ CuCl_2; \ \text{ mass \% } FeCl_3 = 34.94\%$

CHAPTER 5

GASES

Pressure

21. $4.75 \text{ cm} \times \dfrac{10 \text{ mm}}{\text{cm}} = 47.5 \text{ mm Hg or } 47.5 \text{ torr}; \quad 47.5 \text{ torr} \times \dfrac{1 \text{ atm}}{760 \text{ torr}} = 6.25 \times 10^{-2} \text{ atm}$

   $6.25 \times 10^{-2} \text{ atm} \times \dfrac{1.013 \times 10^5 \text{ Pa}}{\text{atm}} = 6.33 \times 10^3 \text{ Pa}$

23.   Suppose we have a column of mercury $1.00 \text{ cm} \times 1.00 \text{ cm} \times 76.0 \text{ cm} = V = 76.0 \text{ cm}^3$:

   $\text{mass} = 76.0 \text{ cm}^3 \times 13.59 \text{ g/cm}^3 = 1.03 \times 10^3 \text{ g} \times \dfrac{1 \text{ kg}}{1000 \text{ g}} = 1.03 \text{ kg}$

   $F = mg = 1.03 \text{ kg} \times 9.81 \text{ m/s}^2 = 10.1 \text{ kg m/s}^2 = 10.1 \text{ N}$

   $\dfrac{\text{Force}}{\text{Area}} = \dfrac{10.1 \text{ N}}{\text{cm}^2} \times \left(\dfrac{100 \text{ cm}}{\text{m}}\right)^2 = 1.01 \times 10^5 \ \dfrac{\text{N}}{\text{m}^2} \text{ or } 1.01 \times 10^5 \text{ Pa}$

   (*Note*: $76.0 \text{ cm Hg} = 1 \text{ atm} = 1.01 \times 10^5 \text{ Pa}$.)

   To exert the same pressure, a column of water will have to contain the same mass as the 76.0-cm column of mercury. Thus the column of water will have to be 13.59 times taller or $76.0 \text{ cm} \times 13.59 = 1.03 \times 10^3 \text{ cm} = 10.3 \text{ m}$.

25.   a. $4.8 \text{ atm} \times \dfrac{760 \text{ mm Hg}}{\text{atm}} = 3.6 \times 10^3 \text{ mm Hg};$ b. $3.6 \times 10^3 \text{ mm Hg} \times \dfrac{1 \text{ torr}}{\text{mm Hg}}$

   $= 3.6 \times 10^3 \text{ torr}$

   c. $4.8 \text{ atm} \times \dfrac{1.013 \times 10^5 \text{ Pa}}{\text{atm}} = 4.9 \times 10^5 \text{ Pa};$ d. $4.8 \text{ atm} \times \dfrac{14.7 \text{ psi}}{\text{atm}} = 71 \text{ psi}$

49

Gas Laws

27.   The decrease in temperature causes the balloon to contract (V and T are directly related). Because weather balloons do expand, the effect of the decrease in pressure must be dominant.

29.   Treat each gas separately, and use the relationship $P_1V_1 = P_2V_2$ (n and T are constant).

For H_2: $P_2 = \dfrac{P_1V_1}{V_2} = 475 \text{ torr} \times \dfrac{2.00 \text{ L}}{3.00 \text{ L}} = 317 \text{ torr}$

For N_2: $P_2 = 0.200 \text{ atm} \times \dfrac{1.00 \text{ L}}{3.00 \text{ L}} = 0.0667 \text{ atm}$; $0.0667 \text{ atm} \times \dfrac{760 \text{ torr}}{\text{atm}} = 50.7 \text{ torr}$

$P_{total} = P_{H_2} + P_{N_2} = 317 + 50.7 = 368 \text{ torr}$

31. $PV = nRT, \quad \dfrac{nT}{P} = \dfrac{V}{R} = \text{constant}, \quad \dfrac{n_1T_1}{P_1} = \dfrac{n_2T_2}{P_2}$; moles × molar mass = mass

$\dfrac{n_1(\text{molar mass})T_1}{P_1} = \dfrac{n_2(\text{molar mass})T_2}{P_2}, \quad \dfrac{\text{mass}_1 \times T_1}{P_1} = \dfrac{\text{mass}_2 \times T_2}{P_2}$

$\text{mass}_2 = \dfrac{\text{mass}_1 \times T_1P_2}{T_2P_1} = \dfrac{1.00 \times 10^3 \text{ g} \times 291 \text{ K} \times 650. \text{ psi}}{299 \text{ K} \times 2050. \text{ psi}} = 309 \text{ g}$

33. $P = P_{CO_2} = \dfrac{n_{CO_2}RT}{V} = \dfrac{\left(22.0 \text{ g} \times \dfrac{1 \text{ mol}}{44.01 \text{ g}}\right) \times \dfrac{0.08206 \text{ L atm}}{\text{K mol}} \times 300. \text{ K}}{4.00 \text{ L}} = 3.08 \text{ atm}$

With air present, the partial pressure of CO_2 will still be 3.08 atm. The total pressure will be the sum of the partial pressures.

$P_{total} = P_{CO_2} + P_{air} = 3.08 \text{ atm} + \left(740. \text{ torr} \times \dfrac{1 \text{ atm}}{760 \text{ torr}}\right) = 3.08 + 0.974 = 4.05 \text{ atm}$

35. $n = \dfrac{PV}{RT} = \dfrac{135 \text{ atm} \times 200.0 \text{ L}}{\dfrac{0.08206 \text{ L atm}}{\text{K mol}} \times (273 + 24) \text{ K}} = 1.11 \times 10^3 \text{ mol}$

For He: $1.11 \times 10^3 \text{ mol} \times \dfrac{4.003 \text{ g He}}{\text{mol}} = 4.44 \times 10^3 \text{ g He}$

For H_2: $1.11 \times 10^3 \text{ mol} \times \dfrac{2.016 \text{ g } H_2}{\text{mol}} = 2.24 \times 10^3 \text{ g } H_2$

37. $\dfrac{PV}{nT} = R$; for a gas at two conditions:

$$\dfrac{P_1 V_1}{n_1 T_1} = \dfrac{P_2 V_2}{n_2 T_2} \ ; \ \text{because n and V are constant:} \ \dfrac{P_1}{T_1} = \dfrac{P_2}{T_2}$$

$$T_2 = \dfrac{P_2 T_1}{P_1} = \dfrac{2500 \ \text{torr} \times 294.2 \ \text{K}}{758 \ \text{torr}} = 970 \ \text{K} = 7.0 \times 10^2 \, ^\circ\text{C}$$

For two-condition problems, units for P and V just need to be the same units for both conditions, not necessarily atm and L. The unit conversions from other P or V units would cancel when applied to both conditions. However, temperature always must be converted to the Kelvin scale. The temperature conversions between other units and Kelvin will not cancel each other.

39.   As NO_2 is converted completely into N_2O_4, the moles of gas present will decrease by a factor of one-half (from the 2 : 1 mol ratio in the balanced equation). Using Avogadro's law:

$$\dfrac{V_1}{n_1} = \dfrac{V_2}{n_2}, \quad V_2 = V_1 \times \dfrac{n_2}{n_1} = 25.0 \ \text{mL} \times \dfrac{1}{2} = 12.5 \ \text{mL}$$

$N_2O_4(g)$ will occupy one-half the original volume of $NO_2(g)$.

41. $PV = nRT$, P is constant. $\dfrac{nT}{V} = \dfrac{P}{R} = \text{constant}, \ \dfrac{n_1 T_1}{V_1} = \dfrac{n_2 T_2}{V_2}$

$$\dfrac{n_2}{n_1} = \dfrac{T_1 V_2}{T_2 V_1} = \dfrac{294 \ \text{K}}{335 \ \text{K}} \times \dfrac{4.20 \times 10^3 \, \text{m}^3}{4.00 \times 10^3 \, \text{m}^3} = 0.921$$

43.   a.   There are 6 He atoms and 4 Ne atoms, and each flask has the same volume. The He flask has 1.5 times as many atoms of gas present as the Ne flask, so the pressure in the He flask will be 1.5 times greater (assuming a constant temperature).

b.   Because the flask volumes are the same, your drawing should have the various atoms equally distributed between the two flasks. So each flask should have 3 He atoms and 2 Ne atoms.

c.   After the stopcock is opened, each flask will have 5 total atoms and the pressures will be equal. If six atoms of He gave an initial pressure of $P_{\text{He, initial}}$, then 5 total atoms will have a pressure of $5/6 \times P_{\text{He, initial}}$.

Using similar reasoning, 4 atoms of Ne gave an initial pressure of $P_{\text{Ne, initial}}$, so 5 total atoms will have a pressure of $5/4 \times P_{\text{Ne, initial}}$. Summarizing:

$$P_{\text{final}} = \dfrac{5}{6} P_{\text{He, initial}} = \dfrac{5}{4} P_{\text{Ne, initial}}$$

d.   For the partial pressures, treat each gas separately. For helium, when the stopcock is opened, the six atoms of gas are now distributed over a larger volume. To solve for the final partial pressures, use Boyle's law for each gas.

$$\text{For He: } P_2 = \frac{P_1 V_1}{V_2} = P_{He, \text{ initial}} \times \frac{X}{2X} = \frac{P_{He, \text{ initial}}}{2}$$

The partial pressure of helium is exactly halved. The same result occurs with neon so that when the volume is doubled, the partial pressure is halved. Summarizing:

$$P_{He, \text{ final}} = \frac{P_{He, \text{ initial}}}{2}; \; P_{Ne, \text{ final}} = \frac{P_{Ne, \text{ initial}}}{2}$$

45. $P_{He} + P_{H_2O} = 1.00 \text{ atm} = 760. \text{ torr} = P_{He} + 23.8 \text{ torr}, \; P_{He} = 736 \text{ torr}$

$$n_{He} = 0.586 \text{ g} \times \frac{1 \text{ mol}}{4.003 \text{ g}} = 0.146 \text{ mol He}$$

$$V = \frac{n_{He} RT}{P_{He}} = \frac{0.146 \text{ mol} \times \dfrac{0.08206 \text{ L atm}}{\text{K mol}} \times 298 \text{ K}}{736 \text{ torr} \times \dfrac{1 \text{ atm}}{760 \text{ torr}}} = 3.69 \text{ L}$$

47.   a.   Mole fraction $CH_4 = \chi_{CH_4} = \dfrac{P_{CH_4}}{P_{total}} = \dfrac{0.175 \text{ atm}}{0.175 \text{ atm} + 0.250 \text{ atm}} = 0.412$

$$\chi_{O_2} = 1.000 - 0.412 = 0.588$$

b. $PV = nRT, \; n_{total} = \dfrac{P_{total} \times V}{RT} = \dfrac{0.425 \text{ atm} \times 10.5 \text{ L}}{\dfrac{0.08206 \text{ L atm}}{\text{K mol}} \times 338 \text{ K}} = 0.161 \text{ mol}$

c. $\chi_{CH_4} = \dfrac{n_{CH_4}}{n_{total}}, \; n_{CH_4} = \chi_{CH_4} \times n_{total} = 0.412 \times 0.161 \text{ mol} = 6.63 \times 10^{-2} \text{ mol CH}_4$

$$6.63 \times 10^{-2} \text{ mol CH}_4 \times \frac{16.04 \text{ g CH}_4}{\text{mol CH}_4} = 1.06 \text{ g CH}_4$$

$$n_{O_2} = 0.588 \times 0.161 \text{ mol} = 9.47 \times 10^{-2} \text{ mol O}_2; \; 9.47 \times 10^{-2} \text{ mol O}_2 \times \frac{32.00 \text{ g O}_2}{\text{mol O}_2}$$

$$= 3.03 \text{ g O}_2$$

49.   We can use the ideal gas law to calculate the partial pressure of each gas or to calculate the total pressure. There will be less math if we calculate the total pressure from the ideal gas law.

$$n_{O_2} = 1.5 \times 10^2 \text{ mg } O_2 \times \frac{1 \text{ g}}{1000 \text{ mg}} \times \frac{1 \text{ mol } O_2}{32.00 \text{ g } O_2} = 4.7 \times 10^{-3} \text{ mol } O_2$$

$$n_{NH_3} = 5.0 \times 10^{21} \text{ molecules } NH_3 \times \frac{1 \text{ mol } NH_3}{6.022 \times 10^{23} \text{ molecules } NH_3} = 8.3 \times 10^{-3} \text{ mol } NH_3$$

$$n_{total} = n_{N_2} + n_{O_3} + n_{NH_3} = 5.0 \times 10^{-2} + 4.7 \times 10^{-3} + 8.3 \times 10^{-3} = 6.3 \times 10^{-2} \text{ mol total}$$

$$P_{total} = \frac{n_{total} \times RT}{V} = \frac{6.3 \times 10^{-2} \text{ mol} \times \frac{0.08206 \text{ L atm}}{\text{K mol}} \times 273 \text{ K}}{1.0 \text{ L}} = 1.4 \text{ atm}$$

$$P_{N_2} = \chi_{N_2} \times P_{total}, \quad \chi_{N_2} = \frac{n_{N_2}}{n_{total}}; \quad P_{N_2} = \frac{5.0 \times 10^{-2} \text{ mol}}{6.3 \times 10^{-2} \text{ mol}} \times 1.4 \text{ atm} = 1.1 \text{ atm}$$

$$P_{O_2} = \frac{4.7 \times 10^{-3}}{6.3 \times 10^{-2}} \times 1.4 \text{ atm} = 0.10 \text{ atm}; \quad P_{NH_3} = \frac{8.3 \times 10^{-3}}{6.3 \times 10^{-2}} \times 1.4 \text{ atm} = 0.18 \text{ atm}$$

Gas Density, Molar Mass, and Reaction Stoichiometry

51.   Rigid container: As temperature is increased, the gas molecules move with a faster average velocity. This results in more frequent and more forceful collisions, resulting in an increase in pressure. Density = mass/volume; the moles of gas are constant, and the volume of the container is constant, so density in this case must be temperature-independent (density is constant).

Flexible container: The flexible container is a constant-pressure container. Therefore, the final internal pressure will be unaffected by an increase in temperature. The density of the gas, however, will be affected because the container volume is affected. As T increases, there is an immediate increase in P inside the container. The container expands its volume to reduce the internal pressure back to the external pressure. We have the same mass of gas in a larger volume. Gas density will decrease in the flexible container as T increases.

53.   Out of 100.0 g of compound, there are:

$$87.4 \text{ g N} \times \frac{1 \text{ mol N}}{14.01 \text{ g N}} = 6.24 \text{ mol N}; \quad \frac{6.24}{6.24} = 1.00$$

$$12.6 \text{ g H} \times \frac{1 \text{ mol H}}{1.008 \text{ g H}} = 12.5 \text{ mol H}; \quad \frac{12.5}{6.24} = 2.00$$

Empirical formula is NH_2. $P \times$ (molar mass) = dRT, where d = density.

$$\text{Molar mass} = \frac{dRT}{P} = \frac{\dfrac{0.977\text{ g}}{L} \times \dfrac{0.08206\text{ L atm}}{K\text{ mol}} \times 373\text{ K}}{710.\text{ torr} \times \dfrac{1\text{ atm}}{760\text{ torr}}} = 32.0\text{ g/mol}$$

Empirical formula mass of $NH_2 = 16.0$ g. Therefore, the molecular formula is N_2H_4.

55.   If Be^{3+}, the formula is $Be(C_5H_7O_2)_3$ and molar mass $\approx 13.5 + 15(12) + 21(1) + 6(16)$ $= 311$ g/mol. If Be^{2+}, the formula is $Be(C_5H_7O_2)_2$ and molar mass $\approx 9.0 + 10(12) + 14(1) + 4(16) = 207$ g/mol.

Data set I (molar mass = dRT/P and d = mass/V):

$$\text{molar mass} = \frac{\text{mass} \times RT}{PV} = \frac{0.2022\text{ g} \times \dfrac{0.08206\text{ L atm}}{K\text{ mol}} \times 286\text{ K}}{(765.2\text{ torr} \times \dfrac{1\text{ atm}}{760\text{ torr}}) \times (22.6 \times 10^{-3}\text{ L})} = 209\text{ g/mol}$$

Data set II:

$$\text{molar mass} = \frac{\text{mass} \times RT}{PV} = \frac{0.2224\text{ g} \times \dfrac{0.08206\text{ L atm}}{K\text{ mol}} \times 290.\text{ K}}{(764.6\text{ torr} \times \dfrac{1\text{ atm}}{760\text{ torr}}) \times (26.0 \times 10^{-3}\text{ L})} = 202\text{ g/mol}$$

These results are close to the expected value of 207 g/mol for $Be(C_5H_7O_2)_2$. Thus we conclude from these data that beryllium is a divalent element with an atomic weight (mass) of 9.0 g/mol.

57. $\text{Molar mass} = \dfrac{dRT}{P} = \dfrac{\dfrac{0.70902\text{ g}}{L} \times \dfrac{0.08206\text{ L atm}}{K\text{ mol}} \times 273.2\text{ K}}{1.000\text{ atm}} = 15.90\text{ g/mol}$

15.90 g/mol is the average molar mass of the mixture of methane and helium. Assume 100.00 mol of total gas present, and let $x =$ mol of CH_4 in the 100.00 mol mixture. This value of x is also equal to the volume percentage of CH_4 in 100.00 L of mixture because T and P are constant.

$$15.90 = \frac{x(16.04) + (100.00 - x)(4.003)}{100.00}, \quad 1590. = (16.04)x + 400.3 - (4.003)x$$

$1190. = (12.04)x, \; x = 98.84\% \; CH_4 \; \text{by volume}; \; \% \; He = 100.00 - x = 1.16\% \; He \; \text{by volume}$

59. $n_{H_2} = \dfrac{PV}{RT} = \dfrac{1.0\ atm \times \left[4800\ m^3 \times \left(\dfrac{100\ cm}{m} \right)^3 \times \dfrac{1\ L}{1000\ cm^3} \right]}{\dfrac{0.08206\ L\ atm}{K\ mol} \times 273\ K} = 2.1 \times 10^5\ mol$

2.1×10^5 mol H_2 are in the balloon. This is 80.% of the total amount of H_2 that had to be generated:

$0.80(\text{total mol } H_2) = 2.1 \times 10^5, \quad \text{total mol } H_2 = 2.6 \times 10^5 \text{ mol } H_2$

$2.6 \times 10^5 \text{ mol } H_2 \times \dfrac{1\ mol\ Fe}{mol\ H_2} \times \dfrac{55.85\ g\ Fe}{mol\ Fe} = 1.5 \times 10^7 \text{ g Fe}$

$2.6 \times 10^5 \text{ mol } H_2 \times \dfrac{1\ mol\ H_2SO_4}{mol\ H_2} \times \dfrac{98.09\ g\ H_2SO_4}{mol\ H_2SO_4} \times \dfrac{100\ g\ reagent}{98\ g\ H_2SO_4} = 2.6 \times 10^7 \text{ g of 98\%}$
sulfuric acid

61.   Because P and T are constant, V and n are directly proportional. The balanced equation requires 2 L of H_2 to react with 1 L of CO (2 : 1 volume ratio due to 2 : 1 mole ratio in the balanced equation). The actual volume ratio present in 1 minute is 16.0 L/25.0 L = 0.640 (0.640 : 1). Because the actual volume ratio present is smaller than the required volume ratio, H_2 is the limiting reactant. The volume of CH_3OH produced at STP will be one-half the volume of H_2 reacted due to the 1 : 2 mole ratio in the balanced equation. In 1 minute, 16.0 L/2 = 8.00 L CH_3OH are produced (theoretical yield).

$n_{CH_3OH} = \dfrac{PV}{RT} = \dfrac{1.00\ atm \times 8.00\ L}{\dfrac{0.08206\ L\ atm}{K\ mol} \times 273\ K} = 0.357 \text{ mol } CH_3OH \text{ in 1 minute}$

$0.357 \text{ mol } CH_3OH \times \dfrac{32.04\ g\ CH_3OH}{mol\ CH_3OH} = 11.4 \text{ g } CH_3OH \text{ (theoretical yield per minute)}$

$\text{Percent yield} = \dfrac{\text{actual yield}}{\text{theoretical yield}} \times 100 = \dfrac{5.30\ g}{11.4\ g} \times 100 = 46.5\% \text{ yield}$

63. $150 \text{ g } (CH_3)_2N_2H_2 \times \dfrac{1\ mol\ (CH_3)_2N_2H_2}{60.10\ g} \times \dfrac{3\ mol\ N_2}{mol\ (CH_3)_2N_2H_2} = 7.5 \text{ mol } N_2 \text{ produced}$

$P_{N_2} = \dfrac{nRT}{V} = \dfrac{7.5\ mol \times \dfrac{0.08206\ L\ atm}{K\ mol} \times 300.\ K}{250\ L} = 0.74\ atm$

We could do a similar calculation for P_{H_2O} and P_{CO_2} and then calculate P_{total} $(= P_{N_2} + P_{H_2O} + P_{CO_2})$. Or we can recognize that 9 total moles of gaseous products form for every mole of $(CH_3)_2N_2H_2$ reacted. This is three times the moles of N_2 produced. Therefore, P_{total} will be three times larger than P_{N_2}. $P_{total} = 3 \times P_{N_2} = 3 \times 0.74 \text{ atm} = 2.2 \text{ atm}$.

65. $2 \text{ NaClO}_3(s) \rightarrow 2 \text{ NaCl}(s) + 3 \text{ O}_2(g)$

$P_{total} = P_{O_2} + P_{H_2O}$, $P_{O_2} = P_{total} - P_{H_2O} = 734 \text{ torr} - 19.8 \text{ torr} = 714 \text{ torr}$

$$n_{O_2} = \frac{P_{O_2} \times V}{RT} = \frac{\left(714 \text{ torr} \times \dfrac{1 \text{ atm}}{760 \text{ torr}}\right) \times 0.0572 \text{ L}}{\dfrac{0.08206 \text{ L atm}}{\text{K mol}} \times (273 + 22) \text{ K}} = 2.22 \times 10^{-3} \text{ mol O}_2$$

$$\text{Mass NaClO}_3 \text{ decomposed} = 2.22 \times 10^{-3} \text{ mol O}_2 \times \frac{2 \text{ mol NaClO}_3}{3 \text{ mol O}_2} \times \frac{106.44 \text{ g NaClO}_3}{\text{mol NaClO}_3}$$

$$= 0.158 \text{ g NaClO}_3$$

$$\text{Mass \% NaClO}_3 = \frac{0.158 \text{ g}}{0.8765 \text{ g}} \times 100 = 18.0\%$$

67. $P_{total} = P_{N_2} + P_{H_2O}$, $P_{N_2} = 726 \text{ torr} - 23.8 \text{ torr} = 702 \text{ torr H} \dfrac{1 \text{ atm}}{760 \text{ torr}} = 0.924 \text{ atm}$

$$n_{N_2} = \frac{P_{N_2} \times V}{RT} = \frac{0.924 \text{ atm} \times 31.8 \times 10^{-3} \text{ L}}{\dfrac{0.08206 \text{ L atm}}{\text{K mol}} \times 298 \text{ K}} = 1.20 \times 10^{-3} \text{ mol N}_2$$

$$\text{Mass of N in compound} = 1.20 \times 10^{-3} \text{ mol N}_2 \times \frac{28.02 \text{ g N}_2}{\text{mol}} = 3.36 \times 10^{-2} \text{ g nitrogen}$$

$$\text{Mass \% N} = \frac{3.36 \times 10^{-2} \text{ g}}{0.253 \text{ g}} \times 100 = 13.3\% \text{ N}$$

69.   For NH$_3$: $P_2 = \dfrac{P_1 V_1}{V_2} = 0.500 \text{ atm} \times \dfrac{2.00 \text{ L}}{3.00 \text{ L}} = 0.333 \text{ atm}$

For O$_2$: $P_2 = \dfrac{P_1 V_1}{V_2} = 1.50 \text{ atm} \times \dfrac{1.00 \text{ L}}{3.00 \text{ L}} = 0.500 \text{ atm}$

After the stopcock is opened, V and T will be constant, so $P \propto n$. The balanced equation requires:

$$\frac{n_{O_2}}{n_{NH_3}} = \frac{P_{O_2}}{P_{NH_3}} = \frac{5}{4} = 1.25$$

The actual ratio present is: $\dfrac{P_{O_2}}{P_{NH_3}} = \dfrac{0.500 \text{ atm}}{0.333 \text{ atm}} = 1.50$

The actual ratio is larger than the required ratio, so NH_3 in the denominator is limiting. Because equal moles of NO will be produced as NH_3 reacted, the partial pressure of NO produced is 0.333 atm (the same as P_{NH_3} reacted).

71. $2\ NH_3(g) \rightarrow N_2(g) + 3\ H_2(g)$; as reactants are converted into products, we go from 2 moles of gaseous reactants to 4 moles of gaseous products (1 mol N_2 + 3 mol H_2). Because the moles of gas doubles as reactants are converted into products, the volume of the gases will double (at constant P and T).

$$PV = nRT,\ P = \left(\frac{RT}{V}\right)n = (constant)n;\ \ \text{pressure is directly related to n at constant T and V.}$$

As the reaction occurs, the moles of gas will double, so the pressure will double. Because 1 mol of N_2 is produced for every 2 mol of NH_3 reacted, $P_{N_2} = 1/2\ P^o_{NH_3}$. Owing to the 3 to 2 mole ratio in the balanced equation, $P_{H_2} = 3/2\ P^o_{NH_3}$.

Note: $P_{total} = P_{H_2} + P_{N_2} = 3/2\ P^o_{NH_3} + 1/2\ P^o_{NH_3} = 2\ P^o_{NH_3}$. As we said earlier, the total pressure doubles as reactants are completely converted into products for this reaction.

Kinetic Molecular Theory and Real Gases

73.   The number of gas particles is constant, so at constant moles of gas, either a temperature change or a pressure change results in the smaller volume. If the temperature is constant, an increase in the external pressure would cause the volume to decrease. Gases are mostly empty space so gases are easily compressible.

If the pressure is constant, a decrease in temperature would cause the volume to decrease. As the temperature is lowered, the gas particles move with a slower average velocity and don't collide with the container walls as frequently and as forcefully. As a result, the internal pressure decreases. In order to keep the pressure constant, the volume of the container must decrease in order to increase the gas particle collisions per unit area.

75.   V, T, and P are all constant, so n must be constant. Because we have equal moles of gas in each container, gas B molecules must be heavier than gas A molecules.

a.   Both gas samples have the same number of molecules present (n is constant).

b.   Because T is constant, KE_{ave} must be the same for both gases ($KE_{ave} = 3/2\ RT$).

c.   The lighter gas A molecules will have the faster average velocity.

d.   The heavier gas B molecules do collide more forcefully, but gas A molecules, with the faster average velocity, collide more frequently. The end result is that P is constant between the two containers.

77. $(KE)_{avg} = 3/2\ RT$; KE depends only on temperature. At each temperature CH_4 and N_2 will have the same average KE. For energy units of joules (J), use $R = 8.3145\ J\ K^{-1}\ mol^{-1}$. To determine average KE per molecule, divide by Avogadro's number, 6.022×10^{23} molecules/mol.

At 273 K: $(KE)_{avg} = \dfrac{3}{2} \times \dfrac{8.3145\ J}{K\ mol} \times 273\ K = 3.40 \times 10^3\ J/mol = 5.65 \times 10^{-21}\ J/molecule$

At 546 K: $(KE)_{avg} = \dfrac{3}{2} \times \dfrac{8.3145\ J}{K\ mol} \times 546\ K = 6.81 \times 10^3\ J/mol = 1.13 \times 10^{-20}\ J/molecule$

79.   No; the numbers calculated in Exercise 77 are the average kinetic energies at the various temperatures. At each temperature, there is a distribution of energies. Similarly, the numbers calculated in Exercise 78 are a special kind of average velocity. There is a distribution of velocities as shown in Figs. 5.15 to 5.17 of the text. Note that the major reason there is a distribution of kinetic energies is because there is a distribution of velocities for any gas sample at some temperature.

81.   a.   They will all have the same average kinetic energy because they are all at the same temperature. Average kinetic energy depends only on temperature.

b.   Flask C; at constant T, $u_{rms} \propto (1/M)^{1/2}$. In general, the lighter the gas molecules, the greater is the root mean square velocity (at constant T).

c.   Flask A: collision frequency is proportional to average velocity $\times$ n/V (as the average velocity doubles, the number of collisions will double, and as the number of molecules in the container doubles, the number of collisions again doubles). At constant T and V, n is proportional to P, and average velocity is proportional to $(1/M)^{1/2}$. We use these relationships and the data in the exercise to determine the following relative values.

	n (relative)	u_{avg} (relative)	Coll. Freq. (relative) $= n \times u_{avg}$
A	1.0	1.0	1.0
B	0.33	1.0	0.33
C	0.13	3.7	0.48

83.   Graham's law of effusion: $\dfrac{Rate_1}{Rate_2} = \left(\dfrac{M_2}{M_1}\right)^{1/2}$

Let Freon-12 = gas 1 and Freon-11 = gas 2:

$\dfrac{1.07}{1.00} = \left(\dfrac{137.4}{M_1}\right)^{1/2}$, $1.14 = \dfrac{137.4}{M_1}$, $M_1 = 121\ g/mol$

The molar mass of CF_2Cl_2 is equal to 121 g/mol, so Freon-12 is CF_2Cl_2.

85. $\dfrac{\text{Rate}_1}{\text{Rate}_2} = \left(\dfrac{M_2}{M_1}\right)^{1/2}$; $\text{rate}_1 = \dfrac{24.0 \text{ mL}}{\text{min}}$; $\text{rate}_2 = \dfrac{47.8 \text{ mL}}{\text{min}}$; $M_2 = \dfrac{16.04 \text{ g}}{\text{mol}}$; $M_1 = ?$

$\dfrac{24.0}{47.8} = \left(\dfrac{16.04}{M_1}\right)^{1/2} = 0.502,\ \ 16.04 = (0.502)^2 \times M_1,\ \ M_1 = \dfrac{16.04}{0.252} = \dfrac{63.7 \text{ g}}{\text{mol}}$

87.   a. $PV = nRT$

$P = \dfrac{nRT}{V} = \dfrac{0.5000 \text{ mol} \times \dfrac{0.08206 \text{ L atm}}{\text{K mol}} \times (25.0 + 273.2) \text{ K}}{1.0000 \text{ L}} = 12.24 \text{ atm}$

b. $\left[P + a\left(\dfrac{n}{V}\right)^2\right](V - nb) = nRT$;   for N_2: $a = 1.39 \text{ atm L}^2/\text{mol}^2$ and $b = 0.0391 \text{ L/mol}$

$\left[P + 1.39\left(\dfrac{0.5000}{1.0000}\right)^2 \text{ atm}\right](1.0000 \text{ L} - 0.5000 \times 0.0391 \text{ L}) = 12.24 \text{ L atm}$

$(P + 0.348 \text{ atm})(0.9805 \text{ L}) = 12.24 \text{ L atm}$

$P = \dfrac{12.24 \text{ L atm}}{0.9805 \text{ L}} - 0.348 \text{ atm} = 12.48 - 0.348 = 12.13 \text{ atm}$

c.   The ideal gas law is high by 0.11 atm, or $\dfrac{0.11}{12.13} \times 100 = 0.91\%$.

89.   The kinetic molecular theory assumes that gas particles do not exert forces on each other and that gas particles are volumeless. Real gas particles do exert attractive forces for each other, and real gas particles do have volumes. A gas behaves most ideally at low pressures and high temperatures. The effect of attractive forces is minimized at high temperatures because the gas particles are moving very rapidly. At low pressure, the container volume is relatively large (P and V are inversely related), so the volume of the container taken up by the gas particles is negligible.

91.   The pressure measured for real gases is too low compared to ideal gases. This is due to the attractions gas particles do have for each other; these attractions "hold" them back from hitting the container walls as forcefully. To make up for this slight decrease in pressure for real gases, a factor is added to the measured pressure. The measured volume is too large. A fraction of the space of the container volume is taken up by the volume of the molecules themselves. Therefore, the actual volume available to real gas molecules is slightly less than the container volume. A term is subtracted from the container volume to correct for the volume taken up by real gas molecules.

93.   The values of a are: H_2, $\dfrac{0.244 \text{ atm L}^2}{\text{mol}^2}$; CO_2, 3.59; N_2, 1.39; CH_4, 2.25

Because a is a measure of intermolecular attractions, the attractions are greatest for CO_2.

95. $u_{rms} = \left(\dfrac{3RT}{M}\right)^{1/2} = \left[\dfrac{3\left(\dfrac{8.3145 \text{ kg m}^2}{s^2 \text{ K mol}}\right)(227+273)\text{ K}}{28.02 \times 10^{-3} \text{ kg/mol}}\right]^{1/2} = 667 \text{ m/s}$

$u_{mp} = \left(\dfrac{2RT}{M}\right)^{1/2} = \left[\dfrac{2\left(\dfrac{8.3145 \text{ kg m}^2}{s^2 \text{ K mol}}\right)(500.\text{ K})}{28.02 \times 10^{-3} \text{ kg/mol}}\right]^{1/2} = 545 \text{ m/s}$

$u_{avg} = \left(\dfrac{8RT}{\pi M}\right)^{1/2} = \left[\dfrac{8\left(\dfrac{8.3145 \text{ kg m}^2}{s^2 \text{ K mol}}\right)(500.\text{ K})}{\pi(28.02 \times 10^{-3} \text{ kg/mol})}\right]^{1/2} = 615 \text{ m/s}$

97.   The force per impact is proportional to $\Delta(mu) = 2mu$. Because $m \propto M$, the molar mass, and $u \propto (1/M)^{1/2}$ at constant T, the force per impact at constant T is proportional to $M \times (1/M)^{1/2} = \sqrt{M}$.

$\dfrac{\text{Impact force } (H_2)}{\text{Impact force } (He)} = \sqrt{\dfrac{M_{H_2}}{M_{He}}} = \sqrt{\dfrac{2.016}{4.003}} = 0.7097$

99. $\Delta(mu) = 2mu =$ change in momentum per impact. Because m is proportional to M, the molar mass, and u is proportional to $(T/M)^{1/2}$:

$$\Delta(mu)_{O_2} \propto 2M_{O_2}\left(\dfrac{T}{M_{O_2}}\right)^{1/2} \text{ and } \Delta(mu)_{He} \propto 2M_{He}\left(\dfrac{T}{M_{He}}\right)^{1/2}$$

$$\dfrac{\Delta(mu)_{O_2}}{\Delta(mu)_{He}} = \dfrac{2M_{O_2}\left(\dfrac{T}{M_{O_2}}\right)^{1/2}}{2M_{He}\left(\dfrac{T}{M_{He}}\right)^{1/2}} = \dfrac{M_{O_2}}{M_{He}}\left(\dfrac{M_{He}}{M_{O_2}}\right)^{1/2} = \dfrac{31.998}{4.003}\left(\dfrac{4.003}{31.998}\right)^{1/2} = 2.827$$

The change in momentum per impact is 2.827 times larger for O_2 molecules than for He atoms.

$$Z_A = A \frac{N}{V}\left(\frac{RT}{2\pi M}\right)^{1/2} = \text{collision rate}$$

$$\frac{Z_{O_2}}{Z_{He}} = \frac{A\left(\dfrac{N}{V}\right)\left(\dfrac{RT}{2\pi M_{O_2}}\right)^{1/2}}{A\left(\dfrac{N}{V}\right)\left(\dfrac{RT}{2\pi M_{He}}\right)^{1/2}} = \frac{\left(\dfrac{1}{M_{O_2}}\right)^{1/2}}{\left(\dfrac{1}{M_{He}}\right)^{1/2}} = 0.3537; \quad \frac{Z_{He}}{Z_{O_2}} = 2.827$$

There are 2.827 times as many impacts per second for He as for O_2.

101. Intermolecular collision frequency $= Z = 4\dfrac{N}{V}d^2\left(\dfrac{\pi RT}{M}\right)^{1/2}$, where d = diameter of He atom

$$\frac{n}{V} = \frac{P}{RT} = \frac{3.0\ \text{atm}}{\dfrac{0.08206\ \text{L atm}}{\text{K mol}} \times 300.\ \text{K}} = 0.12\ \text{mol/L}$$

$$\frac{N}{V} = \frac{0.12\ \text{mol}}{L} \times \frac{6.022 \times 10^{23}\ \text{molecules}}{\text{mol}} \times \frac{1000\ L}{m^3} = \frac{7.2 \times 10^{25}\ \text{molecules}}{m^3}$$

$$Z = 4 \times \frac{7.2 \times 10^{25}\ \text{molecules}}{m^3} \times (50. \times 10^{-12}\ m)^2 \times \left(\frac{\pi(8.3145)(300.)}{4.00 \times 10^{-3}}\right)^{1/2}$$

$$= 1.0 \times 10^9\ \text{collisions/s}$$

Mean free path $= \lambda = \dfrac{u_{avg}}{Z}$; $u_{avg} = \left(\dfrac{8\,RT}{\pi M}\right)^{1/2} = 1260\ \text{m/s}$; $\lambda = \dfrac{1260\ \text{m/s}}{1.0 \times 10^9\ \text{s}^{-1}} = 1.3 \times 10^{-6}\ \text{m}$

Atmospheric Chemistry

103. a. If we have 1.0×10^6 L of air, then there are 3.0×10^2 L of CO.

$$P_{CO} = \chi_{CO}P_{total}; \quad \chi_{CO} = \frac{V_{CO}}{V_{total}}\ \text{because V \% n}; \quad P_{CO} = \frac{3.0 \times 10^2}{1.0 \times 10^6} \times 628\ \text{torr} = 0.19\ \text{torr}$$

b. $n_{CO} = \dfrac{P_{CO}V}{RT}$;   Assuming $1.0\ m^3$ air, $1\ m^3 = 1000$ L:

$$n_{CO} = \frac{\dfrac{0.19}{760}\ \text{atm} \times (1.0 \times 10^3\ L)}{\dfrac{0.08206\ \text{L atm}}{\text{K mol}} \times 273\ \text{K}} = 1.1 \times 10^{-2}\ \text{mol CO}$$

$$1.1 \times 10^{-2} \text{ mol} \times \frac{6.02 \times 10^{23} \text{ molecules}}{\text{mol}} = 6.6 \times 10^{21} \text{ CO molecules in 1.0 m}^3 \text{ of air}$$

c. $$\frac{6.6 \times 10^{21} \text{ molecules}}{\text{m}^3} \times \left(\frac{1 \text{ m}}{100 \text{ cm}}\right)^3 = \frac{6.6 \times 10^{15} \text{ molecules CO}}{\text{cm}^3}$$

105.   For benzene:

$$89.6 \times 10^{-9} \text{ g} \times \frac{1 \text{ mol}}{78.11 \text{ g}} = 1.15 \times 10^{-9} \text{ mol benzene}$$

$$V_{\text{benzene}} = \frac{n_{\text{benzene}} RT}{P} = \frac{1.15 \times 10^{-9} \text{ mol} \times \dfrac{0.08206 \text{ L atm}}{\text{K mol}} \times 296 \text{ K}}{748 \text{ torr} \times \dfrac{1 \text{ atm}}{760 \text{ torr}}} = 2.84 \times 10^{-8} \text{ L}$$

$$\text{Mixing ratio} = \frac{2.84 \times 10^{-8} \text{ L}}{3.00 \text{ L}} \times 10^6 = 9.47 \times 10^{-3} \text{ ppmv}$$

$$\text{or ppbv} = \frac{\text{vol. of X} \times 10^9}{\text{total vol.}} = \frac{2.84 \times 10^{-8} \text{ L}}{3.00 \text{ L}} \times 10^9 = 9.47 \text{ ppbv}$$

$$\frac{1.15 \times 10^{-9} \text{ mol benzene}}{3.00 \text{ L}} \times \frac{1 \text{ L}}{1000 \text{ cm}^3} \times \frac{6.022 \times 10^{23} \text{ molecules}}{\text{mol}}$$

$$= 2.31 \times 10^{11} \text{ molecules benzene/cm}^3$$

For toluene:

$$153 \times 10^{-9} \text{ g C}_7\text{H}_8 \times \frac{1 \text{ mol}}{92.13 \text{ g}} = 1.66 \times 10^{-9} \text{ mol toluene}$$

$$V_{\text{toluene}} = \frac{n_{\text{toluene}} RT}{P} = \frac{1.66 \times 10^{-9} \text{ mol} \times \dfrac{0.08206 \text{ L atm}}{\text{K mol}} \times 296 \text{ K}}{748 \text{ torr} \times \dfrac{1 \text{ atm}}{760 \text{ torr}}} = 4.10 \times 10^{-8} \text{ L}$$

$$\text{Mixing ratio} = \frac{4.10 \times 10^{-8} \text{ L}}{3.00 \text{ L}} \times 10^6 = 1.37 \times 10^{-2} \text{ ppmv (or 13.7 ppbv)}$$

$$\frac{1.66 \times 10^{-9} \text{ mol toluene}}{3.00 \text{ L}} \times \frac{1 \text{ L}}{1000 \text{ cm}^3} \times \frac{6.022 \times 10^{23} \text{ molecules}}{\text{mol}}$$

$$= 3.33 \times 10^{11} \text{ molecules toluene/cm}^3$$

Additional Exercises

107. Processes a, c, and d will all result in a doubling of the pressure. Process a has the effect of halving the volume, which would double the pressure (Boyle's law). Process c doubles the pressure because the absolute temperature is doubled (from 200. K to 400. K). Process d doubles the pressure because the moles of gas are doubled (28 g N_2 is 1 mol of N_2). Process b won't double the pressure since the absolute temperature is not doubled (303 K to 333 K).

109. $0.050 \text{ mL} \times \dfrac{1.149 \text{ g}}{\text{mL}} \times \dfrac{1 \text{ mol O}_2}{32.0 \text{ g}} = 1.8 \times 10^{-3} \text{ mol O}_2$

$$V = \dfrac{nRT}{P} = \dfrac{1.8 \times 10^{-3} \text{ mol} \times \dfrac{0.08206 \text{ L atm}}{\text{K mol}} \times 310. \text{ K}}{1.0 \text{ atm}} = 4.6 \times 10^{-2} \text{ L} = 46 \text{ mL}$$

111. $PV = nRT$, V and T are constant. $\dfrac{P_1}{n_1} = \dfrac{P_2}{n_2}$ or $\dfrac{P_1}{P_2} = \dfrac{n_1}{n_2}$

When V and T are constant, then pressure is directly proportional to moles of gas present, and pressure ratios are identical to mole ratios.

At 25°C: $2 H_2(g) + O_2(g) \rightarrow 2 H_2O(l)$, $H_2O(l)$ is produced at 25°C.

The balanced equation requires 2 mol H_2 for every mol O_2 reacted. The same ratio (2 : 1) holds true for pressure units. The actual pressure ratio present is 2 atm H_2 to 3 atm O_2, well below the required 2 : 1 ratio. Therefore, H_2 is the limiting reagent. The only gas present at 25°C after the reaction goes to completion will be the excess O_2.

$$P_{O_2} \text{ (reacted)} = 2.00 \text{ atm H}_2 \times \dfrac{1 \text{ atm O}_2}{2 \text{ atm H}_2} = 1.00 \text{ atm O}_2$$

$$P_{O_2} \text{ (excess)} = P_{O_2} \text{ (initial)} - P_{O_2} \text{ (reacted)} = 3.00 \text{ atm} - 1.00 \text{ atm} = 2.00 \text{ atm O}_2 = P_{total}$$

At 125°C: $2 H_2(g) + O_2(g) \rightarrow 2 H_2O(g)$, $H_2O(g)$ is produced at 125°C.

The major difference in the problem is that gaseous H_2O is now a product (instead of liquid H_2O), which will increase the total pressure because an additional gas is present.

$$P_{H_2O} \text{ (produced)} = 2.00 \text{ atm H}_2 \times \dfrac{2 \text{ atm H}_2O}{2 \text{ atm H}_2} = 2.00 \text{ atm H}_2O$$

$$P_{total} = P_{O_2} \text{ (excess)} + P_{H_2O} \text{ (produced)} = 2.00 \text{ atm O}_2 + 2.00 \text{ atm H}_2O = 4.00 \text{ atm}$$

113. $Mn(s) + x\ HCl(g) \rightarrow MnCl_x(s) + \dfrac{x}{2} H_2(g)$

$$n_{H_2} = \frac{PV}{RT} = \frac{0.951\ atm \times 3.22\ L}{\dfrac{0.08206\ L\ atm}{K\ mol} \times 373\ K} = 0.100\ mol\ H_2$$

Mol Cl in compound = mol HCl = $0.100\ mol\ H_2 \times \dfrac{x\ mol\ Cl}{\dfrac{x}{2}\ mol\ H_2} = 0.200\ mol\ Cl$

$$\frac{Mol\ Cl}{Mol\ Mn} = \frac{0.200\ mol\ Cl}{2.747\ g\ Mn \times \dfrac{1\ mol\ Mn}{54.94\ g\ Mn}} = \frac{0.200\ mol\ Cl}{0.05000\ mol\ Mn} = 4.00$$

The formula of compound is $MnCl_4$.

115.   At constant T and P, Avogadro's law applies; that is, equal volumes contain equal moles of molecules. In terms of balanced equations, we can say that mole ratios and volume ratios between the various reactants and products will be equal to each other. $Br_2 + 3\ F_2 \rightarrow 2\ X$; 2 moles of X must contain 2 moles of Br and 6 moles of F; X must have the formula BrF_3 for a balanced equation.

117.   PV = nRT, V and T are constant. $\dfrac{P_1}{n_1} = \dfrac{P_2}{n_2}, \ \dfrac{P_2}{P_1} = \dfrac{n_2}{n_1}$

We will do this limiting-reagent problem using an alternative method than described in Chapter 3. Let's calculate the partial pressure of C_3H_3N that can be produced from each of the starting materials assuming each reactant is limiting. The reactant that produces the smallest amount of product will run out first and is the limiting reagent.

$$P_{C_3H_3N} = 0.500\ MPa \times \frac{2\ MPa\ C_3H_3N}{2\ MPa\ C_3H_6} = 0.500\ MPa\ if\ C_3H_6\ is\ limiting$$

$$P_{C_3H_3N} = 0.800\ MPa \times \frac{2\ MPa\ C_3H_3N}{2\ MPa\ NH_3} = 0.800\ MPa\ if\ NH_3\ is\ limiting$$

$$P_{C_3H_3N} = 1.500\ MPa \times \frac{2\ MPa\ C_3H_3N}{3\ MPa\ O_2} = 1.000\ MPa\ if\ O_2\ is\ limiting$$

C_3H_6 is limiting. Although more product could be produced from NH_3 and O_2, there is only enough C_3H_6 to produce 0.500 MPa of C_3H_3N. The partial pressure of C_3H_3N in atmospheres after the reaction is:

$$0.500 \times 10^6 \text{ Pa} \times \frac{1 \text{ atm}}{1.013 \times 10^5 \text{ Pa}} = 4.94 \text{ atm}$$

$$n = \frac{PV}{RT} = \frac{4.94 \text{ atm} \times 150. \text{ L}}{\dfrac{0.08206 \text{ L atm}}{\text{K mol}} \times 298 \text{ K}} = 30.3 \text{ mol } C_3H_3N$$

$$30.3 \text{ mol} \times \frac{53.06 \text{ g}}{\text{mol}} = 1.61 \times 10^3 \text{ g } C_3H_3N \text{ can be produced.}$$

119. $P_{total} = P_{H_2} + P_{H_2O}$, $1.032 \text{ atm} = P_{H_2} + 32 \text{ torr} \times \dfrac{1 \text{ atm}}{760 \text{ torr}}$, $1.032 - 0.042 = 0.990 \text{ atm} = P_{H_2}$

$$n_{H_2} = \frac{P_{H_2} V}{RT} = \frac{0.990 \text{ atm} \times 0.240 \text{ L}}{\dfrac{0.08206 \text{ L atm}}{\text{K mol}} \times 303 \text{ K}} = 9.56 \times 10^{-3} \text{ mol } H_2$$

$$9.56 \times 10^{-3} \text{ mol } H_2 \times \frac{1 \text{ mol Zn}}{\text{mol } H_2} \times \frac{65.38 \text{ g Zn}}{\text{mol Zn}} = 0.625 \text{ g Zn}$$

121. $P_1V_1 = P_2V_2$; the total volume is $1.00 \text{ L} + 1.00 \text{ L} + 2.00 \text{ L} = 4.00 \text{ L}$.

For He: $P_2 = \dfrac{P_1 V_1}{V_2} = 200. \text{ torr} \times \dfrac{1.00 \text{ L}}{4.00 \text{ L}} = 50.0 \text{ torr He}$

For Ne: $P_2 = 0.400 \text{ atm} \times \dfrac{1.00 \text{ L}}{4.00 \text{ L}} = 0.100 \text{ atm}$; $0.100 \text{ atm} \times \dfrac{760 \text{ torr}}{\text{atm}} = 76.0 \text{ torr Ne}$

For Ar: $P_2 = 24.0 \text{ kPa} \times \dfrac{2.00 \text{ L}}{4.00 \text{ L}} = 12.0 \text{ kPa}$; $12.0 \text{ kPa} \times \dfrac{1 \text{ atm}}{101.3 \text{ kPa}} \times \dfrac{760 \text{ torr}}{\text{atm}}$

$$= 90.0 \text{ torr Ar}$$

$P_{total} = 50.0 + 76.0 + 90.0 = 216.0 \text{ torr}$

123.   a. $156 \text{ mL} \times 1.34 \text{ g/mL} = 209 \text{ g } HSiCl_3 = \text{actual yield of } HSiCl_3$

$$n_{HCl} = \frac{PV}{RT} = \frac{10.0 \text{ atm} \times 15.0 \text{ L}}{\dfrac{0.08206 \text{ L atm}}{\text{K mol}} \times 308 \text{ K}} = 5.93 \text{ mol HCl}$$

$$5.93 \text{ mol HCl} \times \frac{1 \text{ mol } HSiCl_3}{3 \text{ mol HCl}} \times \frac{135.45 \text{ g } HSiCl_3}{\text{mol } HSiCl_3} = 268 \text{ g } HSiCl_3$$

$$\text{Percent yield} = \frac{\text{actual yield}}{\text{theoretical yield}} \times 100 = \frac{209 \text{ g}}{268 \text{ g}} \times 100 = 78.0\%$$

b. $209 \text{ g HiSCl}_3 \times \dfrac{1 \text{ mol HSiCl}_3}{135.45 \text{ g HSiCl}_3} \times \dfrac{1 \text{ mol SiH}_4}{4 \text{ mol HSiCl}_3} = 0.386 \text{ mol SiH}_4$

This is the theoretical yield. If the percent yield is 93.1%, then the actual yield is:

$0.386 \text{ mol SiH}_4 \times 0.931 = 0.359 \text{ mol SiH}_4$

$$V_{\text{SiH}_4} = \frac{nRT}{P} = \frac{0.359 \text{ mol} \times \dfrac{0.08206 \text{ L atm}}{\text{K mol}} \times 308 \text{ K}}{10.0 \text{ atm}} = 0.907 \text{ L} = 907 \text{ mL SiH}_4$$

125.   We will apply Boyle's law to solve. $PV = nRT = \text{constant}, \ P_1V_1 = P_2V_2$

Let condition (1) correspond to He from the tank that can be used to fill balloons. We must leave 1.0 atm of He in the tank, so $P_1 = 200. - 1.00 = 199$ atm and $V_1 = 15.0$ L. Condition (2) will correspond to the filled balloons with $P_2 = 1.00$ atm and $V_2 = N(2.00 \text{ L})$, where N is the number of filled balloons, each at a volume of 2.00 L.

199 atm $\times$ 15.0 L = 1.00 atm $\times$ N(2.00 L), N = 1492.5; we can't fill 0.5 of a balloon, so N = 1492 balloons, or to 3 significant figures, 1490 balloons.

127. $n_{\text{Ar}} = \dfrac{228 \text{ g}}{39.95 \text{ g/mol}} = 5.71 \text{ mol Ar}; \ \chi_{\text{CH}_4} = \dfrac{n_{\text{CH}_4}}{n_{\text{CH}_4} + n_{\text{Ar}}}, \ 0.650 = \dfrac{n_{\text{CH}_4}}{n_{\text{CH}_4} + 5.71}$

$0.650(n_{\text{CH}_4} + 5.71) = n_{\text{CH}_4}, \ 3.71 = (0.350)n_{\text{CH}_4}, \ n_{\text{CH}_4} = 10.6 \text{ mol CH}_4$

$KE_{\text{avg}} = \dfrac{3}{2}RT$ for 1 mol

Thus $KE_{\text{total}} = (10.6 + 5.71 \text{ mol}) \times 3/2 \times 8.3145 \text{ J K}^{-1} \text{ mol}^{-1} \times 298 \text{ K} = 6.06 \times 10^4 \text{ J} = 60.6 \text{ kJ}$

129.   Mol of He removed $= \dfrac{PV}{RT} = \dfrac{1.00 \text{ atm} \times (1.75 \times 10^{-3} \text{ L})}{\dfrac{0.08206 \text{ L atm}}{\text{K mol}} \times 298 \text{ K}} = 7.16 \times 10^{-5} \text{ mol He}$

In the original flask, 7.16×10^{-5} mol of He exerted a partial pressure of $1.960 - 1.710 = 0.250$ atm.

$$V = \frac{nRT}{V} = \frac{(7.16 \times 10^{-5} \text{ mol}) \times 0.08206 \text{ L atm K}^{-1} \text{ mol}^{-1} \times 298 \text{ K}}{0.250 \text{ atm}} = 7.00 \times 10^{-3} \text{ L}$$
$$= 7.00 \text{ mL}$$

131.   a. $2 CH_4(g) + 2 NH_3(g) + 3 O_2(g) \rightarrow 2 HCN(g) + 6 H_2O(g)$

b.   Volumes of gases are proportional to moles at constant T and P. Using the balanced equation, methane and ammonia are in stoichiometric amounts and oxygen is in excess. In 1 second:

$$n_{CH_4} = \frac{PV}{RT} = \frac{1.00 \text{ atm} \times 20.0 \text{ L}}{0.08206 \text{ L atm K}^{-1} \text{ mol}^{-1} \times 423 \text{ K}} = 0.576 \text{ mol CH}_4$$

$$\frac{0.576 \text{ mol CH}_4}{s} \times \frac{2 \text{ mol HCN}}{2 \text{ mol CH}_4} \times \frac{27.03 \text{ g HCN}}{\text{mol HCN}} = 15.6 \text{ g HCN/s}$$

Challenge Problems

133.   Initially we have 1.00 mol CH_4 (16.0 g/mol = molar mass) and 2.00 mol O_2 (32.0 g/mol = molar mass).

$$CH_4(g) + a\, O_2(g) \rightarrow b\, CO(g) + c\, CO_2(g) + d\, H_2O(g)$$

$b + c = 1.00$ (C balance); $2a = b + 2c + d$ (O balance)

$2d = 4$ (H balance), $d = 2 = 2.00$ mol H_2O

$$V_{initial} = \frac{nRT}{P} = \frac{3.00 \text{ mol} \times 0.08206 \text{ L atm K}^{-1} \text{ mol}^{-1} \times 425 \text{ K}}{1.00 \text{ atm}} = 104.6 \text{ L (1 extra sig .fig.)}$$

$$\text{Density}_{initial} = \frac{80.0 \text{ g}}{104.6 \text{ L}} = 0.7648 \text{ g/L} \text{ (1 extra significant figure)}$$

Because mass is constant:

$$\text{mass} = V_{initial} \times d_{initial} = V_{final} \times d_{final}, \; V_{final} = V_{initial} \times \frac{d_{initial}}{d_{final}} = 104.6 \text{ L} \times \frac{0.7648 \text{ g/L}}{0.7282 \text{ g/L}}$$

$V_{final} = 109.9$ L (1 extra significant figure)

$$n_{final} = \frac{PV}{RT} = \frac{1.00 \text{ atm} \times 109.9 \text{ L}}{\dfrac{0.08206 \text{ L atm}}{\text{K mol}} \times 425 \text{ K}} = 3.15 \text{ total moles of gas}$$

Assuming an excess of O_2 is present after reaction, an expression for the total moles of gas present at completion is:

$b + c + 2.00 + (2.00 - a) = 3.15$; *Note*: $d = 2.00$ mol H_2O was determined previously.

Because $b + c = 1.00$, solving gives $a = 1.85$ mol O_2 reacted. Indeed, O_2 is in excess.

From the O balance equation:

$$2a = 3.70 = b + 2c + 2.00, \ \ b + 2c = 1.70$$

Because $b + c = 1.00$, solving gives $b = 0.30$ mol CO and $c = 0.70$ mol CO_2.

The fraction of methane that reacts to form CO is 0.30 mol CO/1.00 mol CH_4 = 0.30 (or 30.% by moles of the reacted methane forms CO).

135. $Cr(s) + 3\ HCl(aq) \rightarrow CrCl_3(aq) + 3/2\ H_2(g);\ \ Zn(s) + 2\ HCl(aq) \rightarrow ZnCl_2(aq) + H_2(g)$

$$\text{Mol } H_2 \text{ produced} = n = \frac{PV}{RT} = \frac{\left(750.\ torr \times \dfrac{1\ atm}{760\ torr}\right) \times 0.225\ L}{\dfrac{0.08206\ L\ atm}{K\ mol} \times (273 + 27)\ K} = 9.02 \times 10^{-3} \text{ mol } H_2$$

9.02×10^{-3} mol H_2 = mol H_2 from Cr reaction + mol H_2 from Zn reaction

From the balanced equation: 9.02×10^{-3} mol H_2 = mol Cr $\times$ (3/2) + mol Zn $\times$ 1

Let x = mass of Cr and y = mass of Zn, then:

$$x + y = 0.362 \text{ g and } 9.02 \times 10^{-3} = \frac{(1.5)x}{52.00} + \frac{y}{65.38}$$

We have two equations and two unknowns. Solving by simultaneous equations:

$$9.02 \times 10^{-3} = \ \ (0.02885)x + (0.01530)y$$
$$\underline{-0.01530 \times 0.362 = -(0.01530)x - (0.01530)y}$$
$$3.48 \times 10^{-3} = \ \ (0.01355)x, \qquad x = \text{mass of Cr} = \frac{3.48 \times 10^{-3}}{0.01355} = 0.257 \text{ g}$$

$$y = \text{mass of Zn} = 0.362 \text{ g} - 0.257 \text{ g} = 0.105 \text{ g Zn};\ \ \text{mass \% Zn} = \frac{0.105\ g}{0.362\ g} \times 100$$
$$= 29.0\% \text{ Zn}$$

137. $\text{Molar mass} = \dfrac{dRT}{P}$, P and molar mass are constant; $dT = \dfrac{P \times \text{molar mass}}{R} = \text{constant}$

d = constant(1/T) or $d_1T_1 = d_2T_2$, where T is in kelvin (K).

$T = x + °C;\ 1.2930(x + 0.0) = 0.9460(x + 100.0)$

$(1.2930)x = (0.9460)x + 94.60,\ \ (0.3470)x = 94.60,\ \ x = 272.6$

From these data, absolute zero would be $-272.6°C$. The actual value is $-273.15°C$.

139. $\dfrac{PV}{nRT} = 1 + \beta P$; $\dfrac{n}{V} \times$ molar mass = d

$\dfrac{\text{molar mass}}{RT} \times \dfrac{P}{d} = 1 + \beta P$, $\dfrac{P}{d} = \dfrac{RT}{\text{molar mass}} + \dfrac{\beta RTP}{\text{molar mass}}$

This is in the equation for a straight line: $y = b + mx$. If we plot P/d versus P and extrapolate to P = 0, we get a y intercept = b = 1.398 = RT/molar mass.

At 0.00°C, molar mass = $\dfrac{0.08206 \times 273.15}{1.398}$ = 16.03 g/mol.

141.   Figure 5.16 shows the effect of temperature on the Maxwell-Boltzmann distribution of velocities of molecules. Note that as temperature increases, the probability that a gas particle has the most probable velocity decreases. Thus, since the probability of the gas particle with the most probable velocity decreased by one-half, then the temperature must be higher than 300. K.

The equation that determines the probability that a gas molecule has a certain velocity is:

$$f(u) = 4\pi \left(\dfrac{m}{2\pi k_B T} \right)^{3/2} u^2 e^{-mu^2/2k_B T}$$

Let T_x = the unknown temperature, then:

$$\dfrac{f(u_{mp,x})}{f(u_{mp,300})} = \dfrac{1}{2} = \dfrac{4\pi \left(\dfrac{m}{2\pi k_B T_x} \right)^{3/2} u_{mp,x}^2 \, e^{-mu_{mp,x}^2/2k_B T_x}}{4\pi \left(\dfrac{m}{2\pi k_B T_{300}} \right)^{3/2} u_{mp,300}^2 \, e^{-mu_{mp,300}^2/2k_B T_{300}}}$$

Because $u_{mp} = \sqrt{\dfrac{2k_B T}{m}}$, the equation reduces to:

$$\dfrac{1}{2} = \dfrac{\left(\dfrac{1}{T_x} \right)^{3/2} (T_x)}{\left(\dfrac{1}{T_{300}} \right)^{3/2} (T_{300})} = \left(\dfrac{T_{300}}{T_x} \right)^{1/2}$$

Note that the overall exponent term cancels from the expression when $2k_B T/m$ is substituted for u_{mp}^2 in the exponent term; the temperatures cancel. Solving for T_x:

$$\dfrac{1}{2} = \left(\dfrac{300.\,K}{T_x} \right),   T_x = 1.20 \times 10^3 \, K;   \text{as expected, } T_x \text{ is higher than 300. K.}$$

143.   From the problem, we want $Z_A/Z = 1.00 \times 10^{18}$ where Z_A is the collision frequency of the gas particles with the walls of the container and Z is the intermolecular collision frequency.

From the text: $$\frac{Z_A}{Z} = \frac{A\dfrac{N}{V}\sqrt{\dfrac{RT}{2\pi M}}}{4\dfrac{N}{V}d^2\sqrt{\dfrac{\pi RT}{M}}} = 1.00 \times 10^{18}, \quad 1.00 \times 10^{18} = \frac{A}{4\,d^2\,\pi\sqrt{2}}$$

If l = length of the cube edge container, then the area A of one cube face is l^2 and the total area in the cube is $6l^2$ (6 faces/cube). He diameter = d = $2(3.2 \times 10^{-11}$ m$) = 6.4 \times 10^{-11}$ m.

Solving the above expression for A, and then for l gives l = 0.11 m = 1.1 dm.

Volume = l^3 = (1.1 dm$)^3$ = 1.3 dm^3 = 1.3 L

145.   The reactions are:

$$C(s) + 1/2\ O_2(g) \rightarrow CO(g) \quad \text{and} \quad C(s) + O_2(g) \rightarrow CO_2(g)$$

$$PV = nRT, \quad P = n\left(\frac{RT}{V}\right) = n(\text{constant})$$

Because the pressure has increased by 17.0%, the number of moles of gas has also increased by 17.0%.

$$n_{\text{final}} = (1.170)n_{\text{initial}} = 1.170(5.00) = 5.85 \text{ mol gas} = n_{O_2} + n_{CO} + n_{CO_2}$$

$n_{CO} + n_{CO_2} = 5.00$ (balancing moles of C).   Solving by simultaneous equations:

$$\begin{aligned} n_{O_2} + n_{CO} + n_{CO_2} &= 5.85 \\ -(n_{CO} + n_{CO_2} &= 5.00) \\ \hline n_{O_2} \phantom{+ n_{CO} + n_{CO_2}} &= 0.85 \end{aligned}$$

If all C were converted to CO_2, no O_2 would be left. If all C were converted to CO, we would get 5 mol CO and 2.5 mol excess O_2 in the reaction mixture. In the final mixture, moles of CO equals twice the moles of O_2 present ($n_{CO} = 2n_{O_2}$).

$$n_{CO} = 2n_{O_2} = 1.70 \text{ mol CO}; \quad 1.70 + n_{CO_2} = 5.00, \quad n_{CO_2} = 3.30 \text{ mol } CO_2$$

$$\chi_{CO} = \frac{1.70}{5.85} = 0.291; \quad \chi_{CO_2} = \frac{3.30}{5.85} = 0.564; \quad \chi_{O_2} = 1.000 - 0.291 - 0.564 = 0.145$$

147. a. The reaction is: $CH_4(g) + 2\ O_2(g) \rightarrow CO_2(g) + 2\ H_2O(g)$

$$PV = nRT, \quad \frac{PV}{n} = RT = \text{constant}, \quad \frac{P_{CH_4} V_{CH_4}}{n_{CH_4}} = \frac{P_{air} V_{air}}{n_{air}}$$

The balanced equation requires 2 mol O_2 for every mol of CH_4 that reacts. For three times as much oxygen, we would need 6 mol O_2 per mol of CH_4 reacted ($n_{O_2} = 6n_{CH_4}$). Air is 21% mole percent O_2, so $n_{O_2} = (0.21)n_{air}$. Therefore, the moles of air we would need to deliver the excess O_2 are:

$$n_{O_2} = (0.21)n_{air} = 6n_{CH_4}, \quad n_{air} = 29n_{CH_4}, \quad \frac{n_{air}}{n_{CH_4}} = 29$$

In 1 minute:

$$V_{air} = V_{CH_4} \times \frac{n_{air}}{n_{CH_4}} \times \frac{P_{CH_4}}{P_{air}} = 200.\ L \times 29 \times \frac{1.50\ \text{atm}}{1.00\ \text{atm}} = 8.7 \times 10^3\ L\ \text{air/min}$$

b. If x mol of CH_4 were reacted, then $6x$ mol O_2 were added, producing $(0.950)x$ mol CO_2 and $(0.050)x$ mol of CO. In addition, $2x$ mol H_2O must be produced to balance the hydrogens.

$CH_4(g) + 2\ O_2(g) \rightarrow CO_2(g) + 2\ H_2O(g); \ \ CH_4(g) + 3/2\ O_2(g) \rightarrow CO(g) + 2\ H_2O(g)$

Amount O_2 reacted:

$$(0.950)x\ \text{mol}\ CO_2 \times \frac{2\ \text{mol}\ O_2}{\text{mol}\ CO_2} = (1.90)x\ \text{mol}\ O_2$$

$$(0.050)x\ \text{mol}\ CO \times \frac{1.5\ \text{mol}\ O_2}{\text{mol}\ CO} = (0.075)x\ \text{mol}\ O_2$$

Amount of O_2 left in reaction mixture = $(6.00)x - (1.90)x - (0.075)x = (4.03)x\ \text{mol}\ O_2$

Amount of N_2 = $(6.00)x\ \text{mol}\ O_2 \times \dfrac{79\ \text{mol}\ N_2}{21\ \text{mol}\ O_2} = (22.6)x \approx 23x\ \text{mol}\ N_2$

The reaction mixture contains:

$(0.950)x\ \text{mol}\ CO_2 + (0.050)x\ \text{mol}\ CO + (4.03)x\ \text{mol}\ O_2 + (2.00)x\ \text{mol}\ H_2O$
$$+ 23x\ \text{mol}\ N_2 = (30.)x\ \text{mol of gas total}$$

$$\chi_{CO} = \frac{(0.050)x}{(30.)x} = 0.0017; \quad \chi_{CO_2} = \frac{(0.950)x}{(30.)x} = 0.032; \quad \chi_{O_2} = \frac{(4.03)x}{(30.)x} = 0.13$$

$$\chi_{H_2O} = \frac{(2.00)x}{(30.)x} = 0.067; \quad \chi_{N_2} = \frac{23x}{(30.)x} = 0.77$$

c.   The partial pressures are determined by $P = \chi P_{total}$. Because $P_{total} = 1.00$ atm, $P_{CO} = 0.0017$ atm, $P_{CO_2} = 0.032$ atm, $P_{O_2} = 0.13$ atm, $P_{H_2O} = 0.067$ atm, and $P_{N_2} = 0.77$ atm.

149.   Each stage will give an enrichment of:

$$\frac{\text{Diffusion rate } ^{12}CO_2}{\text{Diffusion rate } ^{13}CO_2} = \left(\frac{M_{^{13}CO_2}}{M_{^{12}CO_2}}\right)^{1/2} = \left(\frac{45.001}{43.998}\right) = 1.0113$$

Because $^{12}CO_2$ moves slightly faster, each successive stage will have less $^{13}CO_2$.

$$\frac{99.90 \; ^{12}CO_2}{0.10 \; ^{13}CO_2} \times 1.0113^N = \frac{99.990 \; ^{12}CO_2}{0.010 \; ^{13}CO_2}$$

$$1.0113^N = \frac{9,999.0}{999.00} = 10.009 \quad \text{(carrying extra significant figures)}$$

$$N \log(1.0113) = \log(10.009), \; N = \frac{1.000391}{4.88 \times 10^{-3}} = 2.05 \times 10^2 \approx 2.1 \times 10^2 \text{ stages are needed.}$$

151.   a.   Average molar mass of air $= 0.790 \times 28.02$ g/mol $+ 0.210 \times 32.00$ g/mol $= 28.9$ g/mol; molar mass of helium $= 4.003$ g/mol

A given volume of air at a given set of conditions has a larger density than helium at those conditions. We need to heat the air to a temperature greater than 25°C in order to lower the air density (by driving air out of the hot air balloon) until the density is the same as that for helium (at 25°C and 1.00 atm).

b.   To provide the same lift as the helium balloon (assume V = 1.00 L), the mass of air in the hot-air balloon (V = 1.00 L) must be the same as that in the helium balloon. Let MM = molar mass:

$$P \bullet MM = dRT, \; \text{mass} = \frac{MM \bullet PV}{RT}; \; \text{solving: mass He} = 0.164 \text{ g}$$

$$\text{Mass air} = 0.164 \text{ g} = \frac{28.9 \text{ g/mol} \times 1.00 \text{ atm} \times 1.00 \text{ L}}{\dfrac{0.08206 \text{ L atm}}{\text{K mol}} \times T}, \; T = 2150 \text{ K} \quad \text{(a very high temperature)}$$

153. $d = $ molar mass(P/RT); at constant P and T, the density of gas is directly proportional to the molar mass of the gas. Thus the molar mass of the gas has a value which is 1.38 times that of the molar mass of O_2.

Molar mass $= 1.38(32.00 \text{ g/mol}) = 44.2$ g/mol

Because H_2O is produced when the unknown binary compound is combusted, the unknown must contain hydrogen. Let A_xH_y be the formula for unknown compound.

$$\text{Mol } A_xH_y = 10.0 \text{ g } A_xH_y \times \frac{1 \text{ mol } A_xH_y}{44.2 \text{ g}} = 0.226 \text{ mol } A_xH_y$$

$$\text{Mol H} = 16.3 \text{ g } H_2O \times \frac{1 \text{ mol } H_2O}{18.02 \text{ g}} \times \frac{2 \text{ mol H}}{\text{mol } H_2O} = 1.81 \text{ mol H}$$

$$\frac{1.81 \text{ mol H}}{0.226 \text{ mol } A_xH_y} = 8 \text{ mol H/mol } A_xH_y ; \quad A_xH_y = A_xH_8$$

The mass of the x moles of A in the A_xH_8 formula is:

$$44.2 \text{ g} - 8(1.008 \text{ g}) = 36.1 \text{ g}$$

From the periodic table and by trial and error, some possibilities for A_xH_8 are ClH_8, F_2H_8, C_3H_8, and Be_4H_8. C_3H_8 and Be_4H_8 fit the data best and because C_3H_8 (propane) is a known substance, C_3H_8 is the best possible identity from the data in this problem.

CHAPTER 6

CHEMICAL EQUILIBRIUM

Characteristics of Chemical Equilibrium

11. $2 NOCl(g) \rightleftharpoons 2 NO(g) + Cl_2(g)$ $K = 1.6 \times 10^{-5}$

The expression for K is the product concentrations divided by the reactant concentrations. When K has a value much less than one, the product concentrations are relatively small, and the reactant concentrations are relatively large.

$2 NO(g) \rightleftharpoons N_2(g) + O_2(g)$ $K = 1 \times 10^{31}$

When K has a value much greater than one, the product concentrations are relatively large, and the reactant concentrations are relatively small. In both cases, however, the rate of the forward reaction equals the rate of the reverse reaction at equilibrium (this is a definition of equilibrium).

13. No, it doesn't matter which direction the equilibrium position is reached. Both experiments will give the same equilibrium position because both experiments started with stoichiometric amounts of reactants or products.

15. When equilibrium is reached, there is no net change in the amount of reactants and products present because the rates of the forward and reverse reactions are equal to each other. The first diagram has 4 A_2B molecules, 2 A_2 molecules, and 1 B_2 molecule present. The second diagram has 2 A_2B molecules, 4 A_2 molecules, and 2 B_2 molecules. Therefore, the first diagram cannot represent equilibrium because there was a net change in reactants and products. Is the second diagram the equilibrium mixture? That depends on whether there is a net change between reactants and products when going from the second diagram to the third diagram. The third diagram contains the same number and type of molecules as the second diagram, so the second diagram is the first illustration that represents equilibrium.

The reaction container initially contained only A_2B. From the first diagram, 2 A_2 molecules and 1 B_2 molecule are present (along with 4 A_2B molecules). From the balanced reaction, these 2 A_2 molecules and 1 B_2 molecule were formed when 2 A_2B molecules decomposed. Therefore, the initial number of A_2B molecules present equals $4 + 2 = 6$ molecules A_2B.

The Equilibrium Constant

17.   K and K_p are equilibrium constants as determined by the law of mass action. For K, concentration units of mol/L are used, and for K_p, partial pressures in units of atm are used (generally). Q is called the reaction quotient. Q has the exact same form as K or K_p, but instead of equilibrium concentrations, initial concentrations are used to calculate the Q value. We use Q to determine if a reaction is at equilibrium. When Q = K (or when $Q_p = K_p$), the reaction is at equilibrium. When Q ≠ K, the reaction is not at equilibrium, and one can deduce the net change that must occur for the system to get to equilibrium.

19.   Solids and liquids do not appear in equilibrium expressions. Only gases and dissolved solutes appear in equilibrium expressions.

$$\text{a.} \quad K = \frac{[H_2O]}{[NH_3]^2[CO_2]}; \quad K_p = \frac{P_{H_2O}}{P_{NH_3}^2 \times P_{CO_2}} \qquad\qquad \text{b.} \quad K = [N_2][Br_2]^3; \quad K_p = P_{N_2} \times P_{Br_2}^3$$

$$\text{c.} \quad K = [O_2]^3; \quad K_p = P_{O_2}^3 \qquad\qquad\qquad\qquad\qquad \text{d.} \quad K = \frac{[H_2O]}{[H_2]}; \quad K_p = \frac{P_{H_2O}}{P_{H_2}}$$

21. $$[NO] = \frac{4.5 \times 10^{-3} \text{ mol}}{3.0 \text{ L}} = 1.5 \times 10^{-3} M; \quad [Cl_2] = \frac{2.4 \text{ mol}}{3.0 \text{ L}} = 0.80 \, M$$

$$[NOCl] = \frac{1.0 \text{ mol}}{3.0 \text{ L}} = 0.33 \, M; \quad K = \frac{[NO]^2[Cl_2]}{[NOCl]^2} = \frac{(1.5 \times 10^{-3})^2(0.80)}{(0.33)^2} = 1.7 \times 10^{-5}$$

23. $K_p = K(RT)^{\Delta n}$, where Δn = sum of gaseous product coefficients − sum of gaseous reactant coefficients. For this reaction, $\Delta n = 3 - 1 = 2$.

$$K = \frac{[CO][H_2]^2}{[CH_3OH]} = \frac{(0.24)(1.1)^2}{(0.15)} = 1.9$$

$$K_p = K(RT)^2 = 1.9(0.08206 \text{ L atm K}^{-1} \text{ mol}^{-1} \times 600. \text{ K})^2 = 4.6 \times 10^3$$

25. $$[N_2O] = \frac{2.00 \times 10^{-2} \text{ mol}}{2.00 \text{ L}}; \quad [N_2] = \frac{2.80 \times 10^{-4} \text{ mol}}{2.00 \text{ L}}; \quad [O_2] = \frac{2.50 \times 10^{-5} \text{ mol}}{2.00 \text{ L}}$$

$$K = \frac{[N_2O]^2}{[N_2]^2[O_2]} = \frac{\left(\dfrac{2.00 \times 10^{-2}}{2.00}\right)^2}{\left(\dfrac{2.80 \times 10^{-4}}{2.00}\right)^2 \left(\dfrac{2.50 \times 10^{-5}}{2.00}\right)} = \frac{(1.00 \times 10^{-2})^2}{(1.40 \times 10^{-4})^2(1.25 \times 10^{-5})}$$

$$= 4.08 \times 10^8$$

If the given concentrations represent equilibrium concentrations, then they should give a value of $K = 4.08 \times 10^8$.

$$\frac{(0.200)^2}{(2.00 \times 10^{-4})^2 (0.00245)} = 4.08 \times 10^8$$

Because the given concentrations when plugged into the equilibrium constant expression give a value equal to K (4.08×10^8), this set of concentrations is a system at equilibrium

27. $K = \dfrac{[H_2]^2 [O_2]}{[H_2O]^2}$, $2.4 \times 10^{-3} = \dfrac{(1.9 \times 10^{-2})^2 [O_2]}{(0.11)^2}$, $[O_2] = 0.080 \, M$

   $\text{Mol } O_2 = 2.0 \, L \times \dfrac{0.080 \, \text{mol } O_2}{L} = 0.16 \, \text{mol } O_2$

29. $K_p = \dfrac{P_{H_2}^4}{P_{H_2O}^4}$; $P_{total} = P_{H_2O} + P_{H_2}$, $36.3 \text{ torr} = 15.0 \text{ torr} + P_{H_2}$, $P_{H_2} = 21.3 \text{ torr}$

   Because 1 atm = 760 torr: $K_p = \dfrac{\left(21.3 \text{ torr} \times \dfrac{1 \text{ atm}}{760 \text{ torr}}\right)^4}{\left(15.0 \text{ torr} \times \dfrac{1 \text{ atm}}{760 \text{ torr}}\right)^4} = 4.07$

Note: Solids and pure liquids are not included in K expressions.

31. $PCl_5(g) \rightleftharpoons PCl_3(g) + Cl_2(g)$ $K_p = \dfrac{P_{PCl_3} \times P_{Cl_2}}{P_{PCl_5}}$

To determine K_p, we must determine the equilibrium partial pressures of each gas. Initially, $P_{PCl_5} = 0.50$ atm and $P_{PCl_3} = P_{Cl_2} = 0$ atm. To reach equilibrium, some of the PCl_5 reacts to produce some PCl_3 and Cl_2, all in a 1 : 1 mole ratio. We must determine the change in partial pressures necessary to reach equilibrium. Because moles $\propto$ P at constant V and T, if we let x = atm of PCl_5 that reacts to reach equilibrium, this will produce x atm of PCl_3 and x atm of Cl_2 at equilibrium. The equilibrium partial pressures of each gas will be the initial partial pressure of each gas plus the change necessary to reach equilibrium. The equilibrium partial pressures are:

   $P_{PCl_5} = 0.50$ atm $- x$, $P_{PCl_3} = P_{Cl_2} = x$

Now we solve for x using the information in the problem:

   $P_{total} = P_{PCl_5} + P_{PCl_3} + P_{Cl_2}$, 0.84 atm $= 0.50 - x + x + x$, 0.84 atm $= 0.50 + x$,

   $x = 0.34$ atm

The equilibrium partial pressures are:

$$P_{PCl_5} = 0.50 - 0.34 = 0.16 \text{ atm}, \ P_{PCl_3} = P_{Cl_2} = 0.34 \text{ atm}$$

$$K_p = \frac{P_{PCl_3} \times P_{Cl_2}}{P_{PCl_5}} = \frac{(0.34)(0.34)}{(0.16)} = 0.72$$

$$K = \frac{K_p}{(RT)^{\Delta n}}, \ \Delta n = 2 - 1 = 1; \ K_p = \frac{0.72}{(0.08206)(523)} = 0.017$$

33. When solving equilibrium problems, a common method to summarize all the information in the problem is to set up a table. We commonly call this table an ICE table because it summarizes *i*nitial concentrations, *c*hanges that must occur to reach equilibrium, and *e*quilibrium concentrations (the sum of the initial and change columns). For the change column, we will generally use the variable x, which will be defined as the amount of reactant (or product) that must react to reach equilibrium. In this problem, the reaction must shift right to reach equilibrium because there are no products present initially. Therefore, x is defined as the amount of reactant SO_3 that reacts to reach equilibrium, and we use the coefficients in the balanced equation to relate the net change in SO_3 to the net change in SO_2 and O_2. The general ICE table for this problem is:

$$2\ SO_3(g) \ \rightleftharpoons \ 2\ SO_2(g) \ + \ O_2(g) \qquad K = \frac{[SO_2]^2[O_2]}{[SO_3]^2}$$

	$2\ SO_3(g)$	$2\ SO_2(g)$	$O_2(g)$
Initial	12.0 mol/3.0 L	0	0
	Let x mol/L of SO_3 react to reach equilibrium.		
Change	$-x$ →	$+x$	$+x/2$
Equil.	$4.0 - x$	x	$x/2$

From the problem, we are told that the equilibrium SO_2 concentration is 3.0 mol/3.0 L = 1.0 M ($[SO_2]_e = 1.0\ M$). From the ICE table setup, $[SO_2]_e = x$, so $x = 1.0$. Solving for the other equilibrium concentrations: $[SO_3]_e = 4.0 - x = 4.0 - 1.0 = 3.0\ M$; $[O_2] = x/2 = 1.0/2 = 0.50\ M$.

$$K = \frac{[SO_2]^2[O_2]}{[SO_3]^2} = \frac{(1.0\ M)^2(0.50\ M)}{(3.0\ M)^2} = 0.056$$

Alternate method: Fractions in the change column can be avoided (if you want) be defining x differently. If we were to let $2x$ mol/L of SO_3 react to reach equilibrium, then the ICE table setup is:

$$2\ SO_3(g) \ \rightleftharpoons \ 2\ SO_2(g) \ + \ O_2(g) \qquad K = \frac{[SO_2]^2[O_2]}{[SO_3]^2}$$

	$2\ SO_3(g)$	$2\ SO_2(g)$	$O_2(g)$
Initial	4.0 M	0	0
	Let $2x$ mol/L of SO_3 react to reach equilibrium.		
Change	$-2x$ →	$+2x$	$+x$
Equil.	$4.0 - 2x$	$2x$	x

Solving: $2x = [SO_2]_e = 1.0\ M$, $x = 0.50\ M$; $[SO_3]_e = 4.0 - 2(0.50) = 3.0\ M$; $[O_2]_e = x$
$= 0.50\ M$

These are exactly the same equilibrium concentrations as solved for previously, thus K will be the same (as it must be). The moral of the story is to define x in a manner that is most comfortable for you. Your final answer is independent of how you define x initially.

Equilibrium Calculations

35. $H_2O(g) + Cl_2O(g) \rightarrow 2\ HOCl(g)$ $K = \dfrac{[HOCl]^2}{[H_2O][Cl_2O]} = 0.0900$

Use the reaction quotient Q to determine which way the reaction shifts to reach equilibrium. For the reaction quotient, initial concentrations given in a problem are used to calculate the value for Q. If $Q < K$, then the reaction shifts right to reach equilibrium. If $Q > K$, then the reaction shifts left to reach equilibrium. If $Q = K$, then the reaction does not shift in either direction because the reaction is already at equilibrium.

a. $Q = \dfrac{[HOCl]_0^2}{[H_2O]_0[Cl_2O]_0} = \dfrac{\left(\dfrac{1.0\ mol}{1.0\ L}\right)^2}{\left(\dfrac{0.10\ mol}{1.0\ L}\right)\left(\dfrac{0.10\ mol}{1.0\ L}\right)} = 1.0 \times 10^2$

$Q > K$, so the reaction shifts left to produce more reactants in order to reach equilibrium.

b. $Q = \dfrac{\left(\dfrac{0.084\ mol}{2.0\ L}\right)^2}{\left(\dfrac{0.98\ mol}{2.0\ L}\right)\left(\dfrac{0.080\ mol}{2.0\ L}\right)} = 0.090 = K$;   at equilibrium

c. $Q = \dfrac{\left(\dfrac{0.25\ mol}{3.0\ L}\right)^2}{\left(\dfrac{0.56\ mol}{3.0\ L}\right)\left(\dfrac{0.0010\ mol}{3.0\ L}\right)} = 110$

$Q > K$, so the reaction shifts to the left to reach equilibrium.

37. $CaCO_3(s) \rightleftharpoons CaO(s) + CO_2(g)$ $K_p = P_{CO_2} = 1.04\ atm$

We only need to calculate the initial partial pressure of CO_2 and compare this value to 1.04 atm. At this temperature, all CO_2 will be in the gas phase.

a. $PV = nRT$, $Q = P_{CO_2} = \dfrac{n_{CO_2}RT}{V} = \dfrac{\dfrac{58.4\ g\ CO_2}{44.01\ g/mol} \times \dfrac{0.08206\ L\ atm}{K\ mol} \times 1173\ K}{50.0\ L} =$

$$2.55\ atm > K_p$$

Reaction will shift to the left because $Q > K_p$; the mass of CaO will decrease.

b. $Q = P_{CO_2} = \dfrac{(23.76)(0.08206)(1173)}{(44.01)(50.0)} = 1.04\ atm = K_p$

At equilibrium because $Q = K_p$; mass of CaO will not change.

c.   Mass of CO_2 is the same as in part b. $P = 1.04\ atm = K_P$. At equilibrium; mass of CaO will not change.

d. $Q = P_{CO_2} = \dfrac{(4.82)(0.08206)(1173)}{(44.01)(50.0)} = 0.211\ atm < K_p$

Reaction will shift to the right because $Q < K_p$; the mass of CaO will increase.

39. $H_2O(g) + Cl_2O(g) \rightleftharpoons 2\ HOCl(g)$ $K = 0.090 = \dfrac{[HOCl]^2}{[H_2O][Cl_2O]}$

a.   The initial concentrations of H_2O and Cl_2O are:

$$\dfrac{1.0\ g\ H_2O}{1.0\ L} \times \dfrac{1\ mol}{18.0\ g} = 5.6 \times 10^{-2}\ mol/L; \quad \dfrac{2.0\ g\ Cl_2O}{1.0\ L} \times \dfrac{1\ mol}{86.9\ g} = 2.3 \times 10^{-2}\ mol/L$$

Because only reactants are present initially, the reaction must proceed to the right to reach equilibrium. Summarizing the problem in a table:

	$H_2O(g)$	+	$Cl_2O(g)$	$\rightleftharpoons$	$2\ HOCl(g)$
Initial	$5.6 \times 10^{-2}\ M$		$2.3 \times 10^{-2}\ M$		0
	x mol/L of H_2O reacts to reach equilibrium				
Change	$-x$		$-x$	$\rightarrow$	$+2x$
Equil.	$5.6 \times 10^{-2} - x$		$2.3 \times 10^{-2} - x$		$2x$

$$K = 0.090 = \dfrac{(2x)^2}{(5.6 \times 10^{-2} - x)(2.3 \times 10^{-2} - x)}$$

$$1.16 \times 10^{-4} - (7.11 \times 10^{-3})x + (0.090)x^2 = 4x^2$$

$$(3.91)x^2 + (7.11 \times 10^{-3})x - 1.16 \times 10^{-4} = 0 \quad \text{(We carried extra significant figures.)}$$

Solving using the quadratic formula (see Appendix 1 of the text):

$$x = \frac{-7.11 \times 10^{-3} \pm (5.06 \times 10^{-5} + 1.81 \times 10^{-3})^{1/2}}{7.82} = 4.6 \times 10^{-3}\ M\ \text{or}\ -6.4 \times 10^{-3}\ M$$

A negative answer makes no physical sense; we can't have less than nothing. Thus $x = 4.6 \times 10^{-3}\ M$.

$[\text{HOCl}] = 2x = 9.2 \times 10^{-3}\ M;\quad [\text{Cl}_2\text{O}] = 2.3 \times 10^{-2} - x = 0.023 - 0.0046 = 1.8 \times 10^{-2}\ M$

$[\text{H}_2\text{O}] = 5.6 \times 10^{-2} - x = 0.056 - 0.0046 = 5.1 \times 10^{-2}\ M$

b. $\text{H}_2\text{O}(g)$ + $\text{Cl}_2\text{O}(g)$ $\rightleftharpoons$ $2\ \text{HOCl}(g)$

Initial	0	0	1.0 mol/2.0 L = 0.50 M

2x mol/L of HOCl reacts to reach equilibrium

Change	$+x$	$+x$ $\leftarrow$	$-2x$
Equil.	x	x	$0.50 - 2x$

$$K = 0.090 = \frac{[\text{HOCl}]^2}{[\text{H}_2\text{O}][\text{Cl}_2\text{O}]} = \frac{(0.50 - 2x)^2}{x^2}$$

The expression is a perfect square, so we can take the square root of each side:

$$0.30 = \frac{0.50 - 2x}{x}, \quad (0.30)x = 0.50 - 2x,\ (2.30)x = 0.50$$

$x = 0.217\ M$ (We carried extra significant figures.)

$x = [\text{H}_2\text{O}] = [\text{Cl}_2\text{O}] = 0.217 = 0.22\ M;\quad [\text{HOCl}] = 0.50 - 2x = 0.50 - 0.434 = 0.07\ M$

41. $2\ \text{SO}_2(g)$ + $\text{O}_2(g)$ $\rightleftharpoons$ $2\ \text{SO}_3(g)$ $K_p = 0.25$

Initial	0.50 atm	0.50 atm	0

2x atm of SO$_2$ reacts to reach equilibrium

Change	$-2x$	$-x$ $\rightarrow$	$+2x$
Equil.	$0.50 - 2x$	$0.50 - x$	$2x$

$$K_p = 0.25 = \frac{P_{\text{SO}_3}^2}{P_{\text{SO}_2}^2 \times P_{\text{O}_2}} = \frac{(2x)^2}{(0.50 - 2x)^2(0.50 - x)}$$

This will give a cubic equation.   Graphing calculators can be used to solve this expression. If you don't have a graphing calculator, an alternative method for solving a cubic equation is to use the method of successive approximations (see Appendix 1 of the text). The first step is to guess a value for x. Because the value of K is small (K < 1), not much of the forward reaction will occur to reach equilibrium. This tells us that x is small. Let's guess that $x =$

0.050 atm. Now we take this estimated value for x and substitute it into the equation everywhere that x appears except for one. For equilibrium problems, we will substitute the estimated value for x into the denominator, and then solve for the numerator value of x. We continue this process until the estimated value of x and the calculated value of x converge on the same number. This is the same answer we would get if we were to solve the cubic equation exactly.   Applying the method of successive approximations and carrying extra significant figures:

$$\frac{4x^2}{[0.50 - 2(0.050)]^2 - [0.50 - (0.050)]} = \frac{4x^2}{(0.40)^2(0.45)} = 0.25, \; x = 0.067$$

$$\frac{4x^2}{[0.50 - 2(0.067)]^2 \, [0.50 - (0.067)]} = \frac{4x^2}{(0.366)^2(0.433)} = 0.25, \; x = 0.060$$

$$\frac{4x^2}{(0.38)^2(0.44)} = 0.25, \; x = 0.063; \quad \frac{4x^2}{(0.374)^2(0.437)} = 0.25, \; x = 0.062$$

The next trial gives the same value for $x = 0.062$ atm. We are done except for determining the equilibrium concentrations. They are:

$$P_{SO_2} = 0.50 - 2x = 0.50 - 2(0.062) = 0.376 = 0.38 \text{ atm}$$

$$P_{O_2} = 0.50 - x = 0.438 = 0.44 \text{ atm}; \quad P_{SO_3} = 2x = 0.124 = 0.12 \text{ atm}$$

43.   The assumption comes from the value of K being much less than 1. For these reactions, the equilibrium mixture will not have a lot of products present; mostly reactants are present at equilibrium. If we define the change that must occur in terms of x as the amount (molarity or partial pressure) of a reactant that must react to reach equilibrium, then x must be a small number because K is a very small number. We want to know the value of x in order to solve the problem, so we don't assume $x = 0$. Instead, we concentrate on the equilibrium row in the ICE table. Those reactants (or products) that have equilibrium concentrations in the form of $0.10 - x$ or $0.25 + x$ or $3.5 - 3x$, etc., is where an important assumption can be made. The assumption is that because $K \ll 1$, x will be small ($x \ll 1$), and when we add x or subtract x from some initial concentration, it will make little or no difference. That is, we assume that $0.10 - x \approx 0.10$ or $0.25 + x \approx 0.25$ or $3.5 - 3x \approx 3.5$, etc.; we assume that the initial concentration of a substance is equal to the final concentration. This assumption makes the math much easier and usually gives a value of x that is well within 5% of the true value of x (we get about the same answer with a lot less work).

We check the assumptions for validity using the 5% rule. From doing a lot of these calculations, it is found that when an assumption such as $0.20 - x \approx 0.20$ is made, if x is less than 5% of the number the assumption was made against, then our final answer is within acceptable error limits of the true value of x (as determined when the equation is solved exactly). For our example above ($0.20 - x \approx 0.20$), if $(x/0.20) \times 100 \leq 5\%$, then our assumption is valid by the 5% rule. If the error is greater than 5%, then we must solve the equation exactly or use a math trick called the method of successive approximations. See Appendix 1 for details regarding the method of successive approximations, as well as for a review in solving quadratic equations exactly.

45. $2 CO_2(g) \rightleftharpoons 2 CO(g) + O_2(g)$ $K = \dfrac{[CO]^2[O_2]}{[CO_2]^2} = 2.0 \times 10^{-6}$

Initial   2.0 mol/5.0 L   0   0
              $2x$ mol/L of CO_2 reacts to reach equilibrium
Change $-2x$ $\rightarrow$ $+2x$ $+x$
Equil. $0.40 - 2x$ $2x$ x

$K = 2.0 \times 10^{-6} = \dfrac{[CO]^2[O_2]}{[CO_2]^2} = \dfrac{(2x)^2(x)}{(0.40 - 2x)^2}$; assuming $2x \ll 0.40$ (K is small, so x is small.):

$2.0 \times 10^{-6} \approx \dfrac{4x^3}{(0.40)^2}$, $2.0 \times 10^{-6} = \dfrac{4x^3}{0.16}$, $x = 4.3 \times 10^{-3} \ M$

Checking assumption: $\dfrac{2(4.3 \times 10^{-3})}{0.40} \times 100 = 2.2\%$; assumption is valid by the 5% rule.

$[CO_2] = 0.40 - 2x = 0.40 - 2(4.3 \times 10^{-3}) = 0.39 \ M$

$[CO] = 2x = 2(4.3 \times 10^{-3}) = 8.6 \times 10^{-3} \ M$; $[O_2] = x = 4.3 \times 10^{-3} \ M$

47. $CaCO_3(s) \rightleftharpoons CaO(s) + CO_2(g)$ $K_p = 1.16 = P_{CO_2}$

Some of the 20.0 g of $CaCO_3$ will react to reach equilibrium. The amount that reacts is the quantity of $CaCO_3$ required to produce a CO_2 pressure of 1.16 atm (from the K_p expression).

$n_{CO_2} = \dfrac{P_{CO_2} V}{RT} = \dfrac{1.16 \text{ atm} \times 10.0 \text{ L}}{\dfrac{0.08206 \text{ L atm}}{\text{K mol}} \times 1073 \text{ K}} = 0.132 \text{ mol } CO_2$

Mass $CaCO_3$ reacted = 0.132 mol $CO_2 \times \dfrac{1 \text{ mol } CaCO_3}{\text{mol } CO_2} \times \dfrac{100.09 \text{ g}}{\text{mol } CaCO_3} = 13.2 \text{ g } CaCO_3$

Mass % $CaCO_3$ reacted = $\dfrac{13.2 \text{ g}}{20.0 \text{ g}} \times 100 = 66.0\%$

49.   a. $K_p = K(RT)^{\Delta n} = 4.5 \times 10^9 \left(\dfrac{0.08206 \text{ L atm}}{\text{K mol}} \times 373 \text{ K} \right)^{-1}$, where $\Delta n = 1 - 2 = -1$

$K_p = 1.5 \times 10^8$

b. K_p is so large that at equilibrium we will have almost all $COCl_2$.   Assume $P_{total} \approx P_{COCl_2} \approx 5.0$ atm.

$$CO(g) + Cl_2(g) \rightleftharpoons COCl_2(g) \qquad K_p = 1.5 \times 10^8$$

Initial	0	0	5.0 atm

x atm $COCl_2$ reacts to reach equilibrium

Change	+x	+x	← −x
Equil.	x	x	5.0 − x

$$K_p = 1.5 \times 10^8 = \frac{5.0 - x}{x^2} \approx \frac{5.0}{x^2} \quad \text{(Assuming } 5.0 - x \approx 5.0.)$$

Solving: $x = 1.8 \times 10^{-4}$ atm. Check assumptions: $5.0 - x = 5.0 - 1.8 \times 10^{-4} = 5.0$ atm. Assumptions are good (well within the 5% rule).

$$P_{CO} = P_{Cl_2} = 1.8 \times 10^{-4} \text{ atm and } P_{COCl_2} = 5.0 \text{ atm}$$

Le Châtelier's Principle

51.   Only statement d is correct. Addition of a catalyst has no effect on the equilibrium position; the reaction just reaches equilibrium more quickly. Statement a is false for reactants that are either solids or liquids (adding more of these has no effect on the equilibrium). Statement b is false always. If temperature remains constant, then the value of K is constant. Statement c is false for exothermic reactions where an increase in temperature decreases the value of K.

53.   a.   No effect; adding more of a pure solid or pure liquid has no effect on the equilibrium position.

   b.   Shifts left; HF(g) will be removed by reaction with the glass. As HF(g) is removed, the reaction will shift left to produce more HF(g).

   c.   Shifts right; as $H_2O(g)$ is removed, the reaction will shift right to produce more $H_2O(g)$.

55. $H^+ + OH^- \rightarrow H_2O$; sodium hydroxide (NaOH) will react with the H^+ on the product side of the reaction. This effectively removes H^+ from the equilibrium, which will shift the reaction to the right to produce more H^+ and CrO_4^{2-}. Because more CrO_4^{2-} is produced, the solution turns yellow.

57.   a.   Right   b.   Right   c.   No effect; He(g) is neither a reactant nor a product.

   d.   Left; because the reaction is exothermic, heat is a product:

   $$CO(g) + H_2O(g) \rightarrow H_2(g) + CO_2(g) + \text{heat}$$

   Increasing T will add heat. The equilibrium shifts to the left to use up the added heat.

   e.   No effect; because the moles of gaseous reactants equals the moles of gaseous products (2 mol versus 2 mol), a change in volume will have no effect on the equilibrium.

59.   a. Left   b. Right   c. Left

d. No effect; the reactant and product concentrations/partial pressures are unchanged.

e. No effect; because there are equal numbers of product and reactant gas molecules, a change in volume has no effect on this equilibrium position.

f. Right; a decrease in temperature will shift the equilibrium to the right because heat is a product in this reaction (as is true in all exothermic reactions).

61.   An endothermic reaction, where heat is a reactant, will shift right to products with an increase in temperature. The amount of $NH_3(g)$ will increase as the reaction shifts right, so the smell of ammonia will increase.

Additional Exercises

63.   a. $N_2(g) + O_2(g) \rightleftharpoons 2\ NO(g)$ $K_p = 1 \times 10^{-31} = \dfrac{P_{NO}^2}{P_{N_2} \times P_{O_2}} = \dfrac{P_{NO}^2}{(0.8)(0.2)}$

$P_{NO} = 1 \times 10^{-16}$ atm

In 1.0 cm^3 of air: $n_{NO} = \dfrac{PV}{RT} = \dfrac{(1 \times 10^{-16}\ \text{atm})(1.0 \times 10^{-3}\ \text{L})}{\left(\dfrac{0.08206\ \text{L atm}}{\text{K mol}}\right)(298\ \text{K})} = 4 \times 10^{-21}$ mol NO

$\dfrac{4 \times 10^{-21}\ \text{mol NO}}{cm^3} \times \dfrac{6.02 \times 10^{23}\ \text{molecules}}{\text{mol NO}} = \dfrac{2 \times 10^3\ \text{molecules NO}}{cm^3}$

b. There is more NO in the atmosphere than we would expect from the value of K. The answer must lie in the rates of the reaction. At 25°C, the rates of both reactions:

$$N_2 + O_2 \rightarrow 2\ NO \text{ and } 2\ NO \rightarrow N_2 + O_2$$

are so slow that they are essentially zero. Very strong bonds must be broken; the activation energy is very high. Therefore, the reaction essentially doesn't occur at low temperatures.   Nitric oxide, however, can be produced in high-energy or high-temperature environments because the production of NO is endothermic. In nature, some NO is produced by lightning and the primary manmade source is automobiles. At these high temperatures, K will increase, and the rates of the reaction will also increase, resulting in a higher production of NO.   Once the NO gets into a more normal temperature environment, it doesn't go back to N_2 and O_2 because of the slow rate.

65. $O(g) + NO(g) \rightleftharpoons NO_2(g)$ $K = 1/6.8 \times 10^{-49} = 1.5 \times 10^{48}$

$NO_2(g) + O_2(g) \rightleftharpoons NO(g) + O_3(g)$ $K = 1/5.8 \times 10^{-34} = 1.7 \times 10^{33}$

$O_2(g) + O(g) \rightleftharpoons O_3(g)$ $K = (1.5 \times 10^{48})(1.7 \times 10^{33}) = 2.6 \times 10^{81}$

67. $3 \, H_2(g)$ $+$ $N_2(g)$ $\rightleftharpoons$ $2 \, NH_3(g)$

Initial $[H_2]_0$ $[N_2]_0$ 0
               x mol/L of N_2 reacts to reach equilibrium
Change $-3x$ $-x$ $\rightarrow$ $+2x$
Equil. $[H_2]_0 - 3x$ $[N_2]_0 - x$ $2x$

From the problem:

$[NH_3]_e = 4.0 \, M = 2x$, $x = 2.0 \, M$; $[H_2]_e = 5.0 \, M = [H_2]_0 - 3x$; $[N_2]_e = 8.0 \, M = [N_2]_0 - x$

$5.0 \, M = [H_2]_0 - 3(2.0 \, M)$, $[H_2]_0 = 11.0 \, M$; $8.0 \, M = [N_2]_0 - 2.0 \, M$, $[N_2]_0 = 10.0 \, M$

69.   a. $PCl_5(g)$ $\rightleftharpoons$ $PCl_3(g)$ $+$ $Cl_2(g)$ $K_p = (P_{PCl_3} \times P_{Cl_2})/P_{PCl_5}$

Initial P_0 0   0 P_0 = initial PCl_5 pressure
Change $-x$ $\rightarrow$ $+x$ $+x$
Equil. $P_0 - x$ x x

$P_{total} = P_0 - x + x + x = P_0 + x = 358.7$ torr

$$P_0 = \frac{n_{PCl_5} RT}{V} = \frac{\dfrac{2.4156 \, g}{208.22 \, g/mol} \times \dfrac{0.08206 \, L \, atm}{K \, mol} \times 523.2 \, K}{2.000 \, L} = 0.2490 \, atm \text{ (or 189.2 torr)}$$

$x = P_{total} - P_0 = 358.7 - 189.2 = 169.5$ torr

$P_{PCl_3} = P_{Cl_2} = 169.5$ torr $\times$ 1 atm/760 torr $= 0.2230$ atm

$P_{PCl_5} = 189.2 - 169.5 = 19.7$ torr $\times$ 1 atm/760 torr $= 0.0259$ atm

$$K_p = \frac{(0.2230)^2}{0.0259} = 1.92$$

b. $P_{Cl_2} = \dfrac{n_{Cl_2} RT}{V} = \dfrac{0.250 \times 0.08206 \times 523.2}{2.000} = 5.37$ atm Cl_2 added

              $PCl_5(g)$ $\rightleftharpoons$ $PCl_3(g)$ $+$ $Cl_2(g)$

Initial   0.0259 atm   0.2230 atm   0.2230 atm   (from a)
             Adding 0.250 mol Cl_2 increases P_{Cl_2} by 5.37 atm.
Initial'   0.0259   0.2230   5.59
Change $+0.2230$ $\leftarrow$ -0.2230 -0.2230 React completely
After   0.2489   0   5.37   New initial
Change $-x$ $\rightarrow$ $+x$ $+x$
Equil. $0.2489 - x$ x $5.37 + x$

$$\frac{(5.37 + x)(x)}{(0.2489 - x)} = 1.92, \quad x^2 + (7.29)x - 0.478 = 0$$

Solving using the quadratic formula: $x = 0.0650$ atm

$P_{PCl_3} = 0.0650$ atm; $P_{PCl_5} = 0.2489 - 0.0650 = 0.1839$ atm; $P_{Cl_2} = 5.37 + 0.0650$
$$= 5.44 \text{ atm}$$

71. $N_2O_4(g) \rightleftharpoons 2\, NO_2(g)$ $K_p = \dfrac{P_{NO_2}^2}{P_{N_2O_4}} = \dfrac{(1.20)^2}{0.34} = 4.2$

Doubling the volume decreases each partial pressure by a factor of 2 (P = nRT/V).
$P_{NO_2} = 0.600$ atm and $P_{N_2O_4} = 0.17$ atm are the new partial pressures.

$$Q = \frac{(0.600)^2}{0.17} = 2.1, \quad Q < K; \quad \text{equilibrium will shift to the right.}$$

	$N_2O_4(g)$	$\rightleftharpoons$	$2\, NO_2(g)$
Initial	0.17 atm		0.600 atm
Equil.	0.17 − x		0.600 + 2x

$$K_p = 4.2 = \frac{(0.600 + 2x)^2}{(0.17 - x)}, \quad 4x^2 + (6.6)x - 0.354 = 0 \quad \text{(carrying extra sig. figs.)}$$

Solving using the quadratic formula: $x = 0.052$ atm

$P_{NO_2} = 0.600 + 2(0.052) = 0.704$ atm; $P_{N_2O_4} = 0.17 - 0.052 = 0.12$ atm

73.   a. $2\, NaHCO_3(s) \rightleftharpoons Na_2CO_3(s) \;+\; CO_2(g) \;+\; H_2O(g)$ $K_p = 0.25$

	$2\, NaHCO_3(s)$	$Na_2CO_3(s)$	$CO_2(g)$	$H_2O(g)$
Initial	−	−	0	0

NaHCO₃(s) decomposes to form x atm each of $CO_2(g)$ and $H_2O(g)$ at equilibrium.

Change	−	$\rightarrow$ −	+x	+x
Equil.	−	−	x	x

$$K_p = 0.25 = P_{CO_2} \times P_{H_2O}, \quad 0.25 = x^2, \quad x = P_{CO_2} = P_{H_2O} = 0.50 \text{ atm}$$

b. $n_{CO_2} = \dfrac{P_{CO_2} V}{RT} = \dfrac{(0.50 \text{ atm})(1.00 \text{ L})}{(0.08206 \text{ L atm K}^{-1} \text{ mol}^{-1})(398 \text{ K})} = 1.5 \times 10^{-2} \text{ mol } CO_2$

Mass of Na_2CO_3 produced:

$$1.5 \times 10^{-2} \text{ mol CO}_2 \times \frac{1 \text{ mol Na}_2\text{CO}_3}{\text{mol CO}_2} \times \frac{106.0 \text{ g Na}_2\text{CO}_3}{\text{mol Na}_2\text{CO}_3} = 1.6 \text{ g Na}_2\text{CO}_3$$

Mass of $NaHCO_3$ reacted:

$$1.5 \times 10^{-2} \text{ mol CO}_2 \times \frac{2 \text{ mol NaHCO}_3}{\text{mol CO}_2} \times \frac{84.01 \text{ g NaHCO}_3}{\text{mol}} = 2.5 \text{ g NaHCO}_3$$

Mass of $NaHCO_3$ remaining = $10.0 - 2.5 = 7.5$ g

c. $10.0 \text{ g NaHCO}_3 \times \dfrac{1 \text{ mol NaHCO}_3}{84.01 \text{ g NaHCO}_3} \times \dfrac{1 \text{ mol CO}_2}{2 \text{ mol NaHCO}_3} = 5.95 \times 10^{-2} \text{ mol CO}_2$

When all the $NaHCO_3$ has just been consumed, we will have 5.95×10^{-2} mol CO_2 gas at a pressure of 0.50 atm (from a).

$$V = \frac{nRT}{P} = \frac{(5.95 \times 10^{-2} \text{ mol})(0.08206 \text{ L atm K}^{-1} \text{ mol}^{-1})(398 \text{ K})}{0.50 \text{ atm}} = 3.9 \text{ L}$$

75. $$NH_3(g) \quad + \quad H_2S(g) \quad \rightleftharpoons \quad NH_4HS(s) \quad K = 400. = \frac{1}{[NH_3][H_2S]}$$

Initial	$\dfrac{2.00 \text{ mol}}{5.00 \text{ L}}$	$\dfrac{2.00 \text{ mol}}{5.00 \text{ L}}$	—

x mol/L of NH_3 reacts to reach equilibrium

Change	$-x$	$-x$	—
Equil.	$0.400 - x$	$0.400 - x$	—

$$K = 400. = \frac{1}{(0.400 - x)(0.400 - x)}, \quad 0.400 - x = \left(\frac{1}{400.}\right)^{1/2} = 0.0500, \quad x = 0.350 \, M$$

Mol $NH_4HS(s)$ produced = $5.00 \text{ L} \times \dfrac{0.350 \text{ mol NH}_3}{\text{L}} \times \dfrac{1 \text{ mol NH}_4\text{HS}}{\text{mol NH}_3} = 1.75$ mol

Total mol $NH_4HS(s)$ = 2.00 mol initially + 1.75 mol produced = 3.75 mol total

$3.75 \text{ mol NH}_4\text{HS} \times \dfrac{51.12 \text{ g NH}_4\text{HS}}{\text{mol NH}_4\text{HS}} = 192 \text{ g NH}_4\text{HS}$

$[H_2S]_e = 0.400 \, M - x = 0.400 \, M - 0.350 \, M = 0.050 \, M \, H_2S$

$$P_{H_2S} = \frac{n_{H_2S}RT}{V} = \frac{n_{H_2S}}{V} \times RT = \frac{0.050 \text{ mol}}{\text{L}} \times \frac{0.08206 \text{ L atm}}{\text{K mol}} \times 308.2 \text{ K} = 1.3 \text{ atm}$$

77. $5.63 \text{ g C}_5\text{H}_6\text{O}_3 \times \dfrac{1 \text{ mol C}_5\text{H}_6\text{O}_3}{114.10 \text{ g}} = 0.0493 \text{ mol C}_5\text{H}_6\text{O}_3$ initially

Total moles of gas = $n_{total} = \dfrac{P_{total}V}{RT} = \dfrac{1.63 \text{ atm} \times 2.50 \text{ L}}{\dfrac{0.08206 \text{ L atm}}{\text{K mol}} \times 473 \text{ K}} = 0.105$ mol
at equilibrium

$$C_5H_6O_3(g) \rightleftharpoons C_2H_6(g) + 3 \, CO(g)$$

Initial   0.0493 mol   0   0
        Let x mol $C_5H_6O_3$ react to reach equilibrium.
Change $-x$ $\rightarrow$ $-x$ $-3x$
Equil. $0.0493 - x$ x $3x$

0.105 mol total = $0.0493 - x + x + 3x = 0.0493 + 3x$, $x = 0.0186$ mol

$$K = \frac{[C_2H_6][CO]^3}{[C_5H_6O_3]} = \frac{\left[\dfrac{0.0186 \text{ mol C}_2\text{H}_6}{2.50 \text{ L}}\right]\left[\dfrac{3(0.0186) \text{ mol CO}}{2.50 \text{ L}}\right]^3}{\left[\dfrac{(0.0493 - 0.0186) \text{ mol C}_5\text{H}_6\text{O}_3}{2.50 \text{ L}}\right]} = 6.74 \times 10^{-6}$$

Challenge Problems

79.   a. $$2 \, NO(g) + Br_2(g) \rightleftharpoons 2 \, NOBr(g)$$

Initial   98.4 torr   41.3 torr   0
        $2x$ torr of NO reacts to reach equilibrium
Change $-2x$ $-x$ $\rightarrow$ $+2x$
Equil. $98.4 - 2x$ $41.3 - x$ $2x$

$P_{total} = P_{NO} + P_{Br_2} + P_{NOBr} = (98.4 - 2x) + (41.3 - x) + 2x = 139.7 - x$

$P_{total} = 110.5 = 139.7 - x$, $x = 29.2$ torr; $P_{NO} = 98.4 - 2(29.2) = 40.0$ torr = 0.0526 atm

$P_{Br_2} = 41.3 - 29.2 = 12.1$ torr = 0.0159 atm; $P_{NOBr} = 2(29.2) = 58.4$ torr = 0.0768 atm

$$K_p = \frac{P_{NOBr}^2}{P_{NO}^2 \times P_{Br_2}} = \frac{(0.0768 \text{ atm})^2}{(0.0526 \text{ atm})^2 (0.0159 \text{ atm})} = 134$$

b. $$2 \, NO(g) + Br_2(g) \rightleftharpoons 2 \, NOBr(g)$$

Initial   0.30 atm   0.30 atm   0
        $2x$ atm of NO reacts to reach equilibrium
Change $-2x$ $-x$ $\rightarrow$ $+2x$
Equil. $0.30 - 2x$ $0.30 - x$ $2x$

This would yield a cubic equation, which can be difficult to solve unless you have a graphing calculator. Because K_p is pretty large, let's approach equilibrium in two steps: Assume the reaction goes to completion, and then solve the back equilibrium problem.

$$2\,NO \quad + \quad Br_2 \quad \rightleftharpoons \quad 2\,NOBr$$

Before	0.30 atm	0.30 atm	0

Let 0.30 atm NO react completely.

Change	−0.30	−0.15	→	+0.30	React completely
After	0	0.15		0.30	New initial

$2y$ atm of NOBr reacts to reach equilibrium

Change	+2y	+y	←	−2y
Equil.	2y	0.15 + y		0.30 − 2y

$$K_p = 134 = \frac{(0.30 - 2y)^2}{(2y)^2(0.15 + y)}, \ \frac{(0.30 - 2y)^2}{(0.15 + y)} = 134 \times 4y^2 = 536y^2$$

If $y \ll 0.15$: $\frac{(0.30)^2}{0.15} \approx 536y^2$ and $y = 0.034$; assumptions are poor (y is 23% of 0.15).

Use 0.034 as an approximation for y, and solve by successive approximations (see Appendix 1):

$$\frac{(0.30 - 0.068)^2}{0.15 + 0.034} = 536y^2, \ y = 0.023; \quad \frac{(0.30 - 0.046)^2}{0.15 + 0.023} = 536y^2, \ y = 0.026$$

$$\frac{(0.30 - 0.052)^2}{0.15 + 0.026} = 536y^2, \ y = 0.026 \text{ atm} \quad \text{(We have converged on the correct answer.)}$$

So: $P_{NO} = 2y = 0.052$ atm; $P_{Br_2} = 0.15 + y = 0.18$ atm; $P_{NOBr} = 0.30 - 2y = 0.25$ atm

81. $P_4(g) \rightleftharpoons 2\,P_2(g)$ $K_p = 0.100 = \frac{P_{P_2}^2}{P_{P_4}}$; $P_{P_4} + P_{P_2} = P_{total} = 1.00$ atm, $P_{P_4} = 1.00 - P_{P_2}$

Let $y = P_{P_2}$ at equilibrium, then $K_p = \frac{y^2}{1.00 - y} = 0.100$

Solving: $y = 0.270$ atm $= P_{P_2}$; $P_{P_4} = 1.00 - 0.270 = 0.73$ atm

To solve for the fraction dissociated, we need the initial pressure of P_4.

$$P_4(g) \quad \rightleftharpoons \quad 2\,P_2(g)$$

Initial	P_0	0

P_0 = initial pressure of P_4

x atm of P_4 reacts to reach equilibrium

Change	−x	→	+2x
Equil.	$P_0 - x$		2x

$P_{total} = P_0 - x + 2x = 1.00$ atm $= P_0 + x$

Solving: 0.270 atm $= P_{P_2} = 2x$, $x = 0.135$ atm; $P_0 = 1.00 - 0.135 = 0.87$ atm

Fraction dissociated $= \dfrac{x}{P_0} = \dfrac{0.135}{0.87} = 0.16$, or 16% of P_4 is dissociated to reach equilibrium.

83. $d =$ density $= \dfrac{P \times (\text{molar mass})}{RT} = \dfrac{P_{O_2}(\text{molar mass}_{O_2}) + P_{O_3}(\text{molar mass}_{O_3})}{RT}$

0.168 g/L $= \dfrac{P_{O_2}(32.00 \text{ g}/\text{mol}) + P_{O_3}(48.00 \text{ g}/\text{mol})}{\dfrac{0.08206 \text{ L atm}}{\text{K mol}} \times 448 \text{ K}}$, $(32.00)P_{O_2} + (48.00)P_{O_3} = 6.18$

$P_{total} = P_{O_2} + P_{O_3} = 128$ torr $\times \dfrac{1 \text{ atm}}{760 \text{ torr}} = 0.168$ atm

We have two equations in two unknowns. Solving using simultaneous equations:

$$(32.00)P_{O_2} + (48.00)P_{O_3} = 6.18$$
$$-(32.00)P_{O_2} - (32.00)P_{O_3} = -5.38$$
$$\overline{\hspace{3cm}(16.00)P_{O_3} = 0.80\hspace{2cm}}$$

$P_{O_3} = \dfrac{0.80}{16.00} = 0.050$ atm and $P_{O_2} = 0.118$ atm; $K_p = \dfrac{P_{O_3}^2}{P_{O_2}^3} = \dfrac{(0.050)^2}{(0.118)^3} = 1.5$

85. 4.72 g $CH_3OH \times \dfrac{1 \text{ mol}}{32.04 \text{ g}} = 0.147$ mol CH_3OH initially

Graham's law of effusion: $\dfrac{\text{Rate}_1}{\text{Rate}_2} = \sqrt{\dfrac{M_2}{M_1}}$

$\dfrac{\text{Rate}_{H_2}}{\text{Rate}_{CH_3OH}} = \sqrt{\dfrac{M_{CH_3OH}}{M_{H_2}}} = \sqrt{\dfrac{32.04}{2.016}} = 3.987$

The effused mixture has 33.0 times as much H_2 as CH_3OH. When the effusion rate ratio is multiplied by the equilibrium mole ratio of H_2 to CH_3OH, the effused mixture will have 33.0 times as much H_2 as CH_3OH. Let n_{H_2} and n_{CH_3OH} equal the equilibrium moles of H_2 and CH_3OH, respectively.

$$33.0 = 3.987 \times \dfrac{n_{H_2}}{n_{CH_3OH}}, \quad \dfrac{n_{H_2}}{n_{CH_3OH}} = 8.28$$

$$CH_3OH(g) \quad \rightleftharpoons \quad CO(g) \quad + 2\ H_2(g)$$

Initial	0.147 mol	0	0
Change	$-x$	$\rightarrow$ $+x$	$+2x$
Equil.	$0.147 - x$	x	$2x$

From the ICE table, $8.28 = \dfrac{n_{H_2}}{n_{CH_3OH}} = \dfrac{2x}{0.147 - x}$

Solving: $x = 0.118$ mol

$$K = \frac{[CO][H_2]^2}{[CH_3OH]} = \frac{\left(\dfrac{0.118\ mol}{1.00\ L}\right)\left(\dfrac{2(0.118\ mol)}{1.00\ L}\right)^2}{\dfrac{(0.147 - 0.118)\ mol}{1.00\ L}} = 0.23$$

87.   a. $P_{PCl_5} = \dfrac{n_{PCl_5} RT}{V} = \dfrac{0.100\ mol \times \dfrac{0.08206\ L\ atm}{K\ mol} \times 480.\ K}{12.0\ L} = 0.328$ atm

$$PCl_5(g) \quad \rightleftharpoons \quad PCl_3(g) \quad + \quad Cl_2(g) \qquad K_p = 0.267$$

Initial	0.328 atm	0	0
Change	$-x$	$\rightarrow$ $+x$	$+x$
Equil.	$0.328 - x$	x	x

$K_p = \dfrac{x^2}{0.328 - x} = 0.267$, $x^2 + (0.267)x - 0.08758 = 0$ (carrying extra sig. figs.)

Solving using the quadratic formula: $x = 0.191$ atm

$P_{PCl_3} = P_{Cl_2} = 0.191$ atm; $P_{PCl_5} = 0.328 - 0.191 = 0.137$ atm

b. $$PCl_5(g) \quad \rightleftharpoons \quad PCl_3(g) \quad + \quad Cl_2(g)$$

Initial	P_0	0	0	$P_0 =$ initial pressure of PCl_5
Change	$-x$	$\rightarrow$ $+x$	$+x$	
Equil.	$P_0 - x$	x	x	

$P_{total} = 2.00$ atm $= (P_0 - x) + x + x = P_0 + x$, $P_0 = 2.00 - x$

$K_p = \dfrac{x^2}{P_0 - x} = 0.267$; $\dfrac{x^2}{2.00 - 2x} = 0.267$, $x^2 = 0.534 - (0.534)x$

$x^2 + (0.534)x - 0.534 = 0$;   solving using the quadratic formula:

$$x = \frac{-0.534 \pm \sqrt{(0.534)^2 + 4(0.534)}}{2} = 0.511\ atm$$

$P_0 = 2.00 - x = 2.00 - 0.511 = 1.49$ atm; the initial pressure of PCl_5 was 1.49 atm.

$$n_{PCl_5} = \frac{P_{PCl_5}V}{RT} = \frac{(1.49 \text{ atm})(5.00 \text{ L})}{(0.08206 \text{ L atm K}^{-1} \text{ mol}^{-1})(480. \text{ K})} = 0.189 \text{ mol } PCl_5$$

0.189 mol $PCl_5 \times 208.22$ g PCl_5/mol = 39.4 g PCl_5 was initially introduced.

89. $N_2(g)$ + $O_2(g)$ $\rightleftharpoons$ 2 $NO(g)$ Let:

equilibrium $P_{O_2} = p$

Equil. $(3.7)p$ p x equilibrium $P_{N_2} = (78/21)P_{O_2} = (3.7)p$

equilibrium $P_{NO} = x$

equilibrium $P_{NO_2} = y$

$$K_p = 1.5 \times 10^{-4} = \frac{P_{NO}^2}{P_{O_2} \times P_{N_2}}$$

N_2 +   2 O_2 $\rightleftharpoons$ 2 NO_2

Equil. $(3.7)p$ p y

$$K_p = 1.0 \times 10^{-5} = \frac{P_{NO_2}^2}{P_{O_2}^2 \times P_{N_2}}$$

We want $P_{NO_2} = P_{NO}$ at equilibrium, so $x = y$.

Taking the ratio of the two K_p expressions:

$$\frac{\dfrac{P_{NO}^2}{P_{O_2} \times P_{N_2}}}{\dfrac{P_{NO_2}^2}{P_{O_2}^2 \times P_{N_2}}} = \frac{1.5 \times 10^{-4}}{1.0 \times 10^{-5}}; \text{ because } P_{NO} = P_{NO_2}: \quad P_{O_2} = \frac{1.5 \times 10^{-4}}{1.0 \times 10^{-5}} = 15 \text{ atm}$$

Air is 21 mol % O_2, so: $P_{O_2} = (0.21)P_{total}$, $P_{total} = \dfrac{15 \text{ atm}}{0.21} = 71$ atm

To solve for the equilibrium concentrations of all gases (not required to answer the question), solve one of the K_p expressions where $p = P_{O_2} = 15$ atm.

$$1.5 \times 10^{-4} = \frac{x^2}{15[3.7(15)]}, \quad x = P_{NO} = P_{NO_2} = 0.35 \text{ atm}$$

Equilibrium pressures:

$P_{O_2} = 15$ atm; $P_{N_2} = 3.7(15) = 55.5 = 56$ atm; $P_{NO} = P_{NO_2} = 0.35$ atm

91.
$$SO_3(g) \; \rightleftharpoons \; SO_2(g) \; + \; 1/2 \; O_2(g)$$

Initial	P_0		0	0	P_0 = initial pressure of SO_3
Change	$-x$	$\rightarrow$	$+x$	$+x/2$
Equil.	$P_0 - x$		x	$x/2$

Average molar mass of the mixture is:

$$\text{average molar mass} = \frac{dRT}{P} = \frac{(1.60 \text{ g}/\text{L})(0.08206 \text{ L atm K}^{-1} \text{ mol}^{-1})(873 \text{ K})}{1.80 \text{ atm}}$$
$$= 63.7 \text{ g/mol}$$

The average molar mass is determined by:

$$\text{average molar mass} = \frac{n_{SO_3}(80.07 \text{ g/mol}) + n_{SO_2}(64.07 \text{ g/mol}) + n_{O_2}(32.00 \text{ g/mol})}{n_{total}}$$

Because χ_A = mol fraction of component A = $n_A/n_{total} = P_A/P_{total}$:

$$63.7 \text{ g/mol} = = \frac{P_{SO_3}(80.07) + P_{SO_2}(64.07) + P_{O_2}(32.00)}{P_{total}}$$

$$P_{total} = P_0 - x + x + x/2 = P_0 + x/2 = 1.80 \text{ atm}, \; P_0 = 1.80 - x/2$$

$$63.7 = \frac{(P_0 - x)(80.07) + x(64.07) + \frac{x}{2}(32.00)}{1.80}$$

$$63.7 = \frac{(1.80 - 3/2 \, x)(80.07) + x(64.07) + \frac{x}{2}(32.00)}{1.80}$$

$$115 = 144 - (120.1)x + (64.07)x + (16.00)x, \; (40.0)x = 29, \; x = 0.73 \text{ atm}$$

$$P_{SO_3} = P_0 - x = 1.80 - (3/2)x = 0.71 \text{ atm}; \; P_{SO_2} = 0.73 \text{ atm}; \; P_{O_2} = x/2 = 0.37 \text{ atm}$$

$$K_p = \frac{P_{SO_2} \times P_{O_2}^{1/2}}{P_{SO_3}} = \frac{(0.73)(0.37)^{1/2}}{(0.71)} = 0.63$$

93. $\quad N_2(g) + 3 \, H_2(g) \rightleftharpoons 2 \, NH_3(g) \qquad K_p = \dfrac{P_{NH_3}^2}{P_{N_2} \times P_{H_2}^3} = 6.5 \times 10^{-3}$

1.0 atm	$N_2(g)$	+	$3 \, H_2(g)$	$\rightleftharpoons$	$2 \, NH_3(g)$
Initial	0.25 atm		0.75 atm		0
Equil.	$0.25 - x$		$0.75 - 3x$		$2x$

$$\frac{(2x)^2}{(0.75 - 3x)^3(0.25 - x)} = 6.5 \times 10^{-3}; \text{ using successive approximations:}$$

$$x = 1.2 \times 10^{-2} \text{ atm}; \quad P_{NH_3} = 2x = 0.024 \text{ atm}$$

<u>10 atm</u> $N_2(g)$ + $3 H_2(g)$ $\rightleftharpoons$ $2 NH_3(g)$

Initial   2.5 atm   7.5 atm   0
Equil. $2.5 - x$ $7.5 - 3x$ $2x$

$$\frac{(2x)^2}{(7.5 - 3x)^3(2.5 - x)} = 6.5 \times 10^{-3}; \text{ using successive approximations:}$$

$$x = 0.69 \text{ atm}; \quad P_{NH_3} = 1.4 \text{ atm}$$

<u>100 atm</u>   Using the same setup as above: $\dfrac{4x^2}{(75 - 3x)^3(25 - x)} = 6.5 \times 10^{-3}$

Solving by successive approximations: $x = 16 \text{ atm}; \quad P_{NH_3} = 32 \text{ atm}$

<u>1000 atm</u>

$$N_2(g)   +   3 H_2(g)   \rightleftharpoons   2 NH_3(g)$$

Initial   250 atm   750 atm   0
            Let 250 atm N_2 react completely.
New initial   0   0 5.0×10^2
Equil. x $3x$ $5.0 \times 10^2 - 2x$

$$\frac{(5.0 \times 10^2 - 2x)^2}{(3x)^3 x} = 6.5 \times 10^{-3}; \text{ using successive approximations:}$$

$$x = 32 \text{ atm}; \quad P_{NH_3} = 5.0 \times 10^2 - 2x = 440 \text{ atm}$$

The results are plotted as $\log P_{NH_3}$ versus $\log P_{total}$. Notice that as P_{total} increases, a larger fraction of N_2 and H_2 is converted to NH_3, that is, as P_{total} increases (V decreases), the reaction shifts further to the right, as predicted by LeChatelier's principle.

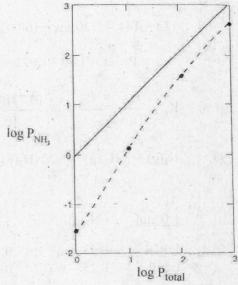

95. $\dfrac{2.00\text{ g}}{165\text{ g/mol}} = 0.0121$ mol XY (initially)

(0.350)(0.0121 mol) = 4.24×10^{-3} mol XY dissociated

	XY	$\rightarrow$	X	+	Y
Initial	0.0121 mol		0		0
Change	−0.00424	$\rightarrow$	+0.00424		+0.00424
Equil.	0.0079 mol		0.00424 mol		0.00424 mol

Total moles of gas = 0.0079 + 0.00424 + 0.00424 = 0.0164 mol

$V \propto n$, so: $\dfrac{V_{final}}{V_{initial}} = \dfrac{n_{final}}{n_{initial}} = \dfrac{0.0164\text{ mol}}{0.0121\text{ mol}} = 1.36$

$V_{initial} = \dfrac{nRT}{P} = \dfrac{(0.0121\text{ mol})(0.008206\text{ L atm K}^{-1}\text{ mol}^{-1})(298\text{ K})}{0.967\text{ atm}} = 0.306$ L

$V_{final} = 0.306$ L(1.36) = 0.416 L

Because mass is conserved in a chemical reaction:

density (final) = $\dfrac{\text{mass}}{\text{volume}} = \dfrac{2.00\text{ g}}{0.416\text{ L}} = 4.81$ g/L

$K = \dfrac{[X][Y]}{[XY]} = \dfrac{\left(\dfrac{0.00424\text{ mol}}{0.416\text{ L}}\right)^2}{\left(\dfrac{0.0079\text{ mol}}{0.416\text{ L}}\right)} = 5.5 \times 10^{-3}$

CHAPTER 7

ACIDS AND BASES

Nature of Acids and Bases

17.　　a.　The first equation is for the reaction of some generic acid, HA, with H_2O.

$$HA + H_2O \rightleftharpoons H_3O^+ + A^-$$

Acid　　Base　　Conjugate　　Conjugate
　　　　　　　Acid of H_2O　　Base of HA

HA is the proton donor (the acid) and H_2O is the proton acceptor (the base). In the reverse reaction, H_3O^+ is the proton donor (the acid) and A^- is the proton acceptor (the base).

The second equation is for some generic base, B, with some generic acid, HX. Note that B has three hydrogens bonded to it.

$$B + HX \rightleftharpoons BH^+ + X^-$$

Base　　Acid　　Conjugate　　Conjugate
　　　　　　　Acid of B　　Base of HX

B is the proton acceptor (the base) and HX is the proton donor (the acid). When B accepts a proton, the central atom goes from having 3 bonded hydrogens to 4 bonded hydrogens. In the reverse reaction, BH^+ is the proton donor (the acid) and X^- is the proton acceptor (the base).

　　b.　Arrhenius acids produce H^+ in solution. So HA in the first equation is an Arrhenius acid. However, in the second equation, H^+ is not a product, so HX is not an Arrhenius acid. Both HA in the first equation and HX in the second equation are proton donors, so both are considered Brønsted-Lowry acids.

For the bases in the two equations, H_2O and B, neither of them produce OH^- in their equations, so neither of them are Arrhenius bases. Both H_2O and B accept protons, so both are Brønsted-Lowry bases.

19.   An acid is a proton (H^+) donor, and a base is a proton acceptor. A conjugate acid-base pair differs by only a proton (H^+) in the formulas.

	Acid	Base	Conjugate Base of Acid	Conjugate Acid of Base
a.	H_2CO_3	H_2O	HCO_3^-	H_3O^+
b.	$C_5H_5NH^+$	H_2O	C_5H_5N	H_3O^+
c.	$C_5H_5NH^+$	HCO_3^-	C_5H_5N	H_2CO_3

21.   The dissociation reaction (the K_a reaction) of an acid in water commonly omits water as a reactant. We will follow this practice. All dissociation reactions produce H^+ and the conjugate base of the acid that is dissociated.

a. $HC_2H_3O_2(aq) \rightleftharpoons H^+(aq) + C_2H_3O_2^-(aq)$ $K_a = \dfrac{[H^+][C_2H_3O_2^-]}{[HC_2H_3O_2]}$

b. $Co(H_2O)_6^{3+}(aq) \rightleftharpoons H^+(aq) + Co(H_2O)_5(OH)^{2+}(aq)$ $K_a = \dfrac{[H^+][Co(H_2O)_5(OH)^{2+}]}{[Co(H_2O)_6^{3+}]}$

c. $CH_3NH_3^+(aq) \rightleftharpoons H^+(aq) + CH_3NH_2(aq)$ $K_a = \dfrac{[H^+][CH_3NH_2]}{[CH_3NH_3^+]}$

23.   The beaker on the left represents a strong acid in solution; the acid HA is 100% dissociated into the H^+ and A^- ions. The beaker on the right represents a weak acid in solution; only a little bit of the acid HB dissociates into ions, so the acid exists mostly as undissociated HB molecules in water.

a. HNO_2: weak acid beaker
b. HNO_3: strong acid beaker
c.   HCl: strong acid beaker
d.   HF: weak acid beaker
e. $HC_2H_3O_2$: weak acid beaker

25.   The K_a value is directly related to acid strength. As K_a increases, acid strength increases. For water, use K_w when comparing the acid strength of water to other species. The K_a values are:

$HClO_4$: strong acid ($K_a \gg 1$); $HClO_2$: $K_a = 1.2 \times 10^{-2}$

NH_4^+: $K_a = 5.6 \times 10^{-10}$; H_2O: $K_a = K_w = 1.0 \times 10^{-14}$

From the K_a values, the ordering is: $HClO_4 > HClO_2 > NH_4^+ > H_2O$

27.   a.   HCl is a strong acid, and water is a very weak acid with $K_a = K_w = 1.0 \times 10^{-14}$. HCl is a much stronger acid than H_2O.

b. H_2O, $K_a = K_w = 1.0 \times 10^{-14}$; HNO_2, $K_a = 4.0 \times 10^{-4}$; HNO_2 is a stronger acid than H_2O because K_a for $HNO_2 > K_w$ for H_2O.

c. HOC_6H_5, $K_a = 1.6 \times 10^{-10}$; HCN, $K_a = 6.2 \times 10^{-10}$; HCN is a slightly stronger acid than HOC_6H_5 because K_a for $HCN > K_a$ for HOC_6H_5.

29. a. H_2O and $CH_3CO_2^-$

b. An acid-base reaction can be thought of as a competition between two opposing bases. Because this equilibrium lies far to the left ($K_a < 1$), $CH_3CO_2^-$ is a stronger base than H_2O.

c. The acetate ion is a better base than water and produces basic solutions in water. When we put acetate ion into solution as the only major basic species, the reaction is:

$$CH_3CO_2^- + H_2O \rightleftharpoons CH_3CO_2H + OH^-$$

Now the competition is between $CH_3CO_2^-$ and OH^- for the proton. Hydroxide ion is the strongest base possible in water. The preceding equilibrium lies far to the left resulting in a K_b value of less than 1. Those species we specifically call weak bases ($10^{-14} < K_b < 1$) lie between H_2O and OH^- in base strength. Weak bases are stronger bases than water but are weaker bases than OH^-.

31. In deciding whether a substance is an acid or a base, strong or weak, you should keep in mind a couple ideas:

1. There are only a few common strong acids and strong bases all of which should be memorized. Common strong acids = HCl, HBr, HI, HNO_3, $HClO_4$, and H_2SO_4. Common strong bases = LiOH, NaOH, KOH, RbOH, CsOH, $Ca(OH)_2$, $Sr(OH)_2$, and $Ba(OH)_2$.

2. All other acids and bases are weak and will have K_a and K_b values of less than 1 but greater than K_w (10^{-14}). Reference Table 7.2 for K_a values for some weak acids and Table 7.3 for K_b values for some weak bases. There are too many weak acids and weak bases to memorize them all. Therefore, use the tables of K_a and K_b values to help you identify weak acids and weak bases. Appendix 5 contains more complete tables of K_a and K_b values.

a. weak acid ($K_a = 4.0 \times 10^{-4}$)   b. strong acid
c. weak base ($K_b = 4.38 \times 10^{-4}$)   d. strong base
e. weak base ($K_b = 1.8 \times 10^{-5}$)   f. weak acid ($K_a = 7.2 \times 10^{-4}$)
g. weak acid ($K_a = 1.8 \times 10^{-4}$)   h. strong base
i. strong acid

Autoionization of Water and pH Scale

33. a. $H_2O(l) \rightleftharpoons H^+(aq) + OH^-(aq)$ $K_w = 2.92 \times 10^{-14} = [H^+][OH^-]$

In pure water: $[H^+] = [OH^-]$, $2.92 \times 10^{-14} = [H^+]^2$, $[H^+] = 1.71 \times 10^{-7}\ M = [OH^-]$

b. $pH = -\log[H^+] = -\log(1.71 \times 10^{-7}) = 6.767$

c. $[H^+] = K_w/[OH^-] = (2.92 \times 10^{-14})/0.10 = 2.9 \times 10^{-13} \ M; \ pH = -\log(2.9 \times 10^{-13}) = 12.54$

35.   At 25°C, the relationship $[H^+][OH^-] = K_w = 1.0 \times 10^{-14}$ always holds for aqueous solutions. When $[H^+]$ is greater than $1.0 \times 10^{-7} \ M$, the solution is acidic; when $[H^+]$ is less than $1.0 \times 10^{-7} \ M$, the solution is basic; when $[H^+] = 1.0 \times 10^{-7} \ M$, the solution is neutral. In terms of $[OH^-]$, an acidic solution has $[OH^-] < 1.0 \times 10^{-7} \ M$, a basic solution has $[\dot{O}H^-] > 1.0 \times 10^{-7} \ M$, and a neutral solution has $[OH^-] = 1.0 \times 10^{-7} \ M$. At 25°C, $pH + pOH = 14.00$.

a. $[H^+] = \dfrac{K_w}{[OH^-]} = \dfrac{1.0 \times 10^{-14}}{1.5} = 6.7 \times 10^{-15} \ M; \ \text{basic}$

$pOH = -\log[OH^-] = -\log(1.5) = -0.18; \ pH = 14.00 - pOH = 14.00 - (-0.18) = 14.18$

b. $[H^+] = \dfrac{1.0 \times 10^{-14}}{3.6 \times 10^{-15}} = 2.8 \ M; \ \text{acidic}$

$pOH = -\log(3.6 \times 10^{-15}) = 14.44; \ pH = 14.00 - 14.44 = -0.44$

c. $[H^+] = \dfrac{1.0 \times 10^{-14}}{1.0 \times 10^{-7}} = 1.0 \times 10^{-7} \ M; \ \text{neutral}$

$pOH = -\log(1.0 \times 10^{-7}) = 7.00; \ pH = 14.00 - 7.00 = 7.00$

d. $[H^+] = \dfrac{1.0 \times 10^{-14}}{7.3 \times 10^{-4}} = 1.4 \times 10^{-11} \ M; \ \text{basic}$

$pOH = -\log(7.3 \times 10^{-4}) = 3.14; \ pH = 14.00 - 3.14 = 10.86$

Note that pH is greater than 14.00 when $[OH^-]$ is greater than 1.0 M (an extremely basic solution). Also note the the pH is negative when $[H^+]$ is greater than 1.0 M (an extremely acidic solution).

37.   a. $[H^+] = 10^{-pH}, \ [H^+] = 10^{-7.40} = 4.0 \times 10^{-8} \ M$

$pOH = 14.00 - pH = 14.00 - 7.40 = 6.60; \ [OH^-] = 10^{-pOH} = 10^{-6.60} = 2.5 \times 10^{-7} \ M$

$\text{or} \ [OH^-] = \dfrac{K_w}{[H^+]} = \dfrac{1.0 \times 10^{-14}}{4.0 \times 10^{-8}} = 2.5 \times 10^{-7} \ M; \ \text{this solution is basic since pH} > 7.00.$

b. $[H^+] = 10^{-15.3} = 5 \times 10^{-16} \ M; \ pOH = 14.00 - 15.3 = -1.3; \ [OH^-] = 10^{-(-1.3)} = 20 \ M; \ \text{basic}$

c. $[H^+] = 10^{-(-1.0)} = 10 \ M; \ pOH = 14.0 - (-1.0) = 15.0; \ [OH^-] = 10^{-15.0} = 1 \times 10^{-15} \ M; \ \text{acidic}$

d. $[H^+] = 10^{-3.20} = 6.3 \times 10^{-4} \, M$; $pOH = 14.00 - 3.20 = 10.80$; $[OH^-] = 10^{-10.80} = 1.6 \times 10^{-11} \, M$; acidic

e. $[OH^-] = 10^{-5.0} = 1 \times 10^{-5} \, M$; $pH = 14.0 - pOH = 14.0 - 5.0 = 9.0$; $[H^+] = 10^{-9.0} = 1 \times 10^{-9} \, M$; basic

f. $[OH^-] = 10^{-9.60} = 2.5 \times 10^{-10} \, M$; $pH = 14.00 - 9.60 = 4.40$; $[H^+] = 10^{-4.40} = 4.0 \times 10^{-5} \, M$; acidic

Solutions of Acids

39.   Strong acids are assumed to completely dissociate in water, for example, $HCl(aq) + H_2O(l) \rightarrow H_3O^+(aq) + Cl^-(aq)$ or $HCl(aq) \rightarrow H^+(aq) + Cl^-(aq)$.

a.   A 0.10 M HCl solution gives 0.10 M H^+ and 0.10 M Cl^- because HCl completely dissociates. The amount of H^+ from H_2O will be insignificant.

$pH = -\log[H^+] = -\log(0.10) = 1.00$

b.   5.0 M H^+ is produced when 5.0 M $HClO_4$ completely dissociates. The amount of H^+ from H_2O will be insignificant. $pH = -\log(5.0) = -0.70$ (Negative pH values just indicate very concentrated acid solutions.)

c. $1.0 \times 10^{-11} \, M$ H^+ is produced when $1.0 \times 10^{-11} \, M$ HI completely dissociates. If you take the negative log of 1.0×10^{-11}, this gives $pH = 11.00$. This is impossible! We dissolved an acid in water and got a basic pH. What we must consider in this problem is that water by itself donates $1.0 \times 10^{-7} \, M$ H^+. We can normally ignore the small amount of H^+ from H_2O except when we have a very dilute solution of an acid (as in the case here). Therefore, the pH is that of neutral water (pH = 7.00) because the amount of HI present is insignificant.

41.   HCl is a strong acid. $[H^+] = 10^{-1.50} = 3.16 \times 10^{-2} \, M$ (carrying one extra sig. fig.)

$$M_1V_1 = M_2V_2, \quad V_1 = \frac{M_2V_2}{M_1} = \frac{3.16 \times 10^{-2} \, \text{mol/L} \times 1.6 \, \text{L}}{12 \, \text{mol/L}} = 4.2 \times 10^{-3} \, \text{L}$$

4.2 mL of 12 M HCl with enough water added to make 1600 mL of solution will result in a solution having $[H^+] = 3.2 \times 10^{-2} \, M$ and pH = 1.50.

43.   a. HNO_2 ($K_a = 4.0 \times 10^{-4}$) and H_2O ($K_a = K_w = 1.0 \times 10^{-14}$) are the major species. HNO_2 is a much stronger acid than H_2O, so it is the major source of H^+. However, HNO_2 is a weak acid ($K_a < 1$), so it only partially dissociates in water. We must solve an equilibrium problem to determine $[H^+]$. In the Solutions Guide, we will summarize the *initial*, *change*, and *equilibrium* concentrations into one table called the ICE table. Solving the weak acid problem:

$$HNO_2 \rightleftharpoons H^+ + NO_2^-$$

Initial	0.250 M	~0	0
	x mol/L HNO_2 dissociates to reach equilibrium		
Change	$-x$ $\rightarrow$	$+x$	$+x$
Equil.	0.250 $- x$	x	x

$$K_a = \frac{[H^+][NO_2^-]}{[HNO_2]} = 4.0 \times 10^{-4} = \frac{x^2}{0.250 - x} ; \text{ if we assume } x << 0.250, \text{ then:}$$

$$4.0 \times 10^{-4} \approx \frac{x^2}{0.250}, \quad x = \sqrt{4.0 \times 10^{-4}(0.250)} = 0.010 \ M$$

We must check the assumption: $\dfrac{x}{0.250} \times 100 = \dfrac{0.010}{0.250} \times 100 = 4.0\%$

All the assumptions are good. The H^+ contribution from water ($1 \times 10^{-7} \ M$) is negligible, and x is small compared to 0.250 (percent error = 4.0%). If the percent error is less than 5% for an assumption, we will consider it a valid assumption (called the 5% rule). Finishing the problem: $x = 0.010 \ M = [H^+]$; pH $= -\log(0.010) = 2.00$

b. CH_3CO_2H ($K_a = 1.8 \times 10^{-5}$) and H_2O ($K_a = K_w = 1.0 \times 10^{-14}$) are the major species. CH_3CO_2H is the major source of H^+. Solving the weak acid problem:

$$CH_3CO_2H \rightleftharpoons H^+ + CH_3CO_2^-$$

Initial	0.250 M	~0	0
	x mol/L CH_3CO_2H dissociates to reach equilibrium		
Change	$-x$ $\rightarrow$	$+x$	$+x$
Equil.	0.250 $- x$	x	x

$$K_a = \frac{[H^+][CH_3CO_2^-]}{[CH_3CO_2H]}, \ 1.8 \times 10^{-5} = \frac{x^2}{0.250 - x} \approx \frac{x^2}{0.250} \text{ (assuming } x << 0.250)$$

$x = 2.1 \times 10^{-3} \ M$; checking assumption: $\dfrac{2.1 \times 10^{-3}}{0.250} \times 100 = 0.84\%$. Assumptions good.

$[H^+] = x = 2.1 \times 10^{-3} M$; pH $= -\log(2.1 \times 10^{-3}) = 2.68$

45.   This is a weak acid in water. Solving the weak acid problem:

$$HF \rightleftharpoons H^+ + F^- \quad K_a = 7.2 \times 10^{-4}$$

Initial	0.020 M	~0	0
	x mol/L HF dissociates to reach equilibrium		
Change	$-x$ $\rightarrow$	$+x$	$+x$
Equil.	0.020 $- x$	x	x

$$K_a = 7.2 \times 10^{-4} = \frac{[H^+][F^-]}{[HF]} = \frac{x^2}{0.020 - x} \approx \frac{x^2}{0.020} \quad (\text{assuming } x << 0.020)$$

$x = [H^+] = 3.8 \times 10^{-3} \, M$; check assumptions:

$$\frac{x}{0.020} \times 100 = \frac{3.8 \times 10^{-3}}{0.020} \times 100 = 19\%$$

The assumption $x << 0.020$ is not good (x is more than 5% of 0.020). We must solve $x^2/(0.020 - x) = 7.2 \times 10^{-4}$ exactly by using either the quadratic formula or the method of successive approximations (see Appendix 1 of the text). Using successive approximations, we let 0.016 M be a new approximation for [HF]. That is, in the denominator try $x = 0.0038$ (the value of x we calculated making the normal assumption) so that $0.020 - 0.0038 = 0.016$; then solve for a new value of x in the numerator.

$$\frac{x^2}{0.020 - x} \approx \frac{x^2}{0.016} = 7.2 \times 10^{-4}, \ x = 3.4 \times 10^{-3}$$

We use this new value of x to further refine our estimate of [HF], that is, $0.020 - x = 0.020 - 0.0034 = 0.0166$ (carrying an extra sig. fig.).

$$\frac{x^2}{0.020 - x} \approx \frac{x^2}{0.0166} = 7.2 \times 10^{-4}, \ x = 3.5 \times 10^{-3}$$

We repeat until we get a self-consistent answer. This would be the same answer we would get solving exactly using the quadratic equation. In this case it is, $x = 3.5 \times 10^{-3}$. Thus:

$$[H^+] = [F^-] = x = 3.5 \times 10^{-3} \, M; \ [OH^-] = K_w/[H^+] = 2.9 \times 10^{-12} \, M$$

$$[HF] = 0.020 - x = 0.020 - 0.0035 = 0.017 \, M; \ pH = 2.46$$

Note: When the 5% assumption fails, use whichever method you are most comfortable with to solve exactly. The method of successive approximations is probably fastest when the percent error is less than ~25% (unless you have a graphing calculator).

47. a. HA is a weak acid. Most of the acid is present as HA molecules; only one set of H^+ and A^- ions is present. In a strong acid, all of the acid would be dissociated into H^+ and A^- ions.

   b. This picture is the result of 1 out of 10 HA molecules dissociating.

   Percent dissociation $= \dfrac{1}{10} \times 100 = 10\%$ (an exact number)

$$HA \rightleftharpoons H^+ + A^- \quad K_a = \frac{[H^+][A^-]}{[HA]}$$

Initial	0.20 M	~0	0

x mol/L HA dissociates to reach equilibrium

Change	$-x$	$\rightarrow$ $+x$	$+x$
Equil.	$0.20 - x$	x	x

$[H^+] = [A^-] = x = 0.10 \times 0.20\ M = 0.020\ M; \quad [HA] = 0.20 - 0.020 = 0.18\ M$

$$K_a = \frac{(0.020)^2}{0.18} = 2.2 \times 10^{-3}$$

49.   Major species: HIO_3, H_2O; major source of H^+: HIO_3 (a weak acid, $K_a = 0.17$)

$$HIO_3 \rightleftharpoons H^+ + IO_3^-$$

Initial	0.010 M	~0	0

x mol/L HIO_3 dissociates to reach equilibrium

Change	$-x$	$\rightarrow$ $+x$	$+x$
Equil.	$0.010 - x$	x	x

$$K_a = 0.17 = \frac{[H^+][IO_3^-]}{[HIO_3]} = \frac{x^2}{0.010 - x} \approx \frac{x^2}{0.010}, \quad x = 0.041; \text{ check assumption.}$$

Assumption is horrible (x is more than 400% of 0.010). When the assumption is this poor, it is generally quickest to solve exactly using the quadratic formula (see Appendix 1 in text). Using the quadratic formula and carrying extra significant figures:

$$0.17 = \frac{x^2}{0.010 - x}, \quad x^2 = 0.17(0.010 - x), \quad x^2 + (0.17)x - 1.7 \times 10^{-3} = 0$$

$$x = \frac{-0.17 \pm [(0.17)^2 - 4(1)(-1.7 \times 10^{-3})]^{1/2}}{2(1)} = \frac{-0.17 \pm 0.189}{2}, \quad x = 9.5 \times 10^{-3}\ M$$
$$\text{(x must be positive)}$$

$$x = 9.5 \times 10^{-3}\ M = [H^+]; \quad pH = -\log(9.5 \times 10^{-3}) = 2.02$$

51.   This is a weak acid in water. We must solve a weak acid problem. Let $HBz = C_6H_5CO_2H$.

$$0.56\ g\ HBz \times \frac{1\ mol\ HBz}{122.1\ g} = 4.6 \times 10^{-3}\ mol; \quad [HBz]_0 = 4.6 \times 10^{-3}\ M$$

$$HBz \rightleftharpoons H^+ + Bz^-$$

Initial	4.6×10^{-3} M	~0	0

x mol/L HBz dissociates to reach equilibrium

Change	$-x$	$\rightarrow$ $+x$	$+x$
Equil.	$4.6 \times 10^{-3} - x$	x	x

$$K_a = 6.4 \times 10^{-5} = \frac{[H^+][Bz^-]}{[HBz]} = \frac{x^2}{(4.6 \times 10^{-3} - x)} \approx \frac{x^2}{4.6 \times 10^{-3}}$$

$$x = [H^+] = 5.4 \times 10^{-4}; \quad \text{check assumptions:} \quad \frac{x}{4.6 \times 10^{-3}} \times 100 = \frac{5.4 \times 10^{-4}}{4.6 \times 10^{-3}} \times 100 = 12\%$$

Assumption is not good (x is 12% of 4.6×10^{-3}). When assumption(s) fail, we must solve exactly using the quadratic formula or the method of successive approximations (see Appendix 1 of text). Using successive approximations:

$$\frac{x^2}{(4.6 \times 10^{-3}) - (5.4 \times 10^{-4})} = 6.4 \times 10^{-5}, \ x = 5.1 \times 10^{-4}$$

$$\frac{x^2}{(4.6 \times 10^{-3}) - (5.1 \times 10^{-4})} = 6.4 \times 10^{-5}, \ x = 5.1 \times 10^{-4} \ M \ \text{(consistent answer)}$$

Thus: $x = [H^+] = [Bz^-] = [C_6H_5CO_2^-] = 5.1 \times 10^{-4} \ M$

$[HBz] = [C_6H_5CO_2H] = 4.6 \times 10^{-3} - x = 4.1 \times 10^{-3} \ M$

$pH = -\log(5.1 \times 10^{-4}) = 3.29; \ pOH = 14.00 - pH = 10.71; \ [OH^-] = 10^{-10.71} = 1.9 \times 10^{-11} \ M$

53. $[HC_9H_7O_4] = \dfrac{2 \text{ tablets} \times \dfrac{0.325 \text{ g } HC_9H_7O_4}{\text{tablet}} \times \dfrac{1 \text{ mol } HC_9H_7O_4}{180.15 \text{ g}}}{0.237 \text{ L}} = 0.0152 \ M$

$$HC_9H_7O_4 \rightleftharpoons H^+ + C_9H_7O_4^-$$

Initial   0.0152 M ~0   0
               x mol/L $HC_9H_7O_4$ dissociates to reach equilibrium
Change $-x$ $\rightarrow$ $-x$ $-x$
Equil. $0.0152 - x$ x x

$$K_a = 3.3 \times 10^{-4} = \frac{[H^+][C_9H_7O_4^-]}{[HC_9H_7O_4]} = \frac{x^2}{0.0152 - x} \approx \frac{x^2}{0.0152}, \ x = 2.2 \times 10^{-3} \ M$$

Assumption that $0.0152 - x \approx 0.0152$ fails the 5% rule: $\dfrac{2.2 \times 10^{-3}}{0.0152} \times 100 = 14\%$

Using successive approximations or the quadratic equation gives an exact answer of $x = 2.1 \times 10^{-3} \ M$.

$[H^+] = x = 2.1 \times 10^{-3} \ M; \ pH = -\log(2.1 \times 10^{-3}) = 2.68$

55.   a.   HCl is a strong acid. It will produce 0.10 M H^+. HOCl is a weak acid. Let's consider the equilibrium:

$$HOCl \quad \rightleftharpoons \quad H^+ \quad + \quad OCl^- \qquad K_a = 3.5 \times 10^{-8}$$

Initial	0.10 M	0.10 M	0

x mol/L HOCl dissociates to reach equilibrium

Change	$-x$	$\rightarrow$	$+x$	$+x$
Equil.	0.10 $- x$		0.10 $+ x$	x

$$K_a = 3.5 \times 10^{-8} = \frac{[H^+][OCl^-]}{[HOCl]} = \frac{(0.10+x)(x)}{0.10-x} \approx x, \; x = 3.5 \times 10^{-8} \, M$$

Assumptions are great (x is 0.000035% of 0.10). We are really assuming that HCl is the only important source of H^+, which it is. The $[H^+]$ contribution from HOCl, x, is negligible. Therefore, $[H^+] = 0.10 \, M$; pH = 1.00.

b. HNO_3 is a strong acid, giving an initial concentration of H^+ equal to 0.050 M. Consider the equilibrium:

$$HC_2H_3O_2 \quad \rightleftharpoons \quad H^+ \quad + \quad C_2H_3O_2^- \qquad K_a = 1.8 \times 10^{-5}$$

Initial	0.50 M	0.050 M	0

x mol/L $HC_2H_3O_2$ dissociates to reach equilibrium

Change	$-x$	$\rightarrow$	$+x$	$+x$
Equil.	0.50 $- x$		0.050 $+ x$	x

$$K_a = 1.8 \times 10^{-5} = \frac{[H^+][C_2H_3O_2^-]}{[HC_2H_3O_2]} = \frac{(0.050+x)x}{(0.50-x)} \approx \frac{(0.050)x}{0.50}$$

$x = 1.8 \times 10^{-4}$; assumptions are good (well within the 5% rule).

$[H^+] = 0.050 + x = 0.050 \, M$ and pH = 1.30

57.

$$HX \quad \rightleftharpoons \quad H^+ \quad + \quad X^-$$

Initial	I	~0	0	where I = $[HX]_0$

x mol/L HX dissociates to reach equilibrium

Change	$-x$	$\rightarrow$	$+x$	$+x$
Equil.	I $- x$		x	x

From the problem, $x = 0.25(I)$ and $I - x = 0.30 \, M$.

$I - 0.25(I) = 0.30 \, M$, $I = 0.40 \, M$ and $x = 0.25(0.40 \, M) = 0.10 \, M$

$$K_a = \frac{[H^+][X^-]}{[HX]} = \frac{x^2}{I-x} = \frac{(0.10)^2}{0.30} = 0.033$$

59. $pH = 2.77$, $[H^+] = 10^{-2.77} = 1.7 \times 10^{-3} \, M$

$$HOCN \quad \rightleftharpoons \quad H^+ \quad + \quad OCN^-$$

Initial   0.0100   ~0   0
Equil.   0.0100 − x   x   x

$x = [H^+] = [OCN^-] = 1.7 \times 10^{-3} \, M$; $[HOCN] = 0.0100 - x = 0.0100 - 0.0017 = 0.0083 \, M$

$$K_a = \frac{[H^+][OCN^-]}{[HOCN]} = \frac{(1.7 \times 10^{-3})^2}{8.3 \times 10^{-3}} = 3.5 \times 10^{-4}$$

61.   Major species: $HC_2H_3O_2$ (acetic acid) and H_2O; major source of H^+: $HC_2H_3O_2$

$$HC_2H_3O_2 \quad \rightleftharpoons \quad H^+ \quad + \quad C_2H_3O_2^-$$

Initial   C   ~0   0   where $C = [HC_2H_3O_2]_0$
                     x mol/L $HC_2H_3O_2$ dissociates to reach equilibrium
Change   −x   →   +x   +x
Equil.   C − x   x   x

$$K_a = 1.8 \times 10^{-5} = \frac{[H^+][C_2H_3O_2^-]}{[HC_2H_3O_2]} = \frac{x^2}{C - x}, \text{ where } x = [H^+]$$

$$1.8 \times 10^{-5} = \frac{[H^+]^2}{C - [H^+]}; \text{ from } pH = 3.0: \; [H^+] = 10^{-3.0} = 1 \times 10^{-3} \, M$$

$$1.8 \times 10^{-5} = \frac{(1 \times 10^{-3})^2}{C - (1 \times 10^{-3})}, \; C - (1 \times 10^{-3}) = \frac{1 \times 10^{-6}}{1.8 \times 10^{-5}}, \; C = 5.7 \times 10^{-2} \approx 6 \times 10^{-2} \, M$$

A $6 \times 10^{-2} \, M$ acetic acid solution will have $pH = 3.0$.

Solutions of Bases

63. NO_3^-: $K_b \ll K_w$ because HNO_3 is a strong acid. All conjugate bases of strong acids have no base strength in water. H_2O: $K_b = K_w = 1.0 \times 10^{-14}$; NH_3: $K_b = 1.8 \times 10^{-5}$; C_5H_5N: $K_b = 1.7 \times 10^{-9}$

Base strength = $NH_3 > C_5H_5N > H_2O > NO_3^-$ (As K_b increases, base strength increases.)

65.   a. $C_6H_5NH_2$ b. $C_6H_5NH_2$ c. OH^- d. CH_3NH_2

The base with the largest K_b value is the strongest base ($K_{b, C_6H_5NH_2} = 3.8 \times 10^{-10}$,

$K_{b, CH_3NH_2} = 4.4 \times 10^{-4}$). OH^- is the strongest base possible in water.

67. $NaOH(aq) \rightarrow Na^+(aq) + OH^-(aq)$; NaOH is a strong base that completely dissociates into Na^+ and OH^-. The initial concentration of NaOH will equal the concentration of OH^- donated by NaOH.

a. $[OH^-] = 0.10\ M$; $pOH = -log[OH^-] = -log(0.10) = 1.00$

pH = 14.00 - pOH = 14.00 - 1.00 = 13.00

Note that H_2O is also present, but the amount of OH^- produced by H_2O will be insignificant compared to the 0.10 M OH^- produced from the NaOH.

b.   The $[OH^-]$ concentration donated by the NaOH is $1.0 \times 10^{-10}\ M$. Water by itself donates $1.0 \times 10^{-7}\ M$. In this exercise, water is the major OH^- contributor, and $[OH^-] = 1.0 \times 10^{-7}\ M$.

$pOH = -log(1.0 \times 10^{-7}) = 7.00$; pH = 14.00 - 7.00 = 7.00

c. $[OH^-] = 2.0\ M$; $pOH = -log(2.0) = -0.30$; pH = 14.00 - (-0.30) = 14.30

69.   pH = 10.50; pOH = 14.00 - 10.50 = 3.50; $[OH^-] = 10^{-3.50} = 3.2 \times 10^{-4}\ M$

$Ba(OH)_2(aq) \rightarrow Ba^{2+}(aq) + 2\ OH^-(aq)$; $Ba(OH)_2$ donates 2 mol OH^- per mol $Ba(OH)_2$.

$[Ba(OH)_2] = 3.2 \times 10^{-4}\ M\ OH^- \times \dfrac{1\ M\ Ba(OH)_2}{2\ M\ OH^-} = 1.6 \times 10^{-4}\ M\ Ba(OH)_2$

A $1.6 \times 10^{-4}\ M\ Ba(OH)_2$ solution will produce a pH = 10.50 solution.

71.   Major species: H_2NNH_2 ($K_b = 3.0 \times 10^{-6}$) and H_2O ($K_b = K_w = 1.0 \times 10^{-14}$); the weak base H_2NNH_2 will dominate OH^- production. We must perform a weak base equilibrium calculation.

	H_2NNH_2	+ H_2O	$\rightleftharpoons$	$H_2NNH_3^+$	+	OH^-	$K_b = 3.0 \times 10^{-6}$
Initial	2.0 M			0		~0	

x mol/L H_2NNH_2 reacts with H_2O to reach equilibrium

Change	$-x$		$\rightarrow$	$+x$		$+x$	
Equil.	$2.0 - x$			x		x	

$K_b = 3.0 \times 10^{-6} = \dfrac{[H_2NNH_3^+][OH^-]}{[H_2NNH_2]} = \dfrac{x^2}{2.0 - x} \approx \dfrac{x^2}{2.0}$ (assuming $x \ll 2.0$)

$x = [OH^-] = 2.4 \times 10^{-3}\ M$; pOH = 2.62; pH = 11.38; assumptions good (x is 0.12% of 2.0).

$[H_2NNH_3^+] = 2.4 \times 10^{-3}\ M$; $[H_2NNH_2] = 2.0\ M$; $[H^+] = 10^{-11.38} = 4.2 \times 10^{-12}\ M$

73.   Neutrally charged organic compounds containing at least one nitrogen atom generally behave as weak bases. The nitrogen atom has an unshared pair of electrons around it. This lone pair of electrons is used to form a bond to H^+.

75.   This is a solution of a weak base in water. We must solve a weak base equilibrium problem.

$$C_2H_5NH_2 \; + \; H_2O \; \rightleftharpoons \; C_2H_5NH_3^+ \; + \; OH^- \qquad K_b = 5.6 \times 10^{-4}$$

Initial	0.20 M	0	~0

x mol/L $C_2H_5NH_2$ reacts with H_2O to reach equilibrium

Change	$-x$	$\rightarrow$	$+x$	$+x$
Equil.	0.20 $-x$		x	x

$$K_b = \frac{[C_2H_5NH_3^+][OH^-]}{[C_2H_5NH_2]} = \frac{x^2}{0.20-x} \approx \frac{x^2}{0.20} \quad \text{(assuming } x \ll 0.20\text{)}$$

$x = 1.1 \times 10^{-2}$; checking assumption: $(1.1 \times 10^{-2}/0.20) \times 100 = 5.5\%$

The assumption fails the 5% rule. We must solve exactly using either the quadratic equation or the method of successive approximations (see Appendix 1 of the text). Using successive approximations and carrying extra significant figures:

$$\frac{x^2}{0.20-0.011} = \frac{x^2}{0.189} = 5.6 \times 10^{-4}, \quad x = 1.0 \times 10^{-2} \, M \quad \text{(consistent answer)}$$

$$x = [OH^-] = 1.0 \times 10^{-2} \, M; \quad [H^+] = \frac{K_w}{[OH^-]} = \frac{1.0 \times 10^{-14}}{1.0 \times 10^{-2}} = 1.0 \times 10^{-12} \, M; \quad pH = 12.00$$

77. $$\frac{5.0 \times 10^{-3} \text{ g}}{0.0100 \text{ L}} \times \frac{1 \text{ mol}}{299.4 \text{ g}} = 1.7 \times 10^{-3} \; M = [\text{codeine}]_0; \quad \text{let cod} = \text{codeine } (C_{18}H_{21}NO_3).$$

Solving the weak base equilibrium problem:

$$\text{cod} \; + \; H_2O \; \rightleftharpoons \; \text{codH}^+ \; + \; OH^- \qquad K_b = 10^{-6.05} = 8.9 \times 10^{-7}$$

Initial	$1.7 \times 10^{-3} \, M$	0	~0

x mol/L codeine reacts with H_2O to reach equilibrium

Change	$-x$	$\rightarrow$	$+x$	$+x$
Equil.	$1.7 \times 10^{-3} -x$		x	x

$$K_b = 8.9 \times 10^{-7} = \frac{x^2}{(1.7 \times 10^{-3} -x)} \approx \frac{x^2}{1.7 \times 10^{-3}}, \quad x = 3.9 \times 10^{-5}; \quad \text{assumptions good.}$$

$$[OH^-] = 3.9 \times 10^{-5} \, M; \quad [H^+] = K_w/[OH^-] = 2.6 \times 10^{-10} \, M; \quad pH = -\log[H^+] = 9.59$$

79.   To solve for percent ionization, just solve the weak base equilibrium problem.

a. $$NH_3 \; + \; H_2O \; \rightleftharpoons \; NH_4^+ \; + \; OH^- \qquad K_b = 1.8 \times 10^{-5}$$

Initial	0.10 M	0	~0
Equil.	0.10 $-x$	x	x

$$K_b = 1.8 \times 10^{-5} = \frac{x^2}{0.10 - x} \approx \frac{x^2}{0.10} \ , \ x = [OH^-] = 1.3 \times 10^{-3} \ M; \text{ assumptions good.}$$

$$\text{Percent ionization} = \frac{[OH^-]}{[NH_3]_0} \times 100 = \frac{1.3 \times 10^{-3} \ M}{0.10 \ M} \times 100 = 1.3\%$$

b. $NH_3 + H_2O \rightleftharpoons NH_4^+ + OH^-$

Initial	0.010 M	0	~0
Equil.	0.010 − x	x	x

$$1.8 \times 10^{-5} = \frac{x^2}{0.010 - x} \approx \frac{x^2}{0.010} \ , \ x = [OH^-] = 4.2 \times 10^{-4} \ M; \text{ assumptions good.}$$

$$\text{Percent ionization} = \frac{4.2 \times 10^{-4}}{0.010} \times 100 = 4.2\%$$

Note: For the same base, the percent ionization increases as the initial concentration of base decreases.

c. $CH_3NH_2 + H_2O \rightleftharpoons CH_3NH_3^+ + OH^- \qquad K_b = 4.38 \times 10^{-4}$

Initial	0.10 M	0	~0
Equil.	0.10 − x	x	x

$$4.38 \times 10^{-4} = \frac{x^2}{0.10 - x} \approx \frac{x^2}{0.10} \ , \ x = 6.6 \times 10^{-3}; \text{ assumption fails the 5\% rule } (x \text{ is}$$

6.6% of 0.10).   Using successive approximations and carrying extra significant figures:

$$\frac{x^2}{0.10 - 0.0066} = \frac{x^2}{0.093} = 4.38 \times 10^{-4}, \ x = 6.4 \times 10^{-3} \quad \text{(consistent answer)}$$

$$\text{Percent ionization} = \frac{6.4 \times 10^{-3}}{0.10} \times 100 = 6.4\%$$

81.   Using the K_b reaction to solve where PT = p-toluidine ($CH_3C_6H_4NH_2$):

$$PT + H_2O \rightleftharpoons PTH^+ + OH^-$$

Initial	0.016 M	0	~0
	x mol/L of PT reacts with H_2O to reach equilibrium		
Change	−x $\rightarrow$	+x	+x
Equil.	0.016 − x	x	x

$$K_b = \frac{[PTH^+][OH^-]}{[PT]} = \frac{x^2}{0.016 - x}$$

Since pH = 8.60: pOH = 14.00 − 8.60 = 5.40 and $[OH^-] = x = 10^{-5.40} = 4.0 \times 10^{-6}$ M

$$K_b = \frac{(4.0 \times 10^{-6})^2}{0.016 - (4.0 \times 10^{-6})} = 1.0 \times 10^{-9}$$

Polyprotic Acids

83. $H_3C_6H_5O_7(aq) \rightleftharpoons H_2C_6H_5O_7^-(aq) + H^+(aq)$ $K_{a_1} = \dfrac{[H_2C_6H_5O_7^-][H^+]}{[H_3C_6H_5O_7]}$

$H_2C_6H_5O_7^-(aq) \rightleftharpoons HC_6H_5O_7^{2-}(aq) + H^+(aq)$ $K_{a_2} = \dfrac{[HC_6H_5O_7^{2-}][H^+]}{[H_2C_6H_5O_7^-]}$

$HC_6H_5O_7^{2-}(aq) \rightleftharpoons C_6H_5O_7^{3-}(aq) + H^+(aq)$ $K_{a_3} = \dfrac{[C_6H_5O_7^{3-}][H^+]}{[HC_6H_5O_7^{2-}]}$

85.   The reactions are:

$$H_3AsO_4 \rightleftharpoons H^+ + H_2AsO_4^-\quad K_{a_1} = 5 \times 10^{-3}$$

$$H_2AsO_4^- \rightleftharpoons H^+ + HAsO_4^{2-}\quad K_{a_2} = 8 \times 10^{-8}$$

$$HAsO_4^{2-} \rightleftharpoons H^+ + AsO_4^{3-}\quad K_{a_3} = 6 \times 10^{-10}$$

We will deal with the reactions in order of importance, beginning with the largest K_a, K_{a_1}.

	H_3AsO_4	$\rightleftharpoons$	H^+	$+$	$H_2AsO_4^-$	$K_{a_1} = 5 \times 10^{-3} = \dfrac{[H^+][H_2AsO_4^-]}{[H_3AsO_4]}$
Initial	0.20 M		~0		0	
Equil.	0.20 - x		x		x	

$$5 \times 10^{-3} = \frac{x^2}{0.20 - x} \approx \frac{x^2}{0.20}, \quad x = 3 \times 10^{-2}\ M;\ \text{assumption fails the 5\% rule.}$$

Solving by the method of successive approximations:

$$5 \times 10^{-3} = x^2/(0.20 - 0.03),\ x = 3 \times 10^{-2}\ \text{(consistent answer)}$$

$[H^+] = [H_2AsO_4^-] = 3 \times 10^{-2}\ M;\ [H_3AsO_4] = 0.20 - 0.03 = 0.17\ M$

Because $K_{a_2} = \dfrac{[H^+][HAsO_4^{2-}]}{[H_2AsO_4^-]} = 8 \times 10^{-8}$ is much smaller than the K_{a_1} value, very little of $H_2AsO_4^-$ (and $HAsO_4^{2-}$) dissociates compared to H_3AsO_4. Therefore, $[H^+]$ and $[H_2AsO_4^-]$ will not change significantly by the K_{a_2} reaction. Using the previously calculated concentrations of H^+ and $H_2AsO_4^-$ to calculate the concentration of $HAsO_4^{2-}$:

$$8 \times 10^{-8} = \frac{(3 \times 10^{-2})[HAsO_4^{2-}]}{3 \times 10^{-2}}, \quad [HAsO_4^{2-}] = 8 \times 10^{-8} \ M$$

The assumption that the K_{a_2} reaction does not change $[H^+]$ and $[H_2AsO_4^-]$ is good. We repeat the process using K_{a_3} to get $[AsO_4^{3-}]$.

$$K_{a_3} = 6 \times 10^{-10} = \frac{[H^+][AsO_4^{3-}]}{[HAsO_4^{2-}]} = \frac{(3 \times 10^{-2})[AsO_4^{3-}]}{8 \times 10^{-8}}$$

$[AsO_4^{3-}] = 1.6 \times 10^{-15} \approx 2 \times 10^{-15} \ M$; assumption good.

So in 0.20 M analytical concentration of H_3AsO_4:

$$[H_3AsO_4] = 0.17 \ M; \quad [H^+] = [H_2AsO_4^-] = 3 \times 10^{-2} \ M;$$

$$[HAsO_4^{2-}] = 8 \times 10^{-8} \ M; \quad [AsO_4^{3-}] = 2 \times 10^{-15} \ M; \quad [OH^-] = K_w/[H^+] = 3 \times 10^{-13} \ M$$

87.   For $H_2C_6H_6O_6$. $K_{a_1} = 7.9 \times 10^{-5}$ and $K_{a_2} = 1.6 \times 10^{-12}$. Because $K_{a_1} \gg K_{a_2}$, the amount of H^+ produced by the K_{a_2} reaction will be negligible.

$$[H_2C_6H_6O_6]_0 = \frac{0.500 \ g \times \dfrac{1 \ mol \ H_2C_6H_6O_6}{176.12 \ g}}{0.2000 \ L} = 0.0142 \ M$$

$$H_2C_6H_6O_6(aq) \rightleftharpoons HC_6H_6O_6^-(aq) \ + \ H^+(aq) \qquad K_{a_1} = 7.9 \times 10^{-5}$$

Initial	0.0142 M	0	~0
Equil.	0.0142 − x	x	x

$$K_{a_1} = 7.9 \times 10^{-5} = \frac{x^2}{0.0142 - x} \approx \frac{x^2}{0.0142}, \ x = 1.1 \times 10^{-3}; \ \text{assumption fails the 5\% rule.}$$

Solving by the method of successive approximations:

$$7.9 \times 10^{-5} = \frac{x^2}{0.0142 - 1.1 \times 10^{-3}}, \ x = 1.0 \times 10^{-3} \ M \ \text{(consistent answer)}$$

Because H^+ produced by the K_{a_2} reaction will be negligible, $[H^+] = 1.0 \times 10^{-3}$ and pH = 3.00.

89.   The dominant H^+ producer is the strong acid H_2SO_4. A 2.0 M H_2SO_4 solution produces 2.0 M HSO_4^- and 2.0 M H^+. However, HSO_4^- is a weak acid that could also add H^+ to the solution.

$$HSO_4^- \quad \rightleftharpoons \quad H^+ \quad + \quad SO_4^{2-}$$

Initial   2.0 M 2.0 M 0

x mol/L HSO_4^- dissociates to reach equilibrium

Change $-x$ $\rightarrow$ $+x$ $+x$

Equil.   2.0 $- x$ 2.0 $+ x$ x

$$K_{a_2} = 1.2 \times 10^{-2} = \frac{[H^+][SO_4^{2-}]}{[HSO_4^-]} = \frac{(2.0 + x)x}{2.0 - x} \approx \frac{2.0(x)}{2.0}, \ x = 1.2 \times 10^{-2} \ M$$

Because x is 0.60% of 2.0, the assumption is valid by the 5% rule. The amount of additional H^+ from HSO_4^- is 1.2×10^{-2} M. The total amount of H^+ present is:

$$[H^+] = 2.0 + (1.2 \times 10^{-2}) = 2.0 \ M; \ pH = -\log(2.0) = -0.30$$

Note: In this problem H^+ from HSO_4^- could have been ignored. However, this is not usually the case in more dilute solutions of H_2SO_4.

Acid-Base Properties of Salts

91.   a.   These are strong acids like HCl, HBr, HI, HNO_3, H_2SO_4, or $HClO_4$.

       b.   These are salts of the conjugate acids of the bases in Table 7.3. These conjugate acids are all weak acids. NH_4Cl, $CH_3NH_3NO_3$, and $C_2H_5NH_3Br$ are three examples. Note that the anions used to form these salts are conjugate bases of strong acids; this is so because they have no acidic or basic properties in water (with the exception of HSO_4^-, which has weak acid properties).

       c.   These are strong ases like LiOH, NaOH, KOH, RbOH, CsOH, $Ca(OH)_2$, $Sr(OH)_2$, and $Ba(OH)_2$.

       d.   These are salts of the conjugate bases of the neutrally charged weak acids in Table 7.2. The conjugate bases of weak acids are weak bases themselves. Three examples are $NaClO_2$, $KC_2H_3O_2$, and CaF_2. The cations used to form these salts are Li^+, Na^+, K^+, Rb^+, Cs^+, Ca^{2+}, Sr^{2+}, or Ba^{2+} because these cations have no acidic or basic properties in water. Notice that these are the cations of the strong bases you should memorize.

       e.   There are two ways to make a neutral salt. The easiest way is to combine a conjugate base of a strong acid (except for HSO_4^-) with one of the cations from a strong base. These ions have no acidic/basic properties in water, so salts of these ions are neutral. Three examples are NaCl, KNO_3, and SrI_2. Another type of strong electrolyte that can produce neutral solutions are salts that contain an ion with weak acid properties combined with an ion of opposite charge having weak base properties. If the K_a for the weak acid ion is equal to the K_b for the weak base ion, then the salt will produce a neutral solution. The most common example of this type of salt is ammonium acetate ($NH_4C_2H_3O_2$). For this salt, K_a for NH_4^+ = K_b for $C_2H_3O_2^-$ = 5.6×10^{-10}. This salt at any concentration produces a neutral solution.

93.   One difficult aspect of acid-base chemistry is recognizing what types of species are present in solution, that is, whether a species is a strong acid, strong base, weak acid, weak base, or a neutral species. Below are some ideas and generalizations to keep in mind that will help in recognizing types of species present.

a.   Memorize the following strong acids: HCl, HBr, HI, HNO_3, $HClO_4$, and H_2SO_4

b.   Memorize the following strong bases: LiOH, NaOH, KOH, RbOH, $Ca(OH)_2$, $Sr(OH)_2$, and $Ba(OH)_2$

c.   Weak acids have a K_a value of less than 1 but greater than K_w. Some weak acids are listed in Table 7.2 of the text. Weak bases have a K_b value of less than 1 but greater than K_w. Some weak bases are listed in Table 7.3 of the text.

d.   Conjugate bases of weak acids are weak bases, that is, all have a K_b value of less than 1 but greater than K_w. Some examples of these are the conjugate bases of the weak acids listed in Table 7.2 of the text.

e.   Conjugate acids of weak bases are weak acids, that is, all have a K_a value of less than 1 but greater than K_w. Some examples of these are the conjugate acids of the weak bases listed in Table 7.3 of the text.

f.   Alkali metal ions (Li^+, Na^+, K^+, Rb^+, Cs^+) and some alkaline earth metal ions (Ca^{2+}, Sr^{2+}, Ba^{2+}) have no acidic or basic properties in water.

g.   Conjugate bases of strong acids (Cl^-, Br^-, I^-, NO_3^-, ClO_4^-, HSO_4^-) have no basic properties in water ($K_b << K_w$), and only HSO_4^- has any acidic properties in water.

Let's apply these ideas to this problem to see what types of species are present.

a.   HI: Strong acid; HF: weak acid ($K_a = 7.2 \times 10^{-4}$)

NaF: F^- is the conjugate base of the weak acid HF, so F^- is a weak base. The K_b value for $F^- = K_w/K_{a, HF} = 1.4 \times 10^{-11}$. Na^+ has no acidic or basic properties.

NaI: Neutral (pH = 7.0); Na^+ and I^- have no acidic/basic properties.

In order of increasing pH, we place the compounds from most acidic (lowest pH) to most basic (highest pH). Increasing pH: HI < HF < NaI < NaF.

b. NH_4Br: NH_4^+ is a weak acid ($K_a = 5.6 \times 10^{-10}$), and Br^- is a neutral species.

HBr: Strong acid

KBr: Neutral; K^+ and Br^- have no acidic/basic properties.

NH_3:   Weak base, $K_b = 1.8 \times 10^{-5}$

Increasing pH: HBr < NH_4Br < KBr < NH_3
                        Most   Most
                        acidic   basic

c. $C_6H_5NH_3NO_3$: $C_6H_5NH_3^+$ is a weak acid $(K_a = K_w/K_{b,\,C_6H_5NH_2} =$
$1.0 \times 10^{-14}/3.8 \times 10^{-10} = 2.6 \times 10^{-5})$, and NO_3^- is a neutral species.

$NaNO_3$:   Neutral; Na^+ and NO_3^- have no acidic/basic properties.

$NaOH$:   Strong base

HOC_6H_5:   Weak acid $(K_a = 1.6 \times 10^{-10})$

KOC_6H_5: $OC_6H_5^-$ is a weak base $(K_b = K_w/K_{a,\,HOC_6H_5} = 6.3 \times 10^{-5})$, and K^+ is
a neutral species.

$C_6H_5NH_2$:   Weak base $(K_b = 3.8 \times 10^{-10})$

HNO_3:   Strong acid

This is a little more difficult than the previous parts of this problem because two weak acids and two weak bases are present. Between the weak acids, $C_6H_5NH_3^+$ is a stronger weak acid than HOC_6H_5 since the K_a value for $C_6H_5NH_3^+$ is larger than the K_a value for HOC_6H_5. Between the two weak bases, because the K_b value for $OC_6H_5^-$ is larger than the K_b value for $C_6H_5NH_2$, $OC_6H_5^-$ is a stronger weak base than $C_6H_5NH_2$.

Increasing pH: $HNO_3 < C_6H_5NH_3NO_3 < HOC_6H_5 < NaNO_3 < C_6H_5NH_2 < KOC_6H_5 < NaOH$
                Most acidic   Most basic

95.   Reference Table 7.6 of the text and the solution to Exercise 93 for some generalizations on acid-base properties of salts.

a. $Sr(NO_3)_2 \rightarrow Sr^{2+} + 2\,NO_3^-$ neutral; Sr^{2+} and NO_3^- have no effect on pH.

b. $C_2H_5NH_3CN \rightarrow C_2H_5NH_3^+ + CN^-$ basic; $C_2H_5NH_3^+$ is a weak acid
$(K_a = K_w/K_{b,C_2H_5NH_2} = 1.0 \times 10^{-14}/5.6 \times 10^{-4} = 1.8 \times 10^{-11})$, and CN^- is a weak base
$(K_b = K_w/K_{a,\,HCN} = 1.0 \times 10^{-14}/6.2 \times 10^{-10} = 1.6 \times 10^{-5})$. Because $K_{b,\,CN^-} > K_{a,\,C_2H_5NH_3^+}$,
the solution of $C_2H_5NH_3CN$ will be basic.

c. $C_5H_5NHF \rightarrow C_5H_5NH^+ + F^-$ acidic; $C_5H_5NH^+$ is a weak acid
$(K_a = K_w/K_{b,C_5H_5N} = 5.9 \times 10^{-6})$, and F^- is a weak base $(K_b = K_w/K_{a,\,HF} = 1.4 \times 10^{-11})$.
Because $K_{a,\,C_5H_5NH^+} > K_{b,\,F^-}$, the solution of C_5H_5NHF will be acidic.

d. $NH_4C_2H_3O_2 \rightarrow NH_4^+ + C_2H_3O_2^-$ neutral; NH_4^+ is a weak acid $(K_a = 5.6 \times 10^{-10})$, and
$C_2H_3O_2^-$ is a weak base $(K_b = K_w/K_{a,\,HC_2H_3O_2} = 5.6 \times 10^{-10})$. Because $K_{a,\,NH_4^+} = K_{b,\,C_2H_3O_2^-}$, the solution of $NH_4C_2H_3O_2$ will have pH = 7.00.

e. $NaHCO_3 \rightarrow Na^+ + HCO_3^-$ basic; ignore Na^+; HCO_3^- is a weak acid $(K_{a_2} = 4.8 \times 10^{-11})$,
and HCO_3^- is a weak base $(K_b = K_w/K_{a_1,\,H_2CO_3} = 2.3 \times 10^{-8})$. HCO_3^- is a stronger base
than an acid because $K_b > K_a$. Therefore, the solution is basic.

97.   a.   KCl is a soluble ionic compound that dissolves in water to produce K^+(aq) and Cl^-(aq). K^+ (like the other alkali metal cations) has no acidic or basic properties. Cl^- is the conjugate base of the strong acid HCl. Cl^- has no basic (or acidic) properties. Therefore, a solution of KCl will be neutral because neither of the ions has any acidic or basic properties. The 1.0 M KCl solution has $[H^+] = [OH^-] = 1.0 \times 10^{-7}$ M and pH = pOH = 7.00.

       b.   KF is also a soluble ionic compound that dissolves in water to produce K^+(aq) and F^-(aq). The difference between the KCl solution and the KF solution is that F^- does have basic properties in water, unlike Cl^-. F^- is the conjugate base of the weak acid HF, and as is true for all conjugate bases of weak acids, F^- is a weak base in water. We must solve an equilibrium problem in order to determine the amount of OH^- this weak base produces in water.

$$F^- + H_2O \rightleftharpoons HF + OH^- \qquad K_b = \frac{K_w}{K_{a,\,HF}} = \frac{1.0 \times 10^{-14}}{7.2 \times 10^{-4}}$$

Initial	1.0 M		0	~0	$K_b = 1.4 \times 10^{-11}$

       x mol/L of F^- reacts with H_2O to reach equilibrium

Change	$-x$	$\rightarrow$	$+x$	$+x$
Equil.	$1.0 - x$		x	x

$$K_b = 1.4 \times 10^{-11} = \frac{[HF][OH^-]}{[F^-]}, \quad 1.4 \times 10^{-11} = \frac{x^2}{1.0 - x} \approx \frac{x^2}{1.0}$$

$x = [OH^-] = 3.7 \times 10^{-6}$ M ; assumptions good

pOH = 5.43; pH = 14.00 − 5.43 = 8.57; $[H^+] = 10^{-8.57} = 2.7 \times 10^{-9}$ M

99. B^- is a weak base. Use the weak base data to determine K_b for B^-.

$$B^- + H_2O \rightleftharpoons HB + OH^-$$

Initial	0.050 M	0	~0
Equil.	$0.050 - x$	x	x

From pH = 9.00: pOH = 5.00, $[OH^-] = 10^{-5.00} = 1.0 \times 10^{-5}$ $M = x$.

$$K_b = \frac{[HB][OH^-]}{[B^-]} = \frac{x^2}{0.050 - x} = \frac{(1.0 \times 10^{-5})^2}{0.050 - (1.0 \times 10^{-5})} = 2.0 \times 10^{-9}$$

Because B^- is a weak base, HB will be a weak acid. Solve the weak acid problem.

$$HB \rightleftharpoons H^+ + B^-$$

Initial	0.010 M	~0	0
Equil.	$0.010 - x$	x	x

$$K_a = \frac{K_w}{K_b} = \frac{1.0 \times 10^{-14}}{2.0 \times 10^{-9}}, \quad 5.0 \times 10^{-6} = \frac{x^2}{0.010 - x} \approx \frac{x^2}{0.010}$$

$x = [H^+] = 2.2 \times 10^{-4}\ M;$ pH = 3.66; assumptions good.

101.   a. $KNO_2 \rightarrow K^+ + NO_2^-$: NO_2^- is a weak base. Ignore K^+.

$$NO_2^- + H_2O \rightleftharpoons HNO_2 + OH^- \quad K_b = \frac{K_w}{K_a} = \frac{1.0 \times 10^{-14}}{4.0 \times 10^{-4}} = 2.5 \times 10^{-11}$$

Initial   0.12 M 0   ~0
Equil.   0.12 – x x x

$$K_b = 2.5 \times 10^{-11} = \frac{[OH^-][HNO_2]}{[NO_2^-]} = \frac{x^2}{0.12 - x} \approx \frac{x^2}{0.12}$$

$x = [OH^-] = 1.7 \times 10^{-6}\ M;$ pOH = 5.77; pH = 8.23; assumptions good.

b. $NaOCl \rightarrow Na^+ + OCl^-$: OCl^- is a weak base. Ignore Na^+.

$$OCl^- + H_2O \rightleftharpoons HOCl + OH^- \quad K_b = \frac{K_w}{K_a} = \frac{1.0 \times 10^{-14}}{3.5 \times 10^{-8}} = 2.9 \times 10^{-7}$$

Initial   0.45 M 0   ~0
Equil.   0.45 – x x x

$$K_b = 2.9 \times 10^{-7} = \frac{[HOCl][OH^-]}{[OCl^-]} = \frac{x^2}{0.45 - x} \approx \frac{x^2}{0.45}$$

$x = [OH^-] = 3.6 \times 10^{-4}\ M;$ pOH = 3.44; pH = 10.56; assumptions good.

c. $NH_4ClO_4 \rightarrow NH_4^+ + ClO_4^-$: NH_4^+ is a weak acid. ClO_4^- is the conjugate base of a strong acid. ClO_4^- has no basic (or acidic) properties.

$$NH_4^+ \rightleftharpoons NH_3 + H^+ \quad K_a = \frac{K_w}{K_b} = \frac{1.0 \times 10^{-14}}{1.8 \times 10^{-5}} = 5.6 \times 10^{-10}$$

Initial   0.40 M 0   ~0
Equil.   0.40 – x x x

$$K_a = 5.6 \times 10^{-10} = \frac{[NH_3][H^+]}{[NH_4^+]} = \frac{x^2}{0.40 - x} \approx \frac{x^2}{0.40}$$

$x = [H^+] = 1.5 \times 10^{-5}\ M;$ pH = 4.82; assumptions good.

103.   All these salts contain Na^+, which has no acidic/basic properties and a conjugate base of a weak acid (except for NaCl, where Cl^- is a neutral species). All conjugate bases of weak acids are weak bases since K_b values for these species are between K_w and 1. To identify the species, we will use the data given to determine the K_b value for the weak conjugate base. From the K_b value and data in Table 7.2 of the text, we can identify the conjugate base present by calculating the K_a value for the weak acid. We will use A^- as an abbreviation for the weak conjugate base.

$$A^- + H_2O \rightleftharpoons HA + OH^-$$

	A^-		HA	OH^-
Initial	0.100 mol/1.00 L		0	~0
	x mol/L A^- reacts with H_2O to reach equilibrium			
Change	$-x$	$\rightarrow$	$+x$	$+x$
Equil.	$0.100 - x$		x	x

$$K_b = \frac{[HA][OH^-]}{[A^-]} = \frac{x^2}{0.100 - x} \; ; \text{ from the problem, pH} = 8.07:$$

$$\text{pOH} = 14.00 - 8.07 = 5.93; \; [OH^-] = x = 10^{-5.93} = 1.2 \times 10^{-6} \, M$$

$$K_b = \frac{(1.2 \times 10^{-6})^2}{0.100 - (1.2 \times 10^{-6})} = 1.4 \times 10^{-11} = K_b \text{ value for the conjugate base of a weak acid.}$$

The K_a value for the weak acid equals K_w/K_b: $K_a = \dfrac{1.0 \times 10^{-14}}{1.4 \times 10^{-11}} = 7.1 \times 10^{-4}$

From Table 7.2 of the text, this K_a value is closest to HF. Therefore, the unknown salt is NaF.

105.   Major species: $Co(H_2O)_6^{3+}$ ($K_a = 1.0 \times 10^{-5}$), Cl^- (neutral), and H_2O ($K_w = 1.0 \times 10^{-14}$); $Co(H_2O)_6^{3+}$ will determine the pH since it is a stronger acid than water. Solving the weak acid problem in the usual manner:

$$Co(H_2O)_6^{3+} \rightleftharpoons Co(H_2O)_5(OH)^{2+} + H^+ \quad K_a = 1.0 \times 10^{-5}$$

	$Co(H_2O)_6^{3+}$			H^+
Initial	0.10 M		0	~0
Equil.	0.10 - x		x	x

$$K_a = 1.0 \times 10^{-5} = \frac{x^2}{0.10 - x} \approx \frac{x^2}{0.10} \, , \; x = [H^+] = 1.0 \times 10^{-3} \, M$$

$$\text{pH} = -\log(1.0 \times 10^{-3}) = 3.00; \text{ assumptions good.}$$

107.   Major species: NH_4^+, OCl^-, and H_2O; K_a for $NH_4^+ = (1.0 \times 10^{-14})/(1.8 \times 10^{-5}) = 5.6 \times 10^{-10}$ and K_b for $OCl^- = (1.0 \times 10^{-14})/(3.5 \times 10^{-8}) = 2.9 \times 10^{-7}$.

Because OCl^- is a better base than NH_4^+ is an acid, the solution will be basic. The dominant equilibrium is the best acid (NH_4^+) reacting with the best base (OCl^-) present.

$$NH_4^+ \quad + \quad OCl^- \quad \rightleftharpoons \quad NH_3 \; + \; HOCl$$

Initial	0.50 M	0.50 M	0	0
Change	$-x$	$-x$ $\rightarrow$	$+x$	$+x$
Equil.	$0.50 - x$	$0.50 - x$	x	x

$$K = K_{a, NH_4^+} \times \frac{1}{K_{a, HOCl}} = (5.6 \times 10^{-10})/(3.5 \times 10^{-8}) = 0.016$$

$$K = 0.016 = \frac{[NH_3][HOCl]}{[NH_4^+][OCl^-]} = \frac{x(x)}{(0.50 - x)(0.50 - x)}$$

$$\frac{x^2}{(0.50 - x)^2} = 0.016, \quad \frac{x}{0.50 - x} = (0.016)^{1/2} = 0.13, \quad x = 0.058 \; M$$

To solve for the H^+, use any pertinent K_a or K_b value. Using K_a for NH_4^+:

$$K_{a, NH_4^+} = 5.6 \times 10^{-10} = \frac{[NH_3][H^+]}{[NH_4^+]} = \frac{(0.058)[H^+]}{0.50 - 0.058}, \quad [H^+] = 4.3 \times 10^{-9} \; M, \quad pH = 8.37$$

Solutions of Dilute Acids and Bases

109. $HBrO \quad \rightleftharpoons \quad H^+ \quad + \quad BrO^- \quad K_a = 2 \times 10^{-9}$

Initial	$1.0 \times 10^{-6} \; M$	~0	0

x mol/L HBrO dissociates to reach equilibrium

Change	$-x$ $\rightarrow$	$+x$	$+x$
Equil.	$1.0 \times 10^{-6} - x$	x	x

$$K_a = 2 \times 10^{-9} = \frac{x^2}{(1.0 \times 10^{-6} - x)} \approx \frac{x^2}{1.0 \times 10^{-6}}, \quad x = [H^+] = 4 \times 10^{-8} \; M; \quad pH = 7.4$$

Let's check the assumptions. This answer is impossible! We can't add a small amount of an acid to a neutral solution and get a basic solution. The highest pH possible for an acid in water is 7.0. In the correct solution we would have to take into account the autoionization of water.

111. $HCN \quad \rightleftharpoons \quad H^+ \quad + \quad CN^- \quad K_a = 6.2 \times 10^{-10}$

Initial	$5.0 \times 10^{-4} \; M$	~0	0
Equil.	$5.0 \times 10^{-4} - x$	x	x

$$K_a = \frac{x^2}{(5.0 \times 10^{-4} - x)} \approx \frac{x^2}{5.0 \times 10^{-4}} = 6.2 \times 10^{-10}, \quad x = 5.6 \times 10^{-7}; \quad \text{check assumptions.}$$

The assumption that the H^+ contribution from water is negligible is poor. Whenever the calculated pH is greater than 6.0 for a weak acid, the water contribution to $[H^+]$ must be considered. From Section 7.9 in text:

if $\dfrac{[H^+]^2 - K_w}{[H^+]} \ll [HCN]_0 = 5.0 \times 10^{-4}$, then we can use $[H^+] = (K_a[HCN]_o + K_w)^{1/2}$.

Using this formula: $[H^+] = [(6.2 \times 10^{-10})(5.0 \times 10^{-4}) + (1.0 \times 10^{-14})]^{1/2}$, $[H^+] = 5.7 \times 10^{-7}\,M$

Checking assumptions: $\dfrac{[H^+]^2 - K_w}{[H^+]} = 5.5 \times 10^{-7} \ll 5.0 \times 10^{-4}$

Assumptions good. $pH = -\log(5.7 \times 10^{-7}) = 6.24$

113.   We can't neglect the $[H^+]$ contribution from H_2O since this is a very dilute solution of the strong acid. Following the strategy developed in Section 7.10 of the text, we first determine the charge balance equation and then manipulate this equation to get into one unknown.

[Positive charge] = [negative charge]

$[H^+] = [Cl^-] + [OH^-] = 7.0 \times 10^{-7} + \dfrac{K_w}{[H^+]}$ (because $[Cl^-] = 7.0 \times 10^{-7}$ and $[OH^-] = \dfrac{K_w}{[H^+]}$)

$\dfrac{[H^+]^2 - K_w}{[H^+]} = 7.0 \times 10^{-7}$, $[H^+]^2 - (7.0 \times 10^{-7})[H^+] - 1.0 \times 10^{-14} = 0$

Using the quadratic formula to solve:

$$[H^+] = \frac{-(-7.0 \times 10^{-7}) \pm [(-7.0 \times 10^{-7})^2 - 4(1)(-1.0 \times 10^{-14})]^{1/2}}{2(1)}$$

$[H^+] = 7.1 \times 10^{-7}\,M$; $pH = -\log(7.1 \times 10^{-7}) = 6.15$

Additional Exercises

115.   a. $NH_3 + H_3O^+ \rightleftharpoons NH_4^+ + H_2O$

$$K_{eq} = \frac{[NH_4^+]}{[NH_3][H^+]} = \frac{1}{K_a \text{ for } NH_4^+} = \frac{K_b}{K_w} = \frac{1.8 \times 10^{-5}}{1.0 \times 10^{-14}} = 1.8 \times 10^9$$

b. $NO_2^- + H_3O^+ \rightleftharpoons H_2O + HNO_2$ $K_{eq} = \dfrac{[HNO_2]}{[NO_2^-][H^+]} = \dfrac{1}{K_a} = \dfrac{1}{4.0 \times 10^{-4}} = 2.5 \times 10^3$

c. $NH_4^+ + CH_3CO_2^- \rightleftharpoons NH_3 + CH_3CO_2H$ $K_{eq} = \dfrac{[NH_3][CH_3CO_2H]}{[NH_4^+][CH_3CO_2^-]} \times \dfrac{[H^+]}{[H^+]}$

$$K_{eq} = \frac{K_a \text{ for } NH_4^+}{K_a \text{ for } CH_3CO_2H} = \frac{K_w}{(K_b \text{ for } NH_3)(K_a \text{ for } CH_3CO_2H)}$$

$$K_{eq} = \frac{1.0 \times 10^{-14}}{(1.8 \times 10^{-5})(1.8 \times 10^{-5})} = 3.1 \times 10^{-5}$$

d. $H_3O^+ + OH^- \rightleftharpoons 2\ H_2O$ $K_{eq} = \dfrac{1}{K_w} = 1.0 \times 10^{14}$

e. $NH_4^+ + OH^- \rightleftharpoons NH_3 + H_2O$ $K_{eq} = \dfrac{1}{K_b \text{ for } NH_3} = 5.6 \times 10^4$

f. $HNO_2 + OH^- \rightleftharpoons H_2O + NO_2^-$

$$K_{eq} = \frac{[NO_2^-]}{[HNO_2][OH^-]} \times \frac{[H^+]}{[H^+]} = \frac{K_a \text{ for } HNO_2}{K_w} = \frac{4.0 \times 10^{-4}}{1.0 \times 10^{-14}} = 4.0 \times 10^{10}$$

117.   At pH = 2.000, $[H^+] = 10^{-2.000} = 1.00 \times 10^{-2}\ M$

At pH = 4.000, $[H^+] = 10^{-4.000} = 1.00 \times 10^{-4}\ M$

$$\text{Mol } H^+ \text{ present} = 0.0100\ L \times \frac{0.0100\ \text{mol } H^+}{L} = 1.00 \times 10^{-4}\ \text{mol } H^+$$

Let V = total volume of solution at pH = 4.000:

$$1.00 \times 10^{-4}\ \text{mol/L} = \frac{1.00 \times 10^{-4}\ \text{mol } H^+}{V}, \quad V = 1.00\ L$$

Volume of water added = 1.00 L − 0.0100 L = 0.99 L = 990 mL

119.   a.   In the lungs there is a lot of O_2, and the equilibrium favors $Hb(O_2)_4$. In the cells there is a lower concentration of O_2, and the equilibrium favors HbH_4^{4+}.

b. CO_2 is a weak acid, $CO_2 + H_2O \rightleftharpoons HCO_3^- + H^+$. Removing CO_2 essentially decreases H^+, which causes the hemoglobin reaction to shift right. $Hb(O_2)_4$ is then favored, and O_2 is not released by hemoglobin in the cells. Breathing into a paper bag increases CO_2 in the blood, thus increasing $[H^+]$, which shifts the hemoglobin reaction left.

c. CO_2 builds up in the blood, and it becomes too acidic, driving the hemoglobin equilibrium to the left. Hemoglobin can't bind O_2 as strongly in the lungs. Bicarbonate ion acts as a base in water and neutralizes the excess acidity.

121.   The light bulb is bright because a strong electrolyte is present; that is, a solute is present that dissolves to produce a lot of ions in solution. The pH meter value of 4.6 indicates that a weak acid is present. (If a strong acid were present, the pH would be close to zero.) Of the possible substances, only HCl (strong acid), NaOH (strong base), and NH_4Cl are strong electrolytes. Of these three substances, only NH_4Cl contains a weak acid (the HCl solution would have a pH close to zero, and the NaOH solution would have a pH close to 14.0). NH_4Cl dissociates into NH_4^+ and Cl^- ions when dissolved in water. Cl^- is the conjugate base of a strong acid, so it has no basic (or acidic properties) in water. NH_4^+, however, is the conjugate acid of the weak base NH_3, so NH_4^+ is a weak acid and would produce a solution with a pH = 4.6 when the concentration is ~1.0 M. NH_4Cl is the solute.

123. $CaO(s) + H_2O(l) \rightarrow Ca(OH)_2(aq);\ \ Ca(OH)_2(aq) \rightarrow Ca^{2+}(aq) + 2\ OH^-(aq)$

$$[OH^-] = \frac{0.25\text{ g CaO} \times \dfrac{1\text{ mol CaO}}{56.08\text{ g}} \times \dfrac{1\text{ mol Ca(OH)}_2}{1\text{ mol CaO}} \times \dfrac{2\text{ mol OH}^-}{\text{mol Ca(OH)}_2}}{1.5\text{ L}} = 5.9 \times 10^{-3}\ M$$

$pOH = -\log(5.9 \times 10^{-3}) = 2.23,\ \ pH = 14.00 - 2.23 = 11.77$

125. $$\frac{30.0\text{ mg papH}^+\text{Cl}^-}{\text{mL soln}} \times \frac{1000\text{ mL}}{\text{L}} \times \frac{1\text{ g}}{1000\text{ mg}} \times \frac{1\text{ mol papH}^+\text{Cl}^-}{378.85\text{ g}} \times \frac{1\text{ mol papH}^+}{\text{mol papH}^+\text{Cl}^-}$$

$$= 0.0792\ M$$

$papH^+ \rightleftharpoons pap + H^+$ $K_a = \dfrac{K_w}{K_{b,\ pap}} = \dfrac{2.1 \times 10^{-14}}{8.33 \times 10^{-9}} = 2.5 \times 10^{-6}$

Initial   0.0792 M 0   ~0
Equil.   0.0792 $- x$ x x

$K_a = 2.5 \times 10^{-6} = \dfrac{x^2}{0.0792 - x} \approx \dfrac{x^2}{0.0792},\ \ x = [H^+] = 4.4 \times 10^{-4}\ M$

$pH = -\log(4.4 \times 10^{-4}) = 3.36;\ \ $ assumptions good.

127.   a. $Fe(H_2O)_6^{3+} + H_2O \rightleftharpoons Fe(H_2O)_5(OH)^{2+} + H_3O^+$

Initial   0.10 M 0   ~0
Equil.   0.10 $- x$ x x

$K_a = \dfrac{[H_3O^+][Fe(H_2O)_5(OH)^{2+}]}{[Fe(H_2O)_6^{3+}]},\ \ 6.0 \times 10^{-3} = \dfrac{x^2}{0.10 - x} \approx \dfrac{x^2}{0.10}$

$x = 2.4 \times 10^{-2}\ M;\ $ assumption is poor (24% error).

Using successive approximations:

$$\frac{x^2}{0.10 - 0.024} = 6.0 \times 10^{-3}, \quad x = 0.021$$

$$\frac{x^2}{0.10 - 0.021} = 6.0 \times 10^{-3}, \quad x = 0.022; \quad \frac{x^2}{0.10 - 0.022} = 6.0 \times 10^{-3}, \quad x = 0.022$$

$$x = [H^+] = 0.022 \ M; \quad pH = 1.66$$

b. $\dfrac{[Fe(H_2O)_5(OH)^{2+}]}{[Fe(H_2O)_6{}^{3+}]} = \dfrac{0.0010}{0.9990}; \quad K_a = 6.0 \times 10^{-3} = \dfrac{[H^+](0.0010)}{0.9990}$

Solving: $[H^+] = 6.0 \ M; \quad pH = -\log(6.0) = -0.78$

c.   Because of the lower charge, $Fe^{2+}(aq)$ will not be as strong an acid as $Fe^{3+}(aq)$. A solution of iron(II) nitrate will be less acidic (have a higher pH) than a solution with the same concentration of iron(III) nitrate.

129. $0.50 \ M$ HA, $K_a = 1.0 \times 10^{-3}$; $0.20 \ M$ HB, $K_a = 1.0 \times 10^{-10}$; $0.10 \ M$ HC, $K_a = 1.0 \times 10^{-12}$

Major source of H^+ is HA because its K_a value is significantly larger than other K_a values.

$$HA \quad \rightleftharpoons \quad H^+ \quad + \quad A^-$$

Initial	$0.50 \ M$	~ 0	0
Equil.	$0.50 - x$	x	x

$K_a = \dfrac{x^2}{0.50 - x}, \quad 1.0 \times 10^{-3} \approx \dfrac{x^2}{0.50}, \quad x = 0.022 \ M = [H^+], \quad \dfrac{0.022}{0.50} \times 100 = 4.4\%$ error

Assumption good. Let's check out the assumption that only HA is an important source of H^+.

For HB: $1.0 \times 10^{-10} = \dfrac{(0.022)[B^-]}{(0.20)}, \quad [B^-] = 9.1 \times 10^{-10} \ M$

At most, HB will produce an additional $9.1 \times 10^{-10} \ M \ H^+$. Even less will be produced by HC. Thus our original assumption was good. $[H^+] = 0.022 \ M$.

131.   Since NH_3 is so concentrated, we need to calculate the OH^- contribution from the weak base NH_3.

$$NH_3 + H_2O \quad \rightleftharpoons \quad NH_4^+ \quad + \quad OH^- \quad\quad K_b = 1.8 \times 10^{-5}$$

Initial	$15.0 \ M$		0	$0.0100 \ M$ (Assume no volume change.)
Equil.	$15.0 - x$		x	$0.0100 + x$

$K_b = 1.8 \times 10^{-5} = \dfrac{x(0.0100 + x)}{15.0 - x} \approx \dfrac{x(0.0100)}{15.0}, \quad x = 0.027;$ assumption is horrible
(x is 270% of 0.0100).

Using the quadratic formula:

$$(1.8 \times 10^{-5})(15.0 - x) = (0.0100)x + x^2, \quad x^2 + (0.0100)x - 2.7 \times 10^{-4} = 0$$

$$x = 1.2 \times 10^{-2} \, M, \quad [OH^-] = (1.2 \times 10^{-2}) + 0.0100 = 0.022 \, M$$

133. a. The initial concentrations are halved since equal volumes of the two solutions are mixed.

$$HC_2H_3O_2 \rightleftharpoons H^+ + C_2H_3O_2^-$$

	$HC_2H_3O_2$	H^+	$C_2H_3O_2^-$
Initial	0.100 M	5.00×10^{-4} M	0
Equil.	0.100 − x	$5.00 \times 10^{-4} + x$	x

$$K_a = 1.8 \times 10^{-5} = \frac{x(5.00 \times 10^{-4} + x)}{0.100 - x} \approx \frac{x(5.00 \times 10^{-4})}{0.100}$$

$x = 3.6 \times 10^{-3}$; assumption is horrible. Using the quadratic formula:

$$x^2 + (5.18 \times 10^{-4})x - 1.8 \times 10^{-6} = 0$$

$$x = 1.1 \times 10^{-3} \, M; \quad [H^+] = 5.00 \times 10^{-4} + x = 1.6 \times 10^{-3} \, M; \quad pH = 2.80$$

b. $x = [C_2H_3O_2^-] = 1.1 \times 10^{-3} \, M$

135. Because the values of K_{a_1} and K_{a_2} are fairly close to each other, we should consider the amount of H^+ produced by the K_{a_1} and K_{a_2} reactions.

$$H_3C_6H_5O_7 \rightleftharpoons H_2C_6H_5O_7^- + H^+ \quad K_{a_1} = 8.4 \times 10^{-4}$$

	$H_3C_6H_5O_7$	$H_2C_6H_5O_7^-$	H^+
Initial	0.15 M	0	~0
Equil.	0.15 - x	x	x

$$8.4 \times 10^{-4} = \frac{x^2}{0.15 - x} \approx \frac{x^2}{0.15}, \quad x = 1.1 \times 10^{-2}; \quad \text{assumption fails the 5\% rule.}$$

Solving more exactly using the method of successive approximations:

$$8.4 \times 10^{-4} = \frac{x^2}{(0.15 - 1.1 \times 10^{-2})}, \quad x = 1.1 \times 10^{-2} \, M \quad \text{(consistent answer)}$$

Now let's solve for the H^+ contribution from the K_{a_2} reaction.

$$H_2C_6H_5O_7^- \rightleftharpoons HC_6H_5O_7^{2-} + H^+ \quad K_{a_2} = 1.8 \times 10^{-5}$$

	$H_2C_6H_5O_7^-$	$HC_6H_5O_7^{2-}$	H^+
Initial	1.1×10^{-2} M	0	1.1×10^{-2} M
Equil.	$1.1 \times 10^{-2} - x$	x	$1.1 \times 10^{-2} + x$

$$1.8 \times 10^{-5} = \frac{x(1.1 \times 10^{-2} + x)}{(1.1 \times 10^{-2} - x)} \approx \frac{x(1.1 \times 10^{-2})}{1.1 \times 10^{-2}}, \quad x = 1.8 \times 10^{-5} \, M; \quad \text{assumption good}$$
$$\text{(0.2\% error).}$$

At most, $1.8 \times 10^{-5}\ M\ H^+$ will be added from the K_{a_2} reaction.

$[H^+]_{total} = (1.1 \times 10^{-2}) + (1.8 \times 10^{-5}) = 1.1 \times 10^{-2}\ M$

Note that the H^+ contribution from the K_{a_2} reaction was negligible compared to the H^+ contribution from the K_{a_1} reaction even though the two K_a values only differed by a factor of 50. Therefore, the H^+ contribution from the K_{a_3} reaction will also be negligible since $K_{a_3} < K_{a_2}$.

Solving: $pH = -\log(1.1 \times 10^{-2}) = 1.96$

Challenge Problems

137.   a. $HCO_3^- + HCO_3^- \rightleftharpoons H_2CO_3 + CO_3^{2-}$

$$K_{eq} = \frac{[H_2CO_3][CO_3^{2-}]}{[HCO_3^-][HCO_3^-]} \times \frac{[H^+]}{[H^+]} = \frac{K_{a_2}}{K_{a_1}} = \frac{4.8 \times 10^{-11}}{4.3 \times 10^{-7}} = 1.1 \times 10^{-4}$$

   b. $[H_2CO_3] = [CO_3^{2-}]$ since the reaction in part a is the principal equilibrium reaction.

   c. $H_2CO_3 \rightleftharpoons 2\ H^+ + CO_3^{2-}$ $K_{eq} = \dfrac{[H^+]^2[CO_3^{2-}]}{[H_2CO_3]} = K_{a_1} \times K_{a_2}$

Because $[H_2CO_3] = [CO_3^{2-}]$ from part b, $[H^+]^2 = K_{a_1} \times K_{a_2}$.

$[H^+] = (K_{a_1} \times K_{a_2})^{1/2}$, or taking the $-\log$ of both sides: $pH = \dfrac{pK_{a_1} + pK_{a_2}}{2}$

   d. $[H^+] = [(4.3 \times 10^{-7}) \times (4.8 \times 10^{-11})]^{1/2}$, $[H^+] = 4.5 \times 10^{-9}\ M$; $pH = 8.35$

139. $HC_2H_3O_2 \rightleftharpoons H^+ + C_2H_3O_2^-$ $K_a = 1.8 \times 10^{-5}$

Initial   1.00 M ~0   0
Equil. $1.00 - x$ x x

$1.8 \times 10^{-5} = \dfrac{x^2}{1.00 - x} \approx \dfrac{x^2}{1.00}$, $x = [H^+] = 4.24 \times 10^{-3}\ M$ (using one extra sig. fig.)

$pH = -\log(4.24 \times 10^{-3}) = 2.37$;   assumptions good.

We want to double the pH to $2(2.37) = 4.74$ by addition of the strong base NaOH. As is true with all strong bases, they are great at accepting protons. In fact, they are so good that we can assume they accept protons 100% of the time. The best acid present will react the strong base. This is $HC_2H_3O_2$. The initial reaction that occurs when the strong base is added is:

$$HC_2H_3O_2 + OH^- \rightarrow C_2H_3O_2^- + H_2O$$

Note that this reaction has the net effect of converting $HC_2H_3O_2$ into its conjugate base, $C_2H_3O_2^-$.

For a pH = 4.74, let's calculate the ratio of $[C_2H_3O_2^-]/[HC_2H_3O_2]$ necessary to achieve this pH.

$$HC_2H_3O_2 \rightleftharpoons H^+ + C_2H_3O_2^- \quad K_a = \frac{[H^+][C_2H_3O_2^-]}{[HC_2H_3O_2]}$$

When pH = 4.74, $[H^+] = 10^{-4.74} = 1.8 \times 10^{-5}$.

$$K_a = 1.8 \times 10^{-5} = \frac{(1.8 \times 10^{-5})[C_2H_3O_2^-]}{[HC_2H_3O_2]}, \quad \frac{[C_2H_3O_2^-]}{[HC_2H_3O_2]} = 1.0$$

For a solution having pH = 4.74, we need to have equal concentrations (equal moles) of $C_2H_3O_2^-$ and $HC_2H_3O_2$. Therefore, we need to add an amount of NaOH that will convert one-half of the $HC_2H_3O_2$ into $C_2H_3O_2^-$. This amount is 0.50 M NaOH.

$$HC_2H_3O_2 + OH^- \rightarrow C_2H_3O_2^- + H_2O$$

Before	1.00 M	0.50 M		0
Change	-0.50	-0.50	$\rightarrow$	$+0.50$
After completion	0.50 M	0		0.50 M

From the preceding stoichiometry problem, adding enough NaOH(s) to produce a 0.50 M OH^- solution will convert one-half the $HC_2H_3O_2$ into $C_2H_3O_2^-$; this results in a solution with pH = 4.74.

$$\text{Mass NaOH} = 1.00 \text{ L} \times \frac{0.50 \text{ mol NaOH}}{\text{L}} \times \frac{40.00 \text{ g NaOH}}{\text{mol}} = 20. \text{ g NaOH}$$

141.   Major species: BH^+, X^-, and H_2O; because BH^+ is the best acid and X^- is the best base in solution, the principal equilibrium is:

$$BH^+ + X^- \rightleftharpoons B + HX$$

Initial	0.100 M	0.100 M	0	0
Equil.	0.100 $- x$	0.100 $- x$	x	x

$$K = \frac{K_{a,BH^+}}{K_{a,HX}} = \frac{[B][HX]}{[BH^+][X^-]}, \text{ where } [B] = [HX] \text{ and } [BH^+] = [X^-]$$

To solve for the K_a of HX, let's use the equilibrium expression to derive a general expression that relates pH to the pK_a for BH^+ and to the pK_a for HX.

$$\frac{K_{a,BH^+}}{K_{a,HX}} = \frac{[HX]^2}{[X^-]^2}; \quad K_{a,HX} = \frac{[H^+][X^-]}{[HX]}, \quad \frac{[HX]}{[X^-]} = \frac{[H^+]}{K_{a,HX}}$$

$$\frac{K_{a,BH^+}}{K_{a,HX}} = \frac{[HX]^2}{[X^-]^2} = \left(\frac{[H^+]}{K_{a,HX}}\right)^2, \quad [H^+]^2 = K_{a,BH^+} \times K_{a,HX}$$

Taking the $-\log$ of both sides: $\quad pH = \dfrac{pK_{a,BH^+} + pK_{a,HX}}{2}$

This is a general equation that applies to all BHX type salts. Solving the problem:

$$K_b \text{ for } B = 1.0 \times 10^{-3}; \quad K_a \text{ for } BH^+ = \frac{K_w}{K_b} = 1.0 \times 10^{-11}$$

$$pH = 8.00 = \frac{11.00 + pK_{a,HX}}{2}, \quad pK_{a,HX} = 5.00 \text{ and } K_a \text{ for } HX = 10^{-5.00} = 1.0 \times 10^{-5}$$

143. $0.0500 \ M \ HCO_2H \ (HA), \ K_a = 1.77 \times 10^{-4}; \ 0.150 \ M \ CH_3CH_2CO_2H \ (HB), \ K_a = 1.34 \times 10^{-5}$

Because two comparable weak acids are present, each contributes to the total pH.

Charge balance: $[H^+] = [A^-] + [B^-] + [OH^-] = [A^-] + [B^-] + K_w/[H^+]$

Mass balance for HA and HB: $0.0500 = [HA] + [A^-]$ and $0.150 = [HB] + [B^-]$

$$\frac{[H^+][A^-]}{[HA]} = 1.77 \times 10^{-4}; \quad \frac{[H^+][B^-]}{[HB]} = 1.34 \times 10^{-5}$$

We have five equations and five unknowns. Manipulate the equations to solve.

$[H^+] = [A^-] + [B^-] + K_w/[H^+]; \quad [H^+]^2 = [H^+][A^-] + [H^+][B^-] + K_w$

$[H^+][A^-] = (1.77 \times 10^{-4})[HA] = (1.77 \times 10^{-4}) \ (0.0500 - [A^-])$

If $[A^-] << 0.0500$, then $[H^+][A^-] \approx (1.77 \times 10^{-4}) \ (0.0500) = 8.85 \times 10^{-6}$.

Similarly, assume $[H^+][B^-] \approx (1.34 \times 10^{-5})(0.150) = 2.01 \times 10^{-6}$.

$[H^+]^2 = 8.85 \times 10^{-6} + 2.01 \times 10^{-6} + 1.00 \times 10^{-14}, \ [H^+] = 3.30 \times 10^{-3} \ mol/L$

Check assumptions: $[H^+][A^-] \approx 8.85 \times 10^{-6}, \ [A^-] \approx \dfrac{8.85 \times 10^{-6}}{3.30 \times 10^{-3}} \approx 2.68 \times 10^{-3}$

Assumed $0.0500 - [A^-] \approx 0.0500$. This assumption is borderline (2.68×10^{-3} is 5.4% of 0.0500). The HB assumption is good (0.4% error).

Using successive approximations to refine the $[H^+][A^-]$ value:

$$[H^+] = 3.22 \times 10^{-3}\ M,\ \ pH = -\log(3.22 \times 10^{-3}) = 2.492$$

Note: If we treat each acid separately:

$$H^+ \text{ from HA} = 2.9 \times 10^{-3}$$
$$H^+ \text{ from HB} = 1.4 \times 10^{-3}$$
$$\overline{}$$
$$4.3 \times 10^{-3}\ M = [H^+]_{total}$$

This assumes the acids did not suppress each other's ionization. They do, and we expect the $[H^+]$ to be less than $4.3 \times 10^{-3}\ M$. We get such an answer.

145.   HA $\rightleftharpoons$ H^+ + A^- $K_a = 5.00 \times 10^{-10}$

Initial	$[HA]_0$	~0	0
Change	$-x$ $\rightarrow$	$+x$	$+x$
Equil.	$[HA]_0 - x$	x	x

From the problem: pH = 5.650, so $[H^+] = x = 10^{-5.650} = 2.24 \times 10^{-6}\ M$

$$5.00 \times 10^{-10} = \frac{x^2}{[HA]_0 - x} = \frac{(2.24 \times 10^{-6})^2}{([HA]_0 - 2.24 \times 10^{-6})},\ \ [HA]_0 = 1.00 \times 10^{-2}\ M$$

After the water is added, the pH of the solution is between 6 and 7, so the water contribution to the $[H^+]$ must be considered. The general expression for a very dilute weak acid solution is:

$$K_a = \frac{[H^+]^2 - K_w}{[HA]_0 - \dfrac{[H^+]^2 - K_w}{[H^+]}}$$

pH = 6.650; $[H^+] = 10^{-6.650} = 2.24 \times 10^{-7}\ M$; let V = volume of water added:

$$5.00 \times 10^{-10} = \frac{(2.24 \times 10^{-7})^2 - (1.00 \times 10^{-14})}{(1.00 \times 10^{-2})\left(\dfrac{0.0500}{0.0500 + V}\right) - \dfrac{(2.24 \times 10^{-7})^2 - (1.00 \times 10^{-14})}{2.24 \times 10^{-7}}}$$

Solving, V = 6.16 L of water were added.

147. $\dfrac{0.135\ \text{mol CO}_2}{2.50\ \text{L}} = 5.40 \times 10^{-2}\ \text{mol CO}_2/\text{L} = 5.40 \times 10^{-2}\ M\ H_2CO_3;\ \ 0.105\ M\ CO_3^{2-}$

The best acid (H_2CO_3) reacts with the best base present (CO_3^{2-}) for the principal equilibrium.

$$H_2CO_3 + CO_3^{2-} \rightarrow 2\,HCO_3^- \quad K = \frac{K_{a_1,\,H_2CO_3}}{K_{a_2,\,H_2CO_3}} = \frac{4.3 \times 10^{-7}}{4.8 \times 10^{-11}} = 9.0 \times 10^3$$

Because $K \gg 1$, assume all CO_2 (H_2CO_3) is converted into HCO_3^-; that is, 5.40×10^{-2} mol/L CO_3^{2-} is converted into HCO_3^-.

$$[HCO_3^-] = 2(5.40 \times 10^{-2}) = 0.108\ M; \quad [CO_3^{2-}] = 0.105 - 0.0540 = 0.051\ M$$

Note: If we solve for the $[H_2CO_3]$ using these concentrations, we get $[H_2CO_3] = 2.5 \times 10^{-5}\ M$; our assumption that the reaction goes to completion is good (2.5×10^{-5} is 0.05% of 0.051). Whenever $K \gg 1$, always assume the reaction goes to completion.

To solve for the $[H^+]$ in equilibrium with HCO_3^- and CO_3^{2-}, use the K_a expression for HCO_3^-.

$$HCO_3^- \rightleftharpoons H^+ + CO_3^{2-} \quad K_{a2} = 4.8 \times 10^{-11}$$

$$4.8 \times 10^{-11} = \frac{[H^+][CO_3^{2-}]}{[HCO_3^-]} = \frac{[H^+](0.051)}{0.108}$$

$[H^+] = 1.0 \times 10^{-10}$; pH = 10.00; assumptions good.

149.   Major species: H_2O, Na^+, and NO_2^-; NO_2^- is a weak base. $NO_2^- + H_2O \rightleftharpoons HNO_2 + OH^-$

Because this is a very dilute solution of a weak base, the OH^- contribution from H_2O must be considered. The weak base equations for dilute solutions are analogous to the weak acid equations derived in Section 7.9 of the text.

For A^- type bases ($A^- + H_2O \rightleftharpoons HA + OH^-$), the general equation is:

$$K_b = \frac{[OH^-]^2 - K_w}{[A^-]_0 - \dfrac{[OH^-]^2 - K_w}{[OH^-]}}$$

When $[A^-]_0 \gg \dfrac{[OH^-]^2 - K_w}{[OH^-]}$, then $K_b = \dfrac{[OH^-]^2 - K_w}{[A^-]_0}$ and:

$$[OH^-] = (K_b[A^-]_0 + K_w)^{1/2}$$

Try: $[OH^-] = \left(\dfrac{1.0 \times 10^{-14}}{4.0 \times 10^{-4}} \times (6.0 \times 10^{-4}) + (1.0 \times 10^{-14}) \right)^{1/2} = 1.6 \times 10^{-7}\ M$

Checking assumption: $6.0 \times 10^{-4} \gg \dfrac{(1.6 \times 10^{-7})^2 - (1.0 \times 10^{-14})}{1.6 \times 10^{-7}} = 9.8 \times 10^{-8}$

Assumption good. $[OH^-] = 1.6 \times 10^{-7}\ M$; pOH = 6.80; pH = 7.20

151.   Major species: H_2O, NH_3, H^+, and Cl^-; the H^+ from the strong acid will react with the best base present (NH_3). Because strong acids are great at donating protons, the reaction between H^+ and NH_3 essentially goes to completion, that is, until one or both of the reactants runs out. The reaction is:

$$NH_3 + H^+ \rightarrow NH_4^+$$

Because equal volumes of 1.0×10^{-4} M NH_3 and 1.0×10^{-4} M H^+ are mixed, both reactants are in stoichiometric amounts, and both reactants will run out at the same time. After reaction, only NH_4^+ and Cl^- remain. Cl^- has no basic properties since it is the conjugate base of a strong acid. Therefore, the only species with acid-base properties is NH_4^+, a weak acid. The initial concentration of NH_4^+ will be exactly one-half of 1.0×10^{-4} M since equal volumes of NH_3 and HCl were mixed. Now we must solve the weak acid problem involving 5.0×10^{-5} M NH_4^+.

	NH_4^+	$\rightleftharpoons$	H^+	$+$	NH_3		$K_a = \dfrac{K_w}{K_b} = 5.6 \times 10^{-10}$
Initial	5.0×10^{-5} M		~0		0		
Equil.	$5.0 \times 10^{-5} - x$		x		x		

$$K_a = \frac{x^2}{(5.0 \times 10^{-5} - x)} \approx \frac{x^2}{5.0 \times 10^{-5}} = 5.6 \times 10^{-10}, \ x = 1.7 \times 10^{-7} \ M; \quad \text{check assumptions.}$$

We cannot neglect $[H^+]$ that comes from H_2O. As discussed in Section 7.9 of the text, assume $5.0 \times 10^{-5} \gg ([H^+]^2 - K_w)/[H^+]$. If this is the case, then:

$$[H^+] = (K_a[HA]_0 + K_w)^{1/2} = 1.9 \times 10^{-7} \ M; \quad \text{checking assumption:}$$

$$\frac{[H^+]^2 - K_w}{[H^+]} = 1.4 \times 10^{-7} \ll 5.0 \times 10^{-5} \quad \text{(assumption good)}$$

So: $[H^+] = 1.9 \times 10^{-7}$ M; pH $= 6.72$

CHAPTER 8

APPLICATIONS OF AQUEOUS EQUILIBRIA

Buffers

15. A buffer solution is one that resists a change in its pH when either hydroxide ions or protons (H^+) are added. Any solution that contains a weak acid and its conjugate base or a weak base and its conjugate acid is classified as a buffer. The pH of a buffer depends on the [base]/[acid] ratio. When H^+ is added to a buffer, the weak base component of the buffer reacts with the H^+ and forms the acid component of the buffer. Even though the concentrations of the acid and base components of the buffer change some, the ratio of [base]/[acid] does not change that much. This translates into a pH that doesn't change much. When OH^- is added to a buffer, the weak acid component is converted into the base component of the buffer. Again, the [base]/[acid] ratio does not change a lot (unless a large quantity of OH^- is added), so the pH does not change much.

$$H^+(aq) + CO_3^{2-}(aq) \rightarrow HCO_3^-(aq); \quad OH^-(aq) + HCO_3^-(aq) \rightarrow H_2O(l) + CO_3^{2-}(aq)$$

17. Only the third beaker represents a buffer solution. A weak acid and its conjugate base must both be present in large quantities in order to have a buffer solution. This is only the case in the third beaker. The first beaker respresents a beaker full of strong acid which is 100°% dissociated. The second beaker represents a weak acid solution. In a weak acid solution, only a small fraction of the acid is dissociated. In this representation, 1/10 of the weak acid has dissociated. The only B^- present in this beaker is from the dissociation of the weak acid. A buffer solution has B^- added from another source.

19. $pH = pK_a + \log\dfrac{[base]}{[acid]}$; when [acid] > [base], then $\dfrac{[base]}{[acid]} < 1$ and $\log\left(\dfrac{[base]}{[acid]}\right) < 0$.

From the Henderson-Hasselbalch equation, if the log term is negative, then $pH < pK_a$. When one has more acid than base in a buffer, the pH will be on the acidic side of the pK_a value; that is, the pH is at a value lower than the pK_a value. When one has more base than acid in a buffer ([conjugate base] > [weak acid]), then the log term in the Henderson-Hasselbalch equation is positive, resulting in $pH > pK_a$. When one has more base than acid in a buffer, the pH is on the basic side of the pK_a value; that is, the pH is at a value greater than the pK_a value. The other scenario you can run across in a buffer is when [acid] = [base]. Here, the log term is equal to zero, and $pH = pK_a$.

21.   a.   This is a weak acid problem. Let $HC_3H_5O_2 = HOPr$ and $C_3H_5O_2^- = OPr^-$.

$$HOPr(aq) \rightleftharpoons H^+(aq) + OPr^-(aq) \qquad K_a = 1.3 \times 10^{-5}$$

Initial	0.100 M	~0	0

x mol/L HOPr dissociates to reach equilibrium

Change	$-x$	$\rightarrow +x$	$+x$
Equil.	$0.100 - x$	x	x

$$K_a = 1.3 \times 10^{-5} = \frac{[H^+][OPr^-]}{[HOPr]} = \frac{x^2}{0.100 - x} \approx \frac{x^2}{0.100}$$

$x = [H^+] = 1.1 \times 10^{-3} \ M;$ pH = 2.96;   assumptions good by the 5% rule.

b.   This is a weak base problem.

$$OPr^-(aq) + H_2O(l) \rightleftharpoons HOPr(aq) + OH^-(aq) \quad K_b = \frac{K_w}{K_a} = 7.7 \times 10^{-10}$$

Initial	0.100 M	0	~0

x mol/L OPr$^-$ reacts with H$_2$O to reach equilibrium

Change	$-x$	$\rightarrow +x$	$+x$
Equil.	$0.100 - x$	x	x

$$K_b = 7.7 \times 10^{-10} = \frac{[HOPr][OH^-]}{[OPr^-]} = \frac{x^2}{0.100 - x} \approx \frac{x^2}{0.100}$$

$x = [OH^-] = 8.8 \times 10^{-6} \ M;$ pOH = 5.06;   pH = 8.94;   assumptions good.

c.   Pure H$_2$O, $[H^+] = [OH^-] = 1.0 \times 10^{-7} \ M;$ pH = 7.00

d.   This solution contains a weak acid and its conjugate base. This is a buffer solution. We will solve for the pH through the weak acid equilibrium reaction.

$$HOPr(aq) \rightleftharpoons H^+(aq) + OPr^-(aq) \qquad K_a = 1.3 \times 10^{-5}$$

Initial	0.100 M	~0	0.100 M

x mol/L HOPr dissociates to reach equilibrium

Change	$-x$	$\rightarrow +x$	$+x$
Equil.	$0.100 - x$	x	$0.100 + x$

$$1.3 \times 10^{-5} = \frac{(0.100 + x)(x)}{0.100 - x} \approx \frac{(0.100)(x)}{0.100} = x = [H^+]$$

$[H^+] = 1.3 \times 10^{-5} \ M;$ pH = 4.89;   assumptions good.

Alternately, we can use the Henderson-Hasselbalch equation to calculate the pH of buffer solutions.

$$pH = pK_a + \log \frac{[base]}{[acid]} = pK_a + \log\left(\frac{0.100}{0.100}\right) = pK_a = -\log(1.3 \times 10^{-5}) = 4.89$$

The Henderson-Hasselbalch equation will be valid when an assumption of the type $0.1 + x \approx 0.1$ that we just made in this problem is valid. From a practical standpoint, this will almost always be true for useful buffer solutions. If the assumption is not valid, the solution will have such a low buffering capacity it will not be of any use to control the pH. *Note*: The Henderson-Hasselbalch equation can <u>only</u> be used to solve for the pH of buffer solutions.

23.   a.   OH⁻ will react completely with the best acid present, HOPr.

	HOPr	+	OH⁻	→	OPr⁻	+	H₂O
Before	0.100 M		0.020 M		0		
Change	−0.020		−0.020	→	+0.020		Reacts completely
After	0.080		0		0.020		

A buffer solution results after the reaction. Using the Henderson-Hasselbalch equation:

$$pH = pK_a + \log \frac{[base]}{[acid]} = 4.89 + \log \frac{(0.020)}{(0.080)} = 4.29; \text{ assumptions good.}$$

b.   We have a weak base and a strong base present at the same time. The amount of OH⁻ added by the weak base will be negligible. To prove it, let's consider the weak base equilibrium:

	OPr⁻	+	H₂O	⇌	HOPr	+	OH⁻		$K_b = 7.7 \times 10^{-10}$
Initial	0.100 M				0		0.020 M		

x mol/L OPr⁻ reacts with H₂O to reach equilibrium

Change	−x			→	+x		+x		
Equil.	0.100 − x				x		0.020 + x		

$[OH^-] = 0.020 + x \approx 0.020$ M; pOH = 1.70; pH = 12.30; assumption good.

Note: The OH⁻ contribution from the weak base OPr⁻ was negligible ($x = 3.9 \times 10^{-9}$ M as compared to 0.020 M OH⁻ from the strong base). The pH can be determined by only considering the amount of strong base present.

c.   This is a strong base in water. $[OH^-] = 0.020$ M; pOH = 1.70; pH = 12.30

d.   OH⁻ will react completely with HOPr, the best acid present.

	HOPr	+	OH⁻	→	OPr⁻	+	H₂O
Before	0.100 M		0.020 M		0.100 M		
Change	−0.020		−0.020	→	+0.020		
After	0.080		0		0.120		

Reacts completely

Using the Henderson-Hasselbalch equation to solve for the pH of the resulting buffer solution:

$$pH = pK_a + \log \frac{[\text{base}]}{[\text{acid}]} = 4.89 + \log \frac{(0.120)}{(0.080)} = 5.07; \text{ assumptions good.}$$

25.   Major species: HF, F⁻, K⁺, and H₂O. K⁺ has no acidic or basic properties. This is a solution containing a weak acid and its conjugate base. This is a buffer solution. One appropriate equilibrium reaction you can use is the K_a reaction of HF, which contains both HF and F⁻. However, you could also use the K_b reaction for F⁻ and come up with the same answer. Alternately, you could use the Henderson-Hasselblach equation to solve for the pH. For this problem, we will use the K_a reaction and set up an ICE table to solve for the pH.

	HF	⇌	F⁻	+	H⁺
Initial	0.60 M		1.00 M		~0

x mol/L HF dissociates to reach equilibrium

| Change | −x | → | +x | | +x |
| Equil. | 0.60 − x | | 1.00 + x | | x |

$$K_a = 7.2 \times 10^{-4} = \frac{[\text{F}^-][\text{H}^+]}{[\text{HF}]} = \frac{(1.00 + x)(x)}{0.60 - x} \approx \frac{(1.00)(x)}{0.60} \quad (\text{assuming } x \ll 0.60)$$

$$x = [\text{H}^+] = 0.60 \times (7.2 \times 10^{-4}) = 4.3 \times 10^{-4} M; \text{ assumptions good.}$$

$$pH = -\log(4.3 \times 10^{-4}) = 3.37$$

27.   Major species after NaOH added: HF, F⁻, K⁺, Na⁺, OH⁻, and H₂O. The OH⁻ from the strong base will react with the best acid present (HF). Any reaction involving a strong base is assumed to go to completion. Because all species present are in the same volume of solution, we can use molarity units to do the stoichiometry part of the problem (instead of moles). The stoichiometry problem is:

	OH⁻	+	HF	→	F⁻	+	H₂O
Before	0.10 mol/1.00 L		0.60 M		1.00 M		
Change	−0.10 M		−0.10 M	→	+0.10 M		
After	0		0.50		1.10		

Reacts completely

After all the OH⁻ reacts, we are left with a solution containing a weak acid (HF) and its conjugate base (F⁻). This is what we call a buffer problem. We will solve this buffer problem using the K_a equilibrium reaction. One could also use the K_b equilibrium reaction or use the Henderson-Hasselbalch equation to solve for the pH.

	HF	$\rightleftharpoons$	F^-	+	H^+
Initial	0.50 M		1.10 M		~0

x mol/L HF dissociates to reach equilibrium

	HF		F^-		H^+
Change	$-x$	$\rightarrow$	$+x$		$+x$
Equil.	$0.50 - x$		$1.10 + x$		x

$$K_a = 7.2 \times 10^{-4} = \frac{(1.10 + x)(x)}{0.50 - x} \approx \frac{(1.10)(x)}{0.50}, \quad x = [H^+] = 3.3 \times 10^{-4}\ M;\quad pH = 3.48;$$

assumptions good.

Note: The added NaOH to this buffer solution changes the pH only from 3.37 to 3.48. If the NaOH were added to 1.0 L of pure water, the pH would change from 7.00 to 13.00.

Major species after HCl added: HF, F^-, H^+, K^+, Cl^-, and H_2O; the added H^+ from the strong acid will react completely with the best base present (F^-).

	H^+	+	F^-	$\rightarrow$	HF
Before	$\dfrac{0.20 \text{ mol}}{1.00 \text{ L}}$		1.00 M		0.60 M
Change	$-0.20\ M$		$-0.20\ M$	$\rightarrow$	$+0.20\ M$
After	0		0.80		0.80

After all the H^+ has reacted, we have a buffer solution (a solution containing a weak acid and its conjugate base). Solving the buffer problem:

	HF	$\rightleftharpoons$	F^-	+	H^+
Initial	0.80 M		0.80 M		0
Equil.	$0.80 - x$		$0.80 + x$		x

$$K_a = 7.2 \times 10^{-4} = \frac{(0.80 + x)(x)}{0.80 - x} \approx \frac{(0.80)(x)}{0.80}, \quad x = [H^+] = 7.2 \times 10^{-4}\ M;\quad pH = 3.14;$$

assumptions good.

Note: The added HCl to this buffer solution changes the pH only from 3.37 to 3.14. If the HCl were added to 1.0 L of pure water, the pH would change from 7.00 to 0.70.

29. K_a for $H_2NNH_3^+ = K_w/K_{b,\ H_2NNH_2} = 1.0 \times 10^{-14}/3.0 \times 10^{-6} = 3.3 \times 10^{-9}$

$$pH = pK_a + \log\frac{[H_2NNH_2]}{[H_2NNH_3^+]} = -\log(3.3 \times 10^{-9}) + \log\left(\frac{0.40}{0.80}\right) = 8.48 + (-0.30) = 8.18$$

$pH = pK_a$ for a buffer when [acid] = [base]. Here, the acid ($H_2NNH_3^+$) concentration needs to decrease, while the base (H_2NNH_2) concentration needs to increase in order for $[H_2NNH_3^+]$ = $[H_2NNH_2]$. Both of these changes are accomplished by adding a strong base (like NaOH) to the original buffer. The added OH^- from the strong base converts the acid component of the buffer into the conjugate base. Here, the reaction is $H_2NNH_3^+ + OH^- \rightarrow H_2NNH_2 + H_2O$.

Because a strong base is reacting, the reaction is assumed to go to completion. The following set-up determines the number of moles of OH^- (x) that must be added so that mol $H_2NNH_3^+$ = mol H_2NNH_2. When mol acid = mol base in a buffer, then [acid] = [base] and $pH = pK_a$.

$$H_2NNH_3^+ \quad + \quad OH^- \quad \rightarrow \quad H_2NNH_2 \quad + \quad H_2O$$

Before	$1.0\ L \times 0.80\ mol/L$	x	$1.0\ L \times 0.40\ mol/L$	
Change	$-x$	$-x$ $\rightarrow$	$+x$	Reacts completely
After	$0.80 - x$	0	$0.40 + x$	

We want mol $H_2NNH_3^+$ = mol H_2NNH_2. So:

$$0.80 - x = 0.40 + x, \ 2x = 0.40, \ x = 0.20 \text{ mol } OH^-$$

When 0.20 mol OH^- is added to the initial buffer, mol $H_2NNH_3^+$ is decreased to 0.60 mol, while mol H_2NNH_2 is increased to 0.60 mol. Therefore, 0.20 mol of NaOH must be added to the initial buffer solution in order to produce a solution where $pH = pK_a$.

31. $$[HC_7H_5O_2] = \frac{21.5 \text{ g } HC_7H_5O_2 \times \dfrac{1 \text{ mol } HC_7H_5O_2}{122.12 \text{ g}}}{0.2000 \text{ L}} = 0.880 \ M$$

$$[C_7H_5O_2^-] = \frac{37.7 \text{ g } NaC_7H_5O_2 \times \dfrac{1 \text{ mol } NaC_7H_5O_2}{144.10 \text{ g}} \times \dfrac{1 \text{ mol } C_7H_5O_2^-}{\text{mol } NaC_7H_5O_2}}{0.2000 \text{ L}} = 1.31 \ M$$

We have a buffer solution since we have both a weak acid and its conjugate base present at the same time. One can use the K_a reaction or the K_b reaction to solve. We will use the K_a reaction for the acid component of the buffer.

$$HC_7H_5O_2 \quad \rightleftharpoons \quad H^+ \quad + \quad C_7H_5O_2^-$$

Initial	$0.880\ M$	~ 0	$1.31\ M$
	x mol/L of $HC_7H_5O_2$ dissociates to reach equilibrium		
Change	$-x$	$\rightarrow +x$	$+x$
Equil.	$0.880 - x$	x	$1.31 + x$

$$K_a = 6.4 \times 10^{-5} = \frac{x(1.31 + x)}{0.880 - x} \approx \frac{x(1.31)}{0.880}, \ x = [H^+] = 4.3 \times 10^{-5} M$$

$pH = -\log(4.3 \times 10^{-5}) = 4.37$; assumptions good.

Alternatively, we can use the Henderson-Hasselbalch equation to calculate the pH of buffer solutions.

$$pH = pK_a + \log\frac{[base]}{[acid]} = pK_a + \log\frac{[C_7H_5O_2^-]}{[HC_7H_5O_2]}$$

$$pH = -\log(6.4 \times 10^{-5}) + \log\left(\frac{1.31}{0.880}\right) = 4.19 + 0.173 = 4.36$$

Within round-off error, this is the same answer we calculated solving the equilibrium problem using the K_a reaction.

The Henderson-Hasselbalch equation will be valid when an assumption of the type $1.31 + x \approx 1.31$ that we just made in this problem is valid. From a practical standpoint, this will almost always be true for useful buffer solutions. If the assumption is not valid, the solution will have such a low buffering capacity that it will be of no use to control the pH. *Note*: The Henderson-Hasselbalch equation can <u>only</u> be used to solve for the pH of buffer solutions.

33. $C_5H_5NH^+ \rightleftharpoons H^+ + C_5H_5N$ $K_a = \dfrac{K_w}{K_b} = \dfrac{1.0 \times 10^{-14}}{1.7 \times 10^{-9}} = 5.9 \times 10^{-6}$

$$pK_a = -\log(5.9 \times 10^{-6}) = 5.23$$

We will use the Henderson-Hasselbalch equation to calculate the concentration ratio necessary for each buffer.

$$pH = pK_a + \log\frac{[\text{base}]}{[\text{acid}]}, \quad pH = 5.23 + \log\frac{[C_5H_5N]}{[C_5H_5NH^+]}$$

a. $4.50 = 5.23 + \log\dfrac{[C_5H_5N]}{[C_5H_5NH^+]}$ b. $5.00 = 5.23 + \log\dfrac{[C_5H_5N]}{[C_5H_5NH^+]}$

$\log\dfrac{[C_5H_5N]}{[C_5H_5NH^+]} = -0.73$ $\log\dfrac{[C_5H_5N]}{[C_5H_5NH^+]} = -0.23$

$\dfrac{[C_5H_5N]}{[C_5H_5NH^+]} = 10^{-0.73} = 0.19$ $\dfrac{[C_5H_5N]}{[C_5H_5NH^+]} = 10^{-0.23} = 0.59$

c. $5.23 = 5.23 + \log\dfrac{[C_5H_5N]}{[C_5H_5NH^+]}$ d. $5.50 = 5.23 + \log\dfrac{[C_5H_5N]}{[C_5H_5NH^+]}$

$\dfrac{[C_5H_5N]}{[C_5H_5NH^+]} = 10^{0.0} = 1.0$ $\dfrac{[C_5H_5N]}{[C_5H_5NH^+]} = 10^{0.27} = 1.9$

35.   When H^+ is added, it converts $C_2H_3O_2^-$ into $HC_2H_3O_2$: $C_2H_3O_2^- + H^+ \rightarrow HC_2H_3O_2$. From this reaction, the moles of $HC_2H_3O_2$ produced must equal the moles of H^+ added and the total concentration of acetate ion + acetic acid must equal $1.0\ M$ (assuming no volume change). Summarizing for each solution:

$$[C_2H_3O_2^-] + [HC_2H_3O_2] = 1.0\ M \text{ and } [HC_2H_3O_2] = [H^+] \text{ added}$$

a. $pH = pK_a + \log \dfrac{[C_2H_3O_2^-]}{[HC_2H_3O_2]}$; for $pH = pK_a$, $[C_2H_3O_2^-] = [HC_2H_3O_2]$.

For this to be true, $[C_2H_3O_2^-] = [HC_2H_3O_2] = 0.50\ M = [H^+]$ added, which means that 0.50 mol of HCl must be added to 1.0 L of the initial solution to produce a solution with $pH = pK_a$.

b. $4.20 = 4.74 + \log \dfrac{[C_2H_3O_2^-]}{[HC_2H_3O_2]}$, $\dfrac{[C_2H_3O_2^-]}{[HC_2H_3O_2]} = 10^{-0.54} = 0.29$

$[C_2H_3O_2^-] = 0.29[HC_2H_3O_2]$; $0.29[HC_2H_3O_2] + [HC_2H_3O_2] = 1.0\ M$

$[HC_2H_3O_2] = 0.78\ M = [H^+]$ added

0.78 mol of HCl must be added to produce a solution with $pH = 4.20$.

c. $5.00 = 4.74 + \log \dfrac{[C_2H_3O_2^-]}{[HC_2H_3O_2]}$, $\dfrac{[C_2H_3O_2^-]}{[HC_2H_3O_2]} = 10^{0.26} = 1.8$

$[C_2H_3O_2^-] = 1.8[HC_2H_3O_2]$; $1.8[HC_2H_3O_2] + [HC_2H_3O_2] = 1.0\ M$

$[HC_2H_3O_2] = 0.36\ M = [H^+]$ added

0.36 mol of HCl must be added to produce a solution with $pH = 5.00$.

37. $pH = pK_a + \log \dfrac{[C_2H_3O_2^-]}{[HC_2H_3O_2]}$; $pK_a = -\log(1.8 \times 10^{-5}) = 4.74$

Because the buffer components, $C_2H_3O_2^-$ and $HC_2H_3O_2$, are both in the same volume of water, the concentration ratio of $[C_2H_3O_2^-]/[HC_2H_3O_2]$ will equal the mole ratio of mol $C_2H_3O_2^-$/mol $HC_2H_3O_2$.

$5.00 = 4.74 + \log \dfrac{\text{mol } C_2H_3O_2^-}{\text{mol } HC_2H_3O_2}$; mol $HC_2H_3O_2 = 0.5000\ L \times \dfrac{0.200\ \text{mol}}{L} = 0.100\ \text{mol}$

$0.26 = \log \dfrac{\text{mol } C_2H_3O_2^-}{0.100\ \text{mol}}$, $\dfrac{\text{mol } C_2H_3O_2^-}{0.100\ \text{mol}} = 10^{0.26} = 1.8$, mol $C_2H_3O_2^- = 0.18\ \text{mol}$

Mass $NaC_2H_3O_2 = 0.18\ \text{mol } NaC_2H_3O_2 \times \dfrac{82.03\ g}{\text{mol}} = 15\ g\ NaC_2H_3O_2$

39. a. pK_b for $C_6H_5NH_2 = -\log(3.8 \times 10^{-10}) = 9.42$; pK_a for $C_6H_5NH_3^+ = 14.00 - 9.42 = 4.58$

$pH = pK_a + \log \dfrac{[C_6H_5NH_2]}{[C_6H_5NH_3^+]}$, $4.20 = 4.58 + \log \dfrac{0.50\ M}{[C_6H_5NH_3^+]}$

$-0.38 = \log \dfrac{0.50\ M}{[C_6H_5NH_3^+]}$, $[C_6H_5NH_3^+] = [C_6H_5NH_3Cl] = 1.2\ M$

b. $4.0 \text{ g NaOH} \times \dfrac{1 \text{ mol NaOH}}{40.00 \text{ g}} \times \dfrac{1 \text{ mol OH}^-}{\text{mol NaOH}} = 0.10 \text{ mol OH}^-$; $[\text{OH}^-] = \dfrac{0.10 \text{ mol}}{1.0 \text{ L}} = 0.10 \text{ M}$

$$\begin{array}{ccccccc} & C_6H_5NH_3^+ & + & OH^- & \rightarrow & C_6H_5NH_2 & + & H_2O \end{array}$$

Before	1.2 M	0.10 M	0.50 M
Change	-0.10	$-0.10 \;\rightarrow$	$+0.10$
After	1.1	0	0.60

A buffer solution exists. $\text{pH} = 4.58 + \log\left(\dfrac{0.60}{1.1}\right) = 4.32$

41. a. $\text{pH} = \text{p}K_a + \log\dfrac{[\text{base}]}{[\text{acid}]}$, $7.15 = -\log(6.2 \times 10^{-8}) + \log\dfrac{[\text{HPO}_4^{2-}]}{[\text{H}_2\text{PO}_4^-]}$

$7.15 = 7.21 + \log\dfrac{[\text{HPO}_4^{2-}]}{[\text{H}_2\text{PO}_4^-]}$, $\dfrac{[\text{HPO}_4^{2-}]}{[\text{H}_2\text{PO}_4^-]} = 10^{-0.06} = 0.9$, $\dfrac{[\text{H}_2\text{PO}_4^-]}{[\text{HPO}_4^{2-}]} = \dfrac{1}{0.9} = 1.1 \approx 1$

b. A best buffer has approximately equal concentrations of weak acid and conjugate base, so pH $\approx$ pK_a for a best buffer.   The pK_a value for a $H_3PO_4/H_2PO_4^-$ buffer is $-\log(7.5 \times 10^{-3}) = 2.12$. A pH of 7.15 is too high for a $H_3PO_4/H_2PO_4^-$ buffer to be effective. At this high of pH, there would be so little H_3PO_4 present that we could hardly consider it a buffer; this solution would not be effective in resisting pH changes, especially when a strong base is added.

43.   At pH = 7.40: $7.40 = -\log(4.3 \times 10^{-7}) + \log\dfrac{[\text{HCO}_3^-]}{[\text{H}_2\text{CO}_3]}$

$\log\dfrac{[\text{HCO}_3^-]}{[\text{H}_2\text{CO}_3]} = 7.40 - 6.37 = 1.03$, $\dfrac{[\text{HCO}_3^-]}{[\text{H}_2\text{CO}_3]} = 10^{1.03}$, $\dfrac{[\text{H}_2\text{CO}_3]}{[\text{HCO}_3^-]} = 10^{-1.03} = 0.093$

At pH = 7.35: $\log\dfrac{[\text{HCO}_3^-]}{[\text{H}_2\text{CO}_3]} = 7.35 - 6.37 = 0.98$, $\dfrac{[\text{HCO}_3^-]}{[\text{H}_2\text{CO}_3]} = 10^{0.98}$

$\dfrac{[\text{H}_2\text{CO}_3]}{[\text{HCO}_3^-]} = 10^{-0.98} = 0.10$

The $[\text{H}_2\text{CO}_3]$: $[\text{HCO}_3^-]$ concentration ratio must increase from 0.093 to 0.10 in order for the onset of acidosis to occur.

45. a. No; a solution of a strong acid (HNO_3) and its conjugate base (NO_3^-) is not generally considered a buffer solution.

b. No; two acids are present (HNO_3 and HF), so it is not a buffer solution.

c. H^+ reacts completely with F^-. Since equal volumes are mixed, the initial concentrations in the mixture are 0.10 M HNO_3 and 0.20 M NaF.

	H^+	$+$	F^-	$\rightarrow$	HF	
Before	0.10 M		0.20 M		0	
Change	-0.10		-0.10	$\rightarrow$	$+0.10$	Reacts completely
After	0		0.10		0.10	

After H^+ reacts completely, a buffer solution results; that is, a weak acid (HF) and its conjugate base (F^-) are both present in solution in large quantities.

d. No; a strong acid (HNO_3) and a strong base (NaOH) do not form buffer solutions. They will neutralize each other to form H_2O.

47. A best buffer has large and equal quantities of weak acid and conjugate base. Because [acid] = [base] for a best buffer, $pH = pK_a + \log \dfrac{[base]}{[acid]} = pK_a + 0 = pK_a$ ($pH \approx pK_a$ for a best buffer).

The best acid choice for a pH = 7.00 buffer would be the weak acid with a pK_a close to 7.0 or $K_a \approx 1 \times 10^{-7}$. HOCl is the best choice in Table 7.2 ($K_a = 3.5 \times 10^{-8}$; $pK_a = 7.46$). To make this buffer, we need to calculate the [base]/[acid] ratio.

$$7.00 = 7.46 + \log \frac{[base]}{[acid]}, \quad \frac{[OCl^-]}{[HOCl]} = 10^{-0.46} = 0.35$$

Any OCl^-/HOCl buffer in a concentration ratio of 0.35 : 1 will have a pH = 7.00. One possibility is [NaOCl] = 0.35 M and [HOCl] = 1.0 M.

49. To solve for [KOCl], we need to use the equation derived in Section 8.3 of the text on the exact treatment of buffered solutions. The equation is:

$$K_a = \frac{[H^+]\left([A^-]_0 + \dfrac{[H^+]^2 - K_w}{[H^+]}\right)}{[HA]_0 - \dfrac{[H^+]^2 - K_w}{[H^+]}}$$

Because pH = 7.20, $[H^+] = 10^{-7.20} = 6.3 \times 10^{-8}$ M.

$$K_a = 3.5 \times 10^{-8} = \frac{6.3 \times 10^{-8}\left([OCl^-] + \dfrac{(6.3 \times 10^{-8})^2 - (1.0 \times 10^{-14})}{6.3 \times 10^{-8}}\right)}{1.0 \times 10^{-6} - \dfrac{(6.3 \times 10^{-8})^2 - (1.0 \times 10^{-14})}{6.3 \times 10^{-8}}}$$

$$3.5 \times 10^{-8} = \frac{6.3 \times 10^{-8}([OCl^-] - 9.57 \times 10^{-8})}{(1.0 \times 10^{-6}) + (9.57 \times 10^{-8})} \quad \text{(Carrying extra significant figures.)}$$

$$3.83 \times 10^{-14} = 6.3 \times 10^{-8}([OCl^-] - 9.57 \times 10^{-8}), \quad [OCl^-] = [KOCl] = 7.0 \times 10^{-7} \, M$$

51.   Using regular procedures, $pH = pK_a = -\log(1.6 \times 10^{-7}) = 6.80$ since $[A^-]_0 = [HA]_0$ in this buffer solution. However, the pH is very close to that of neutral water, so maybe we need to consider the H^+ contribution from water. Another problem with this answer is that $x (= [H^+])$ is not small as compared with $[HA]_0$ and $[A^-]_0$, which was assumed when solving using the regular procedures. Because the concentrations of the buffer components are less than 10^{-6} M, let us use the expression for the exact treatment of buffers to solve.

$$K_a = 1.6 \times 10^{-7} = \frac{[H^+]\left([A^-]_0 + \dfrac{[H^+]^2 - K_w}{[H^+]}\right)}{[HA]_0 - \dfrac{[H^+]^2 - K_w}{[H^+]}} =$$

$$\frac{[H^+]\left(5.0 \times 10^{-7} + \dfrac{[H^+]^2 - (1.0 \times 10^{-14})}{[H^+]}\right)}{5.0 \times 10^{-7} - \dfrac{[H^+]^2 - (1.0 \times 10^{-14})}{[H^+]}}$$

Solving exactly requires solving a cubic equation. Instead, we will use the method of successive approximations where our initial guess for $[H^+] = 1.6 \times 10^{-7} \, M$ (the value obtained using the regular procedures).

$$1.6 \times 10^{-7} = \frac{[H^+]\left(5.0 \times 10^{-7} + \dfrac{(1.6 \times 10^{-7})^2 - (1.0 \times 10^{-14})}{1.6 \times 10^{-7}}\right)}{5.0 \times 10^{-7} - \dfrac{(1.6 \times 10^{-7})^2 - (1.0 \times 10^{-14})}{1.6 \times 10^{-7}}}, \quad [H^+] = 1.1 \times 10^{-7}$$

We continue the process using 1.1×10^{-7} as our estimate for $[H^+]$. This gives $[H^+] = 1.5 \times 10^{-7}$. We continue the process until we get a self consistent answer. After three more iterations, we converge on $[H^+] = 1.3 \times 10^{-7} \, M$. Solving for the pH:

$$pH = -\log(1.3 \times 10^{-7}) = 6.89$$

Note that if we were to solve this problem exactly (using the quadratic formula) while ignoring the H^+ contribution from water, the answer comes out to $[H^+] = 1.0 \times 10^{-7} \, M$. We get a significantly different answer when we consider the H^+ contribution from H_2O.

Acid-Base Titrations

53.   a.   Let's call the acid HB, which is a weak acid. When HB is present in the beakers, it exists in the undissociated form, making it a weak acid. A strong acid would exist as separate H^+ and B^- ions.

   b.   Beaker a contains 4 HB molecules and 2 B^- ions, beaker b contains 6 B^- ions, beaker c contains 6 HB molecules, beaker d contains 6 B^- and 6 OH^- ions, and beaker e contains 3 HB molecules and 3 B^- ions. $HB + OH^- \rightarrow B^- + H_2O$; this is the neutralization reaction that occurs when OH^- is added. We start off the titration with a beaker full of weak acid (beaker c). When some OH^- is added, we convert some weak acid HB into its conjugate base B^- (beaker a). At the halfway point to equivalence, we have converted exactly one-half of the initial amount of acid present into its conjugate base (beaker e). We finally reach the equivalence point when we have added just enough OH^- to convert all of the acid present initially into its conjugate base (beaker b). Past the equivalence point, we have added an excess of OH^-, so we have excess OH^- present as well as the conjugate base of the acid produced from the neutralization reaction (beaker d). The order of the beakers from start to finish is:

   beaker c → beaker a → beaker e → beaker b → beaker d

   c. $pH = pK_a$ when a buffer solution is present that has equal concentrations of the weak acid and conjugate base. This is beaker e.

   d.   The equivalence point is when just enough OH^- has been added to exactly react with all of the acid present initially. This is beaker b.

   e.   Past the equivalence, the pH is dictated by the concentration of excess OH^- added from the strong base. We can ignore the amount of hydroxide added by the weak conjugate base that is also present. This is beaker d.

55.

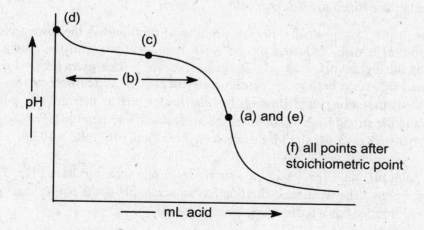

$B + H^+ \rightarrow BH^+$; added H^+ from the strong acid converts the weak base B into its conjugate acid BH^+. Initially, before any H^+ is added (point d), B is the dominant species present. After H^+ is added, both B and BH^+ are present, and a buffered solution results (region b). At the equivalence point (points a and e), exactly enough H^+ has been added to convert all the weak base present initially into its conjugate acid BH^+. Past the equivalence point (region f), excess H^+ is present. For the answer to b, we included almost the entire buffer region. The maximum buffer region is around the halfway point to equivalence (point c), where $[B] = [BH^+]$. Here, $pH = pK_a$, which is a characteristic of a best buffer.

57. Titration i is a strong acid titrated by a strong base. The pH is very acidic until just before the equivalence point; at the equivalence point, $pH = 7.00$; and past the equivalence the pH is very basic. Titration ii is a strong base titrated by a strong acid. Here the pH is very basic until just before the equivalence point; at the equivalence point, $pH = 7.00$; and past the equivalence point, the pH is very acidic. Titration iii is a weak base titrated by a strong acid. The pH starts out basic because a weak base is present. However, the pH will not be as basic as in titration ii, where a strong base is titrated. The pH drops as HCl is added; then at the halfway point to equivalence, $pH = pK_a$. Because $K_b = 4.4 \times 10^{-4}$ for CH_3NH_2, $CH_3NH_3^+$ has $K_a = K_w/K_b = 2.3 \times 10^{-11}$ and $pK_a = 10.64$. So, at the halfway point to equivalence for this weak base-strong acid titration, $pH = 10.64$. The pH continues to drop as HCl is added; then at the equivalence point the pH is acidic ($pH < 7.00$) because the only important major species present is a weak acid (the conjugate acid of the weak base). Past the equivalence point the pH becomes more acidic as excess HCl is added. Titration iv is a weak acid titrated by a strong base. The pH starts off acidic, but not nearly as acidic as the strong acid titration (i). The pH increases as NaOH is added; then, at the halfway point to equivalence, $pH = pK_a$ for $HF = -\log(7.2 \times 10^{-4}) = 3.14$. The pH continues to increase past the halfway point; then at the equivalence point, the pH is basic ($pH > 7.0$) because the only important major species present is a weak base (the conjugate base of the weak acid). Past the equivalence point, the pH becomes more basic as excess NaOH is added.

a. All require the same volume of titrant to reach the equivalence point. At the equivalence point for all these titrations, moles acid = moles base ($M_AV_A = M_BV_B$). Because all the molarities and volumes are the same in the titrations, the volume of titrant will be the same (50.0 mL titrant added to reach equivalence point).

b. Increasing initial pH: i < iv < iii < ii; the strong acid titration has the lowest pH, the weak acid titration is next, followed by the weak base titration, with the strong base titration having the highest pH.

c. i < iv < iii < ii; the strong acid titration has the lowest pH at the halfway point to equivalence, and the strong base titration has the highest halfway point pH. For the weak acid titration, $pH = pK_a = 3.14$, and for the weak base titration, $pH = pK_a = 10.64$.

d. Equivalence point pH: iii < ii = i < iv; the strong-by-strong titrations have $pH = 7.00$ at the equivalence point. The weak base titration has an acidic pH at the equivalence point, and a weak acid titration has a basic equivalence point pH.

The only different answer when the weak acid and weak base are changed would be for part c. This is for the halfway point to equivalence, where $pH = pK_a$.

$$HOC_6H_5; \quad K_a = 1.6 \times 10^{-10}, \quad pK_a = -\log(1.6 \times 10^{-10}) = 9.80$$

$$C_5H_5NH^+, \ K_a = \frac{K_w}{K_{b, C_5H_5N}} = \frac{1.0 \times 10^{-14}}{1.7 \times 10^{-9}} = 5.9 \times 10^{-6}, \ pK_a = 5.23$$

From the pK_a values, the correct ordering at the halfway point to equivalence would be i $<$ iii $<$ iv $<$ ii. Note that for the weak base-strong acid titration using C_5H_5N, the pH is acidic at the halfway point to equivalence, whereas the weak acid-strong base titration using HOC_6H_5 is basic at the halfway point to equivalence. This is fine; this will always happen when the weak base titrated has a $K_b < 1 \times 10^{-7}$ (so K_a of the conjugate acid is greater than 1×10^{-7}) and when the weak acid titrated has a $K_a < 1 \times 10^{-7}$ (so K_b of the conjugate base is greater than 1×10^{-7}).

59.   a.   Because all acids are the same initial concentration, the pH curve with the highest pH at 0 mL of NaOH added will correspond to the titration of the weakest acid. This is curve f.

b.   The pH curve with the lowest pH at 0 mL of NaOH added will correspond to the titration of the strongest acid. This is pH curve a.

The best point to look at to differentiate a strong acid from a weak acid titration (if initial concentrations are not known) is the equivalence point pH. If the pH = 7.00, the acid titrated is a strong acid; if the pH is greater than 7.00, the acid titrated is a weak acid.

c.   For a weak acid-strong base titration, the pH at the halfway point to equivalence is equal to the pK_a value. The pH curve, which represents the titration of an acid with $K_a = 1.0 \times 10^{-6}$, will have a pH $= -\log(1 \times 10^{-6}) = 6.0$ at the halfway point. The equivalence point, from the plots, occurs at 50 mL NaOH added, so the halfway point is 25 mL. Plot d has a pH ≈ 6.0 at 25 mL of NaOH added, so the acid titrated in this pH curve (plot d) has $K_a \approx 1 \times 10^{-6}$.

61.   This is a strong acid ($HClO_4$) titrated by a strong base (KOH). Added OH^- from the strong base will react completely with the H^+ present from the strong acid to produce H_2O.

a.   Only strong acid present. $[H^+] = 0.200 \ M$; pH = 0.699

b.   mmol OH^- added $= 10.0 \ mL \times \dfrac{0.100 \ mmol \ OH^-}{mL} = 1.00 \ mmol \ OH^-$

mmol H^+ present $= 40.0 \ mL \times \dfrac{0.200 \ mmol \ H^+}{mL} = 8.0 \ mmol \ H^+$

Note: The units millimoles are usually easier numbers to work with. The units for molarity are moles per liter but are also equal to millimoles per milliliter.

	H^+	$+$	OH^-	$\rightarrow$	H_2O	
Before	8.00 mmol		1.00 mmol			
Change	−1.00 mmol		−1.00 mmol			Reacts completely
After	7.00 mmol		0			

The excess H^+ determines the pH. $[H^+]_{excess} = \dfrac{7.00 \text{ mmol } H^+}{40.0 \text{ mL} + 10.0 \text{ mL}} = 0.140 \ M$

$pH = -\log(0.140) = 0.854$

c.   mmol OH^- added $= 40.0$ mL $\times 0.100 \ M = 4.00$ mmol OH^-

$$H^+ \quad + \quad OH^- \quad \rightarrow \quad H_2O$$

Before	8.00 mmol	4.00 mmol
After	4.00 mmol	0

$[H^+]_{excess} = \dfrac{4.00 \text{ mmol}}{(40.0 + 40.0) \text{ mL}} = 0.0500 \ M; \ pH = 1.301$

d.   mmol OH^- added $= 80.0$ mL $\times 0.100 \ M = 8.00$ mmol OH^-; this is the equivalence point because we have added just enough OH^- to react with all the acid present. For a strong acid-strong base titration, pH $= 7.00$ at the equivalence point because only neutral species are present (K^+, ClO_4^-, H_2O).

e.   mmol OH^- added $= 100.0$ mL $\times 0.100 \ M = 10.0$ mmol OH^-

$$H^+ \quad + \quad OH^- \quad \rightarrow \quad H_2O$$

Before	8.00 mmol	10.0 mmol
After	0	2.0 mmol

Past the equivalence point, the pH is determined by the excess OH^- present.

$[OH^-]_{excess} = \dfrac{2.0 \text{ mmol}}{(40.0 + 100.0) \text{ mL}} = 0.014 \ M; \ pOH = 1.85; \ pH = 12.15$

63.   This is a weak acid ($HC_2H_3O_2$) titrated by a strong base (KOH).

a.   Only weak acid is present. Solving the weak acid problem:

$$HC_2H_3O_2 \quad \rightleftharpoons \quad H^+ \quad + \quad C_2H_3O_2^-$$

Initial	0.200 M	~0	0
	x mol/L $HC_2H_3O_2$ dissociates to reach equilibrium		
Change	$-x$ $\rightarrow$	$+x$	$+x$
Equil.	$0.200 - x$	x	x

$K_a = 1.8 \times 10^{-5} = \dfrac{x^2}{0.200 - x} \approx \dfrac{x^2}{0.200}, \ x = [H^+] = 1.9 \times 10^{-3} \ M$

pH $= 2.72$; assumptions good.

b. The added OH⁻ will react completely with the best acid present, $HC_2H_3O_2$.

$$\text{mmol } HC_2H_3O_2 \text{ present} = 100.0 \text{ mL} \times \frac{0.200 \text{ mmol } HC_2H_3O_2}{\text{mL}} = 20.0 \text{ mmol } HC_2H_3O_2$$

$$\text{mmol } OH^- \text{ added} = 50.0 \text{ mL} \times \frac{0.100 \text{ mmol } OH^-}{\text{mL}} = 5.00 \text{ mmol } OH^-$$

	$HC_2H_3O_2$	+	OH^-	$\rightarrow$	$C_2H_3O_2^-$	+	H_2O	
Before	20.0 mmol		5.00 mmol		0			
Change	−5.00 mmol		−5.00 mmol	$\rightarrow$	+5.00 mmol			Reacts completely
After	15.0 mmol		0		5.00 mmol			

After reaction of all the strong base, we have a buffer solution containing a weak acid ($HC_2H_3O_2$) and its conjugate base ($C_2H_3O_2^-$). We will use the Henderson-Hasselbalch equation to solve for the pH.

$$pH = pK_a + \log \frac{[C_2H_3O_2^-]}{[HC_2H_3O_2]} = -\log (1.8 \times 10^{-5}) + \log \left(\frac{5.00 \text{ mmol}/V_T}{15.0 \text{ mmol}/V_T} \right), \text{ where } V_T = \text{total volume}$$

$$pH = 4.74 + \log \left(\frac{5.00}{15.0} \right) = 4.74 + (-0.477) = 4.26$$

Note that the total volume cancels in the Henderson-Hasselbalch equation. For the [base]/[acid] term, the mole ratio equals the concentration ratio because the components of the buffer are always in the same volume of solution.

c. mmol OH⁻ added = 100.0 mL × (0.100 mmol OH⁻/mL) = 10.0 mmol OH⁻; the same amount (20.0 mmol) of $HC_2H_3O_2$ is present as before (it doesn't change). As before, let the OH⁻ react to completion, then see what is remaining in solution after this reaction.

	$HC_2H_3O_2$	+	OH^-	$\rightarrow$	$C_2H_3O_2^-$	+	H_2O
Before	20.0 mmol		10.0 mmol		0		
After	10.0 mmol		0		10.0 mmol		

A buffer solution results after reaction. Because $[C_2H_3O_2^-] = [HC_2H_3O_2] = 10.0$ mmol/total volume, $pH = pK_a$. This is always true at the halfway point to equivalence for a weak acid-strong base titration, $pH = pK_a$.

$$pH = -\log(1.8 \times 10^{-5}) = 4.74$$

d. mmol OH⁻ added = 150.0 mL × 0.100 M = 15.0 mmol OH⁻. Added OH⁻ reacts completely with the weak acid.

	$HC_2H_3O_2$	+	OH^-	$\rightarrow$	$C_2H_3O_2^-$	+	H_2O
Before	20.0 mmol		15.0 mmol		0		
After	5.0 mmol		0		15.0 mmol		

We have a buffer solution after all the OH⁻ reacts to completion. Using the Henderson-Hasselbalch equation:

$$pH = 4.74 + \log \frac{[C_2H_3O_2^-]}{[HC_2H_3O_2]} = 4.74 + \log \left(\frac{15.0 \text{ mmol}}{5.0 \text{ mmol}} \right)$$

$$pH = 4.74 + 0.48 = 5.22$$

e. mmol OH⁻ added = 200.00 mL × 0.100 M = 20.0 mmol OH⁻; as before, let the added OH⁻ react to completion with the weak acid; then see what is in solution after this reaction.

$$HC_2H_3O_2 \quad + \quad OH^- \quad \rightarrow \quad C_2H_3O_2^- \quad + \quad H_2O$$

Before	20.0 mmol	20.0 mmol	0
After	0	0	20.0 mmol

This is the equivalence point. Enough OH⁻ has been added to exactly neutralize all the weak acid present initially. All that remains that affects the pH at the equivalence point is the conjugate base of the weak acid ($C_2H_3O_2^-$). This is a weak base equilibrium problem.

$$C_2H_3O_2^- + H_2O \rightleftharpoons HC_2H_3O_2 + OH^- \qquad K_b = \frac{K_w}{K_b} = \frac{1.0 \times 10^{-14}}{1.8 \times 10^{-5}}$$

Initial   20.0 mmol/300.0 mL   0   0 $K_b = 5.6 \times 10^{-9}$
          x mol/L $C_2H_3O_2^-$ reacts with H_2O to reach equilibrium

Change	$-x$	$\rightarrow$ $+x$	$+x$
Equil.	$0.0667 - x$	x	x

$$K_b = 5.6 \times 10^{-10} = \frac{x^2}{0.0667 - x} \approx \frac{x^2}{0.0667} \, , \; x = [OH^-] = 6.1 \times 10^{-6} \, M$$

pOH = 5.21; pH = 8.79; assumptions good.

f. mmol OH⁻ added = 250.0 mL × 0.100 M = 25.0 mmol OH⁻

$$HC_2H_3O_2 \quad + \quad OH^- \quad \rightarrow \quad C_2H_3O_2^- \quad + \quad H_2O$$

Before	20.0 mmol	25.0 mmol	0
After	0	5.0 mmol	20.0 mmol

After the titration reaction, we have a solution containing excess OH⁻ and a weak base $C_2H_3O_2^-$. When a strong base and a weak base are both present, assume that the amount of OH⁻ added from the weak base will be minimal; that is, the pH past the equivalence point is determined by the amount of excess strong base.

$$[OH^-]_{excess} = \frac{5.0 \text{ mmol}}{100.0 \text{ mL} + 250.0 \text{ mL}} = 0.014 \, M; \; pOH = 1.85; \; pH = 12.15$$

65. We will do sample calculations for the various parts of the titration. All results are summarized in Table 8.1 at the end of Exercise 67.

At the beginning of the titration, only the weak acid $HC_3H_5O_3$ is present. Let HLac = $HC_3H_5O_3$ and $Lac^- = C_3H_5O_3^-$.

$$HLac \rightleftharpoons H^+ + Lac^- \quad K_a = 10^{-3.86} = 1.4 \times 10^{-4}$$

	HLac	H$^+$	Lac$^-$
Initial	0.100 M	~0	0

x mol/L HLac dissociates to reach equilibrium

Change	$-x$ $\rightarrow$	$+x$	$+x$
Equil.	$0.100 - x$	x	x

$$1.4 \times 10^{-4} = \frac{x^2}{0.100 - x} \approx \frac{x^2}{0.100}, \quad x = [H^+] = 3.7 \times 10^{-3} \, M; \quad pH = 2.43; \quad \text{assumptions good.}$$

Up to the stoichiometric point, we calculate the pH using the Henderson-Hasselbalch equation. This is the buffer region. For example, at 4.0 mL of NaOH added:

$$\text{initial mmol HLac present} = 25.0 \text{ mL} \times \frac{0.100 \text{ mmol}}{\text{mL}} = 2.50 \text{ mmol HLac}$$

$$\text{mmol OH}^- \text{ added} = 4.0 \text{ mL} \times \frac{0.100 \text{ mmol}}{\text{mL}} = 0.40 \text{ mmol OH}^-$$

Note: The units millimoles are usually easier numbers to work with. The units for molarity are moles per liter but are also equal to millimoles per milliliter.

The 0.40 mmol of added OH$^-$ converts 0.40 mmol HLac to 0.40 mmol Lac$^-$ according to the equation:

$$HLac + OH^- \rightarrow Lac^- + H_2O \qquad \text{Reacts completely since a strong base is added.}$$

mmol HLac remaining = 2.50 – 0.40 = 2.10 mmol; mmol Lac$^-$ produced = 0.40 mmol

We have a buffer solution. Using the Henderson-Hasselbalch equation where pK$_a$ = 3.86:

$$pH = pK_a + \log \frac{[\text{Lac}^-]}{[\text{HLac}]} = 3.86 + \log \frac{(0.40)}{(2.10)} \qquad \text{(Total volume cancels, so we can use use the ratio of moles or millimoles.)}$$

$$pH = 3.86 - 0.72 = 3.14$$

Other points in the buffer region are calculated in a similar fashion. Perform a stoichiometry problem first, followed by a buffer problem. The buffer region includes all points up to and including 24.9 mL OH$^-$ added.

At the stoichiometric point (25.0 mL OH⁻ added), we have added enough OH⁻ to convert all of the HLac (2.50 mmol) into its conjugate base (Lac⁻). All that is present is a weak base. To determine the pH, we perform a weak base calculation.

$$[\text{Lac}^-]_0 = \frac{2.50 \text{ mmol}}{25.0 \text{ mL} + 25.0 \text{ mL}} = 0.0500 \ M$$

$$\text{Lac}^- + \text{H}_2\text{O} \rightleftharpoons \text{HLac} + \text{OH}^- \qquad K_b = \frac{1.0 \times 10^{-14}}{1.4 \times 10^{-4}} = 7.1 \times 10^{-11}$$

Initial	0.0500 M	0	0
	x mol/L Lac⁻ reacts with H₂O to reach equilibrium		
Change	$-x$	$\rightarrow$ $+x$	$+x$
Equil.	0.0500 $-x$	x	x

$$K_b = \frac{x^2}{0.0500 - x} \approx \frac{x^2}{0.0500} = 7.1 \times 10^{-11}$$

$x = [\text{OH}^-] = 1.9 \times 10^{-6} \ M$; pOH = 5.72; pH = 8.28; assumptions good.

Past the stoichiometric point, we have added more than 2.50 mmol of NaOH. The pH will be determined by the excess OH⁻ ion present. An example of this calculation follows.

At 25.1 mL: OH⁻ added = 25.1 mL $\times \dfrac{0.100 \text{ mmol}}{\text{mL}} = 2.51$ mmol OH⁻

2.50 mmol OH⁻ neutralizes all the weak acid present. The remainder is excess OH⁻.

Excess OH⁻ = 2.51 − 2.50 = 0.01 mmol OH⁻

$$[\text{OH}^-]_{excess} = \frac{0.01 \text{ mmol}}{(25.0 + 25.1) \text{ mL}} = 2 \times 10^{-4} \ M; \quad \text{pOH} = 3.7; \quad \text{pH} = 10.3$$

All results are listed in Table 8.1 at the end of the solution to Exercise 67.

67.   At beginning of the titration, only the weak base NH₃ is present. As always, solve for the pH using the K_b reaction for NH₃.

$$\text{NH}_3 + \text{H}_2\text{O} \rightleftharpoons \text{NH}_4^+ + \text{OH}^- \qquad K_b = 1.8 \times 10^{-5}$$

Initial	0.100 M	0	~0
Equil.	0.100 $-x$	x	x

$$K_b = \frac{x^2}{0.100 - x} \approx \frac{x^2}{0.100} = 1.8 \times 10^{-5}$$

$x = [\text{OH}^-] = 1.3 \times 10^{-3} \ M$; pOH = 2.89; pH = 11.11; assumptions good.

In the buffer region (4.0 – 24.9 mL), we can use the Henderson-Hasselbalch equation:

$$K_a = \frac{1.0 \times 10^{-14}}{1.8 \times 10^{-5}} = 5.6 \times 10^{-10}; \quad pK_a = 9.25; \quad pH = 9.25 + \log\frac{[NH_3]}{[NH_4^+]}$$

We must determine the amounts of NH_3 and NH_4^+ present after the added H^+ reacts completely with the NH_3. For example, after 8.0 mL HCl added:

$$\text{initial mmol } NH_3 \text{ present} = 25.0 \text{ mL} \times \frac{0.100 \text{ mmol}}{mL} = 2.50 \text{ mmol } NH_3$$

$$\text{mmol } H^+ \text{ added} = 8.0 \text{ mL} \times \frac{0.100 \text{ mmol}}{mL} = 0.80 \text{ mmol } H^+$$

Added H^+ reacts with NH_3 to completion: $NH_3 + H^+ \rightarrow NH_4^+$

mmol NH_3 remaining = 2.50 – 0.80 = 1.70 mmol; mmol NH_4^+ produced = 0.80 mmol

$$pH = 9.25 + \log\frac{1.70}{0.80} = 9.58 \text{(Mole ratios can be used since the total volume cancels.)}$$

Other points in the buffer region are calculated in similar fashion. Results are summarized in Table 8.1 on the next page.

At the stoichiometric point (25.0 mL H^+ added), just enough HCl has been added to convert all the weak base (NH_3) into its conjugate acid (NH_4^+). Perform a weak acid calculation.

$[NH_4^+]_0$ = 2.50 mmol/50.0 mL = 0.0500 M

$$\begin{array}{ccccccc} & NH_4^+ & \rightleftharpoons & H^+ & + & NH_3 & \quad K_a = 5.6 \times 10^{-10} \end{array}$$

Initial   0.0500 M 0   0
Equil.   0.0500 - x x x

$$5.6 \times 10^{-10} = \frac{x^2}{0.0500 - x} \approx \frac{x^2}{0.0500}, \quad x = [H^+] = 5.3 \times 10^{-6} \ M; \quad pH = 5.28; \quad \text{assumptions}$$
$$\text{good.}$$

Beyond the stoichiometric point, the pH is determined by the excess H^+. For example, at 28.0 mL of H^+ added:

$$H^+ \text{ added} = 28.0 \text{ mL} \times \frac{0.100 \text{ mmol}}{mL} = 2.80 \text{ mmol } H^+$$

Excess H^+ = 2.80 mmol – 2.50 mmol = 0.30 mmol excess H^+

$$[H^+]_{excess} = \frac{0.30 \text{ mmol}}{(25.0 + 28.0) \text{ mL}} = 5.7 \times 10^{-3} \ M; \quad pH = 2.24$$

Table 8.1 Summary of pH Results for Exercises 65 and 67   (Graph follows)

Titrant mL	Exercise 65	Exercise 67
0.0	2.43	11.11
4.0	3.14	9.97
8.0	3.53	9.58
12.5	3.86	9.25
20.0	4.46	8.65
24.0	5.24	7.87
24.5	5.6	7.6
24.9	6.3	6.9
25.0	8.28	5.28
25.1	10.3	3.7
26.0	11.30	2.71
28.0	11.75	2.24
30.0	11.96	2.04

Note: The following figure includes the pH curves for Exercises 66 and 68 in addition to the pH curves for Exercises 65 and 67. The solid circles are the data points for Exercises 65 and 67, and the open circles are the data points for Exercises 66 and 68.

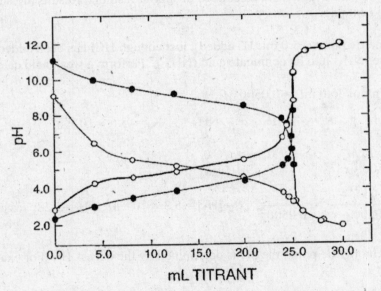

69.   a.   This is a weak acid-strong base titration. At the halfway point to equivalence, [weak acid] = [conjugate base], so pH = pK$_a$ (always for a weak acid-strong base titration).

pH = $-\log(6.4 \times 10^{-5})$ = 4.19

mmol HC$_7$H$_5$O$_2$ present = 100.0 mL $\times$ 0.10 M = 10. mmol HC$_7$H$_5$O$_2$. For the equivalence point, 10. mmol of OH$^-$ must be added. The volume of OH$^-$ added to reach the equivalence point is:

$$10. \text{ mmol OH}^- \times \frac{1 \text{ mL}}{0.10 \text{ mmol OH}^-} = 1.0 \times 10^2 \text{ mL OH}^-$$

At the equivalence point, 10. mmol of $HC_7H_5O_2$ is neutralized by 10. mmol of OH^- to produce 10. mmol of $C_7H_5O_2^-$. This is a weak base. The total volume of the solution is 100.0 mL + 1.0×10^2 mL = 2.0×10^2 mL. Solving the weak base equilibrium problem:

$$C_7H_5O_2^- + H_2O \rightleftharpoons HC_7H_5O_2 + OH^- \quad K_b = \frac{1.0 \times 10^{-14}}{6.4 \times 10^{-5}} = 1.6 \times 10^{-10}$$

Initial   10. mmol/2.0×10^2 mL   0   0
Equil. $0.050 - x$ x x

$$K_b = 1.6 \times 10^{-10} = \frac{x^2}{0.050 - x} \approx \frac{x^2}{0.050}, \quad x = [OH^-] = 2.8 \times 10^{-6} \, M$$

pOH = 5.55; pH = 8.45; assumptions good.

b. At the halfway point to equivalence for a weak base-strong acid titration, pH = pK_a because [weak base] = [conjugate acid].

$$K_a = \frac{K_w}{K_b} = \frac{1.0 \times 10^{-14}}{5.6 \times 10^{-4}} = 1.8 \times 10^{-11}; \quad pH = pK_a = -\log(1.8 \times 10^{-11}) = 10.74$$

For the equivalence point (mmol acid added = mmol base present):

mmol $C_2H_5NH_2$ present = 100.0 mL $\times$ 0.10 M = 10. mmol $C_2H_5NH_2$

$$\text{mL H}^+ \text{ added} = 10. \text{ mmol H}^+ \times \frac{1 \text{ mL}}{0.20 \text{ mmol H}^+} = 50. \text{ mL H}^+$$

The strong acid added completely converts the weak base into its conjugate acid. Therefore, at the equivalence point, $[C_2H_5NH_3^+]_0$ = 10. mmol/(100.0 + 50.) mL = 0.067 M. Solving the weak acid equilibrium problem:

$$C_2H_5NH_3^+ \rightleftharpoons H^+ + C_2H_5NH_2$$

Initial   0.067 M 0   0
Equil. $0.067 - x$ x x

$$K_a = 1.8 \times 10^{-11} = \frac{x^2}{0.067 - x} \approx \frac{x^2}{0.067}, \quad x = [H^+] = 1.1 \times 10^{-6} \, M$$

pH = 5.96; assumptions good.

c. In a strong acid-strong base titration, the halfway point has no special significance other than that exactly one-half of the original amount of acid present has been neutralized.

mmol H^+ present = 100.0 mL $\times$ 0.50 M = 50. mmol H^+

$$\text{mL OH}^- \text{ added} = 25 \text{ mmol OH}^- \times \frac{1\,\text{mL}}{0.25\,\text{mmol}} = 1.0 \times 10^2 \text{ mL OH}^-$$

$$\begin{array}{cccc} & \text{H}^+ & + & \text{OH}^- & \rightarrow & \text{H}_2\text{O} \end{array}$$

Before	50. mmol	25 mmol
After	25 mmol	0

$$[\text{H}^+]_{\text{excess}} = \frac{25\,\text{mmol}}{(100.0 + 1.0 \times 10^2)\,\text{mL}} = 0.13\,M; \quad \text{pH} = 0.89$$

At the equivalence point of a strong acid-strong base titration, only neutral species are present (Na^+, Cl^-, and H_2O), so the pH = 7.00.

71. $\text{Mol H}^+ \text{ added} = 0.0400 \text{ L} \times 0.100 \text{ mol/L} = 0.00400 \text{ mol H}^+$

The added strong acid reacts to completion with the weak base to form the conjugate acid of the weak base and H_2O. Let B = weak base:

$$\begin{array}{ccccc} \text{B} & + & \text{H}^+ & \rightarrow & \text{BH}^+ \end{array}$$

Before	0.0100 mol	0.00400 mol	0
After	0.0060	0	0.0400 mol

After the H^+ reacts to completion, we have a buffer solution. Using the Henderson-Hasselbalch equation:

$$\text{pH} = \text{p}K_a + \log\frac{[\text{base}]}{[\text{acid}]}, \quad 8.00 = \text{p}K_a + \log\frac{(0.0060/\text{V}_\text{T})}{(0.00400/\text{V}_\text{T})}, \quad \text{where V}_\text{T} = \text{total volume of solution}$$

$$\text{p}K_a = 8.00 - \log\frac{(0.0060)}{(0.00400)} = 8.00 - 0.18, \quad \text{p}K_a = 7.82$$

For a conjugate acid-base pair, $\text{p}K_a + \text{p}K_b = 14,00$, so:

$$\text{p}K_b = 14.00 - 7.82 = 6.18; \quad K_b = 10^{-6.18} = 6.6 \times 10^{-7}$$

Indicators

73. $\text{HIn} \rightleftharpoons \text{In}^- + \text{H}^+ \quad K_a = \dfrac{[\text{In}^-][\text{H}^+]}{[\text{HIn}]} = 1.0 \times 10^{-9}$

a.   In a very acid solution, the HIn form dominates, so the solution will be yellow.

b.   The color change occurs when the concentration of the more dominant form is approximately ten times as great as the less dominant form of the indicator.

$$\frac{[\text{HIn}]}{[\text{In}^-]} = \frac{10}{1}; \quad K_a = 1.0 \times 10^{-9} = \left(\frac{1}{10}\right)[\text{H}^+], \quad [\text{H}^+] = 1 \times 10^{-8}\,M; \quad \text{pH} = 8.0 \text{ at color change}$$

    c.   This is way past the equivalence point (100.0 mL OH⁻ added), so the solution is very basic and the In⁻ form of the indicator dominates. The solution will be blue.

75.   pH > 5 for bromcresol green to be blue. pH < 8 for thymol blue to be yellow. The pH is between 5 and 8.

77.   When choosing an indicator, we want the color change of the indicator to occur approximately at the pH of the equivalence point. Since the pH generally changes very rapidly at the equivalence point, we don't have to be exact. This is especially true for strong acid-strong base titrations. The following are some indicators where the color change occurs at about the pH of the equivalence point.

Exercise	pH at Eq. Pt.	Indicator
61	7.00	bromthymol blue or phenol red
63	8.79	o-cresolphthalein or phenolphthalein

79.

Exercise	pH at Eq. Pt.	Indicator
65	8.28	o-cresolphthalein or phenolphthalein
67	5.28	bromcresol green

81.   The color of the indicator will change over the approximate range of pH = $pK_a \pm 1 = 5.3 \pm 1$. Therefore, the useful pH range of methyl red where it changes color would be about 4.3 (red) to 6.3 (yellow). Note that at pH < 4.3, the HIn form of the indicator dominates, and the color of the solution is the color of HIn (red). At pH > 6.3, the In⁻ form of the indicator dominates, and the color of the solution is the color of In⁻ (yellow). In titrating a weak acid with base, we start off with an acidic solution with pH < 4.3, so the color would change from red to reddish orange at pH ≈ 4.3. In titrating a weak base with acid, the color change would be from yellow to yellowish orange at pH ≈ 6.3. Only a weak base-strong acid titration would have an acidic pH at the equivalence point, so only in this type of titration would the color change of methyl red indicate the approximate endpoint.

Polyprotic Acid Titrations

83.   The first titration plot (from 0 – 100.0 mL) corresponds to the titration of H_2A by OH⁻. The reaction is $H_2A + OH^- \rightarrow HA^- + H_2O$. After all the H_2A has been reacted, the second titration (from 100.0 – 200.0 mL) corresponds to the titration of HA⁻ by OH⁻. The reaction is $HA^- + OH^- \rightarrow A^{2-} + H_2O$.

    a.   At 100.0 mL of NaOH, just enough OH⁻ has been added to react completely with all of the H_2A present (mol OH⁻ added = mol H_2A present initially). From the balanced equation, the mol of HA⁻ produced will equal the mol of H_2A present initially. Because mol of HA⁻ present at 100.0 mL OH⁻ added equals the mol of H_2A present initially, exactly 100.0 mL more of NaOH must be added to react with all of the HA⁻. The volume of NaOH added to reach the second equivalence point equals 100.0 mL + 100.0 mL = 200.0 mL.

b. $H_2A + OH^- \rightarrow HA^- + H_2O$ is the reaction occurring from 0 – 100.0 mL NaOH added.

   i.   No reaction has taken place, so H_2A and H_2O are the major species.

   ii. Adding OH^- converts H_2A into HA^-. The major species between 0 mL and 100.0 mL NaOH added are H_2A, HA^-, H_2O, and Na^+.

   iii. At 100.0 mL NaOH added, mol of OH^- = mol H_2A, so all of the H_2A present initially has been converted into HA^-. The major species are HA^-, H_2O, and Na^+.

   iv. Between 100.0 and 200.0 mL NaOH added, the OH^- converts HA^- into A^{2-}. The major species are HA^-, A^{2-}, H_2O, and Na^+.

   v.   At the second equivalence point (200.0 mL), just enough OH^- has been added to convert all of the HA^- into A^{2-}. The major species are A^{2-}, H_2O, and Na^+.

   vi. Past 200.0 mL NaOH added, excess OH^- is present. The major species are OH^-, A^{2-}, H_2O, and Na^+.

c. 50.0 mL of NaOH added corresponds to the first halfway point to equivalence. Exactly one-half of the H_2A present initially has been converted into its conjugate base HA^-, so $[H_2A] = [HA^-]$ in this buffer solution.

$$H_2A \rightleftharpoons HA^- + H^+ \qquad K_{a_1} = \frac{[HA^-][H^+]}{[H_2A]}$$

When $[HA^-] = [H_2A]$, then $K_{a_1} = [H^+]$ or $pK_{a_1} = pH$.

Here, pH = 4.0, so $pK_{a_1} = 4.0$ and $K_{a_1} = 10^{-4.0} = 1 \times 10^{-4}$.

150.0 mL of NaOH added correspond to the second halfway point to equivalence, where $[HA^-] = [A^{2-}]$ in this buffer solution.

$$HA^- \rightleftharpoons A^{2-} + H^+ \qquad K_{a_2} = \frac{[A^{2-}][H^+]}{[HA^-]}$$

When $[A^{2-}] = [HA^-]$, then $K_{a_2} = [H^+]$ or $pK_{a_2} = pH$.

Here, pH = 8.0, so $pK_{a_2} = 8.0$ and $K_{a_2} = 10^{-8.0} = 1 \times 10^{-8}$.

85.   100.0 mL × 0.0500 M = 5.00 mmol H_3X initially

a. Because $K_{a_1} \gg K_{a_2} \gg K_{a_3}$, pH initially is determined by H_3X equilibrium reaction.

$$H_3X \quad \rightleftharpoons \quad H^+ \; + \; H_2X^-$$

Initial	0.0500 M	~0	0
Equil.	0.0500 − x	x	x

$$K_{a_1} = 1.0 \times 10^{-3} = \frac{x^2}{0.0500 - x} \approx \frac{x^2}{0.0500}, \quad x = 7.1 \times 10^{-3}; \quad \text{assumption poor.}$$

Using the quadratic formula:

$$x^2 + (1.0 \times 10^{-3})x - 5.0 \times 10^{-5} = 0, \quad x = 6.6 \times 10^{-3} \ M = [H^+]; \ pH = 2.18$$

b. 1.00 mmol OH^- added converts H_3X into H_2X^-. After this reaction goes to completion, 4.00 mmol H_3X and 1.00 mmol H_2X^- are in a total volume of 110.0 mL. Solving the buffer problem:

	H_3X	$\rightleftharpoons$	H^+	+	H_2X^-
Initial	0.0364 M		~0		0.00909 M
Equil.	0.0364 − x		x		0.00909 + x

$$K_{a_1} = 1.0 \times 10^{-3} = \frac{x(0.00909 + x)}{0.0364 - x}; \quad \text{assumption that } x \text{ is small does not work here.}$$

Using the quadratic formula and carrying extra significant figures:

$$x^2 + (1.01 \times 10^{-2})x - 3.64 \times 10^{-5} = 0, \ x = 2.8 \times 10^{-3} \ M = [H^+]; \ pH = 2.55$$

c. 2.50 mmol OH^- added results in 2.50 mmol H_3X and 2.50 mmol H_2X^- after OH^- reacts completely with H_3X. This is the first halfway point to equivalence. $pH = pK_{a_1} = 3.00$; assumptions good (5% error).

d. 5.00 mmol OH^- added results in 5.00 mmol H_2X^- after OH^- reacts completely with H_3X. This is the first stoichiometric point.

$$pH = \frac{pK_{a_1} + pK_{a_2}}{2} = \frac{3.00 + 7.00}{2} = 5.00$$

e. 6.00 mmol OH^- added results in 4.00 mmol H_2X^- and 1.00 mmol HX^{2-} after OH^- reacts completely with H_3X and then reacts completely with H_2X^-.

Using the $H_2X^- \rightleftharpoons H^+ + HX^{2-}$ reaction:

$$pH = pK_{a_2} + \log\frac{[HX^{2-}]}{[H_2X^-]} = 7.00 - \log(1.00/4.00) = 6.40; \quad \text{assumptions good.}$$

f. 7.50 mmol KOH added results in 2.50 mmol H_2X^- and 2.50 mmol HX^{2-} after OH^- reacts completely. This is the second halfway point to equivalence.

$$pH = pK_{a_2} = 7.00; \quad \text{assumptions good.}$$

g. 10.0 mmol OH^- added results in 5.0 mmol HX^{2-} after OH^- reacts completely. This is the second stoichiometric point.

$$pH = \frac{pK_{a_2} + pK_{a_3}}{2} = \frac{7.00 + 12.00}{2} = 9.50$$

h.   12.5 mmol OH^- added results in 2.5 mmol HX^{2-} and 2.5 mmol X^{3-} after OH^- reacts completely with H_3X first, then H_2X^-, and finally HX^{2-}. This is the third halfway point to equivalence. Usually $pH = pK_{a_3}$ but normal assumptions don't hold. We must solve for the pH exactly.

$[X^{3-}] = [HX^{2-}] = 2.5 \text{ mmol}/225.0 \text{ mL} = 1.1 \times 10^{-2} \, M$

$$X^{3-} + H_2O \rightleftharpoons HX^{2-} + OH^- \qquad K_b = \frac{K_w}{K_{a_3}} = 1.0 \times 10^{-2}$$

Initial	0.011 M	0.011 M	0
Equil.	0.011 $- x$	0.011 $+ x$	x

$K_b = 1.0 \times 10^{-2} = \dfrac{x(0.011 + x)}{0.011 - x}$; using the quadratic formula:

$x^2 + (2.1 \times 10^{-2})x - 1.1 \times 10^{-4} = 0, \quad x = 4.3 \times 10^{-3} \, M = OH^-; \quad pH = 11.63$

i.   15.0 mmol OH^- added results in 5.0 mmol X^{3-} after OH^- reacts completely. This is the third stoichiometric point.

$$X^{3-} + H_2O \rightleftharpoons HX^{2-} + OH^- \qquad K_b = \frac{K_w}{K_{a_3}} = 1.0 \times 10^{-2}$$

Initial	$\dfrac{5.0 \text{ mmol}}{250.0 \text{ mL}} = 0.020 \, M$	0	0
Equil.	0.020 $- x$	x	x

$K_b = \dfrac{x^2}{0.020 - x}, \quad 1.0 \times 10^{-2} \approx \dfrac{x^2}{0.020}, \quad x = 1.4 \times 10^{-2}; \quad$ assumption poor.

Using the quadratic formula: $x^2 + (1.0 \times 10^{-2})x - 2.0 \times 10^{-4} = 0$

$x = [OH^-] = 1.0 \times 10^{-2} \, M; \quad pH = 12.00$

j.   20.0 mmol OH^- added results in 5.0 mmol X^{3-} and 5.0 mmol OH^- excess after OH^- reacts completely. Because K_b for X^{3-} is fairly large for a weak base, we have to worry about the OH^- contribution from X^{3-}.

$[X^{3-}] = [OH^-] = 5.0 \text{ mmol}/300.0 \text{ mL} = 1.7 \times 10^{-2} \, M$

$$X^{3-} + H_2O \rightleftharpoons OH^- + HX^{2-}$$

Initial	$1.7 \times 10^{-2} \, M$	$1.7 \times 10^{-2} \, M$	0
Equil.	$1.7 \times 10^{-2} - x$	$1.7 \times 10^{-2} + x$	x

$K_b = 1.0 \times 10^{-2} = \dfrac{(1.7 \times 10^{-2} + x)x}{(1.7 \times 10^{-2} - x)}$

Using the quadratic formula: $x^2 + (2.7 \times 10^{-2})x - 1.7 \times 10^{-4} = 0$, $x = 5.3 \times 10^{-3}\,M$

$[OH^-] = (1.7 \times 10^{-2}) + x = (1.7 \times 10^{-2}) + (5.3 \times 10^{-3}) = 2.2 \times 10^{-2}\,M$; pH = 12.34

87. $\dfrac{0.200\ \text{g}}{165.0\ \text{g/mol}} = 1.212 \times 10^{-3}\ \text{mol} = 1.212\ \text{mmol H}_3\text{A}$ (carrying extra sig. figs.)

   a. $10.50\ \text{mL} \times 0.0500\,M = 0.525\ \text{mmol OH}^-$ added; $H_3A + OH^- \rightarrow H_2A^- + H_2O$;
   $1.212 - 0.525 = 0.687$ mmol H_3A remains after OH^- reacts completely and 0.525 mmol
   H_2A^- formed. Solving the buffer problem using the K_{a_1} reaction gives:

$$K_{a_1} = \frac{(10^{-3.73})\left(\dfrac{0.525}{60.50} + 10^{-3.73}\right)}{\dfrac{0.687}{60.50} - 10^{-3.73}} = 1.5 \times 10^{-4};\ pK_{a_1} = -\log(1.5 \times 10^{-4}) = 3.82$$

First stoichiometric point: $pH = \dfrac{pK_{a_1} + pK_{a_2}}{2} = 5.19 = \dfrac{3.82 + pK_{a_2}}{2}$

$pK_{a_2} = 6.56;\ K_{a_2} = 10^{-6.56} = 2.8 \times 10^{-7}$

Second stoichiometric point: $pH = \dfrac{pK_{a_2} + pK_{a_3}}{2},\ 8.00 = \dfrac{6.56 + pK_{a_3}}{2}$

$pK_{a_3} = 9.44;\ K_{a_3} = 10^{-9.44} = 3.6 \times 10^{-10}$

   b. 1.212 mmol $H_3A = 0.0500\,M\ OH^- \times V_{OH^-}$, $V_{OH^-} = 24.2$ mL; 24.2 mL of OH^- are
   necessary to reach the first stoichiometric point. It will require 60.5 mL to reach the third
   halfway point to equivalence, where $pH = pK_{a_3} = 9.44$. The pH at 59.0 mL of NaOH
   added should be a little lower than 9.44.

   c. 59.0 mL of $0.0500\,M\ OH^- = 2.95$ mmol OH^- added

	H_3A	+	OH^-	$\rightarrow$	H_2A^-	+	H_2O
Before	1.212 mmol		2.95 mmol		0		
After	0		1.74		1.212		

	H_2A^-	+	OH^-	$\rightarrow$	HA^{2-}	+	H_2O
Before	1.212		1.74		0		
After	0		0.53		1.212		

	HA^{2-}	+	OH^-	$\rightarrow$	A^{3-}	+	H_2O
Before	1.212		0.53		0		
After	0.68 mmol		0		0.53 mmol		

Use the K_{a_3} reaction to solve for the $[H^+]$ in this buffer solution and make the normal assumptions.

$$K_{a_3} = 3.6 \times 10^{-10} = \frac{\left(\dfrac{0.53 \text{ mmol}}{109 \text{ mL}}\right)[H^+]}{\left(\dfrac{0.68 \text{ mmol}}{109 \text{ mL}}\right)}, \quad [H^+] = 4.6 \times 10^{-10} \, M; \quad pH = 9.34$$

Assumptions good.

89.   a. Na^+ is present in all solutions. The added H^+ from HCl reacts completely with CO_3^{2-} to convert it into HCO_3^-. After all CO_3^{2-} is reacted (after point C, the first equivalence point), H^+ then reacts completely with the next best base present, HCO_3^-. Point E represents the second equivalence point. The major species present at the various points after H^+ reacts completely follow.

A. CO_3^{2-}, H_2O, Na^+

B. CO_3^{2-}, HCO_3^-, H_2O, Cl^-, Na^+

C. HCO_3^-, H_2O, Cl^-, Na^+

D. HCO_3^-, CO_2 (H_2CO_3), H_2O, Cl^-, Na^+

E. CO_2 (H_2CO_3), H_2O, Cl^-, Na^+

F. H^+ (excess), CO_2 (H_2CO_3), H_2O, Cl^-, Na^+

b.   Point A (initially):

$$CO_3^{2-} + H_2O \rightleftharpoons HCO_3^- + OH^- \qquad K_{b, CO_3^{2-}} = \frac{K_w}{K_{a_2}} = \frac{1.0 \times 10^{-14}}{4.8 \times 10^{-11}}$$

Initial	$0.100 \, M$	0	~0	$K_b = 2.1 \times 10^{-4}$
Equil.	$0.100 - x$	x	x	

$$K_b = 2.1 \times 10^{-4} = \frac{[HCO_3^-][OH^-]}{[CO_3^{2-}]} = \frac{x^2}{0.100 - x} \approx \frac{x^2}{0.100}$$

$x = 4.6 \times 10^{-3} \, M = [OH^-]; \quad pH = 11.66; \quad$ assumptions good.

Point B: The first halfway point where $[CO_3^{2-}] = [HCO_3^-]$.

$pH = pK_{a_2} = -\log(4.8 \times 10^{-11}) = 10.32; \quad$ assumptions good.

Point C: First equivalence point (25.00 mL of 0.100 M HCl added). The amphoteric HCO_3^- is the major acid-base species present.

$$pH = \frac{pK_{a_1} + pK_{a_2}}{2}; \quad pK_{a_1} = -\log(4.3 \times 10^{-7}) = 6.37$$

$$pH = \frac{6.37 + 10.32}{2} = 8.35$$

Point D: The second halfway point where $[HCO_3^-] = [H_2CO_3]$.

$pH = pK_{a_1} = 6.37$; assumptions good.

Point E: This is the second equivalence point, where all of the CO_3^{2-} present initially has been converted into H_2CO_3 by the added strong acid. 50.0 mL HCl added.

$[H_2CO_3] = 2.50$ mmol/75.0 mL $= 0.0333$ M

$$H_2CO_3 \quad \rightleftharpoons \quad H^+ + HCO_3^- \qquad K_{a_1} = 4.3 \times 10^{-7}$$

Initial	0.0333 M	0	0
Equil.	0.0333 $-x$	x	x

$$K_{a_1} = 4.3 \times 10^{-7} = \frac{x^2}{0.0333 - x} \approx \frac{x^2}{0.0333}$$

$x = [H^+] = 1.2 \times 10^{-4}$ M; $pH = 3.92$; assumptions good.

Solubility Equilibria

91. $MX(s) \rightleftharpoons M^{n+}(aq) + X^{n-}(aq)$ $K_{sp} = [M^{n+}][X^{n-}]$; the K_{sp} reaction always refers to a solid breaking up into its ions. The representations all show 1 : 1 salts, i.e., the formula of the solid contains 1 cation for every 1 anion (either +1 and −1, or +2 and −2, or +3 and −3). The solution with the largest number of ions (largest $[M^{n+}]$ and $[X^{n-}]$) will have the largest K_{sp} value. From the representations, the second beaker has the largest number of ions present, so this salt has the largest K_{sp} value. Conversely, the third beaker, with the fewest number of hydrated ions, will have the smallest K_{sp} value.

93.   In our setups, s = solubility in mol/L. Because solids do not appear in the K_{sp} expression, we do not need to worry about their initial or equilibrium amounts.

a. $Ag_3PO_4(s) \rightleftharpoons 3\,Ag^+(aq) + PO_4^{3-}(aq)$

Initial		0	0
	s mol/L of $Ag_3PO_4(s)$ dissolves to reach equilibrium		
Change	$-s$ $\rightarrow$	$+3s$	$+s$
Equil.		$3s$	s

$K_{sp} = 1.8 \times 10^{-18} = [Ag^+]^3[PO_4^{3-}] = (3s)^3(s) = 27s^4$

$27s^4 = 1.8 \times 10^{-18}$, $s = (6.7 \times 10^{-20})^{1/4} = 1.6 \times 10^{-5}$ mol/L = molar solubility

$$\frac{1.6 \times 10^{-5} \text{ mol } Ag_3PO_4}{L} \times \frac{418.7 \text{ g } Ag_3PO_4}{\text{mol } Ag_3PO_4} = 6.7 \times 10^{-3} \text{ g/L}$$

b. $CaCO_3(s)$ $\rightleftharpoons$ $Ca^{2+}(aq)$ + $CO_3^{2-}(aq)$

Initial	s = solubility (mol/L)	0	0
Equil.		s	s

$K_{sp} = 8.7 \times 10^{-9} = [Ca^{2+}][CO_3^{2-}] = s^2,\ s = 9.3 \times 10^{-5}$ mol/L

$$\frac{9.3 \times 10^{-5}\ \text{mol}}{L} \times \frac{100.1\ \text{g}}{\text{mol}} = 9.3 \times 10^{-3}\ \text{g/L}$$

c. $Hg_2Cl_2(s)$ $\rightleftharpoons$ $Hg_2^{2+}(aq)$ + $2\ Cl^-(aq)$

Initial	s = solubility (mol/L)	0	0
Equil.		s	$2s$

$K_{sp} = 1.1 \times 10^{-18} = [Hg_2^{2+}][Cl^-]^2 = (s)(2s)^2 = 4s^3,\ s = 6.5 \times 10^{-7}$ mol/L

$$\frac{6.5 \times 10^{-7}\ \text{mol}}{L} \times \frac{472.1\ \text{g}}{\text{mol}} = 3.1 \times 10^{-4}\ \text{g/L}$$

95.   In our setup, s = solubility of the ionic solid in mol/L. This is defined as the maximum amount of a salt that can dissolve. Because solids do not appear in the K_{sp} expression, we do not need to worry about their initial and equilibrium amounts.

a. $CaC_2O_4(s)$ $\rightleftharpoons$ $Ca^{2+}(aq)$ + $C_2O_4^{2-}(aq)$

Initial		0	0
	s mol/L of $CaC_2O_4(s)$ dissolves to reach equilibrium		
Change	$-s$ $\rightarrow$	$+s$	$+s$
Equil.		s	s

From the problem, $s = \dfrac{6.1 \times 10^{-3}\ \text{g}}{L} \times \dfrac{1\ \text{mol } CaC_2O_4}{128.10\ \text{g}} = 4.8 \times 10^{-5}$ mol/L.

$K_{sp} = [Ca^{2+}][C_2O_4^{2-}] = (s)(s) = s^2,\ K_{sp} = (4.8 \times 10^{-5})^2 = 2.3 \times 10^{-9}$

b. $BiI_3(s)$ $\rightleftharpoons$ $Bi^{3+}(aq)$ + $3\ I^-(aq)$

Initial		0	0
	s mol/L of $BiI_3(s)$ dissolves to reach equilibrium		
Change	$-s$ $\rightarrow$	$+s$	$+3s$
Equil.		s	$3s$

$K_{sp} = [Bi^{3+}][I^-]^3 = (s)(3s)^3 = 27s^4,\ K_{sp} = 27(1.32 \times 10^{-5})^4 = 8.20 \times 10^{-19}$

97.
$$Ag_2C_2O_4(s) \rightleftharpoons 2\,Ag^+(aq) \;+\; C_2O_4^{2-}(aq)$$

Initial	s = solubility (mol/L)	0	0
Equil.		$2s$	s

From problem, $[Ag^+] = 2s = 2.2 \times 10^{-4}\,M$, $s = 1.1 \times 10^{-4}\,M$

$K_{sp} = [Ag^+]^2[C_2O_4^{2-}] = (2s)^2(s) = 4s^3 = 4(1.1 \times 10^{-4})^3 = 5.3 \times 10^{-12}$

99.   a.   Because both solids dissolve to produce three ions in solution, we can compare values of K_{sp} to determine relative solubility. Because the K_{sp} for CaF_2 is the smallest, $CaF_2(s)$ has the smallest molar solubility.

   b.   We must calculate molar solubilities because each salt yields a different number of ions when it dissolves.

$$Ca_3(PO_4)_2(s) \;\rightleftharpoons\; 3\,Ca^{2+}(aq) \;+\; 2\,PO_4^{3-}(aq) \qquad K_{sp} = 1.3 \times 10^{-32}$$

Initial	s = solubility (mol/L)	0	0
Equil.		$3s$	$2s$

$K_{sp} = [Ca^{2+}]^3[PO_4^{3-}]^2 = (3s)^3(2s)^2 = 108s^5$, $s = (1.3 \times 10^{-32}/108)^{1/5} = 1.6 \times 10^{-7}$ mol/L

$$FePO_4(s) \;\rightleftharpoons\; Fe^{3+}(aq) \;+\; PO_4^{3-}(aq) \qquad K_{sp} = 1.0 \times 10^{-22}$$

Initial	s = solubility (mol/L)	0	0
Equil.		s	s

$K_{sp} = [Fe^{3+}][PO_4^{3-}] = s^2$, $s = \sqrt{1.0 \times 10^{-22}} = 1.0 \times 10^{-11}$ mol/L

$FePO_4$ has the smallest molar solubility.

101.   a.
$$Fe(OH)_3(s) \;\rightleftharpoons\; Fe^{3+}(aq) \;+\; 3\,OH^-(aq)$$

Initial		0	$1 \times 10^{-7}\,M$ (from water)

s mol/L of $Fe(OH)_3(s)$ dissolves to reach equilibrium = molar solubility

Change	$-s$	$\rightarrow$ $+s$	$+3s$
Equil.		s	$1 \times 10^{-7} + 3s$

$K_{sp} = 4 \times 10^{-38} = [Fe^{3+}][OH^-]^3 = (s)(1 \times 10^{-7} + 3s)^3 \approx s(1 \times 10^{-7})^3$

$s = 4 \times 10^{-17}$ mol/L; assumption good ($3s \ll 1 \times 10^{-7}$)

   b.
$$Fe(OH)_3(s) \;\rightleftharpoons\; Fe^{3+}(aq) \;+\; 3\,OH^-(aq) \qquad pH = 5.0,\; [OH^-] = 1 \times 10^{-9}\,M$$

Initial		0	$1 \times 10^{-9}\,M$ (buffered)

s mol/L dissolves to reach equilibrium

Change	$-s$	$\rightarrow$ $+s$	(assume no pH change in buffer)
Equil.		s	1×10^{-9}

$$K_{sp} = 4 \times 10^{-38} = [Fe^{3+}][OH^-]^3 = (s)(1 \times 10^{-9})^3, \quad s = 4 \times 10^{-11} \text{ mol/L} = \text{molar solubility}$$

c. $$Fe(OH)_3(s) \rightleftharpoons Fe^{3+}(aq) + 3 OH^-(aq) \quad pH = 11.0, \quad [OH^-] = 1 \times 10^{-3} M$$

Initial		0	0.001 M (buffered)

s mol/L dissolves to reach equilibrium

Change	$-s$	$\rightarrow$ $+s$	(assume no pH change)
Equil.		s	0.001

$$K_{sp} = 4 \times 10^{-38} = [Fe^{3+}][OH^-]^3 = (s)(0.001)^3, \quad s = 4 \times 10^{-29} \text{ mol/L} = \text{molar solubility}$$

Note: As $[OH^-]$ increases, solubility decreases. This is the common ion effect.

103. $$ZnS(s) \rightleftharpoons Zn^{2+}(aq) + S^{2-}(aq) \qquad K_{sp} = [Zn^{2+}][S^{2-}]$$

Initial	s = solubility (mol/L)	0.050 M	0
Equil.		0.050 + s	s

$$K_{sp} = 2.5 \times 10^{-22} = (0.050 + s)(s) \approx (0.050)s, \quad s = 5.0 \times 10^{-21} \text{ mol/L}; \quad \text{assumption good.}$$

$$\text{Mass ZnS that dissolves} = 0.3000 \text{ L} \times \frac{5.0 \times 10^{-21} \text{ mol ZnS}}{L} \times \frac{97.45 \text{ g ZnS}}{\text{mol}} = 1.5 \times 10^{-19} \text{ g}$$

105.   If the anion in the salt can act as a base in water, then the solubility of the salt will increase as the solution becomes more acidic. Added H^+ will react with the base, forming the conjugate acid. As the basic anion is removed, more of the salt will dissolve to replenish the basic anion. The salts with basic anions are Ag_3PO_4, $CaCO_3$, $CdCO_3$ and $Sr_3(PO_4)_2$. Hg_2Cl_2 and PbI_2 do not have any pH dependence because Cl^- and I^- are terrible bases (the conjugate bases of a strong acids).

$$Ag_3PO_4(s) + H^+(aq) \rightarrow 3 Ag^+(aq) + HPO_4^{2-}(aq) \xrightarrow{\text{excess } H^+} 3 Ag^+(aq) + H_3PO_4(aq)$$

$$CaCO_3(s) + H^+ \rightarrow Ca^{2+} + HCO_3^- \xrightarrow{\text{excess } H^+} Ca^{2+} + H_2CO_3 \; [H_2O(l) + CO_2(g)]$$

$$CdCO_3(s) + H^+ \rightarrow Cd^{2+} + HCO_3^- \xrightarrow{\text{excess } H^+} Cd^{2+} + H_2CO_3 \; [H_2O(l) + CO_2(g)]$$

$$Sr_3(PO_4)_2(s) + 2 H^+ \rightarrow 3 Sr^{2+} + 2 HPO_4^{2-} \xrightarrow{\text{excess } H^+} 3 Sr^{2+} + 2 H_3PO_4$$

107.   a. AgF   b. $Pb(OH)_2$ c. $Sr(NO_2)_2$ d. $Ni(CN)_2$

All these salts have anions that are bases. The anions of the other choices are conjugate bases of strong acids. They have no basic properties in water and, therefore, do not have solubilities that depend on pH.

109. $[BaBr_2]_0 = \dfrac{0.150\ L(1.0 \times 10^{-4}\ mol/L)}{0.250\ L} = 6.0 \times 10^{-5}\ M$

$[K_2C_2O_4]_0 = \dfrac{0.100\ L(6.0 \times 10^{-4}\ mol/L)}{0.250\ L} = 2.4 \times 10^{-4}\ M$

$Q = [Ba^{2+}]_0[C_2O_4^{2-}]_0 = (6.0 \times 10^{-5})(2.4 \times 10^{-4}) = 1.5 \times 10^{-8}\ M$

Because $Q < K_{sp}$, $BaC_2O_4(s)$ will not precipitate. The final concentration of ions will be:

$[Ba^{2+}] = 6.0 \times 10^{-5}\ M$, $[Br^-] = 1.2 \times 10^{-4}\ M$

$[K^+] = 4.8 \times 10^{-4}\ M$, $[C_2O_4^{2-}] = 2.4 \times 10^{-4}\ M$

111. $Al(OH)_3(s) \rightleftharpoons Al^{3+}(aq) + 3\ OH^-(aq)$ $\qquad\qquad K_{sp} = 2 \times 10^{-32}$

$Q = 2 \times 10^{-32} = [Al^{3+}]_0[OH^-]_0^3 = (0.2)[OH^-]_0^3$, $[OH^-]_0 = 4.6 \times 10^{-11}$ (carrying extra sig. fig.)

$pOH = -\log(4.6 \times 10^{-11}) = 10.3$; when the pOH of the solution equals 10.3, $K_{sp} = Q$. For precipitation, we want $Q > K_{sp}$. This will occur when $[OH^-]_0 > 4.6 \times 10^{-11}$ or when pOH < 10.3. Because pH + pOH = 14.00, precipitation of $Al(OH)_3(s)$ will begin when pH > 3.7 because this corresponds to a solution with pOH < 10.3.

113. $Ag_3PO_4(s) \rightleftharpoons 3\ Ag^+(aq) + PO_4^{3-}(aq)$; when Q is greater than K_{sp}, precipitation will occur. We will calculate the $[Ag^+]_0$ necessary for $Q = K_{sp}$. Any $[Ag^+]_0$ greater than this calculated number will cause precipitation of $Ag_3PO_4(s)$. In this problem, $[PO_4^{3-}]_0 = [Na_3PO_4]_0 = 1.0 \times 10^{-5}\ M$.

$K_{sp} = 1.8 \times 10^{-18}$; $Q = 1.8 \times 10^{-18} = [Ag^+]_0^3[PO_4^{3-}]_0 = [Ag^+]_0^3(1.0 \times 10^{-5}\ M)$

$[Ag^+]_0 = \left(\dfrac{1.8 \times 10^{-18}}{1.0 \times 10^{-5}} \right)^{1/3}$, $[Ag^+]_0 = 5.6 \times 10^{-5}\ M$

When $[Ag^+]_0 = [AgNO_3]_0$ is greater than $5.6 \times 10^{-5}\ M$, $Ag_3PO_4(s)$ will precipitate.

115.

a.

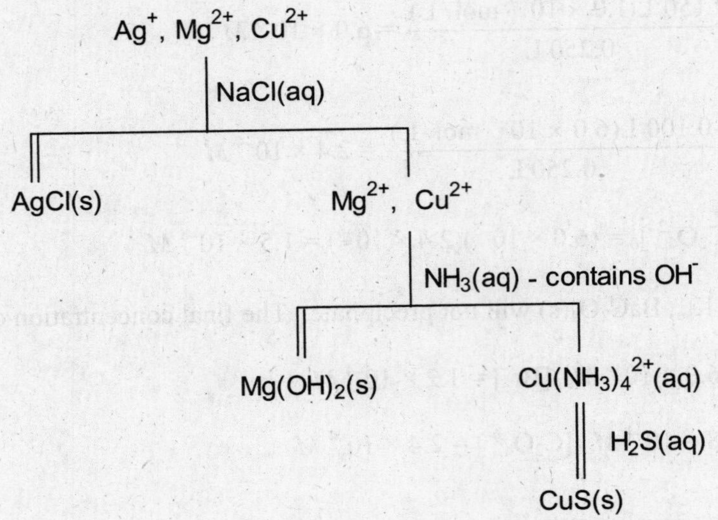

Ag^+, Mg^{2+}, Cu^{2+}

NaCl(aq)

AgCl(s) Mg^{2+}, Cu^{2+}

$NH_3(aq)$ - contains OH^-

$Mg(OH)_2(s)$ $Cu(NH_3)_4^{2+}(aq)$

$H_2S(aq)$

CuS(s)

b.

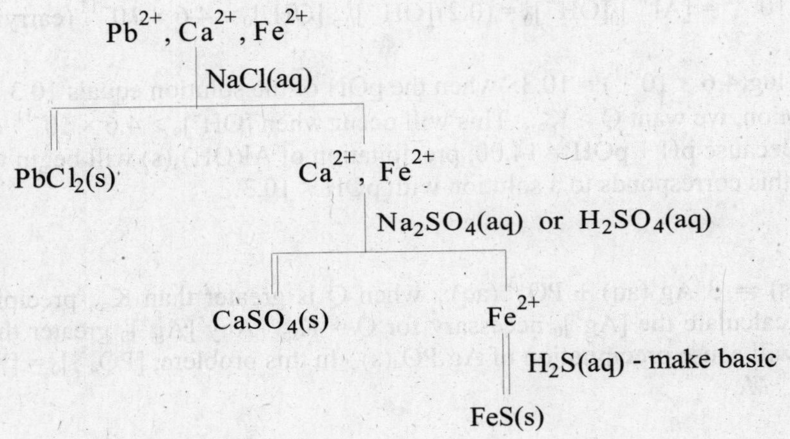

$Pb^{2+}, Ca^{2+}, Fe^{2+}$

NaCl(aq)

$PbCl_2(s)$ Ca^{2+}, Fe^{2+}

$Na_2SO_4(aq)$ or $H_2SO_4(aq)$

$CaSO_4(s)$ Fe^{2+}

$H_2S(aq)$ - make basic

FeS(s)

c.

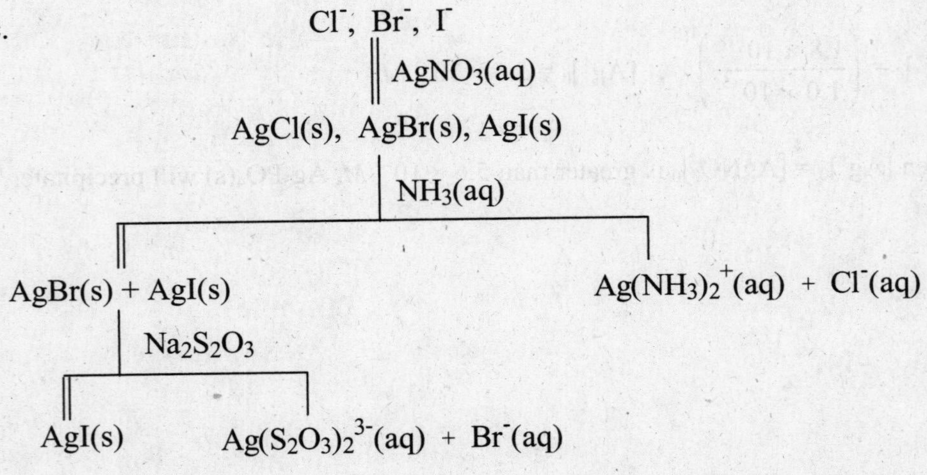

Cl^-, Br^-, I^-

$AgNO_3(aq)$

AgCl(s), AgBr(s), AgI(s)

$NH_3(aq)$

AgBr(s) + AgI(s) $Ag(NH_3)_2^+(aq) + Cl^-(aq)$

$Na_2S_2O_3$

AgI(s) $Ag(S_2O_3)_2^{3-}(aq) + Br^-(aq)$

d.

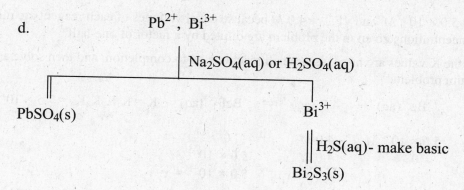

$$Pb^{2+}, \ Bi^{3+}$$

Na₂SO₄(aq) or H₂SO₄(aq)

PbSO₄(s)   Bi³⁺

H₂S(aq)- make basic

Bi₂S₃(s)

117. S^{2-} is a very basic anion and reacts significantly with H^+ to form HS^- ($S^{2-} + H^+ \rightleftharpoons HS^-$). Thus, the actual concentration of S^{2-} in solution depends on the amount of H^+ present. In basic solutions, little H^+ is present, which shifts the above equilibrium to the left. In basic solutions, the S^{2-} concentration is relatively high. So, in basic solutions, a wider range of sulfide salts will precipitate. However, in acidic solutions, added H^+ shifts the equilibrium to the right resulting in a lower S^{2-} concentration. In acidic solutions, only the least soluble sulfide salts will precipitate out of solution.

Complex Ion Equilibria

119. $Hg^{2+}(aq) + 2 \ I^-(aq) \rightarrow HgI_2(s)$, orange ppt; $HgI_2(s) + 2 \ I^-(aq) \rightarrow HgI_4^{2-}(aq)$

Soluble complex ion

121. $\dfrac{65 \text{ g KI}}{0.500 \text{ L}} \times \dfrac{1 \text{ mol KI}}{166.0 \text{ g KI}} = 0.78 \ M \text{ KI}$

The formation constant for HgI_4^{2-} is an extremely large number. Because of this, we will let the Hg^{2+} and I^- ions present initially react to completion and then solve an equilibrium problem to determine the Hg^{2+} concentration.

	$Hg^{2+}(aq)$	+	$4 \ I^-(aq)$	$\rightleftharpoons$	$HgI_4^{2-}(aq)$	$K = 1.0 \times 10^{30}$
Before	0.010 M		0.78 M		0	
Change	−0.010		−0.040	$\rightarrow$	+0.010	Reacts completely (K large)
After	0		0.74		0.010	New initial

x mol/L HgI_4^{2-} dissociates to reach equilibrium

Change	$+x$	$+4x$	$\leftarrow$ $-x$
Equil.	x	$0.74 + 4x$	$0.010 - x$

$K = 1.0 \times 10^{30} = \dfrac{[HgI_4^{2-}]}{[Hg^{2+}][I^-]^4} = \dfrac{(0.010 - x)}{(x)(0.74 + 4x)^4}$; making usual assumptions:

$1.0 \times 10^{30} \approx \dfrac{(0.010)}{(x)(0.74)^4}$, $x = [Hg^{2+}] = 3.3 \times 10^{-32} \ M$; assumptions good.

Note: 3.3×10^{-32} mol/L corresponds to one Hg^{2+} ion per 5×10^7 L. It is very reasonable to approach the equilibrium in two steps. The reaction does essentially go to completion.

123. $[Be^{2+}]_0 = 5.0 \times 10^{-5} M$ and $[F^-]_0 = 4.0 M$ because equal volumes of each reagent are mixed, so all concentrations given in the problem are diluted by a factor of one-half.

Because the K values are large, assume all reactions go to completion, and then solve an equilibrium problem.

$$Be^{2+}(aq) \quad + \quad 4 F^-(aq) \quad \rightleftharpoons \quad BeF_4^{2-}(aq) \qquad K = K_1K_2K_3K_4 = 7.5 \times 10^{12}$$

Before	$5.0 \times 10^{-5} M$	$4.0 M$	0
After	0	$4.0 M$	$5.0 \times 10^{-5} M$
Equil.	x	$4.0 + 4x$	$5.0 \times 10^{-5} - x$

$$K = 7.5 \times 10^{12} = \frac{[BeF_4^{2-}]}{[Be^{2+}][F^-]^4} = \frac{5.0 \times 10^{-5} - x}{x(4.0 + 4x)^4} \approx \frac{5.0 \times 10^{-5}}{x(4.0)^4}$$

$x = [Be^{2+}] = 2.6 \times 10^{-20} M$;   assumptions good. $[F^-] = 4.0 M$; $[BeF_4^{2-}] = 5.0 \times 10^{-5} M$

Now use the stepwise K values to determine the other concentrtations.

$$K_1 = 7.9 \times 10^4 = \frac{[BeF^+]}{[Be^{2+}][F^-]} = \frac{[BeF^+]}{(2.6 \times 10^{-20})(4.0)}, \quad [BeF^+] = 8.2 \times 10^{-15} M$$

$$K_2 = 5.8 \times 10^3 = \frac{[BeF_2]}{[BeF^+][F^-]} = \frac{[BeF_2]}{(8.2 \times 10^{-15})(4.0)}, \quad [BeF_2] = 1.9 \times 10^{-10} M$$

$$K_3 = 6.1 \times 10^2 = \frac{[BeF_3^-]}{[BeF_2][F^-]} = \frac{[BeF_3^-]}{(1.9 \times 10^{-10})(4.0)}, \quad [BeF_3^-] = 4.6 \times 10^{-7} M$$

125.   a.
$$Cu(OH)_2 \rightleftharpoons Cu^{2+} + 2 OH^- \qquad\qquad K_{sp} = 1.6 \times 10^{-19}$$
$$Cu^{2+} + 4 NH_3 \rightleftharpoons Cu(NH_3)_4^{2+} \qquad\qquad K_f = 1.0 \times 10^{13}$$

$$\overline{Cu(OH)_2(s) + 4 NH_3(aq) \rightleftharpoons Cu(NH_3)_4^{2+}(aq) + 2 OH^-(aq) \qquad K = K_{sp}K_f = 1.6 \times 10^{-6}}$$

b.
$$Cu(OH)_2(s) \quad + \quad 4 NH_3 \quad \rightleftharpoons \quad Cu(NH_3)_4^{2+} + \quad 2 OH^- \qquad K = 1.6 \times 10^{-6}$$

Initial		$5.0 M$	0	$0.0095 M$
	s mol/L Cu(OH)$_2$ dissolves to reach equilibrium		
Equil.		$5.0 - 4s$	s	$0.0095 + 2s$

$$K = 1.6 \times 10^{-6} = \frac{[Cu(NH_3)_4^{2+}][OH^-]^2}{[NH_3]^4} = \frac{s(0.0095 + 2s)^2}{(5.0 - 4s)^4}$$

If s is small: $1.6 \times 10^{-6} = \frac{s(0.0095)^2}{(5.0)^4}$, $s = 11.$ mol/L

Assumptions are not good. We will solve the problem by successive approximations.

$$s_{calc} = \frac{1.6 \times 10^{-6}(5.0 - 4s_{guess})^4}{(0.0095 + 2s_{guess})^2}; \quad \text{the results from six trials are:}$$

s_{guess}:   0.10, 0.050, 0.060, 0.055, 0.056

s_{calc}: 1.6×10^{-2}, 0.071, 0.049, 0.058, 0.056

Thus the solubility of $Cu(OH)_2$ is 0.056 mol/L in 5.0 M NH_3.

127.
$$AgBr(s) \rightleftharpoons Ag^+ + Br^- \qquad K_{sp} = 5.0 \times 10^{-13}$$
$$Ag^+ + 2\,S_2O_3^{2-} \rightleftharpoons Ag(S_2O_3)_2^{3-} \qquad K_f = 2.9 \times 10^{13}$$

$$AgBr(s) + 2\,S_2O_3^{2-} \rightleftharpoons Ag(S_2O_3)_2^{3-} + Br^- \qquad K = K_{sp} \times K_f = 14.5 \quad \text{(Carry extra sig. figs.)}$$

$$AgBr(s) + 2\,S_2O_3^{2-}(aq) \rightleftharpoons Ag(S_2O_3)_2^{3-}(aq) + Br^-\,aq)$$

Initial	0.500 M	0	0

s mol/L $AgBr(s)$ dissolves to reach equilibrium

Change	$-s$	$-2s$	$\rightarrow$	$+s$	$+s$
Equil.		0.500 $-2s$	s	s

$$K = \frac{s^2}{(0.500 - 2s)^2} = 14.5; \quad \text{taking the square root of both sides:}$$

$$\frac{s}{0.500 - 2s} = 3.81, \quad s = 1.91 - (7.62)s, \quad s = 0.222 \text{ mol/L}$$

$$1.00\ L \times \frac{0.222 \text{ mol AgBr}}{L} \times \frac{187.8 \text{ g AgBr}}{\text{mol AgBr}} = 41.7 \text{ g AgBr} = 42 \text{ g AgBr}$$

129.   Test tube 1: Added Cl^- reacts with Ag^+ to form a silver chloride precipitate. The net ionic equation is $Ag^+(aq) + Cl^-(aq) \rightarrow AgCl(s)$. Test tube 2: Added NH_3 reacts with Ag^+ ions to form a soluble complex ion, $Ag(NH_3)_2^+$. As this complex ion forms, Ag^+ is removed from the solution, which causes the $AgCl(s)$ to dissolve. When enough NH_3 is added, all the silver chloride precipitate will dissolve. The equation is $AgCl(s) + 2\,NH_3(aq) \rightarrow Ag(NH_3)_2^+(aq) + Cl^-(aq)$. Test tube 3: Added H^+ reacts with the weak base, NH_3, to form NH_4^+. As NH_3 is removed from the $Ag(NH_3)_2^+$ complex ion, Ag^+ ions are released to solution and can then react with Cl^- to re-form $AgCl(s)$. The equations are $Ag(NH_3)_2^+(aq) + 2\,H^+(aq) \rightarrow Ag^+(aq) + 2\,NH_4^+(aq)$, and $Ag^+(aq) + Cl^-(aq) \rightarrow AgCl(s)$.

Additional Exercises

131.   a.   The optimum pH for a buffer is when pH = pK_a. At this pH a buffer will have equal neutralization capacity for both added acid and base. As shown next, because the pK_a for $TRISH^+$ is 8.1, the optimal buffer pH is about 8.1.

$$K_b = 1.19 \times 10^{-6}; \quad K_a = K_w/K_b = 8.40 \times 10^{-9}; \quad pK_a = -\log(8.40 \times 10^{-9}) = 8.076$$

b. $pH = pK_a + \log\dfrac{[TRIS]}{[TRISH^+]}, \quad 7.00 = 8.076 + \log\dfrac{[TRIS]}{[TRISH^+]}$

$$\dfrac{[TRIS]}{[TRISH^+]} = 10^{-1.08} = 0.083 \quad \text{(at pH = 7.00)}$$

$$9.00 = 8.076 + \log\dfrac{[TRIS]}{[TRISH^+]}, \quad \dfrac{[TRIS]}{[TRISH^+]} = 10^{0.92} = 8.3 \quad \text{(at pH = 9.00)}$$

c. $\dfrac{50.0 \text{ g TRIS}}{2.0 \text{ L}} \times \dfrac{1 \text{ mol}}{121.14 \text{ g}} = 0.206\,M = 0.21\,M = [TRIS]$

$\dfrac{65.0 \text{ g TRISHCl}}{2.0 \text{ L}} \times \dfrac{1 \text{ mol}}{157.60 \text{ g}} = 0.206\,M = 0.21\,M = [TRISHCl] = [TRISH^+]$

$$pH = pK_a + \log\dfrac{[TRIS]}{[TRISH^+]} = 8.076 + \log\dfrac{(0.21)}{(0.21)} = 8.08$$

The amount of H^+ added from HCl is: $(0.50 \times 10^{-3} \text{ L}) \times 12 \text{ mol/L} = 6.0 \times 10^{-3} \text{ mol } H^+$

The H^+ from HCl will convert TRIS into $TRISH^+$. The reaction is:

	TRIS	+	H^+	→	$TRISH^+$	
Before	0.21 M		$\dfrac{6.0 \times 10^{-3}}{0.2005} = 0.030\,M$		0.21 M	
Change	−0.030		−0.030	→	+0.030	Reacts completely
After	0.18		0		0.24	

Now use the Henderson-Hasselbalch equation to solve this buffer problem.

$$pH = 8.076 + \log\left(\dfrac{0.18}{0.24}\right) = 7.95$$

133.   A best buffer is when pH ≈ pK_a; these solutions have about equal concentrations of weak acid and conjugate base. Therefore, choose combinations that yield a buffer where pH ≈ pK_a; that is, look for acids whose pK_a is closest to the pH.

a.   Potassium fluoride + HCl will yield a buffer consisting of HF ($pK_a = 3.14$) and F^-.

b. Benzoic acid + NaOH will yield a buffer consisting of benzoic acid (pKa = 4.19) and benzoate anion.

c. Sodium acetate + acetic acid (pKa = 4.74) is the best choice for pH = 5.0 buffer since acetic acid has a pKa value closest to 5.0.

d. HOCl and NaOH: This is the best choice to produce a conjugate acid-base pair with pH = 7.0. This mixture would yield a buffer consisting of HOCl (pKa = 7.46) and OCl⁻. Actually, the best choice for a pH = 7.0 buffer is an equimolar mixture of ammonium chloride and sodium acetate. NH_4^+ is a weak acid ($K_a = 5.6 \times 10^{-10}$), and $C_2H_3O_2^-$ is a weak base ($K_b = 5.6 \times 10^{-10}$). A mixture of the two will give a buffer at pH = 7.0 because the weak acid and weak base are the same strengths (K_a for $NH_4^+ = K_b$ for $C_2H_3O_2^-$). $NH_4C_2H_3O_2$ is commercially available, and its solutions are used for pH = 7.0 buffers.

e. Ammonium chloride + NaOH will yield a buffer consisting of NH_4^+ (pKa = 9.26) and NH_3.

135. NaOH added = $50.0 \text{ mL} \times \dfrac{0.500 \text{ mmol}}{\text{mL}} = 25.0$ mmol NaOH

NaOH left unreacted = $31.92 \text{ mL HCl} \times \dfrac{0.289 \text{ mmol}}{\text{mL}} \times \dfrac{1 \text{ mmol NaOH}}{\text{mmol HCl}} = 9.22$ mmol NaOH

NaOH reacted with aspirin = $25.0 - 9.22 = 15.8$ mmol NaOH

$15.8 \text{ mmol NaOH} \times \dfrac{1 \text{ mmol aspirin}}{2 \text{ mmol NaOH}} \times \dfrac{180.2 \text{ mg}}{\text{mmol}} = 1420 \text{ mg} = 1.42$ g aspirin

Purity = $\dfrac{1.42 \text{ g}}{1.427 \text{ g}} \times 100 = 99.5\%$

Here, a strong base is titrated by a strong acid. The equivalence point will be at pH = 7.0. Bromthymol blue would be the best indicator since it changes color at pH ≈ 7 (from base color to acid color). See Fig. 8.8 of the text.

137. At the equivalence point, P^{2-} is the major species. P^{2-} is a weak base in water because it is the conjugate base of a weak acid.

	P^{2-}	+	H_2O	⇌	HP^-	+	OH^-
Initial	$\dfrac{0.5 \text{ g}}{0.1 \text{ L}} \times \dfrac{1 \text{ mol}}{204.2 \text{ g}} = 0.024\ M$				0		~0 (carry extra sig. fig.)
Equil.	$0.024 - x$				x		x

$K_b = \dfrac{[HP^-][OH^-]}{P^{2-}} = \dfrac{K_w}{K_a} = \dfrac{1.0 \times 10^{-14}}{10^{-5.51}},\ 3.2 \times 10^{-9} = \dfrac{x^2}{0.024 - x} \approx \dfrac{x^2}{0.024}$

$x = [OH^-] = 8.8 \times 10^{-6} \, M$; pOH = 5.1; pH = 8.9; assumptions good.

Phenolphthalein would be the best indicator for this titration because it changes color at pH $\approx$ 9 (from acid color to base color).

139. $CaF_2(s) \rightleftharpoons Ca^{2+}(aq) + 2 \, F^-(aq)$ $K_{sp} = [Ca^{2+}][F^-]^2$

We need to determine the F^- concentration present in a 1.0 M HF solution. Solving the weak acid equilibrium problem:

$$HF(aq) \rightleftharpoons H^+(aq) + F^-(aq) \quad K_a = \frac{[H^+][F^-]}{[HF]}$$

Initial   1.0 M ~0   0
Equil.   1.0 − x x x

$K_a = 7.2 \times 10^{-4} = \dfrac{x(x)}{1.0 - x} \approx \dfrac{x^2}{1.0}$, $x = [F^-] = 2.7 \times 10^{-2} \, M$; assumption good.

Next, calculate the Ca^{2+} concentration necessary for $Q = K_{sp, CaF_2}$.

$Q = [Ca^{2+}]_0[F^-]_0^2$, $4.0 \times 10^{-11} = [Ca^{2+}]_0(2.7 \times 10^{-2})^2$, $[Ca^{2+}]_0 = 5.5 \times 10^{-8}$ mol/L

$$\text{Mass Ca(NO}_3)_2 = 1.0 \, L \times \frac{5.5 \times 10^{-8} \, \text{mol Ca}^{2+}}{L} \times \frac{1 \, \text{mol Ca(NO}_3)_2}{\text{mol Ca}^{2+}} \times \frac{164.10 \, \text{g Ca(NO}_3)_2}{\text{mol}}$$

$$= 9.0 \times 10^{-6} \, \text{g Ca(NO}_3)_2$$

For precipitation of $CaF_2(s)$ to occur, we need $Q > K_{sp}$. When 9.0×10^{-6} g Ca(NO$_3$)$_2$ has been added to 1.0 L of solution, $Q = K_{sp}$. So precipitation of $CaF_2(s)$ will begin to occur when just more than 9.0×10^{-6} g Ca(NO$_3$)$_2$ has been added.

141.   a. $Pb(OH)_2(s) \rightleftharpoons Pb^{2+} + 2 \, OH^-$

Initial s = solubility (mol/L)   0 $1.0 \times 10^{-7} \, M$ (from water)
Equil. s $1.0 \times 10^{-7} + 2s$

$K_{sp} = 1.2 \times 10^{-15} = [Pb^{2+}][OH^-]^2 = s(1.0 \times 10^{-7} + 2s)^2 \approx s(2s^2) = 4s^3$

$s = [Pb^{2+}] = 6.7 \times 10^{-6} \, M$; assumption is good by the 5% rule.

b. $Pb(OH)_2(s) \rightleftharpoons Pb^{2+} + 2 \, OH^-$

Initial   0   0.10 M pH = 13.00, $[OH^-] = 0.10 \, M$
s mol/L Pb(OH)$_2$(s) dissolves to reach equilibrium
Equil. s 0.10   (Buffered solution)

$1.2 \times 10^{-15} = (s)(0.10)^2$, $s = [Pb^{2+}] = 1.2 \times 10^{-13} \, M$

c. We need to calculate the Pb^{2+} concentration in equilibrium with $EDTA^{4-}$. Since K is large for the formation of $PbEDTA^{2-}$, let the reaction go to completion, and then solve an equilibrium problem to get the Pb^{2+} concentration.

$$Pb^{2+} \quad + \quad EDTA^{4-} \quad \rightleftharpoons \quad PbEDTA^{2-} \quad K = 1.1 \times 10^{18}$$

Before	0.010 M	0.050 M	0

0.010 mol/L Pb^{2+} reacts completely (large K)

Change	-0.010	-0.010	$\rightarrow$ +0.010	Reacts completely
After	0	0.040	0.010	New initial

x mol/L $PbEDTA^{2-}$ dissociates to reach equilibrium

Equil.	x	$0.040 + x$	$0.010 - x$

$$1.1 \times 10^{18} = \frac{(0.010 - x)}{(x)(0.040 + x)} \approx \frac{0.010}{x(0.040)}, \quad x = [Pb^{2+}] = 2.3 \times 10^{-19} \ M; \text{ assumptions good.}$$

Now calculate the solubility quotient for $Pb(OH)_2$ to see if precipitation occurs. The concentration of OH^- is 0.10 M since we have a solution buffered at pH = 13.00.

$$Q = [Pb^{2+}]_0[OH^-]_0^2 = (2.3 \times 10^{-19})(0.10)^2 = 2.3 \times 10^{-21} < K_{sp} \ (1.2 \times 10^{-15})$$

$Pb(OH)_2(s)$ will not form since Q is less than K_{sp}.

143. $HC_2H_3O_2 \rightleftharpoons H^+ + C_2H_3O_2^-$; let C_0 = initial concentration of $HC_2H_3O_2$

From normal weak acid setup: $K_a = 1.8 \times 10^{-5} = \dfrac{[H^+][C_2H_3O_2^-]}{[HC_2H_3O_2]} = \dfrac{[H^+]^2}{C_0 - [H^+]}$

$[H^+] = 10^{-2.68} = 2.1 \times 10^{-3} \ M; \ 1.8 \times 10^{-5} = \dfrac{(2.1 \times 10^{-3})^2}{C_0 - (2.1 \times 10^{-3})}, \ C_0 = 0.25 \ M$

25.0 mL $\times$ 0.25 mmol/mL = 6.3 mmol $HC_2H_3O_2$

Need 6.3 mmol KOH = $V_{KOH} \times 0.0975$ mmol/mL, $V_{KOH} = 65$ mL

145.   0.400 mol/L $\times V_{NH_3}$ = mol NH_3 = mol NH_4^+ after reaction with HCl at the equivalence point.

At the equivalence point: $[NH_4^+]_0 = \dfrac{\text{mol } NH_4^+}{\text{total volume}} = \dfrac{0.400 \times V_{NH_3}}{1.50 \times V_{NH_3}} = 0.267 \ M$

$$NH_4^+ \quad \rightleftharpoons \quad H^+ \quad + \quad NH_3$$

Initial	0.267 M	0	0
Equil.	$0.267 - x$	x	x

$$K_a = \frac{K_w}{K_b} = \frac{1.0 \times 10^{-14}}{1.8 \times 10^{-5}}, \quad 5.6 \times 10^{-10} = \frac{x^2}{0.267 - x} \approx \frac{x^2}{0.267}$$

$x = [H^+] = 1.2 \times 10^{-5}\ M;\ \ pH = 4.92;\ \ \text{assumptions good.}$

147. $HA + OH^- \rightarrow A^- + H_2O$, where HA = acetylsalicylic acid

$$\text{mmol HA present} = 27.36\ \text{mL OH}^- \times \frac{0.5106\ \text{mmol OH}^-}{\text{mL OH}^-} \times \frac{1\ \text{mmol HA}}{\text{mmol OH}^-} = 13.97\ \text{mmol HA}$$

$$\text{Molar mass of HA} = \frac{2.51\ \text{g HA}}{13.97 \times 10^{-3}\ \text{mol HA}} = 180.\ \text{g/mol}$$

To determine the K_a value, use the pH data. After complete neutralization of acetylsalicylic acid by OH^-, we have 13.97 mmol of A^- produced from the neutralization reaction. A^- will react completely with the added H^+ and re-form acetylsalicylic acid HA.

$$\text{mmol H}^+ \text{added} = 15.44\ \text{mL} \times \frac{0.4524\ \text{mmol H}^+}{\text{mL}} = 6.985\ \text{mmol H}^+$$

	A^-	+	H^+	$\rightarrow$	HA	
Before	13.97 mmol		6.985 mmol		0	
Change	−6.985		−6.985	$\rightarrow$	+6.985	Reacts completely
After	6.985 mmol		0		6.985 mmol	

We have back titrated this solution to the halfway point to equivalence, where pH = pK_a (assuming HA is a weak acid). This is true because after H^+ reacts completely, equal milliliters of HA and A^- are present, which only occurs at the halfway point to equivalence. Assuming acetylsalicylic acid is a weak acid, then pH = pK_a = 3.48. $K_a = 10^{-3.48} = 3.3 \times 10^{-4}$.

149. K_{a_3} is so small (4.8×10^{-13}) that a break is not seen at the third stoichiometric point.

Challenge Problems

151. $\text{mmol HC}_3\text{H}_5\text{O}_2 \text{ present initially} = 45.0\ \text{mL} \times \frac{0.750\ \text{mmol}}{\text{mL}} = 33.8\ \text{mmol HC}_3\text{H}_5\text{O}_2$

$\text{mmol C}_3\text{H}_5\text{O}_2^- \text{ present initially} = 55.0\ \text{mL} \times \frac{0.700\ \text{mmol}}{\text{mL}} = 38.5\ \text{mmol C}_3\text{H}_5\text{O}_2^-$

The initial pH of the buffer is:

$$pH = pK_a + \log \frac{[C_3H_5O_2^-]}{[HC_3H_5O_2]} = -\log(1.3 \times 10^{-5}) + \log \frac{\dfrac{38.5\ \text{mmol}}{100.0\ \text{mL}}}{\dfrac{33.8\ \text{mmol}}{100.0\ \text{mL}}} = 4.89 + \log \frac{38.5}{33.8} = 4.95$$

Note: Because the buffer components are in the same volume of solution, we can use the mole (or millimole) ratio in the Henderson-Hasselbalch equation to solve for pH instead of using the concentration ratio of $[C_3H_5O_2^-]/[HC_3H_5O_2]$. The total volume always cancels for buffer solutions.

When NaOH is added, the pH will increase, and the added OH^- will convert $HC_3H_5O_2$ into $C_3H_5O_2^-$. The pH after addition of OH^- increases by 2.5%, so the resulting pH is:

$$4.95 + 0.025(4.95) = 5.07$$

At this pH, a buffer solution still exists, and the millimole ratio between $C_3H_5O_2^-$ and $HC_3H_5O_2$ is:

$$pH = pK_a + \log \frac{\text{mmol } C_3H_5O_2^-}{\text{mmol } HC_3H_5O_2} , \quad 5.07 = 4.89 + \log \frac{\text{mmol } C_3H_5O_2^-}{\text{mmol } HC_3H_5O_2}$$

$$\frac{\text{mmol } C_3H_5O_2^-}{\text{mmol } HC_3H_5O_2} = 10^{0.18} = 1.5$$

Let x = mmol OH^- added to increase pH to 5.07. Because OH^- will essentially react to completion with $HC_3H_5O_2$, the setup to the problem using millimoles is:

	$HC_3H_5O_2$	+	OH^-	→	$C_3H_5O_2^-$	
Before	33.8 mmol		x mmol		38.5 mmol	
Change	$-x$		$-x$	→	$+x$	Reacts completely
After	$33.8 - x$		0		$38.5 + x$	

$$\frac{\text{mmol } C_3H_5O_2^-}{\text{mmol } HC_3H_5O_2} = 1.5 = \frac{38.5 + x}{33.8 - x}, \quad 1.5(33.8 - x) = 38.5 + x, \quad x = 4.9 \text{ mmol } OH^- \text{ added}$$

The volume of NaOH necessary to raise the pH by 2.5% is:

$$4.9 \text{ mmol NaOH} \times \frac{1 \text{ mL}}{0.10 \text{ mmol NaOH}} = 49 \text{ mL}$$

49 mL of 0.10 *M* NaOH must be added to increase the pH by 2.5%.

153.   a.   Best acid will react with the best base present, so the dominate equilibrium is:

$$NH_4^+ + X^- \rightleftharpoons NH_3 + HX \qquad K_{eq} = \frac{[NH_3][HX]}{[NH_4^+][X^-]} = \frac{K_{a, NH_4^+}}{K_{a, HX}}$$

Because initially $[NH_4^+]_0 = [X^-]_0$ and $[NH_3]_0 = [HX]_0 = 0$, at equilibrium $[NH_4^+] = [X^-]$ and $[NH_3] = [HX]$.   Therefore:

$$K_{eq} = \frac{K_{a,\,NH_4^+}}{K_{a,\,HX}} = \frac{[HX]^2}{[X^-]^2}$$

The K_a expression for HX is: $K_{a,\,HX} = \frac{[H^+][X^-]}{[HX]}$, $\frac{[HX]}{[X^-]} = \frac{[H^+]}{K_{a,\,HX}}$

Substituting into the K_{eq} expression: $K_{eq} = \frac{K_{a,\,NH_4^+}}{K_{a,\,HX}} = \frac{[HX]^2}{[X^-]^2} = \left(\frac{[H^+]}{K_{a,\,HX}}\right)^2$

Rearranging: $[H^+]^2 = K_{a,\,NH_4^+} \times K_{a,\,HX}$, or taking the $-\log$ of both sides:

$$pH = \frac{pK_{a,\,NH_4^+} + pK_{a,\,HX}}{2}$$

b. Ammonium formate = $NH_4(HCO_2)$

$$K_{a,\,NH_4^+} = \frac{1.0 \times 10^{-14}}{1.8 \times 10^{-5}} = 5.6 \times 10^{-10},\ pK_a = 9.25;\ K_{a,\,HCO_2H} = 1.8 \times 10^{-4},\ pK_a = 3.74$$

$$pH = \frac{pK_{a,\,NH_4^+} + pK_{a,\,HCO_2H}}{2} = \frac{9.25 + 3.74}{2} = 6.50$$

Ammonium acetate = $NH_4(C_2H_3O_2)$; $K_{a,\,HC_2H_3O_2} = 1.8 \times 10^{-5}$; $pK_a = 4.74$

$$pH = \frac{9.25 + 4.74}{2} = 7.00$$

Ammonium bicarbonate = $NH_4(HCO_3)$; $K_{a,\,H_2CO_3} = 4.3 \times 10^{-7}$; $pK_a = 6.37$

$$pH = \frac{9.25 + 6.37}{2} = 7.81$$

c. $NH_4^+(aq) + OH^-(aq) \rightarrow NH_3(aq) + H_2O(l)$; $C_2H_3O_2^-(aq) + H^+(aq) \rightarrow HC_2H_3O_2(aq)$

155. a.
$$CuBr(s) \rightleftharpoons Cu^+ + Br^- \qquad K_{sp} = 1.0 \times 10^{-5}$$
$$Cu^+ + 3\,CN^- \rightleftharpoons Cu(CN)_3^{2-} \qquad K_f = 1.0 \times 10^{11}$$

$$CuBr(s) + 3\,CN \rightleftharpoons Cu(CN)_3^{2-} + Br^- \quad K = 1.0 \times 10^6$$

Because K is large, assume that enough CuBr(s) dissolves to completely use up the 1.0 M CN^-; then solve the back equilibrium problem to determine the equilibrium concentrations.

$$CuBr(s) + 3\,CN^- \rightleftharpoons Cu(CN)_3^{2-} + Br^-$$

Before	x	$1.0\,M$	0	0

x mol/L of CuBr(s) dissolves to react completely with $1.0\,M\,CN^-$

Change	$-x$	$-3x \rightarrow$	$+x$	$+x$
After	0	$1.0 - 3x$	x	x

For reaction to go to completion, $1.0 - 3x = 0$ and $x = 0.33$ mol/L. Now solve the back equilibrium problem.

$$CuBr(s) + 3\,CN^- \rightleftharpoons Cu(CN)_3^{2-} + Br^-$$

Initial		0	$0.33\,M$	$0.33\,M$

Let y mol/L of $Cu(CN)_3^{2-}$ react to reach equilibrium.

Change		$+3y \leftarrow$	$-y$	$-y$
Equil.		$3y$	$0.33 - y$	$0.33 - y$

$$K = 1.0 \times 10^6 = \frac{(0.33 - y)^2}{(3y)^3} \approx \frac{(0.33)^2}{27y^3}, \quad y = 1.6 \times 10^{-3}\,M; \text{ assumptions good.}$$

Of the initial $1.0\,M\,CN^-$, only $3(1.6 \times 10^{-3}) = 4.8 \times 10^{-3}\,M$ is present at equilibrium. Indeed, enough CuBr(s) did dissolve to essentially remove the initial $1.0\,M\,CN^-$. This amount, 0.33 mol/L, is the solubility of CuBr(s) in $1.0\,M$ NaCN.

b. $[Br^-] = 0.33 - y = 0.33 - 1.6 \times 10^{-3} = 0.33\,M$

c. $[CN^-] = 3y = 3(1.6 \times 10^{-3}) = 4.8 \times 10^{-3}\,M$

157. a.

$$SrF_2(s) \rightleftharpoons Sr^{2+}(aq) + 2\,F^-(aq)$$

Initial		0	0

s mol/L SrF$_2$ dissolves to reach equilibrium

Equil.		s	$2s$

$$[Sr^{2+}][F^-]^2 = K_{sp} = 7.9 \times 10^{-10} = 4s^3, \quad s = 5.8 \times 10^{-4} \text{ mol/L in pure water}$$

b. Greater, because some of the F^- would react with water:

$$F^- + H_2O \rightleftharpoons HF + OH^- \quad K_b = \frac{K_w}{K_{a,\,HF}} = 1.4 \times 10^{-11}$$

This lowers the concentration of F^-, forcing more SrF$_2$ to dissolve.

c. $SrF_2(s) \rightleftharpoons Sr^{2+} + 2\,F^- \quad K_{sp} = 7.9 \times 10^{-10} = [Sr^{2+}][F^-]^2$

Let s = solubility = $[Sr^{2+}]$; then $2s$ = total F^- concentration.

Since F^- is a weak base, some of the F^- is converted into HF. Therefore:

$$\text{total } F^- \text{ concentration} = 2s = [F^-] + [HF]$$

$$HF \rightleftharpoons H^+ + F^- \quad K_a = 7.2 \times 10^{-4} = \frac{[H^+][F^-]}{[HF]} = \frac{1.0 \times 10^{-2}[F^-]}{[HF]} \quad \text{(since pH = 2.00 buffer)}$$

$$7.2 \times 10^{-2} = \frac{[F^-]}{[HF]}, \quad [HF] = 14[F^-]; \text{ Solving:}$$

$$[Sr^{2+}] = s; \quad 2s = [F^-] + [HF] = [F^-] + 14[F^-], \quad 2s = 15[F^-], \quad [F^-] = 2s/15$$

$$K_{sp} = 7.9 \times 10^{-10} = [Sr^{2+}][F^-]^2 = (s)\left(\frac{2s}{15}\right)^2, \quad s = 3.5 \times 10^{-3} \text{ mol/L in pH = 2.00 solution}$$

159.   We need to determine $[S^{2-}]_0$ that will cause precipitation of CuS(s) but not MnS(s).
For CuS(s):

$$CuS(s) \rightleftharpoons Cu^{2+}(aq) + S^{2-}(aq) \quad K_{sp} = [Cu^{2+}][S^{2-}] = 8.5 \times 10^{-45}$$

$$[Cu^{2+}]_0 = 1.0 \times 10^{-3} M, \quad \frac{K_{sp}}{[Cu^{2+}]_0} = \frac{8.5 \times 10^{-45}}{1.0 \times 10^{-3}} = 8.5 \times 10^{-42} M = [S^{2-}]$$

This $[S^{2-}]$ represents the concentration that we must exceed to cause precipitation of CuS because if $[S^{2-}]_0 > 8.5 \times 10^{-42} M$, $Q > K_{sp}$.

For MnS(s):

$$MnS(s) \rightleftharpoons Mn^{2+}(aq) + S^{2-}(aq) \quad K_{sp} = [Mn^{2+}][S^{2-}] = 2.3 \times 10^{-13}$$

$$[Mn^{2+}]_0 = 1.0 \times 10^{-3} M, \quad \frac{K_{sp}}{[Mn^{2+}]} = \frac{2.3 \times 10^{-13}}{1.0 \times 10^{-3}} = 2.3 \times 10^{-10} M = [S^{2-}]$$

This value of $[S^{2-}]$ represents the largest concentration of sulfide that can be present without causing precipitation of MnS. That is, for this value of $[S^{2-}]$, $Q = K_{sp}$, and no precipitatation of MnS occurs. However, for any $[S^{2-}]_0 > 2.3 \times 10^{-10} M$, MnS(s) will form.

We must have $[S^{2-}]_0 > 8.5 \times 10^{-42} M$ to precipitate CuS, but $[S^{2-}]_0 < 2.3 \times 10^{-10} M$ to prevent precipitation of MnS.

The question asks for a pH that will precipitate CuS(s) but not MnS(s). We need to first choose an initial concentration of S^{2-} that will do this. Let's choose $[S^{2-}]_0 = 1.0 \times 10^{-10} M$ because this will clearly cause CuS(s) to precipitate but is still less than the $[S^{2-}]_0$ required for MnS(s) to precipitate. The problem now is to determine the pH necessary for a 0.1 M H_2S solution to have $[S^{2-}] = 1.0 \times 10^{-10} M$. Let's combine the K_{a_1} and K_{a_2} equations for H_2S to determine the required $[H^+]$.

$$H_2S(aq) \rightleftharpoons H^+(aq) + HS^-(aq) \qquad K_{a_1} = 1.0 \times 10^{-7}$$

$$HS^-(aq) \rightleftharpoons H^+(aq) + S^{2-}(aq) \qquad K_{a_2} = 1 \times 10^{-19}$$

$$\overline{H_2S(aq) \rightleftharpoons 2H^+(aq) + S^{2-}(aq) \qquad K = K_{a_1} \times K_{a_2} = 1.0 \times 10^{-26}}$$

$$1 \times 10^{-26} = \frac{[H^+]^2[S^{2-}]}{[H_2S]} = \frac{[H^+]^2(1 \times 10^{-10})}{0.10}, \quad [H^+] = 3 \times 10^{-9} \ M$$

$pH = -\log(3 \times 10^{-9}) = 8.5$. So, if $pH = 8.5$, $[S^{2-}] = 1 \times 10^{-10} \ M$, which will cause precipitation of $CuS(s)$ but not $MnS(s)$.

Note: Any pH less than 8.7 would be a correct answer to this problem.

161.
$$AgCN(s) \rightleftharpoons Ag^+(aq) + CN^-(aq) \qquad K_{sp} = 2.2 \times 10^{-12}$$

$$H^+(aq) + CN^-(aq) \rightleftharpoons HCN(aq) \qquad K = 1/K_{a, \ HCN} = 1.6 \times 10^9$$

$$\overline{AgCN(s) + H^+(aq) \rightleftharpoons Ag^+(aq) + HCN(aq) \qquad K = 2.2 \times 10^{-12}(1.6 \times 10^9) = 3.5 \times 10^{-3}}$$

$$AgCN(s) + H^+(aq) \rightleftharpoons Ag^+(aq) + HCN(aq)$$

Initial	1.0 M	0	0
	s mol/L AgCN(s) dissolves to reach equilibrium		
Equil.	$1.0 - s$	s	s

$$3.5 \times 10^{-3} = \frac{[Ag^+][HCN]}{[H^+]} = \frac{s(s)}{1.0 - s} \approx \frac{s^2}{1.0}, \quad s = 5.9 \times 10^{-2}$$

Assumption fails the 5% rule (s is 5.9% of 1.0 M). Using the method of successive approximations:

$$3.5 \times 10^{-3} = \frac{s^2}{1.0 - 0.059}, \quad s = 5.7 \times 10^{-2}$$

$$3.5 \times 10^{-3} = \frac{s^2}{1.0 - 0.057}, \quad s = 5.7 \times 10^{-2} \quad \text{(consistent answer)}$$

The molar solubility of AgCN(s) in 1.0 M H$^+$ is 5.7×10^{-2} mol/L.

163.   For HOCl, $K_a = 3.5 \times 10^{-8}$ and $pK_a = -\log(3.5 \times 10^{-8}) = 7.46$. This will be a buffer solution because the pH is close to the pK_a value.

$$pH = pK_a + \log\frac{[OCl^-]}{[HOCl]}, \quad 8.00 = 7.46 + \log\frac{[OCl^-]}{[HOCl]}, \quad \frac{[OCl^-]}{[HOCl]} = 10^{0.54} = 3.5$$

1.00 L $\times$ 0.0500 M = 0.0500 mol HOCl initially. Added OH$^-$ converts HOCl into OCl$^-$. The total moles of OCl$^-$ and HOCl must equal 0.0500 mol. Solving where n = moles:

$$n_{OCl^-} + n_{HOCl} = 0.0500 \text{ and } n_{OCl^-} = (3.5)n_{HOCl}$$

$$(4.5)n_{HOCl} = 0.0500, \ n_{HOCl} = 0.011 \text{ mol}; \ n_{OCl^-} = 0.039 \text{ mol}$$

Need to add 0.039 mol NaOH to produce 0.039 mol OCl$^-$.

0.039 mol = V $\times$ 0.0100 M, V = 3.9 L NaOH; *note:* Normal buffer assumptions hold.

165.   a.   200.0 mL $\times$ 0.250 mmol Na$_3$PO$_4$/mL = 50.0 mmol Na$_3$PO$_4$

135.0 mL $\times$ 1.000 mmol HCl/mL = 135.0 mmol HCl

100.0 mL $\times$ 0.100 mmol NaCN/mL = 10.0 mmol NaCN

Let H$^+$ from the HCl react to completion with the bases in solution. In general, react the strongest base first and so on. Here, 110.0 mmol of HCl reacts to convert all CN$^-$ to HCN and all PO$_4^{3-}$ to H$_2$PO$_4^-$. At this point 10.0 mmol HCN, 50.0 mmol H$_2$PO$_4^-$, and 25.0 mmol HCl are in solution. The remaining HCl reacts completely with H$_2$PO$_4^-$, converting 25.0 mmol to H$_3$PO$_4$. The final solution contains 25.0 mmol H$_3$PO$_4$, (50.0 − 25.0 =) 25.0 mmol H$_2$PO$_4^-$, and 10.0 mmol HCN. HCN (K$_a$ = 6.2 $\times$ 10^{-10}) is a much weaker acid than either H$_3$PO$_4$ (K$_{a_1}$ = 7.5 $\times$ 10^{-3}) or H$_2$PO$_4^-$ (K$_{a_2}$ = 6.2 $\times$ 10^{-8}), so ignore it. We have a buffer solution. Principal equilibrium reaction is:

$$H_3PO_4 \ \rightleftharpoons \ H^+ \ + \ H_2PO_4^- \qquad K_{a_1} = 7.5 \times 10^{-3}$$

	H$_3$PO$_4$	H$^+$	H$_2$PO$_4^-$
Initial	25.0 mmol/435.0 mL	0	25.0/435.0
Equil.	0.0575 − x	x	0.0575 + x

$$K_{a_1} = 7.5 \times 10^{-3} = \frac{x(0.0575 + x)}{0.0575 - x}; \text{ normal assumptions don't hold here.}$$

Using the quadratic formula and carrying extra sig. figs.:

$$x^2 + (0.0650)x - 4.31 \times 10^{-4} = 0, \ x = 0.0061 \ M = [H^+]; \ pH = 2.21$$

b.   [HCN] = $\dfrac{10.0 \text{ mmol}}{435.0 \text{ mL}}$ = 2.30 $\times$ 10^{-2} M; HCN dissociation will be minimal.

167.   Major species PO$_4^{3-}$, H$^+$, HSO$_4^-$, H$_2$O, and Na$^+$; let the best base (PO$_4^{3-}$) react with the best acid (H$^+$). Assume the reaction goes to completion because H$^+$ is reacting. Note that the concentrations are halved when equal volumes of the two reagents are mixed.

$$PO_4^{3-} \ + \ H^+ \ \rightarrow \ HPO_4^{2-}$$

Before 0.25 M 0.050 M 0
After   0.20 M 0   0.050 M

Major species: PO_4^{3-}, HPO_4^{2-}, HSO_4^-, H_2O, and Na^+; react the best base (PO_4^{3-}) with the best acid (HSO_4^-). Because K for this reaction is very large, assume the reaction goes to completion.

$$PO_4^{3-} \ + \ HSO_4^- \ \rightarrow \ HPO_4^{2-} \ + \ SO_4^{2-} \qquad K = \frac{K_{a,\,HSO_4^-}}{K_{a,\,HPO_4^{2-}}} = 2.5 \times 10^{10}$$

Before 0.20 M 0.050 M 0.050 M 0
After   0.15 M 0   0.100 M 0.050 M

Major species: PO_4^{3-}, HPO_4^{2-}, SO_4^{2-} (a very weak base with $K_b = 8.3 \times 10^{-13}$), H_2O, and Na^+; because the best base present (PO_4^{3-}) and best acid present (HPO_4^{2-}) are conjugate acid-base pairs, a buffer solution exists. Because K_b for PO_4^{3-} is a relatively large value ($K_b = K_w/K_{a,\,HPO_4^{2-}} = 0.021$), the usual assumptions that the amount of base that reacts to each equilibrium is negligible compared with the initial concentration of base will not hold. Solving using the K_b reaction for PO_4^{3-}:

$$PO_4^{3-} \ + \ H_2O \ \rightleftharpoons \ HPO_4^{2-} \ + \ OH^- \qquad K_b = 0.021$$

Initial 0.15 M 0.100 M 0
Change $-x$ $\rightarrow$ $+x$ $+x$
Equil. $0.15 - x$ $0.100 + x$ x

$$K_b = 0.021 = \frac{(0.100 + x)(x)}{0.15 - x}; \text{ using quadratic equation:}$$

$$x = [OH^-] = 0.022 \ M; \ pOH = 1.66; \ pH = 12.34$$

169. H_3A: $pK_{a_1} = 3.00$, $pK_{a_2} = 7.30$, $pK_{a_3} = 11.70$

The pH at the second stoichiometric point is:

$$pH = \frac{pK_{a_2} + pK_{a_3}}{2} = \frac{7.30 + 11.70}{2} = 9.50$$

Thus to reach a pH of 9.50, we must go to the second stoichiometric point. 100.0 mL × 0.0500 M = 5.00 mmol H_3A initially. To reach the second stoichiometric point, we need 10.0 mmol OH^- = 1.00 mmol/mL × V_{NaOH}. Solving for V_{NaOH}:

$$V_{NaOH} = 10.0 \text{ mL} \ (\text{to reach pH} = 9.50)$$

pH = 4.00 is between the first halfway point to equivalence (pH = pK_{a_1} = 3.00) and the first stoichiometric point (pH = $\dfrac{pK_{a_1} + pK_{a_2}}{2}$ = 5.15).

This is the buffer region controlled by $H_3A \rightleftharpoons H_2A^- + H^+$.

$$pH = pK_{a_1} + \log\frac{[H_2A^-]}{[H_3A]}, \quad 4.00 = 3.00 + \log\frac{[H_2A^-]}{[H_3A]}, \quad \frac{[H_2A^-]}{[H_3A]} = 10.$$

Because both species are in the same volume, the mole ratio also equals 10. Let n = mmol:

$$\frac{n_{H_2A^-}}{n_{H_3A}} = 10. \text{ and } n_{H_2A^-} + n_{H_3A} = 5.00 \text{ mmol (mole balance)}$$

$$11n_{H_3A} = 5.00, \quad n_{H_3A} = 0.45 \text{ mmol}; \quad n_{H_2A^-} = 4.55 \text{ mmol}$$

We need to add 4.55 mmol OH^- to get 4.55 mmol H_2A^- from the original H_3A present.

4.55 mmol = 1.00 mmol/mL $\times$ V_{NaOH}, $\quad$ V_{NaOH} = 4.55 mL of NaOH (to reach pH = 4.00)

Note: Normal buffer assumptions are good.

CHAPTER 9

ENERGY, ENTHALPY, AND THERMOCHEMISTRY

The Nature of Energy

15. Ball A: $PE = mgz = 2.00 \text{ kg} \times \dfrac{9.81 \text{ m}}{s^2} \times 10.0 \text{ m} = \dfrac{196 \text{ kg m}^2}{s^2} = 196 \text{ J}$

At point I: All this energy is transferred to ball B. All of B's energy is kinetic energy at this point. $E_{total} = KE = 196 \text{ J}$. At point II, the sum of the total energy will equal 196 J.

At point II: $PE = mgz = 4.00 \text{ kg} \times \dfrac{9.81 \text{ m}}{s^2} \times 3.00 \text{ m} = 118 \text{ J}$

$KE = E_{total} - PE = 196 \text{ J} - 118 \text{ J} = 78 \text{ J}$

17. Path-dependent functions for a trip from Chicago to Denver are those quantities that depend on the route taken. One can fly directly from Chicago to Denver or one could fly from Chicago to Atlanta to Los Angeles and then to Denver. Some path-dependent quantities are miles traveled, fuel consumption of the airplane, time traveling, airplane snacks eaten, etc. State functions are path independent; they only depend on the initial and final states. Some state functions for an airplane trip from Chicago to Denver would be longitude change, latitude change, elevation change, and overall time zone change.

19. Step 1: $\Delta E_1 = q + w = 72 \text{ J} + 35 \text{ J} = 107 \text{ J}$; step 2: $\Delta E_2 = 35 \text{ J} - 72 \text{ J} = -37 \text{ J}$

$\Delta E_{overall} = \Delta E_1 + \Delta E_2 = 107 \text{ J} - 37 \text{ J} = 70. \text{ J}$

21. $q = \text{molar heat capacity} \times \text{mol} \times \Delta T = \dfrac{20.8 \text{ J}}{°\text{C mol}} \times 39.1 \text{ mol} \times (38.0 - 0.0)°\text{C} = 30,900 \text{ J}$

$= 30.9 \text{ kJ}$

$w = -P\Delta V = -1.00 \text{ atm} \times (998 \text{ L} - 876 \text{ L}) = -122 \text{ L atm} \times \dfrac{101.3 \text{ J}}{\text{L atm}} = -12,400 \text{ J} = -12.4 \text{ kJ}$

$\Delta E = q + w = 30.9 \text{ kJ} + (-12.4 \text{ kJ}) = 18.5 \text{ kJ}$

181

23. $H_2O(g) \rightarrow H_2O(l)$; $\Delta E = q + w$; $q = -40.66$ kJ; $w = -P\Delta V$

Volume of 1 mol $H_2O(l)$ = 1.000 mol $H_2O(l) \times \dfrac{18.02 \text{ g}}{\text{mol}} \times \dfrac{1 \text{ cm}^3}{0.996 \text{ g}} = 18.1 \text{ cm}^3 = 18.1 \text{ mL}$

$w = -P\Delta V = -1.00 \text{ atm} \times (0.0181 \text{ L} - 30.6 \text{ L}) = 30.6 \text{ L atm} \times \dfrac{101.3 \text{ J}}{\text{L atm}} = 3.10 \times 10^3 \text{ J}$

$= 3.10 \text{ kJ}$

$\Delta E = q + w = -40.66 \text{ kJ} + 3.10 \text{ kJ} = -37.56 \text{ kJ}$

Properties of Enthalpy

25. $\Delta H = \Delta E + P\Delta V$ at constant P; from the definition of enthalpy, the difference between ΔH and ΔE at constant P is the quantity $P\Delta V$. Thus when a system at constant P can do pressure-volume work, then $\Delta H \neq \Delta E$. When the system cannot do PV work, then $\Delta H = \Delta E$ at constant pressure. An important way to differentiate ΔH from ΔE is to concentrate on q, the heat flow; the heat flow by a system at constant pressure equals ΔH, and the heat flow by a system at constant volume equals ΔE.

27.   One should try to cool the reaction mixture or provide some means of removing heat since the reaction is very exothermic (heat is released). The $H_2SO_4(aq)$ will get very hot and possibly boil unless cooling is provided.

29. $4 \text{ Fe(s)} + 3 \text{ O}_2(g) \rightarrow 2 \text{ Fe}_2O_3(s)$ $\Delta H = -1652$ kJ; note that 1652 kJ of heat is released when 4 mol Fe reacts with 3 mol O_2 to produce 2 mol Fe_2O_3.

a. $4.00 \text{ mol Fe} \times \dfrac{-1652 \text{ kJ}}{4 \text{ mol Fe}} = -1650 \text{ kJ}$; 1650 kJ of heat released

b. $1.00 \text{ mol Fe}_2O_3 \times \dfrac{-1652 \text{ kJ}}{2 \text{ mol Fe}_2O_3} = -826 \text{ kJ}$; 826 kJ of heat released

c. $1.00 \text{ g Fe} \times \dfrac{1 \text{ mol Fe}}{55.85 \text{ g}} \times \dfrac{-1652 \text{ kJ}}{4 \text{ mol Fe}} = -7.39 \text{ kJ}$; 7.39 kJ of heat released

d. $10.0 \text{ g Fe} \times \dfrac{1 \text{ mol Fe}}{55.85 \text{ g}} = 0.179 \text{ mol Fe}$; $2.00 \text{ g O}_2 \times \dfrac{1 \text{ mol O}_2}{32.00 \text{ g}} = 0.0625 \text{ mol O}_2$

0.179 mol Fe/0.0625 mol O_2 = 2.86; the balanced equation requires a 4 mol Fe/3 mol O_2 = 1.33 mole ratio. O_2 is limiting since the actual mole Fe/mole O_2 ratio is greater than the required mole ratio.

$0.0625 \text{ mol O}_2 \times \dfrac{-1652 \text{ kJ}}{3 \text{ mol O}_2} = -34.4 \text{ kJ}$; 34.4 kJ of heat released

31.   When a liquid is converted into gas, there is an increase in volume. The 2.5 kJ/mol quantity is the work done by the vaporization process in pushing back the atmosphere.

The Thermodynamics of Ideal Gases

33.   Consider the constant volume process first.

$$n = 1.00 \times 10^3 \text{ g} \times \frac{1 \text{ mol}}{30.07 \text{ g}} = 33.3 \text{ mol C}_2\text{H}_6; \quad C_v = \frac{44.60 \text{ J}}{\text{K mol}} = \frac{44.60 \text{ J}}{{}^\circ\text{C mol}}$$

$$\Delta E = nC_v\Delta T = (33.3 \text{ mol})(44.60 \text{ J }{}^\circ\text{C}^{-1}\text{ mol}^{-1})(75.0 - 25.0{}^\circ\text{C}) = 74,300 \text{ J} = 74.3 \text{ kJ}$$

$$\Delta E = q + w; \text{ since } \Delta V = 0, w = 0; \quad \Delta E = q_v = 74.3 \text{ kJ}$$

$$\Delta H = \Delta E + \Delta PV = \Delta E + nR\Delta T$$

$$\Delta H = 74.3 \text{ kJ} + (33.3 \text{ mol})(8.3145 \text{ J K}^{-1}\text{ mol}^{-1})(50.0 \text{ K})(1 \text{ kJ}/1000 \text{ J})$$

$$\Delta H = 74.3 \text{ kJ} + 13.8 \text{ kJ} = 88.1 \text{ kJ}$$

Now consider the constant pressure process.

$$q_p = \Delta H = nC_p\Delta T = (33.3 \text{ mol})(52.92 \text{ J K}^{-1}\text{ mol}^{-1})(50.0 \text{ K})$$

$$q_p = 88,100 \text{ J} = 88.1 \text{ kJ} = \Delta H$$

$$w = -P\Delta V = -nR\Delta T = -(33.3 \text{ mol})(8.3145 \text{ J K}^{-1}\text{ mol}^{-1})(50.0 \text{ K}) = -13,800 \text{ J} = -13.8 \text{ kJ}$$

$$\Delta E = q + w = 88.1 \text{ kJ} - 13.8 \text{ kJ} = 74.3 \text{ kJ}$$

Summary	Constant V	Constant P
q	74.3 kJ	88.1 kJ
ΔE	74.3 kJ	74.3 kJ
ΔH	88.1 kJ	88.1 kJ
w	0	−13.8 kJ

35.   Pathway I:

Step 1: (5.00 mol, 3.00 atm, 15.0 L) $\rightarrow$ (5.00 mol, 3.00 atm, 55.0 L)

$$w = -P\Delta V = -(3.00 \text{ atm})(55.0 - 15.0 \text{ L}) = -120. \text{ L atm}$$

$$w = -120. \text{ L atm} \times \frac{101.3 \text{ J}}{\text{L atm}} \times \frac{1 \text{ kJ}}{1000 \text{ J}} = -12.2 \text{ kJ}$$

$$\Delta H = q_p = nC_p\Delta T = nC_p \times \frac{\Delta(PV)}{nR} = \frac{C_p\Delta(PV)}{R}; \quad \Delta(PV) = (P_2V_2 - P_1V_1)$$

For an ideal monatomic gas: $C_p = \frac{5}{2}R$; $\Delta H = \left(\frac{5}{2}R\right)\frac{\Delta(PV)}{R} = \frac{5}{2}\Delta(PV)$

$$\Delta H = q_p = \frac{5}{2}\Delta(PV) = \frac{5}{2}(165 - 45.0) \text{ L atm} = 300. \text{ L atm}$$

$$\Delta H = q_p = 300. \text{ L atm} \times \frac{101.3 \text{ J}}{\text{L atm}} \times \frac{1 \text{ kJ}}{1000 \text{ J}} = 30.4 \text{ kJ}$$

$$\Delta E = q + w = 30.4 \text{ kJ} - 12.2 \text{ kJ} = 18.2 \text{ kJ}$$

Step 2: (5.00 mol, 3.00 atm, 55.0 L) $\rightarrow$ (5.00 mol, 6.00 atm, 20.0 L)

$$\Delta E = nC_v\Delta T = n\left(\frac{3}{2}R\right)\left(\frac{\Delta(PV)}{nR}\right) = \frac{3}{2}\Delta PV$$

$$\Delta E = \frac{3}{2}(120. - 165) \text{ L atm} = -67.5 \text{ L atm}\quad \text{(Carry an extra significant figure.)}$$

$$\Delta E = -67.5 \text{ L atm} \times \frac{101.3 \text{ J}}{\text{L atm}} \times \frac{1 \text{ kJ}}{1000 \text{ J}} = -6.8 \text{ kJ}$$

$$\Delta H = nC_p\Delta T = n\left(\frac{5}{2}R\right)\left(\frac{\Delta(PV)}{nR}\right) = \frac{5}{2}\Delta PV$$

$$\Delta H = \frac{5}{2}(-45 \text{ L atm}) = -113 \text{ L atm}\quad \text{(Carry an extra significant figure.)}$$

$$\Delta H = -113 \text{ L atm} \times \frac{101.3 \text{ J}}{\text{L atm}} \times \frac{1 \text{ kJ}}{1000 \text{ J}} = -11.4 = -11 \text{ kJ}$$

$$w = -P_{ext}\Delta V = -(6.00 \text{ atm})(20.0 - 55.0) \text{ L} = 210. \text{ L atm}$$

$$w = 210. \text{ L atm} \times \frac{101.3 \text{ J}}{\text{L atm}} \times \frac{1 \text{ kJ}}{1000 \text{ J}} = 21.3 \text{ kJ}$$

$$\Delta E = q + w, \; -6.8 \text{ kJ} = q + 21.3 \text{ kJ}, \; q = -28.1 \text{ kJ}$$

Summary:

	Path I Step 1	Step 2	Total
q	30.4 kJ	−28.1 kJ	2.3 kJ
w	−12.2 kJ	21.3 kJ	9.1 kJ
ΔE	18.2 kJ	−6.8 kJ	11.4 kJ
ΔH	30.4 kJ	−11 kJ	19 kJ

Pathway II:

Step 3: (5.00 mol, 3.00 atm, 15.0 L) $\rightarrow$ (5.00 mol, 6.00 atm, 15.0 L)

$$\Delta E = q_v = \frac{3}{2}\Delta(PV) = \frac{5}{2}(90.0 - 45.0) \text{ L atm} = 67.5 \text{ L atm}$$

$$\Delta E = q_v = 67.5 \text{ L atm} \times \frac{101.3 \text{ J}}{\text{L atm}} \times \frac{1 \text{ kJ}}{1000 \text{ J}} = 6.84 \text{ kJ}$$

$$w = -P\Delta V = 0 \text{ because } \Delta V = 0$$

$$\Delta H = \Delta E + \Delta(PV) = 67.5 \text{ L atm} + 45.0 \text{ L atm} = 112.5 \text{ L atm} = 11.40 \text{ kJ}$$

Step 4: (5.00 mol, 6.00 atm, 15.0 L) → (5.00 mol, 6.00 atm, 20.0 L)

$$\Delta H = q_p = nC_p\Delta T = \left(\frac{5}{2}R\right)\left(\frac{\Delta(PV)}{nR}\right) = \frac{5}{2}\Delta PV$$

$$\Delta H = \frac{5}{2}(120. - 90.0) \text{ L atm} = 75 \text{ L atm}$$

$$\Delta H = q_p = 75 \text{ L atm} \times \frac{101.3 \text{ J}}{\text{L atm}} \times \frac{1 \text{ kJ}}{1000 \text{ J}} = 7.6 \text{ kJ}$$

$$w = -P\Delta V = -(6.00 \text{ atm})(20.0 - 15.0) \text{ L} = -30. \text{ L atm}$$

$$w = -30. \text{ L atm} \times \frac{101.3 \text{ J}}{\text{L atm}} \times \frac{1 \text{ kJ}}{1000 \text{ J}} = -3.0 \text{ kJ}$$

$$\Delta E = q + w = 7.6 \text{ kJ} - 3.0 \text{ kJ} = 4.6 \text{ kJ}$$

Summary:

Path II	Step 3	Step 4	Total
q	6.84 kJ	7.6 kJ	14.4 kJ
w	0	−3.0 kJ	−3.0 kJ
ΔE	6.84 kJ	4.6 kJ	11.4 kJ
ΔH	11.40 kJ	7.6 kJ	19.0 kJ

State functions are independent of the particular pathway taken between two states; path functions are dependent on the particular pathway. In this problem, the overall values of ΔH and ΔE for the two pathways are the same; hence ΔH and ΔE are state functions. The overall values of q and w for the two pathways are different; hence q and w are path functions.

Calorimetry and Heat Capacity

37.   In calorimetry, heat flow is determined into or out of the surroundings. Because $\Delta E_{univ} = 0$ by the first law of thermodynamics, $\Delta E_{sys} = -\Delta E_{surr}$; what happens to the surroundings is the exact opposite of what happens to the system. To determine heat flow, we need to know the heat capacity of the surroundings, the mass of the surroundings that accepts/donates the heat, and the change in temperature. If we know these quantities, q_{surr} can be calculated and then equated to q_{sys} ($-q_{surr} = q_{sys}$). For an endothermic reaction, the surroundings (the calorimeter contents) donates heat to the system. This is accompanied by a decrease in temperature of the surroundings. For an exothermic reaction, the system donates heat to the surroundings (the calorimeter), so temperature increases.

$q_P = \Delta H$; $q_V = \Delta E$; a coffee-cup calorimeter is at constant (atmospheric) pressure. The heat released or gained at constant pressure is ΔH. A bomb calorimeter is at constant volume. The heat released or gained at constant volume is ΔE.

39.   Specific heat capacity is defined as the amount of heat necessary to raise the temperature of one gram of substance by one degree Celsius. Therefore, $H_2O(l)$ with the largest heat capacity value requires the largest amount of heat for this process. The amount of heat for $H_2O(l)$ is:

$$\text{energy} = s \times m \times \Delta T = \frac{4.18 \text{ J}}{°C \text{ g}} \times 25.0 \text{ g} \times (37.0°C - 15.0°C) = 2.30 \times 10^3 \text{ J}$$

The largest temperature change when a certain amount of energy is added to a certain mass of substance will occur for the substance with the smallest specific heat capacity. This is $Hg(l)$, and the temperature change for this process is:

$$\Delta T = \frac{\text{energy}}{s \times m} = \frac{10.7 \text{ kJ} \times \dfrac{1000 \text{ J}}{\text{kJ}}}{\dfrac{0.14 \text{ J}}{°C \text{ g}} \times 550. \text{ g}} = 140°C$$

41.   Heat loss by hot water = heat gain by cold water; keeping all quantities positive to avoid sign errors:

$$\frac{4.18 \text{ J}}{°C \text{ g}} \times m_{hot} \times (55.0°C - 37.0°C) = \frac{4.18 \text{ J}}{°C \text{ g}} \times 90.0 \text{ g} \times (37.0 °C - 22.0°C)$$

$$m_{hot} = \frac{90.0 \text{ g} \times 15.0°C}{18.0°C} = 75.0 \text{ g hot water needed}$$

43.   Heat gained by water = heat lost by nickel = $s \times m \times \Delta T$, where s = specific heat capacity.

$$\text{Heat gain} = \frac{4.18 \text{ J}}{°C \text{ g}} \times 150.0 \text{ g} \times (25.0°C - 23.5°C) = 940 \text{ J}$$

A common error in calorimetry problems is sign errors. Keeping all quantities positive helps to eliminate sign errors.

$$\text{Heat loss} = 940 \text{ J} = \frac{0.444 \text{ J}}{°C \text{ g}} \times \text{mass} \times (99.8 - 25.0) °C, \quad \text{mass} = \frac{940}{0.444 \times 74.8} = 28 \text{ g}$$

45. $50.0 \times 10^{-3} \text{ L} \times 0.100 \text{ mol/L} = 5.00 \times 10^{-3} \text{ mol}$ of both $AgNO_3$ and HCl are reacted. Thus 5.00×10^{-3} mol of AgCl will be produced because there is a 1 : 1 mole ratio between reactants.

Heat lost by chemicals = heat gained by solution

$$\text{Heat gain} = \frac{4.18 \text{ J}}{°C \text{ g}} \times 100.0 \text{ g} \times (23.40 - 22.60)°C = 330 \text{ J}$$

Heat loss = 330 J; this is the heat evolved (exothermic reaction) when 5.00×10^{-3} mol of AgCl is produced. So q = –330 J and ΔH (heat per mol AgCl formed) is negative with a value of:

$$\Delta H = \frac{-330 \text{ J}}{5.00 \times 10^{-3} \text{ mol}} \times \frac{1 \text{ kJ}}{1000 \text{ J}} = -66 \text{ kJ/mol}$$

Note: Sign errors are common with calorimetry problems. However, the correct sign for ΔH can be determined easily from the ΔT data; i.e., if ΔT of the solution increases, then the reaction is exothermic because heat was released, and if ΔT of the solution decreases, then the reaction is endothermic because the reaction absorbed heat from the water. For calorimetry problems, keep all quantities positive until the end of the calculation and then decide the sign for ΔH. This will help eliminate sign errors.

47.   Heat lost by solution = heat gained by KBr; mass of solution = 125 g + 10.5 g = 136 g

Note: Sign errors are common with calorimetry problems. However, the correct sign for ΔH can easily be obtained from the ΔT data. When working calorimetry problems, keep all quantities positive (ignore signs). When finished, deduce the correct sign for ΔH. For this problem, T decreases as KBr dissolves, so ΔH is positive; the dissolution of KBr is endothermic (absorbs heat).

$$\text{Heat lost by solution} = \frac{4.18 \text{ J}}{°C \text{ g}} \times 136 \text{ g} \times (24.2°C - 21.1°C) = 1800 \text{ J} = \text{heat gained by KBr}$$

$$\Delta H \text{ in units of J/g} = \frac{1800 \text{ J}}{10.5 \text{ g KBr}} = 170 \text{ J/g}$$

$$\Delta H \text{ in units of kJ/mol} = \frac{170 \text{ J}}{\text{g KBr}} \times \frac{119.0 \text{ g KBr}}{\text{mol KBr}} \times \frac{1 \text{ kJ}}{1000 \text{ J}} = 20. \text{ kJ/mol}$$

49.   Because ΔH is exothermic, the temperature of the solution will increase as $CaCl_2(s)$ dissolves.   Keeping all quantities positive:

$$\text{heat loss as } CaCl_2 \text{ dissolves} = 11.0 \text{ g } CaCl_2 \times \frac{1 \text{ mol } CaCl_2}{110.98 \text{ g } CaCl_2} \times \frac{81.5 \text{ kJ}}{\text{mol } CaCl_2} = 8.08 \text{ kJ}$$

$$\text{heat gained by solution} = 8.08 \times 10^3 \text{ J} = \frac{4.18 \text{ J}}{°C \text{ g}} \times (125 + 11.0) \text{ g} \times (T_f - 25.0°C)$$

$$T_f - 25.0°C = \frac{8.08 \times 10^3}{4.18 \times 136} = 14.2°C, \ T_f = 14.2°C + 25.0°C = 39.2°C$$

51.   a.   Heat gain by calorimeter = heat loss by CH_4 = 6.79 g $CH_4 \times \dfrac{1 \text{ mol } CH_4}{16.04 \text{ g}} \times \dfrac{802 \text{ kJ}}{\text{mol}}$

$$= 340. \text{ kJ}$$

$$\text{Heat capacity of calorimeter} = \frac{340. \text{ kJ}}{10.8 °C} = 31.5 \text{ kJ/°C}$$

b. Heat loss by C_2H_2 = heat gain by calorimeter = $16.9°C \times \dfrac{31.5\ kJ}{°C} = 532\ kJ$

A bomb calorimeter is at constant volume, so heat released = $q_v = \Delta E$:

$$\Delta E_{comb} = \dfrac{-532\ kJ}{12.6\ g\ C_2H_2} \times \dfrac{26.04\ g}{mol\ C_2H_2} = -1.10 \times 10^3\ kJ/mol$$

53. a. $C_{12}H_{22}O_{11}(s) + 12\ O_2(g) \rightarrow 12\ CO_2(g) + 11\ H_2O(l)$

b. A bomb calorimeter is at constant volume, so heat released = $q_v = \Delta E$:

$$\Delta E = \dfrac{-24.00\ kJ}{1.46\ g} \times \dfrac{342.30\ g}{mol} = -5630\ kJ/mol\ C_{12}H_{22}O_{11}$$

c. $\Delta H = \Delta E + \Delta(PV) = \Delta E + \Delta(nRT) = \Delta E + \Delta nRT$, where Δn = moles of gaseous products − moles of gaseous reactants.

For this reaction, $\Delta n = 12 - 12 = 0$, so $\Delta H = \Delta E = -5630\ kJ/mol$.

Hess's Law

55.

$$
\begin{array}{ll}
2\ N_2(g) + 6\ H_2(g) \rightarrow 4\ NH_3(g) & \Delta H = -2(92\ kJ) \\
6\ H_2O(g) \rightarrow 6\ H_2(g) + 3\ O_2(g) & \Delta H = -3(-484\ kJ) \\
\hline
2\ N_2(g) + 6\ H_2O(g) \rightarrow 3\ O_2(g) + 4\ NH_3(g) & \Delta H = 1268\ kJ
\end{array}
$$

No, because the reaction is very endothermic (requires a lot of heat to react), it would not be a practical way of making ammonia because of the high energy costs required.

57.

$$
\begin{array}{ll}
2\ C + 2\ O_2 \rightarrow 2\ CO_2 & \Delta H = 2(-394\ kJ) \\
H_2 + 1/2\ O_2 \rightarrow H_2O & \Delta H = -286\ kJ \\
2\ CO_2 + H_2O \rightarrow C_2H_2 + 5/2\ O_2 & \Delta H = -(-1300.kJ) \\
\hline
2\ C(s) + H_2(g) \rightarrow C_2H_2(g) & \Delta H = 226\ kJ
\end{array}
$$

Note: The enthalpy change for a reaction that is reversed is the negative quantity of the enthalpy change for the original reaction. If the coefficients in a balanced reaction are multiplied by an integer, then the value of ΔH is multiplied by the same integer.

59.

$$
\begin{array}{ll}
CaC_2 \rightarrow Ca + 2\ C & \Delta H = -(-62.8\ kJ) \\
CaO + H_2O \rightarrow Ca(OH)_2 & \Delta H = -653.1\ kJ \\
2\ CO_2 + H_2O \rightarrow C_2H_2 + 5/2\ O_2 & \Delta H = -(-1300.\ kJ) \\
Ca + 1/2\ O_2 \rightarrow CaO & \Delta H = -635.5\ kJ \\
2\ C + 2\ O_2 \rightarrow 2\ CO_2 & \Delta H = 2(-393.5\ kJ) \\
\hline
CaC_2(s) + 2\ H_2O(l) \rightarrow Ca(OH)_2(aq) + C_2H_2(g) & \Delta H = -713\ kJ
\end{array}
$$

61. $C_4H_4(g) + 5 O_2(g) \rightarrow 4 CO_2(g) + 2 H_2O(l)$ $\Delta H_{comb} = -2341$ kJ

        $C_4H_8(g) + 6 O_2(g) \rightarrow 4 CO_2(g) + 4 H_2O(l)$ $\Delta H_{comb} = -2755$ kJ

        $H_2(g) + 1/2 O_2(g) \rightarrow H_2O(l)$ $\Delta H_{comb} = -286$ kJ

By convention, $H_2O(l)$ is produced when enthalpies of combustion are given, and because per mole quantities are given, the combustion reaction refers to 1 mole of that quantity reacting with $O_2(g)$.

Using Hess's Law to solve:

$$C_4H_4(g) + 5 O_2(g) \rightarrow 4 CO_2(g) + 2 H_2O(l) \qquad \Delta H_1 = -2341 \text{ kJ}$$
$$4 CO_2(g) + 4 H_2O(l) \rightarrow C_4H_8(g) + 6 O_2(g) \qquad \Delta H_2 = -(-2755 \text{ kJ})$$
$$2 H_2(g) + O_2(g) \rightarrow 2 H_2O(l) \qquad \Delta H_3 = 2(-286 \text{ kJ})$$

$$C_4H_4(g) + 2 H_2(g) \rightarrow C_4H_8(g) \qquad \Delta H = \Delta H_1 + \Delta H_2 + \Delta H_3 = -158 \text{ kJ}$$

63.
$$C_6H_4(OH)_2 \rightarrow C_6H_4O_2 + H_2 \qquad \Delta H = 177.4 \text{ kJ}$$
$$H_2O_2 \rightarrow H_2 + O_2 \qquad \Delta H = -(-191.2 \text{ kJ})$$
$$2 H_2 + O_2 \rightarrow 2 H_2O(g) \qquad \Delta H = 2(-241.8 \text{ kJ})$$
$$2 H_2O(g) \rightarrow 2 H_2O(l) \qquad \Delta H = 2(-43.8 \text{ kJ})$$

$$C_6H_4(OH)_2(aq) + H_2O_2(aq) \rightarrow C_6H_4O_2(aq) + 2 H_2O(l) \qquad \Delta H = -202.6 \text{ kJ}$$

Standard Enthalpies of Formation

65.   The change in enthalpy that accompanies the formation of one mole of a compound from its elements, with all substances in their standard states, is the standard enthalpy of formation for a compound. The reactions that refer to ΔH_f° are:

$$Na(s) + 1/2 Cl_2(g) \rightarrow NaCl(s); \quad H_2(g) + 1/2 O_2(g) \rightarrow H_2O(l)$$

$$6 C(graphite, s) + 6 H_2(g) + 3 O_2(g) \rightarrow C_6H_{12}O_6(s); \quad Pb(s) + S(s) + 2 O_2(g) \rightarrow PbSO_4(s)$$

67.   In general: $\Delta H^\circ = \sum n_p \Delta H_{f, products}^\circ - \sum n_r \Delta H_{f, reactants}^\circ$, and all elements in their standard state have $\Delta H_f^\circ = 0$ by definition.

a. The balanced equation is: $2 NH_3(g) + 3 O_2(g) + 2 CH_4(g) \rightarrow 2 HCN(g) + 6 H_2O(g)$

$$\Delta H^\circ = (2 \text{ mol HCN} \times \Delta H_{f, HCN}^\circ + 6 \text{ mol } H_2O(g) \text{ H } \Delta H_{f, H_2O}^\circ)$$
$$- (2 \text{ mol } NH_3 \times \Delta H_{f, NH_3}^\circ + 2 \text{ mol } CH_4 \times \Delta H_{f, CH_4}^\circ)$$

$$\Delta H^\circ = [2(135.1) + 6(-242)] - [2(-46) + 2(-75)] = -940. \text{ kJ}$$

b. $Ca_3(PO_4)_2(s) + 3\ H_2SO_4(l) \rightarrow 3\ CaSO_4(s) + 2\ H_3PO_4(l)$

$$\Delta H° = \left[3\ mol\ CaSO_4(s)\left(\frac{-1433\ kJ}{mol}\right) + 2\ mol\ H_3PO_4(l)\left(\frac{-1267\ kJ}{mol}\right) \right]$$

$$- \left[1\ mol\ Ca_3(PO_4)_2(s)\left(\frac{-4126\ kJ}{mol}\right) + 3\ mol\ H_2SO_4(l)\left(\frac{-814\ kJ}{mol}\right) \right]$$

$\Delta H° = -6833\ kJ - (-6568\ kJ) = -265\ kJ$

c. $NH_3(g) + HCl(g) \rightarrow NH_4Cl(s)$

$$\Delta H° = (1\ mol\ NH_4Cl \times \Delta H°_{f,\ NH_4Cl}) - (1\ mol\ NH_3 \times \Delta H°_{f,\ NH_3} + 1\ mol\ HCl \times \Delta H°_{f,\ HCl})$$

$$\Delta H° = \left[1\ mol\left(\frac{-314\ kJ}{mol}\right) \right] - \left[1\ mol\left(\frac{-46\ kJ}{mol}\right) + 1\ mol\left(\frac{-92\ kJ}{mol}\right) \right]$$

$\Delta H° = -314\ kJ + 138\ kJ = -176\ kJ$

d.   The balanced equation is: $C_2H_5OH(l) + 3\ O_2(g) \rightarrow 2\ CO_2(g) + 3\ H_2O(g)$

$$\Delta H° = \left[2\ mol\left(\frac{-393.5\ kJ}{mol}\right) + 3\ mol\left(\frac{-242\ kJ}{mol}\right) \right] - \left[1\ mol\left(\frac{-278\ kJ}{mol}\right) \right]$$

$\Delta H° = -1513\ kJ - (-278\ kJ) = -1235\ kJ$

e. $SiCl_4(l) + 2\ H_2O(l) \rightarrow SiO_2(s) + 4\ HCl(aq)$

Because $HCl(aq)$ is $H^+(aq) + Cl^-(aq)$, $\Delta H°_f = 0 - 167 = -167\ kJ/mol$.

$$\Delta H° = \left[4\ mol\left(\frac{-167\ kJ}{mol}\right) + 1\ mol\left(\frac{-911\ kJ}{mol}\right) \right]$$

$$- \left[1\ mol\left(\frac{-687\ kJ}{mol}\right) + 2\ mol\left(\frac{-286\ kJ}{mol}\right) \right]$$

$\Delta H° = -1579\ kJ - (-1259\ kJ) = -320.\ kJ$

f. $MgO(s) + H_2O(l) \rightarrow Mg(OH)_2(s)$

$$\Delta H° = \left[1\ mol\left(\frac{-925\ kJ}{mol}\right) \right] - \left[1\ mol\left(\frac{-602\ kJ}{mol}\right) + 1\ mol\left(\frac{-286\ kJ}{mol}\right) \right]$$

$\Delta H° = -925\ kJ - (-888\ kJ) = -37\ kJ$

69. $4 Na(s) + O_2(g) \rightarrow 2 Na_2O(s), \quad \Delta H^\circ = 2 \text{ mol}\left(\dfrac{-416 \text{ kJ}}{\text{mol}}\right) = -832 \text{ kJ}$

$2 Na(s) + 2 H_2O(l) \rightarrow 2 NaOH(aq) + H_2(g)$

$\Delta H^\circ = \left[2 \text{ mol}\left(\dfrac{-470. \text{ kJ}}{\text{mol}}\right)\right] - \left[2 \text{ mol}\left(\dfrac{-286 \text{ kJ}}{\text{mol}}\right)\right] = -368 \text{ kJ}$

$2Na(s) + CO_2(g) \rightarrow Na_2O(s) + CO(g)$

$\Delta H^\circ = \left[1 \text{ mol}\left(\dfrac{-416 \text{ kJ}}{\text{mol}}\right) + 1 \text{ mol}\left(\dfrac{-110.5 \text{ kJ}}{\text{mol}}\right)\right] - \left[1 \text{ mol}\left(\dfrac{-393.5 \text{ kJ}}{\text{mol}}\right)\right] = -133 \text{ kJ}$

In reactions 2 and 3, sodium metal reacts with the "extinguishing agent." Both reactions are exothermic and each reaction produces a flammable gas, H_2 and CO, respectively.

71. $5 N_2O_4(l) + 4 N_2H_3CH_3(l) \rightarrow 12 H_2O(g) + 9 N_2(g) + 4 CO_2(g)$

$\Delta H^\circ = \left[12 \text{ mol}\left(\dfrac{-242 \text{ kJ}}{\text{mol}}\right) + 4 \text{ mol}\left(\dfrac{-393.5 \text{ kJ}}{\text{mol}}\right)\right]$

$- \left[5 \text{ mol}\left(\dfrac{-20. \text{ kJ}}{\text{mol}}\right) + 4 \text{ mol}\left(\dfrac{54 \text{ kJ}}{\text{mol}}\right)\right] = -4594 \text{ kJ}$

73.   a. $\Delta H^\circ = 3 \text{ mol}(227 \text{ kJ/mol}) - 1 \text{ mol}(49 \text{ kJ/mol}) = 632 \text{ kJ}$

b.   Because $3 C_2H_2(g)$ is higher in energy than $C_6H_6(l)$, acetylene will release more energy per gram when burned in air.

75. $2 ClF_3(g) + 2 NH_3(g) \rightarrow N_2(g) + 6 HF(g) + Cl_2(g) \quad \Delta H^\circ = -1196 \text{ kJ}$

$\Delta H^\circ = (6 \Delta H^\circ_{f, \, HF}) - (2 \Delta H^\circ_{f, \, ClF_3} + 2 \Delta H^\circ_{f, \, NH_3})$

$-1196 \text{ kJ} = 6 \text{ mol}\left(\dfrac{-271 \text{ kJ}}{\text{mol}}\right) - 2 \Delta H^\circ_{f, \, ClF_3} - 2 \text{ mol}\left(\dfrac{-46 \text{ kJ}}{\text{mol}}\right)$

$-1196 \text{ kJ} = -1626 \text{ kJ} - 2 \Delta H^\circ_{f, \, ClF_3} + 92 \text{ kJ}, \quad \Delta H^\circ_{f, \, ClF_3} = \dfrac{(-1626 + 92 + 1196) \text{ kJ}}{2 \text{ mol}} = \dfrac{-169 \text{ kJ}}{\text{mol}}$

Energy Consumption and Sources

77.   Mass of H_2O = 1.00 gal $\times \dfrac{3.785\ L}{gal} \times \dfrac{1000\ mL}{L} \times \dfrac{1.00\ g}{mL}$ = 3790 g H_2O

Energy required (theoretical) = s $\times$ m $\times \Delta T = \dfrac{4.18\ J}{^\circ C\ g} \times$ 3790 g $\times$ 10.0 °C = 1.58 $\times 10^5$ J

For an actual (80.0% efficient) process, more than this quantity of energy is needed since heat is always lost in any transfer of energy. The energy required is:

$$1.58 \times 10^5\ J \times \dfrac{100.\ J}{80.0\ J} = 1.98 \times 10^5\ J$$

$$\text{Mass of } C_2H_2 = 1.98 \times 10^5\ J \times \dfrac{1\ mol\ C_2H_2}{1300. \times 10^3\ J} \times \dfrac{26.04\ g\ C_2H_2}{mol\ C_2H_2} = 3.97\ g\ C_2H_2$$

79. $CO(g) + 2\ H_2(g) \rightarrow CH_3OH(l)$ ΔH° = –239 kJ – (–110.5 kJ) = –129 kJ

81. $CH_3OH(l) + 3/2\ O_2(g) \rightarrow CO_2(g) + 2\ H_2O(l)$

ΔH° = [–393.5 kJ + 2(–286 kJ)] – (– 239 kJ) = –727 kJ/mol CH_3OH

$$\dfrac{-727\ kJ}{mol} \times \dfrac{1\ mol}{32.04\ g} = -22.7\ kJ/g \text{ versus } -29.67\ kJ/g \text{ for ethanol}$$

Ethanol has a higher fuel value than methanol.

Additional Exercises

83. $\Delta E_{overall} = \Delta E_{step\ 1} + \Delta E_{step\ 2}$; this is a cyclic process, which means that the overall initial state and final state are the same. Because ΔE is a state function, $\Delta E_{overall}$ = 0 and $\Delta E_{step\ 1}$ = $-\Delta E_{step\ 2}$.

$\Delta E_{step\ 1}$ = q + w = 45 J + (–10. J) = 35 J

$\Delta E_{step\ 2} = -\Delta E_{step\ 1}$ = –35 J = q + w,   –35 J = –60. J + w,   w = 25 J

85. $H_2(g) + 1/2\ O_2(g) \rightarrow H_2O(l)$ $\Delta H^\circ = \Delta H^\circ_{f,\ H_2O(l)}$ = –285.8 kJ

$H_2O(l) \rightarrow H_2(g) + 1/2\ O_2(g)$ ΔH° = 285.8 kJ

$\Delta E^\circ = \Delta H^\circ - P\Delta V = \Delta H^\circ - \Delta nRT$

ΔE° = 285.8 kJ – (1.50 – 0 mol)(8.3145 J K^{-1} mol^{-1})(298 K)$\left(\dfrac{1\ kJ}{1000\ J}\right)$

ΔE° = 285.8 kJ – 3.72 kJ = 282.1 kJ

87.   The specific heat of water is 4.18 J °C^{-1} g^{-1}, which is equal to 4.18 kJ °C^{-1} kg^{-1}

We have 1.00 kg of H_2O, so: 1.00 kg $\times \dfrac{4.18 \text{ kJ}}{°C \text{ kg}}$ = 4.18 kJ/°C

This is the portion of the heat capacity that can be attributed to H_2O.

Total heat capacity = C_{cal} + C_{H_2O}, C_{cal} = 10.84 − 4.18 = 6.66 kJ/°C

89. $Na_2SO_4(aq) + Ba(NO_3)_2(aq) \rightarrow BaSO_4(s) + 2 NaNO_3(aq)$ $\Delta H = ?$

1.00 L $\times \dfrac{2.00 \text{ mol}}{L}$ = 2.00 mol Na_2SO_4; 2.00 L $\times \dfrac{0.750 \text{ mol}}{L}$ = 1.50 mol $Ba(NO_3)_2$

The balanced equation requires a 1 : 1 mole ratio between Na_2SO_4 and $Ba(NO_3)_2$. Because we have fewer moles of $Ba(NO_3)_2$ present, it is limiting and 1.50 mol $BaSO_4$ will be produced [there is a 1 : 1 mole ratio between $Ba(NO_3)_2$ and $BaSO_4$].

Heat gain by solution = heat loss by reaction

Mass of solution = 3.00 L $\times \dfrac{1000 \text{ mL}}{1 \text{ L}} \times \dfrac{2.00 \text{ g}}{\text{mL}}$ = 6.00 $\times 10^3$ g

Heat gain by solution = $\dfrac{6.37 \text{ J}}{°C \text{ g}} \times$ 6.00 $\times 10^3$ g $\times$ (42.0 − 30.0) °C = 4.59 $\times 10^5$ J

Because the solution gained heat, the reaction is exothermic;   q = −4.59 $\times 10^5$ J for the reaction.

$\Delta H = \dfrac{-4.59 \times 10^5 \text{ J}}{1.50 \text{ mol } BaSO_4}$ = −3.06 $\times 10^5$ J/mol = −306 kJ/mol

91. $HNO_3(aq) + KOH(aq) \rightarrow H_2O(l) + KNO_3(aq)$ $\Delta H = -56$ kJ

0.2000 L $\times \dfrac{0.400 \text{ mol } HNO_3}{L}$ = 8.00 $\times 10^{-2}$ mol HNO_3

0.1500 L $\times \dfrac{0.500 \text{ mol KOH}}{L}$ = 7.50 $\times 10^{-2}$ mol KOH

Because the balanced reaction requires a 1 : 1 mole ratio between HNO_3 and KOH, and because fewer moles of KOH are actually present as compared with HNO_3, KOH is the limiting reagent.

7.50 $\times 10^{-2}$ mol KOH $\times \dfrac{-56 \text{ kJ}}{\text{mol KOH}}$ = −4.2 kJ; 4.2 kJ of heat is released.

93. $w = -P\Delta V$; Δn = moles of gaseous products – moles of gaseous reactants. Only gases can do PV work (we ignore solids and liquids). When a balanced reaction has more moles of product gases than moles of reactant gases (Δn positive), the reaction will expand in volume (ΔV positive), and the system will do work on the surroundings. For example, in reaction c, $\Delta n = 2 - 0 = 2$ moles, and this reaction would do expansion work against the surroundings. When a balanced reaction has a decrease in the moles of gas from reactants to products (Δn negative), the reaction will contract in volume (ΔV negative), and the surroundings will do compression work on the system, e.g., reaction a, where $\Delta n = 0 - 1 = -1$. When there is no change in the moles of gas from reactants to products, $\Delta V = 0$ and $w = 0$, e.g., reaction b, where $\Delta n = 2 - 2 = 0$.

When $\Delta V > 0$ ($\Delta n > 0$), then $w < 0$, and the system does work on the surroundings (c and e).

When $\Delta V < 0$ ($\Delta n < 0$), then $w > 0$, and the surroundings do work on the system (a and d).

When $\Delta V = 0$ ($\Delta n = 0$), then $w = 0$ (b).

95.

$I(g) + Cl(g) \rightarrow ICl(g)$	$\Delta H = -(211.3 \text{ kJ})$
$1/2\ Cl_2(g) \rightarrow Cl(g)$	$\Delta H = 1/2(242.3 \text{ kJ})$
$1/2\ I_2(g) \rightarrow I(g)$	$\Delta H = 1/2(151.0 \text{ kJ})$
$1/2\ I_2(s) \rightarrow 1/2\ I_2(g)$	$\Delta H = 1/2(62.8 \text{ kJ})$

$1/2\ I_2(s) + 1/2\ Cl_2(g) \rightarrow ICl(g)$ $\Delta H = 16.8 \text{ kJ/mol} = \Delta H^\circ_{f,\,ICl}$

97.   a.   aluminum oxide = Al_2O_3; $2\ Al(s) + 3/2\ O_2(g) \rightarrow Al_2O_3(s)$

b. $C_2H_5OH(l) + 3\ O_2(g) \rightarrow 2\ CO_2(g) + 3\ H_2O(l)$

c. $Ba(OH)_2(aq) + 2\ HCl(aq) \rightarrow 2\ H_2O(l) + BaCl_2(aq)$

d. $2\ C(\text{graphite, s}) + 3/2\ H_2(g) + 1/2\ Cl_2(g) \rightarrow C_2H_3Cl(g)$

e. $C_6H_6(l) + 15/2\ O_2(g) \rightarrow 6\ CO_2(g) + 3\ H_2O(l)$

Note: ΔH_{comb} values assume 1 mole of compound combusted.

f. $NH_4Br(s) \rightarrow NH_4^+(aq) + Br^-(aq)$

Challenge Problems

99. $H_2O(s) \rightarrow H_2O(l)$ $\Delta H = \Delta H_{fus}$; for 1 mol of supercooled water at $-15.0°C$ (or 258.2 K), $\Delta H_{fus,\ 258.2\ K} = 10.9 \text{ kJ}/2.00 \text{ mol} = 5.45 \text{ kJ/mol}$. Using Hess's law and the equation $\Delta H = nC_p\Delta T$:

$H_2O(s, 273.2\ K) \rightarrow H_2O(s, 258.2\ K)$ $\Delta H_1 = 1\ \text{mol}(37.5\ \text{J K}^{-1}\ \text{mol}^{-1})(-15.0\ K)$
$= -563\ J = -0.563\ kJ$

$H_2O(s, 258.2\ K) \rightarrow H_2O(l, 258.2\ K)$ $\Delta H_2 = 1\ \text{mol}(5.45\ \text{kJ/mol}) = 5.45\ kJ$

$H_2O(l, 258.2\ K) \rightarrow H_2O(l, 273.2\ K)$ $\Delta H_3 = 1\ \text{mol}(75.3\ \text{J K}^{-1}\ \text{mol}^{-1})(15.0\ K)$
$= 1130\ J = 1.13\ kJ$

___

$H_2O(s, 273.\ 2\ K) \rightarrow H_2O(l, 273.2\ K)$ $\Delta H_{fus, 273.2} = \Delta H_1 + \Delta H_2 + \Delta H_3$

$\Delta H_{fus, 273.2} = -0.563\ kJ + 5.45\ kJ + 1.13\ kJ,\quad \Delta H_{fus, 273.2} = 6.02\ kJ/mol$

101.   Molar heat capacity of $H_2O(l) = 4.184\ \text{J K}^{-1}\ \text{g}^{-1}(18.015\ \text{g/mol}) = 75.37\ \text{J K}^{-1}\ \text{mol}^{-1}$

Molar heat capacity of $H_2O(g) = 2.02\ \text{J K}^{-1}\ \text{g}^{-1}(18.015\ \text{g/mol}) = 36.4\ \text{J K}^{-1}\ \text{mol}^{-1}$

Using Hess's law and the equation $\Delta H = nC_p\Delta T$:

$H_2O(l, 298.2\ K) \rightarrow H_2O(l, 373.2\ K)$ $\Delta H_1 = 1\ \text{mol}(75.37\ \text{J K}^{-1}\ \text{mol}^{-1})(75.0\ K)(1\ \text{kJ}/1000\ J)$
$= 5.65\ kJ$

$H_2O(l, 373.2\ K) \rightarrow H_2O(g, 373.2\ K)$ $\Delta H_2 = 1\ \text{mol}(40.66\ \text{kJ/mol}) = 40.66\ kJ$

$H_2O(g, 373.2\ K) \rightarrow H_2O(g, 298.2\ K)$ $\Delta H_3 = 1\ \text{mol}(36.4\ \text{J K}^{-1}\ \text{mol}^{-1})(-75.0\ K)(1\ \text{kJ}/1000\ J)$
$= -2.73\ kJ$

___

$H_2O(l, 298.2\ K) \rightarrow H_2O(g, 298.2\ K)$ $\Delta H_{vap, 298.2\ K} = \Delta H_1 + \Delta H_2 + \Delta H_3 = 43.58\ kJ/mol$

Using ΔH_f° values in Appendix 4 (which are determined at 25° C):

$\Delta H^\circ{}_{vap} = -242\ kJ - (-286\ kJ) = 44\ kJ$

To two significant figures, the two calculated ΔH_{vap} values agree (as they should).

103.   Energy used in 8.0 hours = 40. kWh = $\dfrac{40.\ \text{kJ h}}{s} \times \dfrac{3600\ s}{h} = 1.4 \times 10^5\ kJ$

Energy from the sun in 8.0 hours = $\dfrac{10.\ \text{kJ}}{s\ m^2} \times \dfrac{60\ s}{\min} \times \dfrac{60\ \min}{h} \times 8.0\ h = 2.9 \times 10^4\ kJ/m^2$

Only 15% of the sunlight is converted into electricity:

$0.15 \times (2.9 \times 10^4\ kJ/m^2) \times \text{area} = 1.4 \times 10^5\ kJ,\quad \text{area} = 32\ m^2$

105.   For an isothermal (constant T) process involving the expansion or compression of a gas, $\Delta E = nC_v\Delta T = 0$ (ΔH is also zero). Because $\Delta E = q + w = 0$, $q = -w = -(-P_{ex}\Delta V)$. So if the expansion or compression occurs against some nonzero external pressure, $w \neq 0$ and $q \neq 0$. Instead, $q = -w = P_{ex}\Delta V$ (if the gas expands or contracts against some constant pressure).

CHAPTER 10

SPONTANEITY, ENTROPY, AND FREE ENERGY

Spontaneity and Entropy

13.   Possible arrangements for one molecule:

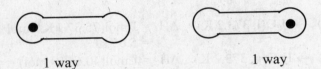

1 way   1 way

Both are equally probable.

Possible arrangements for two molecules:

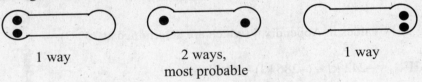

1 way   2 ways,   1 way
            most probable

Possible arrangement for three molecules:

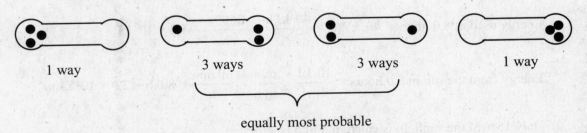

1 way   3 ways   3 ways   1 way

equally most probable

15.   We draw all the possible arrangements of the two particles in the three levels.

2 kJ	__	__	x	__	x	xx
1 kJ	__	x	__	xx	x	__
0 kJ	xx	x	x	__	__	__

Total E =   0 kJ   1 kJ   2 kJ   2 kJ   3 kJ   4 kJ

The most likely total energy is 2 kJ.

196

17.   Processes a, b, d, and g are spontaneous. Processes c, e, and f require an external source of energy in order to occur since they are nonspontaneous.

19.   a.   Positional probability increases; there is a greater volume accessible to the randomly moving gas molecules, which increases disorder.

       b.   The positional probability doesn't change. There is no change in volume and thus no change in the numbers of positions of the molecules.

       c.   Positional probability decreases because the volume decreases (P and V are inversely related).

21.   There are more ways to roll a seven. We can consider all the possible throws by constructing a table.

One die	1	2	3	4	5	6	
1	2	3	4	5	6	7	
2	3	4	5	6	7	8	
3	4	5	6	7	8	9	Sum of the two dice
4	5	6	7	8	9	10	
5	6	7	8	9	10	11	
6	7	8	9	10	11	12	

There are six ways to get a seven, more than any other number. The seven is not favored by energy; rather, it is favored by probability. To change the probability, we would have to expend energy (do work).

Energy, Enthalpy, and Entropy Changes Involving Ideal Gases and Physical Changes

23. 1.00×10^3 g $C_2H_6 \times 1$ mol/30.07 g = 33.3 mol

$q_v = \Delta E = nC_v\Delta T = 33.3$ mol$(44.60$ J K^{-1}mol$^{-1})(48.4$ K$) = 7.19 \times 10^4$ J = 71.9 kJ

At constant volume, 71.9 kJ of energy is required, and $\Delta E = 71.9$ kJ.

At constant pressure (assuming ethane acts as an ideal gas):

$C_p = C_v + R = 44.60 + 8.31 = 52.91$ J K^{-1}mol^{-1}

Energy required = $q_p = \Delta H = nC_p\Delta T = 33.3$ mol$(52.91$ J K^{-1}mol$^{-1})(48.4$ K$) = 8.53 \times 10^4$ J
$= 85.3$ kJ

For the constant pressure process, $\Delta E = 71.9$ kJ, as calculated previously (ΔE is unchanged).

25.   Heat gain by He = heat loss by N_2; because ΔT in °C = ΔT in K, the units on the heat capacities also could be J °C^{-1} mol^{-1}.

$(0.400$ mol$)(12.5$ J °C^{-1} mol$^{-1})(T_f - 20.0$ °C$) = (0.600$ mol$)(20.7$ J °C^{-1} mol$^{-1})(100.0$ °C $- T_f)$

$(5.00)T_f - 100. = 1240 - (12.4)T_f$, $T_f = \dfrac{1340}{17.4} = 77.0$ °C

27. $P_1V_1 = P_2V_2$ since n and T are constant; $P_2 = \dfrac{P_1V_1}{V_2} = \dfrac{(5.0)(1.0)}{(2.0)} = 2.5$ atm

Gas expands isothermally against no pressure, so $\Delta E = 0$, w = 0, and q = 0.

$\Delta E = 0$, so $q_{rev} = -w_{rev} = nRT \ln(V_2/V_1)$; $T = \dfrac{PV}{nR} = 61$ K

$q_{rev} = (1.0$ mol$)(8.3145$ J K^{-1} mol$^{-1})(61$ K$) \ln(2.0/1.0) = 350$ J

29.   a. $q_v = \Delta E = nC_v\Delta T = (1.000$ mol$)(28.95$ J K^{-1} mol$^{-1})(350.0 - 298.0$ K$)$

$q_v = 1.51 \times 10^3$ J = 1.51 kJ

$q_p = \Delta H = nC_p\Delta T = (1.000)(37.27)(350.0 - 298.0) = 1.94 \times 10^3$ J = 1.94 kJ

b. $\Delta S = S_{350} - S_{298} = nC_p \ln(T_2/T_1)$

$S_{350} - 213.64$ J/K $= (1.000$ mol$)(37.27$ J K^{-1} mol$^{-1}) \ln(350.0/298.0)$

$S_{350} = 213.64$ J/K + 5.994 J/K = 219.63 J/K = molar entropy at 350.0 K and 1.000 atm

c. $\Delta S = nR \ln(V_2/V_1)$, $V = nRT/P$, $\Delta S = nR \ln(P_1/P_2) = S_{(350,\ 1.174)} - S_{(350,\ 1.000)}$

$\Delta S = S_{(350,\ 1.174)} - 219.63$ J/K $= (1.000$ mol$)(8.3145$ J K^{-1} mol$^{-1}) \ln(1.000$ atm$/1.174$ atm$)$

$\Delta S = -1.334$ J/K $= S_{(350,\ 1.174)} - 219.63$, $S_{(350,\ 1.174)} = 218.30$ J K^{-1} mol^{-1}

31.   For A(l, 125°C) → A(l, 75°C):

$\Delta S = nC_p \ln(T_2/T_1) = 1.00$ mol$(75.0$ J K^{-1} mol$^{-1}) \ln(348$ K$/398$ K$) = -10.1$ J/K

For A(l, 75°C) → A(g, 155°C): $\Delta S = 75.0$ J K^{-1} mol^{-1}

For A(g, 155°C) → A(g,125°C):

$\Delta S = nC_p \ln(T_2/T_1) = 1.00$ mol$(29.0$ J K^{-1} mol$^{-1}) \ln(398$ K$/428$ K$) = -2.11$ J/K

The sum of the three step gives A(l, 125°C) → A(g, 125°C). ΔS for this process is the sum of ΔS for each of the three steps.

$$\Delta S = -10.1 + 75.0 - 2.11 = 62.8 \text{ J/K}$$

For a phase change, $\Delta S = \Delta H/T$. At 125°C: $\Delta H_{vap} = T\Delta S = 398 \text{ K}(62.8 \text{ J/K}) = 2.50 \times 10^4 \text{ J}$

33.   Calculate the final temperature by equating heat loss to heat gain. Keep all quantities positive to avoid sign errors.

$$(3.00 \text{ mol})(75.3 \text{ J °C}^{-1} \text{ mol}^{-1})(T_f - 0°C) = (1.00 \text{ mol})(75.3 \text{ J °C}^{-1} \text{ mol}^{-1})(100.°C - T_f)$$

Solving: $T_f = 25°C = 298$ K

Now we can calculate ΔS for the various changes using $\Delta S = nC_p \ln(T_2/T_1)$.

Heat 3 mol H_2O: $\Delta S_1 = (3.00 \text{ mol})(75.3 \text{ J K}^{-1} \text{ mol}^{-1}) \ln(298 \text{ K}/273 \text{ K}) = 19.8 \text{ J/K}$

Cool 1 mol H_2O: $\Delta S_2 = (1.00 \text{ mol})(75.3 \text{ J K}^{-1} \text{ mol}^{-1}) \ln(298/373) = -16.9 \text{ J/K}$

$\Delta S_{total} = \Delta S_{heat} + \Delta S_{cool} = 19.8 - 16.9 = 2.9 \text{ J/K}$

Entropy and the Second Law of Thermodynamics: Free Energy

35.   Living organisms need an external source of energy to carry out these processes. Green plants use the energy from sunlight to produce glucose from carbon dioxide and water by photosynthesis. In the human body the energy released from the metabolism of glucose helps drive the synthesis of proteins. For all processes combined, ΔS_{univ} must be greater than zero (second law).

37. ΔS_{surr} is primarily determined by heat flow. This heat flow into or out of the surroundings comes from the heat flow out of or into the system. In an exothermic process ($\Delta H < 0$), heat flows into the surroundings from the system. The heat flow into the surroundings increases the random motions in the surroundings and increases the entropy of the surroundings ($\Delta S_{surr} > 0$). This is a favorable driving force for spontaneity. In an endothermic reaction ($\Delta H > 0$), heat is transferred from the surroundings into the system. This heat flow out of the surroundings decreases the random motions in the surroundings and decreases the entropy of the surroundings ($\Delta S_{surr} < 0$). This is unfavorable. The magnitude of ΔS_{surr} also depends on the temperature. The relationship is inverse; at low temperatures, a specific amount of heat exchange makes a larger percent change in the surroundings than the same amount of heat flow at a higher temperature. The negative sign in the $\Delta S_{surr} = -\Delta H/T$ equation is necessary to get the signs correct. For an exothermic reaction where ΔH is negative, this increases ΔS_{surr}, so the negative sign converts the negative ΔH value into a positive quantity. For an endothermic process where ΔH is positive, the sign of ΔS_{surr} is negative, and the negative sign converts the positive ΔH value into a negative quantity.

39. a. $\Delta S_{surr} = \dfrac{-\Delta H}{T} = \dfrac{-(-2221\,kJ)}{298\,K} = 7.45\,kJ/K = 7.45 \times 10^3\,J/K$

    b. $\Delta S_{surr} = \dfrac{-\Delta H}{T} = \dfrac{-112\,kJ}{298\,K} = -0.376\,kJ/K = -376\,J/K$

41. a. Decrease in positional probability; $\Delta S°(-)$

    b. Increase in positional probability; $\Delta S°(+)$

    c. Decrease in positional probability $(\Delta n < 0)$; $\Delta S°(-)$

    d. Decrease in positional probability $(\Delta n < 0)$; $\Delta S°(-)$

    e. HCl(g) has a greater positional probability due to the huge volume of gas; $\Delta S°(-)$.

    f. Increase in positional probability; $\Delta S°(+)$

For c, d, and e, concentrate on the gaseous products and reactants. When there are more gaseous product molecules than gaseous reactant molecules $(\Delta n > 0)$, then $\Delta S°$ will be positive. When Δn is negative, then $\Delta S°$ is negative.

43. $C_2H_2(g) + 4\,F_2(g) \rightarrow 2\,CF_4(g) + H_2(g)$; $\Delta S° = 2S°_{CF_4} + S°_{H_2} - [S°_{C_2H_2} + 4S°_{F_2}]$

    $-358\,J/K = (2\,mol)S°_{CF_4} + 131\,J/K - [201\,J/K + 4(203\,J/K)]$, $S°_{CF_4} = 262\,J\,K^{-1}\,mol^{-1}$

45. $-144\,J/K = (2\,mol)\,S°_{AlBr_3} - [2(28\,J/K) + 3(152\,J/K)]$, $S°_{AlBr_3} = 184\,J\,K^{-1}\,mol^{-1}$

47. At the boiling point, $\Delta G = 0$ so $\Delta H = T\Delta S$. $T = \dfrac{\Delta H}{\Delta S} = \dfrac{58.51 \times 10^3\,J/mol}{92.92\,J\,K^{-1}\,mol^{-1}} = 629.7\,K$

49. a. $NH_3(s) \rightarrow NH_3(l)$; $\Delta G = \Delta H - T\Delta S = 5650\,J/mol - 200.\,K(28.9\,J\,K^{-1}\,mol^{-1})$

      $\Delta G = 5650\,J/mol - 5780\,J/mol = -130\,J/mol$

      Yes, NH_3 will melt because $\Delta G < 0$ at this temperature.

    b. At the melting point, $\Delta G = 0$, so $T = \dfrac{\Delta H}{\Delta S} = \dfrac{5650\,J/mol}{28.9\,J\,K^{-1}\,mol^{-1}} = 196\,K$.

51. Solid I $\rightarrow$ solid II; equilibrium occurs when $\Delta G = 0$.

    $\Delta G = \Delta H - T\Delta S$, $\Delta H = T\Delta S$, $T = \Delta H/\Delta S = \dfrac{-743.1\,J/mol}{-17.0\,J\,K^{-1}\,mol^{-1}} = 43.7\,K = -229.5°C$

Free Energy and Chemical Reactions

53. $-5490. \text{ kJ} = 8(-394 \text{ kJ}) + 10(-237 \text{ kJ}) - 2 \Delta G^\circ_{f, C_4H_{10}}, \quad \Delta G^\circ_{f, C_4H_{10}} = -16 \text{ kJ/mol}$

55. $\Delta G^\circ = -58.03 \text{ kJ} - (298 \text{ K})(-0.1766 \text{ kJ/K}) = -5.40 \text{ kJ}$

$$\Delta G^\circ = 0 = \Delta H^\circ - T \Delta S^\circ, \quad T = \frac{\Delta H^\circ}{\Delta S^\circ} = \frac{-58.03 \text{ kJ}}{-0.1766 \text{ kJ/K}} = 328.6 \text{ K}$$

ΔG° is negative below 328.6 K, where the favorable ΔH° term dominates.

57.   a. $\Delta G^\circ = 2(-270. \text{ kJ}) - 2(-502 \text{ kJ}) = 464 \text{ kJ}$

b.   Because ΔG° is positive, this reaction is not spontaneous at standard conditions at 298 K.

c. $\Delta G^\circ = \Delta H^\circ - T \Delta S^\circ, \quad \Delta H^\circ = \Delta G^\circ + T \Delta S^\circ = 464 \text{ kJ} + 298 \text{ K}(0.179 \text{ kJ/K}) = 517 \text{ kJ}$

We need to solve for the temperature when $\Delta G^\circ = 0$:

$$\Delta G^\circ = 0 = \Delta H^\circ - T \Delta S^\circ, \quad T = \frac{\Delta H^\circ}{\Delta S^\circ} = \frac{517 \text{ kJ}}{0.179 \text{ kJ/K}} = 2890 \text{ K}$$

This reaction will be spontaneous at standard conditions ($\Delta G^\circ < 0$) when T > 2890 K. At these temperatures, the favorable entropy term will dominate.

59. $CH_4(g) + CO_2(g) \rightarrow CH_3CO_2H(l)$

$\Delta H^\circ = -484 - [-75 + (-393.5)] = -16 \text{ kJ}; \quad \Delta S^\circ = 160. - (186 + 214) = -240. \text{ J/K}$

$\Delta G^\circ = \Delta H^\circ - T \Delta S^\circ = -16 \text{ kJ} - (298 \text{ K})(-0.240 \text{ kJ/K}) = 56 \text{ kJ}$

At standard concentrations, where $\Delta G = \Delta G^\circ$, this reaction is spontaneous only at temperatures below T = $\Delta H^\circ / \Delta S^\circ$ = 67 K (where the favorable ΔH° term will dominate, giving a negative ΔG° value). This is not practical. Substances will be in condensed phases and rates will be very slow at this extremely low temperature.

$CH_3OH(g) + CO(g) \rightarrow CH_3CO_2H(l)$

$\Delta H^\circ = -484 - [-110.5 + (-201)] = -173 \text{ kJ}; \quad \Delta S^\circ = 160. - (198 + 240.) = -278 \text{ J/K}$

$\Delta G^\circ = -173 \text{ kJ} - (298 \text{ K})(-0.278 \text{ kJ/K}) = -90. \text{ kJ}$

This reaction also has a favorable enthalpy and an unfavorable entropy term. This reaction is spontaneous at temperatures below T = $\Delta H^\circ / \Delta S^\circ$ = 622 K (assuming standard concentrations). The reaction of CH_3OH and CO will be preferred at standard conditions. It is spontaneous at high enough temperatures that the rates of reaction should be reasonable.

61.   Enthalpy is not favorable, so ΔS must provide the driving force for the change. Thus ΔS is positive. There is an increase in positional probability, so the original enzyme has the more ordered structure (has the smaller positional probability).

63.   Because there are more product gas molecules than reactant gas molecules ($\Delta n > 0$), ΔS will be positive. From the signs of ΔH and ΔS, this reaction is spontaneous at all temperatures. It will cost money to heat the reaction mixture. Because there is no thermodynamic reason to do this, the purpose of the elevated temperature must be to increase the rate of the reaction, i.e., kinetic reasons.

Free Energy: Pressure Dependence and Equilibrium

65.   At constant temperature and pressure, the sign of ΔG (positive or negative) tells us which reaction is spontaneous (the forward or reverse reaction). If $\Delta G < 0$, then the forward reaction is spontaneous, and if $\Delta G > 0$, then the reverse reaction is spontaneous. If $\Delta G = 0$, then the reaction is at equilibrium (neither the forward nor reverse reaction is spontaneous). $\Delta G°$ gives the reaction equilibrium position by determining K through the equation $\Delta G° = -RT \ln K$. $\Delta G°$ can only be used to predict spontaneity when all reactants and products are present at standard pressures of 1 atm and/or standard concentrations of 1 M.

67. $\Delta G = \Delta G° + RT \ln Q = \Delta G° + RT \ln \dfrac{P_{N_2O_4}}{P_{NO_2}^2}$

       $\Delta G° = 1 \text{ mol}(98 \text{ kJ/mol}) - 2 \text{ mol}(52 \text{ kJ/mol}) = -6 \text{ kJ}$

   a.   These are standard conditions, so $\Delta G = \Delta G°$ since $Q = 1$ and $\ln Q = 0$. Because $\Delta G°$ is negative, the forward reaction is spontaneous. The reaction shifts right to reach equilibrium.

   b. $\Delta G = -6 \times 10^3 \text{ J} + 8.3145 \text{ J K}^{-1} \text{ mol}^{-1}(298 \text{ K}) \ln \dfrac{0.50}{(0.21)^2}$

       $\Delta G = -6 \times 10^3 \text{ J} + 6.0 \times 10^3 \text{ J} = 0$

       Since $\Delta G = 0$, this reaction is at equilibrium (no shift).

   c. $\Delta G = -6 \times 10^3 \text{ J} + 8.3145 \text{ J K}^{-1} \text{ mol}^{-1}(298 \text{ K}) \ln \dfrac{1.6}{(0.29)^2}$

       $\Delta G = -6 \times 10^3 \text{ J} + 7.3 \times 10^3 \text{ J} = 1.3 \times 10^3 \text{ J} = 1 \times 10^3 \text{ J}$

       Since ΔG is positive, the reverse reaction is spontaneous, so the reaction shifts to the left to reach equilibrium.

69. $\Delta H° = 2\,\Delta H°_{f,\,NH_3} = 2(-46) = -92$ kJ; $\Delta G° = 2\Delta G°_{f,\,NH_3} = 2(-17) = -34$ kJ

$\Delta S° = 2(193$ J/K$) - [192$ J/K $+ 3(131$ J/K$)] = -199$ J/K; $\Delta G° = -RT \ln K$

$$K = \frac{-\Delta G°}{RT} = \exp\left(\frac{-(-34{,}000 \text{ J})}{(8.3145 \text{ J K}^{-1}\text{mol}^{-1})(298 \text{ K})}\right) = e^{13.72} = 9.1 \times 10^5$$

Note: When determining exponents, we will round off after the calculation is complete. This helps eliminate excessive round off error.

a. $\Delta G = \Delta G° + RT \ln \dfrac{P_{NH_3}^2}{P_{N_2} \times P_{H_2}^3} = -34$ kJ $+ \dfrac{(8.3145 \text{ J K}^{-1}\text{mol}^{-1})(298 \text{ K})}{1000 \text{ J/kJ}} \ln \dfrac{(50.)^2}{(200.)(200.)^3}$

$\Delta G = -34$ kJ $- 33$ kJ $= -67$ kJ

b. $\Delta G = -34$ kJ $+ \dfrac{(8.3145 \text{ J K}^{-1}\text{mol}^{-1})(298 \text{ K})}{1000 \text{ J/kJ}} \ln \dfrac{(200.)^2}{(200.)(600.)^3}$

$\Delta G = -34$ kJ $- 34.4$ kJ $= -68$ kJ

c. Assume $\Delta H°$ and $\Delta S°$ are temperature independent.

$\Delta G°_{100} = \Delta H° - T\Delta S°$, $\Delta G°_{100} = -92$ kJ $- (100.$ K$)(-0.199$ kJ/K$) = -72$ kJ

$\Delta G_{100} = \Delta G°_{100} + RT \ln Q = -72$ kJ $+ \dfrac{(8.3145 \text{ J K}^{-1}\text{mol}^{-1})(100.\text{K})}{1000 \text{ J/kJ}} \ln \dfrac{(10.)^2}{(50.)(200.)^3}$

$\Delta G_{100} = -72$ kJ $- 13$ kJ $= -85$ kJ

d. $\Delta G°_{700} = -92$ kJ $- (700.$ K$)(-0.199$ kJ/K$) = 47$ kJ

$\Delta G_{700} = 47$ kJ $+ \dfrac{(8.3145 \text{ J K}^{-1}\text{mol}^{-1})(700.\text{K})}{1000 \text{ J/kJ}} \ln \dfrac{(10.)^2}{(50.)(200.)^3}$

$\Delta G_{700} = 47$ kJ $- 88$ kJ $= -41$ kJ

71. $\Delta G° = 2$ mol$(-229$ kJ/mol$) - [2$ mol$(-34$ kJ/mol$) + 1$ mol$(-300.$ kJ/mol$)] = -90.$ kJ

$$K = \exp\frac{-\Delta G°}{RT} = \exp\left[\frac{-(-9.0 \times 10^4 \text{ J})}{8.3145 \text{ J K}^{-1}\text{mol}^{-1}(298 \text{ K})}\right] = e^{36.32} = 5.9 \times 10^{15}$$

$\Delta G° = \Delta H° - T\Delta S°$; because there is a decrease in the number of moles of gaseous particles, $\Delta S°$ is negative. Because $\Delta G°$ is negative, $\Delta H°$ must be negative. The reaction will be spontaneous at low temperatures (the favorable $\Delta H°$ term dominates at low temperatures).

73. $\Delta G° = -RT \ln K$; to determine K at a temperature other than 25°C, one needs to know $\Delta G°$ at that temperature. We assume $\Delta H°$ and $\Delta S°$ are temperature-independent and use the equation $\Delta G° = \Delta H° - T\Delta S°$ to estimate $\Delta G°$ at the different temperature. For $K = 1$, we want $\Delta G° = 0$, which occurs when $\Delta H° = T\Delta S°$. Again, assume $\Delta H°$ and $\Delta S°$ are temperature independent, and then solve for T $(= \Delta H° /\Delta S°)$. At this temperature, $K = 1$ because $\Delta G° = 0$. This only works for reactions where the signs of $\Delta H°$ and $\Delta S°$ are the same (either both positive or both negative). When the signs are opposite, K will always be greater than one (when $\Delta H°$ is negative and $\Delta S°$ is positive) or K will always be less than one (when $\Delta H°$ is positive and $\Delta S°$ is negative). When the signs of $\Delta H°$ and $\Delta S°$ are opposite, K can never equal one.

75. $\Delta G° = - RT \ln K$; when $K = 1.00$, $\Delta G° = 0$ since $\ln(1.00) = 0$. $\Delta G° = 0 = \Delta H° - T\Delta S°$

   $\Delta H° = 3(-242 \text{ kJ}) - [- 826 \text{ kJ}] = 100. \text{ kJ}$; $\Delta S° = [2(27 \text{ J/K}) + 3(189 \text{ J/K})] -$
   $$[90. \text{ J/K} + 3(131 \text{ J/K})] = 138 \text{ J/K}$$

   $$\Delta H° = T\Delta S°, \quad T = \frac{\Delta H°}{\Delta S°} = \frac{100. \text{ kJ}}{0.138 \text{ kJ/K}} = 725 \text{ K}$$

77. $$K = \frac{P_{NF_3}^2}{P_{N_2}^2 \times P_{F_2}^3} = \frac{(0.48)^2}{0.021(0.063)^3} = 4.4 \times 10^4$$

   $\Delta G_{800}^o = -RT \ln K = -8.3145 \text{ J K}^{-1} \text{ mol}^{-1}(800. \text{ K}) \ln(4.4 \times 10^4) = -7.1 \times 10^4 \text{ J/mol}$
   $$= -71 \text{ kJ/mol}$$

79. Because the partial pressure of C(g) decreased, the net change that occurs for this reaction to reach equilibrium is for products to convert to reactants.

	A(g)	+	2 B(g)	⇌	C(g)
Initial	0.100 atm		0.100 atm		0.100 atm
Change	+x		+2x	←	-x
Equil.	0.100 + x		0.100 + 2x		0.100 - x

   From the problem, $P_C = 0.040 \text{ atm} = 0.100 - x$, $x = 0.060 \text{ atm}$.

   The equilibrium partial pressures are: $P_A = 0.100 + x = 0.100 + 0.060 = 0.160 \text{ atm}$, $P_B = 0.100 + 2(0.60) = 0.220 \text{ atm}$, and $P_C = 0.040 \text{ atm}$.

   $$K = \frac{0.040}{0.160(0.220)^2} = 5.2$$

   $\Delta G° = -RT \ln K = -8.3145 \text{ J K}^{-1} \text{ mol}^{-1}(298 \text{ K}) \ln(5.2) = -4.1 \times 10^3 \text{ J/mol} = -4.1 \text{ kJ/mol}$

81.   When reactions are added together, the equilibrium constants are multiplied together to determine the K value for the final reaction.

$$H_2(g) + O_2(g) \rightleftharpoons H_2O_2(g) \qquad K = 2.3 \times 10^6$$
$$H_2O(g) \rightleftharpoons H_2(g) + 1/2\,O_2(g) \qquad K = (1.8 \times 10^{37})^{-1/2}$$

$$H_2O(g) + 1/2\,O_2(g) \rightleftharpoons H_2O_2(g) \qquad K = 2.3 \times 10^6 (1.8 \times 10^{37})^{-1/2} = 5.4 \times 10^{-13}$$

$$\Delta G^\circ = -RT \ln K = \frac{-8.3145\ J}{K\ mol} (600.\ K) \ln(5.4 \times 10^{-13}) = 1.4 \times 10^5\ J/mol = 140\ kJ/mol$$

83.
$$HgbO_2 \rightarrow Hgb + O_2 \qquad \Delta G^\circ = -(-70\ kJ)$$
$$Hgb + CO \rightarrow HgbCO \qquad \Delta G^\circ = -80\ kJ$$

$$HgbO_2 + CO \rightarrow HgbCO + O_2 \qquad \Delta G^\circ = -10\ kJ$$

$$\Delta G^\circ = -RT \ln K, \quad K = \exp\left(\frac{-\Delta G^\circ}{RT}\right) = \exp\left(\frac{-(-10 \times 10^3\ J)}{(8.3145\ J\ K^{-1}\ mol^{-1})(298\ K)}\right) = 60$$

85.   At 25.0°C: $\Delta G^\circ = \Delta H^\circ - T\Delta S^\circ = -58.03 \times 10^3\ J/mol - (298.2\ K)(-176.6\ J\ K^{-1}\ mol^{-1})$
$$= -5.37 \times 10^3\ J/mol$$

$$\Delta G^\circ = -RT \ln K, \ \ln K = \frac{-\Delta G^\circ}{RT} = \frac{-(-5.37 \times 10^3\ J/mol)}{(8.3145\ J\ K^{-1}\ mol^{-1})(298.2\ K)} = 2.166; \ \ K = e^{2.166} = 8.72$$

At 100.0°C: $\Delta G^\circ = -58.03 \times 10^3\ J/mol - (373.2\ K)(-176.6\ J\ K^{-1}\ mol^{-1}) = 7.88 \times 10^3\ J/mol$

$$\ln K = \frac{-(7.88 \times 10^3\ J/mol)}{(8.3145\ J\ K^{-1}\ mol^{-1})(373.2\ K)} = -2.540, \ K = e^{-2.540} = 0.0789$$

87.   From the equation in Exercise 10.86, a graph of ln K vs. 1/T will yield a straight line with slope equal to $-\Delta H^\circ/R$ and y intercept equal to $\Delta S^\circ/R$.

a.

Temp (°C)	T(K)	1000/T (K^{-1})	K_w	ln K_w
0	273	3.66	1.14×10^{-15}	−34.408
25	298	3.36	1.00×10^{-14}	−32.236
35	308	3.25	2.09×10^{-14}	−31.499
40.	313	3.19	2.92×10^{-14}	−31.165
50.	323	3.10	5.47×10^{-14}	−30.537

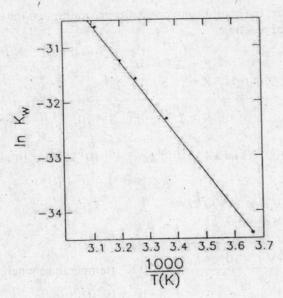

The straight-line equation (from a calculator) is: $\ln K = -6.91 \times 10^3 \left(\dfrac{1}{T}\right) - 9.09$

Slope $= -6.91 \times 10^3 \text{ K} = \dfrac{-\Delta H^\circ}{R}$

$\Delta H^\circ = -(-6.91 \times 10^3 \text{ K} \times 8.3145 \text{ J K}^{-1} \text{ mol}^{-1}) = 5.75 \times 10^4 \text{ J/mol} = 57.5 \text{ kJ/mol}$

y intercept $= -9.09 = \dfrac{\Delta S^\circ}{R}$, $\Delta S^\circ = -9.09 \times 8.3145 \text{ J K}^{-1} \text{ mol}^{-1} = -75.6 \text{ J K}^{-1} \text{ mol}^{-1}$

b. From part a, $\Delta H^\circ = 57.5$ kJ/mol and $\Delta S^\circ = -75.6$ J K^{-1} mol^{-1}. Assuming that ΔH° and ΔS° are temperature independent:

$$\Delta G^\circ = 57{,}500 \text{ J/mol} - 647 \text{ K}(-75.6 \text{ J K}^{-1} \text{ mol}^{-1}) = 106{,}400 \text{ J/mol} = 106.4 \text{ kJ/mol}$$

89.   From Exercise 10.86: $\ln K = \dfrac{-\Delta H^\circ}{RT} + \dfrac{\Delta S^\circ}{R}$, $R = 8.3145$ J K^{-1} mol^{-1}

For two sets of K and T data:

$$\ln K_1 = \dfrac{-\Delta H^\circ}{R}\left(\dfrac{1}{T_1}\right) + \dfrac{\Delta S^\circ}{R}; \quad \ln K_2 = \dfrac{-\Delta H^\circ}{R}\left(\dfrac{1}{T_2}\right) + \dfrac{\Delta S^\circ}{R}$$

Subtracting the first expression from the second:

$$\ln K_2 - \ln K_1 = \dfrac{\Delta H^\circ}{R}\left(\dfrac{1}{T_1} - \dfrac{1}{T_2}\right) \text{ or } \ln \dfrac{K_2}{K_1} = \dfrac{\Delta H^\circ}{R}\left(\dfrac{1}{T_1} - \dfrac{1}{T_2}\right)$$

$$\ln\left(\dfrac{3.25 \times 10^{-2}}{8.84}\right) = \dfrac{\Delta H^\circ}{8.3145 \text{ J K}^{-1} \text{ mol}^{-1}}\left(\dfrac{1}{298 \text{ K}} - \dfrac{1}{348 \text{ K}}\right)$$

$-5.61 = (5.8 \times 10^{-5} \text{ mol/J})(\Delta H°), \ \Delta H° = -9.7 \times 10^4 \text{ J/mol}$

For K = 8.84 at T = 25°C:

$$\ln(8.84) = \frac{-(-9.7 \times 10^4 \text{ J/mol})}{(8.3145 \text{ J K}^{-1} \text{ mol}^{-1})(298 \text{ K})} + \frac{\Delta S°}{8.3145 \text{ J K}^{-1} \text{ mol}^{-1}}$$

$$\frac{\Delta S°}{8.3145} = -37, \ \Delta S° = -310 \text{ J K}^{-1} \text{ mol}^{-1}$$

We get the same value for $\Delta S°$ using $K = 3.25 \times 10^{-2}$ at T = 348 K data. $\Delta G° = -RT \ln K$; when K = 1.00, then $\Delta G° = 0$ since $\ln(1.00) = 0$. $\Delta G° = 0 = \Delta H° - T\Delta S°$. Assuming that $\Delta H°$ and $\Delta S°$ do not depend on temperature:

$$\Delta H° = T\Delta S°, \ T = \frac{\Delta H°}{\Delta S°} = \frac{-9.7 \times 10^4 \text{ J/mol}}{-310 \text{ J K}^{-1} \text{ mol}^{-1}} = 310 \text{ K} \quad \text{(temperature where K = 1.00)}$$

Adiabatic Processes

91.   For reversible adaiabatic process, $P_1V_1^{\gamma} = P_2V_2^{\gamma}$, where $\gamma = C_p/C_v$. For an ideal gas, $C_p = C_v + R$.

$C_p = C_v + R = 20.5 \text{ J K}^{-1} \text{ mol}^{-1} + 8.3145 \text{ J K}^{-1} \text{ mol}^{-1} = 28.8 \text{ J K}^{-1} \text{ mol}^{-1}$

$$V_2^{\gamma} = \frac{P_1 V_1^{\gamma}}{P_2}, \ \text{where } \gamma = \frac{C_p}{C_v} = \frac{28.8}{20.5} = 1.40$$

$$V_1 = \frac{nRT_1}{P_1} = \frac{1.75 \text{ mol}(0.08206 \text{ L atm K}^{-1} \text{ mol}^{-1})(294 \text{ K})}{1.50 \text{ atm}} = 28.1 \text{ L}$$

$$V_2^{1.40} = \frac{1.50 \text{ atm}(28.1 \text{ L})^{1.40}}{4.50 \text{ atm}} = 35.6, \ V_2 = (35.6)^{1/1.40} = 12.8 \text{ L}$$

To calculate $w = \Delta E = nC_V\Delta T$, we need T_2 in order to calculate ΔT.

$$T_2 = \frac{P_2 V_2}{nR} = \frac{4.50 \text{ atm}(12.8 \text{ L})}{1.75 \text{ mol}(0.08206 \text{ L atm K}^{-1} \text{ mol}^{-1})} = 401 \text{ K}$$

Note: we also could have used the formula $T_2/T_1 = (V_1/V_2)^{\gamma-1}$ to calculate T_2.

$w = \Delta E = nC_V\Delta T = 1.75 \text{ mol}(20.5 \text{ J K}^{-1} \text{ mol}^{-1})(401 \text{ K} - 294 \text{ K}) = 3.84 \times 10^3 \text{ J}$

Note that for the adiabatic reversible compression, T increases, so ΔE increases. As work is added during the compression, the internal energy of the system increases ($\Delta E = w$).

93.   a.   Isothermal reversible expansion: $\Delta T = 0$ so $\Delta E = 0$

$w = -nRT \ln(V_2/V_1) = -nRT \ln(P_1/P_2)$

$w = -1.00 \text{ mol}(8.3145 \text{ J K}^{-1} \text{ mol}^{-1})(300. \text{ K}) \ln(5.00/1.00) = -4.01 \times 10^3 \text{ J}$

Since $\Delta E = 0$, $q = -w = 4.01 \times 10^3$ J.

b.   Isothermal irreversible expansion against a constant external pressure of 1.00 atm, so $\Delta E = 0$, and $q = -w$.

$$w = -P_{ex}\Delta V; \; V_1 = \frac{1.00 \text{ mol}(0.08206 \text{ L atm K}^{-1} \text{ mol}^{-1})300. \text{ K}}{5.00 \text{ atm}} = 4.92 \text{ L}$$

$$V_f = \frac{1.00 \text{ mol}(0.08206)(300. \text{ K})}{1.00 \text{ atm}} = 24.6 \text{ L}$$

$w = -(1.00 \text{ atm})(24.6 \text{ L} - 4.92 \text{ L}) \times 101.3 \text{ J L}^{-1} \text{ atm}^{-1} = -1.99 \times 10^3 \text{ J}$

$q = -w = 1.99 \times 10^3$ J

As expected, we get more work from the reversible process.

c.   Adiabatic reversible expansion: $q = 0$ so $\Delta E = w$, $\Delta E = nC_v\Delta T$; we need C_v and ΔT to calculate ΔE and, in turn, w. In order to calculate T_2, we need to determine V_2.

$P_1 V_1^{\gamma} = P_2 V_2^{\gamma}$, where $\gamma = C_p/C_v$. From part b, $V_1 = 4.92$ L.

For a gas behaving ideally:

$C_p = C_v + R$, $C_v = C_p - R = 37.1 \text{ J K}^{-1} \text{ mol}^{-1} - 8.3145 \text{ J K}^{-1} \text{ mol}^{-1} = 28.8 \text{ J K}^{-1} \text{ mol}^{-1}$

$$\frac{C_p}{C_v} = \frac{37.1}{28.8} = 1.29; \; V_2^{1.29} = \frac{P_1 V_1^{1.29}}{P_2}$$

$$V_2^{1.29} = \frac{5.00 \text{ atm}(4.92 \text{ L})^{1.29}}{1.00 \text{ atm}} = 39.0, \; V_2 = 17.1 \text{ L}$$

$$T_2 = \frac{P_2 V_2}{nR} = \frac{1.00 \text{ atm}(17.1 \text{ L})}{1.00 \text{ mol}(0.08206 \text{ L atm K}^{-1} \text{ mol}^{-1})} = 208 \text{ K}$$

$\Delta E = w = nC_v\Delta T = 1.00 \text{ mol}(28.8 \text{ J K}^{-1} \text{ mol}^{-1})(208 \text{ K} - 300. \text{ K}) = -2.6 \times 10^3 \text{ J}$

As expected, we do not get as much work in the adiabatic reversible process as when the gas expands isothermally and reversibly.

Additional Exercises

95.   The light source for the first reaction is necessary for kinetic reasons. The first reaction is just too slow to occur unless a light source is available. The kinetics of a reaction are independent of the thermodynamics of a reaction. Even though the first reaction is more favorable thermodynamically (assuming standard conditions), it is unfavorable for kinetic reasons. The second reaction has a negative $\Delta G°$ value and is a fast reaction, so the second reaction which occurs very quickly is favored both kinetically and thermodynamically. When considering if a reaction will occur, thermodynamics and kinetics must both be considered.

97.   When an ionic solid dissolves, positional probability increases, so ΔS_{sys} is positive. Since temperature increased as the solid dissolved, this is an exothermic process, and ΔS_{surr} is positive ($\Delta S_{surr} = -\Delta H/T$). Since the solid did dissolve, the dissolving process is spontaneous, so ΔS_{univ} is positive (as it must be when ΔS_{sys} and ΔS_{surr} are both positive).

99. $w_{max} = \Delta G$; when ΔG is negative, the magnitude of ΔG is equal to the maximum possible useful work obtainable from the process (at constant T and P). When ΔG is positive, the magnitude of ΔG is equal to the minimum amount of work that must be expended to make the process spontaneous. Due to waste energy (heat) in any real process, the amount of useful work obtainable from a spontaneous process is always less than w_{max}, and for a nonspontaneous reaction, an amount of work greater than w_{max} must be applied to make the process spontaneous.

101.   At the boiling point, $\Delta S_{univ} = 0$ because the system is at equilibrium.

$\Delta S_{univ} = 0 = \Delta S_{sys} + \Delta S_{surr}, \quad \Delta S_{sys} = -\Delta S_{surr}$

Because we are at the boiling point, $\Delta G = 0$. So for 1.00 mol of $CHCl_3$:

$$\Delta H = T\Delta S, \quad \Delta S = \frac{\Delta H}{T} = \frac{31.4 \times 10^3 \text{ J}}{(273.2 + 61.7) \text{ K}} = 93.8 \text{ J K}^{-1}$$

$\Delta S_{sys} = -\Delta S_{surr}$ (at the boiling point), $\Delta S_{surr} = -\Delta S_{sys} = -93.8 \text{ J K}^{-1}$

103. $HF(aq) \rightleftharpoons H^+(aq) + F^-(aq); \quad \Delta G = \Delta G° + RT \ln \dfrac{[H^+][F^-]}{[HF]}$

$\Delta G° = -RT \ln K = -(8.3145 \text{ J K}^{-1} \text{ mol}^{-1})(298 \text{ K}) \ln(7.2 \times 10^{-4}) = 1.8 \times 10^4 \text{ J/mol}$

a.   The concentrations are all at standard conditions, so $\Delta G = \Delta G = 1.8 \times 10^4$ J/mol (since $Q = 1.0$ and $\ln Q = 0$). Because $\Delta G°$ is positive, the reaction shifts left to reach equilibrium.

b. $\Delta G = 1.8 \times 10^4 \text{ J/mol} + (8.3145 \text{ J K}^{-1} \text{ mol}^{-1})(298 \text{ K}) \ln \dfrac{(2.7 \times 10^{-2})^2}{0.98}$

$\Delta G = 1.8 \times 10^4 \text{ J/mol} - 1.8 \times 10^4 \text{ J/mol} = 0$

Because $\Delta G = 0$, the reaction is at equilibrium (no shift).

c. $\Delta G = 1.8 \times 10^4 + 8.3145(298) \ln \dfrac{(1.0 \times 10^{-5})^2}{1.0 \times 10^{-5}} = -1.1 \times 10^4$ J/mol; shifts right

d. $\Delta G = 1.8 \times 10^4 + 8.3145(298) \ln \dfrac{7.2 \times 10^{-4}(0.27)}{0.27} = 1.8 \times 10^4 - 1.8 \times 10^4 = 0$;
at equilibrium

e. $\Delta G = 1.8 \times 10^4 + 8.3145(298) \ln \dfrac{1.0 \times 10^{-3}(0.67)}{0.52} = 2 \times 10^3$ J/mol; shifts left

105. $\Delta S°$ will be negative because 2 moles of gaseous reactants forms 1 mole of gaseous product. For $\Delta G°$ to be negative, $\Delta H°$ must be negative (exothermic). For this sign combination of $\Delta H°$ and $\Delta S°$, K decreases as T increases because $\Delta G°$ becomes more positive ($\Delta G° = -RT \ln K$). Therefore, the ratio of the partial pressure of PCl_5 (a product) to the partial pressure of PCl_3 (a reactant) will decrease when T is raised.

107. Using Le Chatelier's principle: A decrease in pressure (volume increases) will favor the side with the greater number of particles. Thus 2 I(g) will be favored at low pressure.

Looking at ΔG: $\Delta G = \Delta G° + RT \ln (P_I^2 / P_{I_2})$; $\ln(P_I^2 / P_{I_2}) > 0$ when $P_I = P_{I_2} = 10$ atm, and ΔG is positive (not spontaneous). But at $P_I = P_{I_2} = 0.10$ atm, the logarithm term is negative. If $|RT \ln Q| > \Delta G°$, then ΔG becomes negative, and the reaction is spontaneous.

109. $NaCl(s) \rightleftharpoons Na^+(aq) + Cl^-(aq)$ $K = K_{sp} = [Na^+][Cl^-]$

$\Delta G° = [(-262$ kJ$) + (-131$ kJ$)] - (-384$ kJ$) = -9$ kJ $= -9000$ J

$\Delta G° = -RT \ln K_{sp}$, $K_{sp} = \exp\left[\dfrac{-(-9000 \text{ J})}{8.3145 \text{ J K}^{-1} \text{ mol}^{-1} \times 298 \text{ K}}\right] = 38 = 40$

$NaCl(s) \rightleftharpoons Na^+(aq) + Cl^-(aq)$ $K_{sp} = 40$

Initial s = solubility (mol/L)   0   0
Equil. s s

$K_{sp} = 40 = s(s)$, $s = (40)^{1/2} = 6.3 = 6$ $M = [Cl^-]$

111. ΔS is more favorable (less negative) for reaction 2 than for reaction 1, resulting in $K_2 > K_1$. In reaction 1, seven particles in solution are forming one particle in solution. In reaction 2, four particles are forming one, which results in a smaller decrease in positional probability than for reaction 1.

113. Note that these substances are not in the solid state but are in the aqueous state; water molecules are also present. There is an apparent increase in ordering (decrease in positional probability) when these ions are placed in water as compared to the separated state. The hydrating water molecules must be in a highly ordered arrangement when surrounding these anions.

115. $S = k \ln \Omega$; S has units of $J K^{-1} mol^{-1}$, and k has units of J/K ($k = 1.38 \times 10^{-23}$ J/K).   To make units match: $S (J K^{-1} mol^{-1}) = N_A k \ln \Omega$ when N_A = Avogadro's number

$$189 \, J K^{-1} mol^{-1} = 8.31 \, J K^{-1} mol^{-1} \ln \Omega_g$$
$$-(70. \, J K^{-1} mol^{-1} = 8.31 \, J K^{-1} mol^{-1} \ln \Omega_l)$$

_____

Subtracting: $119 \, J K^{-1} mol^{-1} = 8.31 \, J K^{-1} mol^{-1}(\ln \Omega_g - \ln \Omega_l)$

$14.3 = \ln(\Omega_g/\Omega_l)$, $\dfrac{\Omega_g}{\Omega_l} = e^{14.3} = 1.6 \times 10^6$

117.   Isothermal: $\Delta H = 0$ (assume ideal gas)

$$\Delta S = nR \ln\left(\frac{V_2}{V_1}\right) = (1.00 \text{ mol})(8.3145 \, J K^{-1} mol^{-1}) \ln\left(\frac{1.00 \text{ L}}{100.0 \text{ L}}\right) = -38.3 \text{ J/K}$$

$\Delta G = \Delta H - T\Delta S = 0 - (300. \text{ K})(-38.3 \text{ J/K}) = +11,500 \text{ J} = 11.5 \text{ kJ}$

119.   a.   Isothermal: $\Delta E = 0$ and $\Delta H = 0$ if gas is ideal.

$\Delta S = nR \ln(P_1/P_2) = (1.00 \text{ mol})(8.3145 \, J K^{-1} mol^{-1}) \ln(5.00 \text{ atm}/2.00 \text{ atm}) = 7.62 \text{ J/K}$

$$T = \frac{PV}{nR} = \frac{5.00 \text{ atm} \times 5.00 \text{ L}}{1.00 \text{ mol} \times 0.08206 \text{ L atm } K^{-1} mol^{-1}} = 305 \text{ K}$$

$\Delta G = \Delta H - T\Delta S = 0 - (305 \text{ K})(7.62 \text{ J/K}) = -2320 \text{ J}$

$w = -P\Delta V = -(2.00 \text{ atm})\Delta V$, where $V_f = \dfrac{nRT}{2.00 \text{ atm}} = 12.5 \text{ L}$

$w = -2.00 \text{ atm} (12.5 - 5.00 \text{ L}) \times (101.3 \, J L^{-1} atm^{-1}) = -1500 \text{ J}$

$\Delta E = 0 = q + w$, $q = 1500 \text{ J}$

b.   Second law, $\Delta S_{univ} > 0$ for spontaneous processes:

$$\Delta S_{univ} = \Delta S_{sys} + \Delta S_{surr} = \Delta S_{sys} - \frac{q_{actual}}{T}$$

$\Delta S_{univ} = 7.62 \text{ J/K} - \dfrac{1500 \text{ J}}{305 \text{ K}} = 7.62 - 4.9 = 2.7 \text{ J/K}$; thus the process is spontaneous.

Challenge Problems

121.   a. $V_1 = \dfrac{nRT_1}{P_1} = \dfrac{2.00 \text{ mol} \times 0.08206 \text{ L atm } K^{-1} mol^{-1} \times 298 \text{ K}}{2.00 \text{ atm}} = 24.5 \text{ L}$

For an adiabatic reversible process, $P_1V_1^\gamma = P_2V_2^\gamma$ and $T_1V_1^{\gamma-1} = T_2V_2^{\gamma-1}$, where $\gamma = C_p/C_v$. Because argon is a monoatomic gas, $C_p = (5/2)R$ and $C_v = (3/2)R$, so $\gamma = 5/3$.

$$V_2^\gamma = \frac{P_1V_1^\gamma}{P_2} = \frac{2.00 \text{ atm} (24.5 \text{ L})^{5/3}}{1.00 \text{ atm}} = 413, \quad V_2 = (413)^{3/5} = 37.1 \text{ L}$$

We can either use the ideal gas law or the $T_1V^{\gamma-1} = T_2V_2^{\gamma-1}$ equation to calculate the final temperature. Using the ideal gas law:

$$T_2 = \frac{P_2V_2}{nR} = \frac{1.00 \text{ atm} \times 37.1 \text{ L}}{2.00 \text{ mol} \times 0.08206 \text{ L atm K}^{-1} \text{ mol}^{-1}} = 226 \text{ K}$$

b. For an adiabatic process ($q = 0$), $\Delta E = w = nC_v\Delta T$. For an expansion against a fixed external pressure, $w = -P\Delta V$. From the ideal gas equation (see part a), $V_1 = 24.5$ L.

$$w = -P\Delta V = nC_v\Delta T$$

$$1.00 \text{ atm}(V_2 - 24.5 \text{ L})\left(\frac{101.3 \text{ J}}{\text{L atm}}\right) = 2.00 \text{ mol}(3/2)\left(\frac{8.3145 \text{ J}}{\text{mol K}}\right)(T_2 - 298 \text{ K})$$

We will ignore units from here. Note that both sides of the equation are in units of J.

$$(-101)V_2 + 2480 = (24.9)T_2 - 7430$$

To solve for T_2, we need to find an expression for V_2. Using the ideal gas equation:

$$V_2 = \frac{nRT_2}{P_2} = \frac{2.00 \text{ mol}(0.08206)T_2}{1.00} = (0.164)T_2; \quad \text{substituting:}$$

$$-101(0.164 \text{ } T_2) + 2480 = (24.9)T_2 - 7430, \quad (41.5)T_2 = 9910, \quad T_2 = 239 \text{ K}$$

123. a. $\Delta S = \dfrac{q_{rev}}{T}$; isothermal: $\Delta T = 0$ so $\Delta E = 0$ and $q = -w$.

Reversible expansion: $q_{rev} = 855$ J; $\Delta S = \dfrac{q_{rev}}{T} = \dfrac{855 \text{ J}}{298 \text{ K}} = 2.87$ J/K

For the compression, we go back to the initial state for the overall process (expansion then compression). Because the initial and overall final state are the same, all state functions like ΔS must equal zero.

$$\Delta S_{overall} = 0 = \Delta S_{exp} + \Delta S_{comp}, \quad \Delta S_{comp} = -\Delta S_{exp} = -2.87 \text{ J/K}$$

Note: Even though the compression step is not a reversible process, it still must have $q_{rev} = -855$ J.

$$\Delta S_{comp} = \frac{q_{rev}}{T} = \frac{-855 \text{ J}}{298 \text{ K}} = -2.87 \text{ J/K}$$

b. $\Delta S_{univ, \, overall} = \Delta S_{sys, \, overall} + \Delta S_{surr, \, overall}$

$\Delta S_{sys, \, overall} = 2.87 \text{ J/K} - 2.87 \text{ J/K} = 0$

$$\Delta S_{surr, \, exp} = \frac{-q_{actual}}{T} = \frac{-855 \text{ J}}{298 \text{ K}} = -2.87 \text{ J/K}$$

From the problem, the compression step is isothermal, so q = −w. The work done on the system is 2(855) J, so q = −2(855) J.

$$\Delta S_{surr, \, comp} = \frac{-q_{actual}}{T} = \frac{-[-2(855) \text{ J}]}{298 \text{ K}} = 5.74 \text{ J/K}$$

$\Delta S_{surr, \, overall} = -2.87 \text{ J/K} + 5.74 \text{ J/K} = 2.87 \text{ J/K}$

$\Delta S_{univ, \, overall} = 0 + 2.87 \text{ J/K} = 2.87 \text{ J/K}$

125.   The system is the 1.00-L sample of water. The process is the cooling of the water from 90.°C to 25°C. The surroundings are the room (at 25°C) and everything else.

$$1.00 \times 10^3 \text{ mL} \times \frac{1.00 \text{ g}}{\text{mL}} \times \frac{1 \text{ mol}}{18.02 \text{ g}} = 55.5 \text{ mol H}_2\text{O}$$

$\Delta S = nC_p \ln(T_2/T_1) = 55.5 \text{ mol}(75.3 \text{ J K}^{-1} \text{ mol}^{-1}) \ln(298/363) = -825 \text{ J/K}$

$\Delta S_{surr} = \dfrac{-q_{actual}}{T}$; $q_{actual} = nC_p \Delta T = 55.5 \text{ mol}(75.3 \text{ J K}^{-1} \text{ mol}^{-1})(-65 \text{ K}) = -2.72 \times 10^5 \text{ J}$

$$\Delta S_{surr} = \frac{-q_{actual}}{T} = \frac{-(-2.72 \times 10^5 \text{ J})}{298 \text{ K}} = 913 \text{ J/K}$$

$\Delta S_{univ} = \Delta S + \Delta S_{surr} = -825 \text{ J/K} + 913 \text{ J/K} = 88 \text{ J/K}$

Not too surprising, this is a spontaneous process due to the favorable ΔS_{surr} term.

127. $K = P_{CO_2}$; to prevent Ag_2CO_3 from decomposing, P_{CO_2} should be greater than K.

From Exercise 10.86, $\ln K = \dfrac{-\Delta H^\circ}{RT} + \dfrac{\Delta S^\circ}{R}$. For two conditions of K and T, the equation is:

$$\ln \frac{K_2}{K_1} = \frac{\Delta H^\circ}{R} \left(\frac{1}{T_1} - \frac{1}{T_2} \right)$$

Let $T_1 = 25°C = 298 \text{ K}$, $K_1 = 6.23 \times 10^{-3}$ torr; $T_2 = 110.°C = 383 \text{ K}$, $K_2 = ?$

$$\ln \frac{K_2}{6.23 \times 10^{-3} \text{ torr}} = \frac{79.14 \times 10^3 \text{ J/mol}}{8.3145 \text{ J K}^{-1} \text{ mol}^{-1}} \left(\frac{1}{298 \text{ K}} - \frac{1}{383 \text{ K}} \right)$$

$$\ln \frac{K_2}{6.23 \times 10^{-3}} = 7.1, \quad \frac{K_2}{6.23 \times 10^{-3}} = e^{7.1} = 1.2 \times 10^3, \quad K_2 = 7.5 \text{ torr}$$

To prevent decomposition of Ag_2CO_3, the partial pressure of CO_2 should be greater than 7.5 torr.

129. $3 O_2(g) \rightleftharpoons 2 O_3(g)$; $\Delta H^\circ = 2(143) = 286$ kJ; $\Delta G^\circ = 2(163) = 326$ kJ

$$\ln K = \frac{-\Delta G^\circ}{RT} = \frac{-326 \times 10^3 \text{ J}}{(8.3145 \text{ J K}^{-1} \text{ mol}^{-1})(298 \text{ K})} = -131.573, \quad K = e^{-131.573} = 7.22 \times 10^{-58}$$

We need the value of K at 230. K. From Exercise 10.86: $\ln K = \frac{-\Delta H^\circ}{RT} + \frac{\Delta S^\circ}{R}$

For two sets of K and T:

$$\ln \frac{K_2}{K_1} = \frac{\Delta H^\circ}{R} \left(\frac{1}{T_1} - \frac{1}{T_2} \right)$$

Let $K_2 = 7.22 \times 10^{-58}$, $T_2 = 298$; $K_1 = K_{230}$, $T_1 = 230.$ K; $\Delta H^\circ = 286 \times 10^3$ J

$$\ln \frac{7.22 \times 10^{-58}}{K_{230}} = \frac{286 \times 10^3}{8.3145} \left(\frac{1}{230.} - \frac{1}{298} \right) = 34.13$$

$$\frac{7.22 \times 10^{-58}}{K_{230}} = e^{34.13} = 6.6 \times 10^{14}, \quad K_{230} = 1.1 \times 10^{-72}$$

$$K_{230} = 1.1 \times 10^{-72} = \frac{P_{O_3}^2}{P_{O_2}^3} = \frac{P_{O_3}^2}{(1.0 \times 10^{-3} \text{ atm})^3}, \quad P_{O_3} = 3.3 \times 10^{-41} \text{ atm}$$

The volume occupied by one molecule of ozone is:

$$V = \frac{nRT}{P} = \frac{(1/6.022 \times 10^{23} \text{ mol})(0.8206 \text{ L atm K}^{-1} \text{ mol}^{-1})(230. \text{ K})}{(3.3 \times 10^{-41} \text{ atm})}, \quad V = 9.5 \times 10^{17} \text{ L}$$

Equilibrium is probably not maintained under these conditions. When only two ozone molecules are in a volume of 9.5×10^{17} L, the reaction is not at equilibrium. Under these conditions, Q > K and the reaction shifts left. But with only two ozone molecules in this huge volume, it is extremely unlikely that they will collide with each other. At these conditions, the concentration of ozone is not large enough to maintain equilibrium.

131.   a. $\Delta G° = G_B^° - G_A^° = 11{,}718 - 8996 = 2722$ J

$$K = \exp\left(\frac{-\Delta G°}{RT}\right) = \exp\left[\frac{-2722 \text{ J}}{(8.3145 \text{ J K}^{-1} \text{ mol}^{-1})(298 \text{ K})}\right] = 0.333$$

   b.   Since $Q = 1.00 > K$, reaction shifts left.   Let x = atm of B(g) that reacts to reach equilibrium.

	A(g)	$\rightleftharpoons$	B(g)	$K = P_B/P_A$
Initial	1.00 atm		1.00 atm	
Equil.	$1.00 + x$		$1.00 - x$	

$$K = \frac{1.00 - x}{1.00 + x} = 0.333, \quad 1.00 - x = 0.333 + (0.333)x, \quad x = 0.50 \text{ atm}$$

$P_B = 1.00 - 0.50 = 0.50$ atm; $P_A = 1.00 + 0.50 = 1.50$ atm

   c. $\Delta G = \Delta G° + RT \ln Q = \Delta G° + RT \ln(P_B/P_A)$

$\Delta G = 2722 \text{ J} + (8.3145)(298) \ln(0.50/1.50) = 2722 \text{ J} - 2722 \text{ J} = 0$

(carrying extra sig. figs.)

133.   Step 1: $\Delta E = 0$ and $\Delta H = 0$ since $\Delta T = 0$

$$w = -P\Delta V = -(9.87 \times 10^{-3} \text{ atm})\Delta V; \quad V = \frac{nRT}{P}, \quad R = 0.08206 \text{ L atm K}^{-1} \text{ mol}^{-1}$$

$$\Delta V = V_f - V_i = nRT\left(\frac{1}{P_f} - \frac{1}{P_i}\right)$$

$$\Delta V = 1.00 \text{ mol}(0.08206)(298 \text{ K})\left(\frac{1}{9.87 \times 10^{-3} \text{ atm}} - \frac{1}{2.45 \times 10^{-2} \text{ atm}}\right)$$

$\Delta V = 1480$ L (we will carry all values to three sig. figs.)

$w = -(9.87 \times 10^{-3} \text{ atm})(1480 \text{ L}) = -14.6 \text{ L atm}(101.3 \text{ J L}^{-1} \text{ atm}^{-1}) = -1480$ J

$$\Delta E = q + w = 0, \quad q = -w = +1480 \text{ J}; \quad \Delta S = nR \ln\left(\frac{P_1}{P_2}\right)$$

$$\Delta S = 1.00 \text{ mol}(8.3145 \text{ J K}^{-1} \text{ mol}^{-1}) \ln\left(\frac{2.45 \times 10^{-2} \text{ atm}}{9.87 \times 10^{-3} \text{ atm}}\right), \quad \Delta S = 7.56 \text{ J/K}$$

$\Delta G = \Delta H - T\Delta S = 0 - 298 \text{ K}(7.56 \text{ J/K}) = -2250$ J

Step 2: $\Delta E = 0$, $\Delta H = 0$

$$w = -(4.93 \times 10^{-3} \text{ atm})\left(\frac{nRT}{4.93 \times 10^{-3}} - \frac{nRT}{9.87 \times 10^{-3} \text{ atm}}\right) L \times \frac{101.3 \text{ J}}{L \text{ atm}} = -1240 \text{ J}$$

$$q = -w = 1240 \text{ J}; \quad \Delta S = nR \ln\left(\frac{9.87 \times 10^{-3} \text{ atm}}{4.93 \times 10^{-3} \text{ atm}}\right) = 5.77 \text{ J/K}$$

$$\Delta G = 0 - 298K(5.77 \text{ J/K}) = -1720 \text{ J}$$

Step 3: $\Delta E = 0$, $\Delta H = 0$

$$w = -(2.45 \times 10^{-3} \text{ atm})\left(\frac{nRT}{2.45 \times 10^{-3}} - \frac{nRT}{4.93 \times 10^{-3}}\right) L \times \frac{101.3 \text{ J}}{L \text{ atm}} = -1250 \text{ J}$$

$$q = -w = 1250 \text{ J}; \quad \Delta S = nR \ln\left(\frac{4.93 \times 10^{-3} \text{ atm}}{2.45 \times 10^{-3} \text{ atm}}\right) = 5.81 \text{ J/K}$$

$$\Delta G = 0 - 298 \text{ K}(5.81 \text{ J/K}) = -1730 \text{ J}$$

	q	w	ΔE	ΔS	ΔH	ΔG
Step 1	1480 J	−1480 J	0	7.56 J/K	0	−2250 J
Step 2	1240 J	−1240 J	0	5.77 J/K	0	−1720 J
Step 3	1250 J	−1250 J	0	5.81 J/K	0	−1730 J
Total	3970 J	−3970 J	0	19.14 J/K	0	-5.70×10^3 J

135.   a. $\Delta G° = 2 \text{ mol}(-394 \text{ kJ/mol}) - 2 \text{ mol}(-137 \text{ kJ/mol}) = -514 \text{ kJ}$

$$K = \exp\left(\frac{-\Delta G°}{RT}\right) = \exp\left(\frac{-(-514{,}000 \text{ J})}{(8.3145 \text{ J K}^{-1} \text{ mol}^{-1})(298 \text{ K})}\right) = 1.24 \times 10^{90}$$

b. $\Delta S° = 2(214 \text{ J/K}) - [2(198 \text{ J/K}) + 205 \text{ J/K}] = -173 \text{ J/K}$

$2 \text{ CO}(1.00 \text{ atm}) + O_2(1.00 \text{ atm}) \rightarrow 2 \text{ CO}_2(1.00 \text{ atm})$	$\Delta S° = -173 \text{ J/K}$
$2 \text{ CO}_2(1.00 \text{ atm}) \rightarrow 2 \text{ CO}_2(10.0 \text{ atm})$	$\Delta S = nR \ln(P_1/P_2) = -38.3 \text{ J/K}$
$2 \text{ CO}(10.0 \text{ atm}) \rightarrow 2 \text{ CO}(1.00 \text{ atm})$	$\Delta S = nR \ln(P_1/P_2) = 38.3 \text{ J/K}$
$O_2(10.0 \text{ atm}) \rightarrow O_2(1.00 \text{ atm})$	$\Delta S = nR \ln(P_1/P_2) = 19.1 \text{ J/K}$

$2 \text{ CO}(10.0 \text{ atm}) + O_2(10.0 \text{ atm}) \rightarrow \text{CO}_2(10.0 \text{ atm})$	$\Delta S = -173 + 19.1 = -154 \text{ J/K}$

137.

T(°C)	T(K)	C_p (J K^{-1} mol^{-1})	C_p/T (J K^{-2} mol^{-1})
−200.	73	12	0.16
-180.	93	15	0.16
-160.	113	17	0.15
-140.	133	19	0.14
-100.	173	24	0.14
-60.	213	29	0.14
-30.	243	33	0.14
-10.	263	36	0.14
0	273	37	0.14

Total area of C_p/T versus T plot = ΔS = I + II + III   (See following plot.)

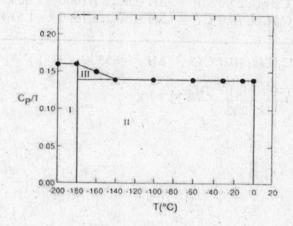

ΔS = (0.16 J K^{-2} mol^{-1})(20. K) + (0.14 J K^{-2} mol^{-1})(180. K) + 1/2 (0.02 J K^{-2} mol^{-1})(40. K)

ΔS = 3.2 + 25 + 0.4 = 29 J K^{-1} mol^{-1}

139.   To calculate ΔS_{sys} at 10.0°C, we need a place to start. From the data in the problem, we can calculate ΔS_{sys} at the melting point (5.5°C). For a phase change, $\Delta S_{sys} = q_{rev}/T = \Delta H/T$, where ΔH is determined at the melting point (5.5°C). Solving for ΔH at 5.5°C (using a thermochemical cycle):

C_6H_6(l, 25.0°C) → C_6H_6(s, 25.0°C)	ΔH = −10.04 kJ
C_6H_6(l, 5.5°C) → C_6H_6(l, 25.0°C)	ΔH = $nC_p\Delta T/1000$ = 2.59 kJ
C_6H_6(s, 25.0°C) → C_6H_6(s, 5.5°C)	ΔH = $nC_p\Delta T/1000$ = −1.96 kJ

C_6H_6(l, 5.5°C) → C_6H_6(s, 5.5°C) ΔH = −9.41 kJ

At the melting point, $\Delta S_{sys} = \dfrac{\Delta H}{T} = \dfrac{-9.41 \times 10^3 \text{ J}}{278.7 \text{ K}} = -33.8$ J/K.

For the phase change at 10.0°C (283.2 K):

C_6H_6(l, 278.7 K) → C_6H_6(s, 278.7 K) $\Delta S = -33.8$ J/K

C_6H_6(l, 283.2 K) → C_6H_6(l, 278.7 K) $\Delta S = nC_p \ln(T_2/T_1) = -2.130$ J/K

C_6H_6(s, 278.7 K) → C_6H_6(s, 283.2 K) $\Delta S = nC_p \ln(T_2/T_1) = 1.608$ J/K

C_6H_6(l, 283.2 K) → C_6H_6(s, 283.2 K) $\Delta S_{sys} = -34.3$ J/K

To calculate ΔS_{surr}, we need ΔH at 10.0°C ($\Delta S_{surr} = \dfrac{-\Delta H}{T}$).

C_6H_6(l, 25.0°C) → C_6H_6(s, 25.0°C) $\Delta H = -10.04$ kJ

C_6H_6(l, 10.0°C) → C_6H_6(l, 25.0°C) $\Delta H = nC_p \Delta T/1000 = 2.00$ kJ

C_6H_6(s, 25.0°C) → C_6H_6(s, 10.0°C) $\Delta H = nC_p \Delta T/1000 = -1.51$ kJ

C_6H_6(l, 10.0°C) → C_6H_6(s, 10.0°C) $\Delta H = -9.55$ kJ

$$\Delta S_{surr} = \frac{-\Delta H}{T} = \frac{-(-9.55 \times 10^3 \text{ J})}{283.2 \text{ K}} = 33.7 \text{ J/K}$$

CHAPTER 11

ELECTROCHEMISTRY

Galvanic Cells, Cell Potentials, and Standard Reduction Potentials

15.   Electrochemistry is the study of the interchange of chemical and electrical energy. A redox (oxidation-reduction) reaction is a reaction in which one or more electrons are transferred. In a galvanic cell, a spontaneous redox reaction occurs that produces an electric current. In an electrolytic cell, electricity is used to force a nonspontaneous redox reaction to occur.

17.   A galvanic cell at standard conditions must have a positive overall standard cell potential $(E^{\circ}_{cell} > 0)$. The only combination of the half-reactions that gives a positive cell potential is:

$$Cu^{2+} + 2e^- \rightarrow Cu \qquad\qquad E°(\text{cathode}) = 0.34 \text{ V}$$
$$Zn \rightarrow Zn^{2+} + 2e^- \qquad -E°(\text{anode}) = 0.76 \text{ V}$$

$$Cu^{2+}(aq) + Zn(s) \rightarrow Cu(s) + Zn^{2+}(aq) \qquad E^{\circ}_{cell} = 1.10 \text{ V}$$

a. The reducing agent causes reduction to occur since it always contains the species which is oxidized. Zn is oxidized in the galvanic cell, so Zn is the reducing agent. The oxidizing agent causes oxidation to occur since it always contains the species which is reduced. Cu^{2+} is reduced in the galvanic cell, so Cu^{2+} is the oxidizing agent. Electrons will flow from the zinc compartment (the anode) to the copper compartment (the cathode).

b. From the work above, $E^{\circ}_{cell} = 1.10$ V.

c. The pure metal that is a product in the spontaneous reaction is copper. So the copper electrode will increase in mass as $Cu^{2+}(aq)$ is reduced to $Cu(s)$. The zinc electrode will decrease in mass for this galvanic cell as $Zn(s)$ is oxidized to $Zn^{2+}(aq)$.

19.   A typical galvanic cell diagram is:

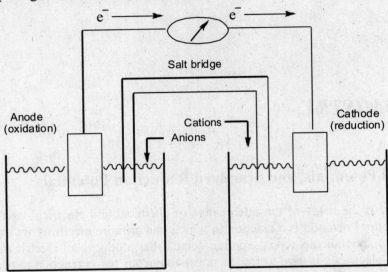

The diagram for all cells will look like this. The contents of each half-cell will be identified for each reaction, with all concentrations at 1.0 M and partial pressures at 1.0 atm. Note that cations always flow into the cathode compartment and anions always flow into the anode compartment. This is required to keep each compartment electrically neutral.

a.   Reference Table 11.1 for standard reduction potentials.   Remember that E^o_{cell} = $E°$(cathode) $-$ $E°$(anode); in the *Solutions Guide*, we will represent $E°$(cathode) as E^o_c and represent $-E°$(anode) as $-E^o_a$.   Also remember that standard potentials are *not* multiplied by the integer used to obtain the overall balanced equation.

$$(Cl_2 + 2\ e^- \rightarrow 2\ Cl^-) \times 3 \qquad\qquad\qquad E^o_c = 1.36\ V$$
$$7\ H_2O + 2\ Cr^{3+} \rightarrow Cr_2O_7^{2-} + 14\ H^+ + 6\ e^- \qquad -E^o_a = -1.33\ V$$

$$7\ H_2O(l) + 2\ Cr^{3+}(aq) + 3\ Cl_2(g) \rightarrow Cr_2O_7^{2-}\ (aq) + 6\ Cl^-\ (aq) + 14\ H^+(aq) \quad E^o_{cell} = 0.03\ V$$

The contents of each compartment are:

Cathode: Pt electrode; Cl_2 bubbled into solution, Cl^- in solution

Anode: Pt electrode; Cr^{3+}, H^+, and $Cr_2O_7^{2-}$ in solution

We need a nonreactive metal to use as the electrode in each case since all the reactants and products are in solution. Pt or graphite are the most common choices.

b. $Cu^{2+} + 2\ e^- \rightarrow Cu$ $\qquad\qquad\qquad\qquad\qquad E^o_c = 0.34\ V$
            $Mg \rightarrow Mg^{2+} + 2\ e^-$ $\qquad\qquad\qquad\qquad -E^o_a = 2.37\ V$

$Cu^{2+}(aq) + Mg(s) \rightarrow Cu(s) + Mg^{2+}(aq)$ $\qquad\qquad E^o_{cell} = 2.71\ V$

Cathode: Cu electrode; Cu^{2+} in solution

Anode: Mg electrode; Mg^{2+} in solution

c.

$$5 \; e^- + 6 \; H^+ + IO_3^- \rightarrow 1/2 \; I_2 + 3 \; H_2O \qquad E_c^o = 1.20 \; V$$

$$(Fe^{2+} \rightarrow Fe^{3+} + e^-) \times 5 \qquad -E_a^o = -0.77 \; V$$

$$6 \; H^+ + IO_3^- + 5 \; Fe^{2+} \rightarrow 5 \; Fe^{3+} + 1/2 \; I_2 + 3 \; H_2O \qquad E_{cell}^o = 0.43 \; V$$

or $12 \; H^+(aq) + 2 \; IO_3^-(aq) + 10 \; Fe^{2+}(aq) \rightarrow 10 \; Fe^{3+}(aq) + I_2(s) + 6 \; H_2O(l)$ $E_{cell}^o = 0.43 \; V$

Cathode: Pt electrode; IO_3^-, I_2, and H_2SO_4 (H^+ source) in solution

Anode: Pt electrode; Fe^{2+} and Fe^{3+} in solution

Note: $I_2(s)$ would make a poor electrode since it sublimes.

d.

$$(Ag^+ + e^- \rightarrow Ag) \times 2 \qquad E_c^o = 0.80 \; V$$

$$Zn \rightarrow Zn^{2+} + 2 \; e^- \qquad -E_a^o = 0.76 \; V$$

$$Zn(s) + 2 \; Ag^+(aq) \rightarrow 2 \; Ag(s) + Zn^{2+}(aq) \qquad E_{cell}^o = 1.56 \; V$$

Cathode: Ag electrode; Ag^+ in solution

Anode: Zn electrode; Zn^{2+} in solution

21. Reference Exercise 11.19 for a typical galvanic cell design. The contents of each half-cell compartment are identified below, with all solute concentrations at 1.0 *M* and all gases at 1.0 atm. For each pair of half-reactions, the half-reaction with the largest standard reduction potential will be the cathode reaction, and the half-reaction with the smallest reduction potential will be reversed to become the anode reaction. Only this combination gives a spontaneous overall reaction, i.e., a reaction with a positive overall standard cell potential.

a.

$$Cl_2 + 2 \; e^- \rightarrow 2 \; Cl^- \qquad E_c^o = 1.36 \; V$$

$$2 \; Br^- \rightarrow Br_2 + 2 \; e^- \qquad -E_a^o = -1.09 \; V$$

$$Cl_2(g) + 2 \; Br^-(aq) \rightarrow Br_2(aq) + 2 \; Cl^-(aq) \qquad E_{cell}^o = 0.27 \; V$$

The contents of each compartment are:

Cathode: Pt electrode; $Cl_2(g)$ bubbled in, Cl^- in solution

Anode: Pt electrode; Br_2 and Br^- in solution

b. $\qquad (2\ e^- + 2\ H^+ + IO_4^- \rightarrow IO_3^- + H_2O) \times 5 \qquad\qquad E_c^\circ = 1.60\ V$

$\qquad\qquad (4\ H_2O + Mn^{2+} \rightarrow MnO_4^- + 8\ H^+ + 5\ e^-) \times 2 \qquad -E_a^\circ = -1.51\ V$

$\overline{10\ H^+ + 5\ IO_4^- + 8\ H_2O + 2\ Mn^{2+} \rightarrow 5\ IO_3^- + 5\ H_2O + 2\ MnO_4^- + 16\ H^+ \qquad E_{cell}^\circ = 0.09\ V}$

This simplifies to:

$$3\ H_2O(l) + 5\ IO_4^-(aq) + 2\ Mn^{2+}(aq) \rightarrow 5\ IO_3^-(aq) + 2\ MnO_4^-(aq) + 6\ H^+(aq)$$

$$E_{cell}^\circ = 0.09\ V$$

Cathode: Pt electrode; IO_4^-, IO_3^-, and H_2SO_4 (as a source of H^+) in solution

Anode: Pt electrode; Mn^{2+}, MnO_4^-, and H_2SO_4 in solution

c. $H_2O_2 + 2\ H^+ + 2\ e^- \rightarrow 2\ H_2O \qquad\qquad E_c^\circ = 1.78\ V$

$\qquad\qquad H_2O_2 \rightarrow O_2 + 2\ H^+ + 2\ e^- \qquad -E_a^\circ = -0.68\ V$

$\overline{\qquad\qquad 2\ H_2O_2(aq) \rightarrow 2\ H_2O(l) + O_2(g) \qquad\qquad E_{cell}^\circ = 1.10\ V}$

Cathode: Pt electrode; H_2O_2 and H^+ in solution

Anode: Pt electrode; $O_2(g)$ bubbled in, H_2O_2 and H^+ in solution

d. $\qquad\qquad (Fe^{3+} + 3\ e^- \rightarrow Fe) \times 2 \qquad\qquad E_c^\circ = -0.036\ V$

$\qquad\qquad (Mn \rightarrow Mn^{2+} + 2\ e^-) \times 3 \qquad -E_a^\circ = 1.18\ V$

$\overline{\qquad 2\ Fe^{3+}(aq) + 3\ Mn(s) \rightarrow 2\ Fe(s) + 3\ Mn^{2+}(aq) \qquad E_{cell}^\circ = 1.14\ V}$

Cathode: Fe electrode; Fe^{3+} in solution; Anode: Mn electrode; Mn^{2+} in solution

23.   In standard line notation, the anode is listed first and the cathode is listed last. A double line separates the two compartments. By convention, the electrodes are on the ends, with all solutes and gases toward the middle. A single line is used to indicate a phase change. We also included all concentrations.

21a. $Pt \mid Br^- (1.0\ M), Br_2 (1.0\ M) \parallel Cl_2 (1.0\ atm) \mid Cl^- (1.0\ M) \mid Pt$

21b. $Pt \mid Mn^{2+} (1.0\ M), MnO_4^- (1.0\ M), H^+ (1.0\ M) \parallel IO_4^- (1.0\ M), IO_3^- (1.0\ M),$

$H^+ (1.0\ M) \mid Pt$

21c. $Pt \mid H_2O_2 (1.0\ M), H^+ (1.0\ M) \mid O_2 (1.0\ atm) \parallel H_2O_2 (1.0\ M), H^+ (1.0\ M) \mid Pt$

21d. $Mn \mid Mn^{2+} (1.0\ M) \parallel Fe^{3+} (1.0\ M) \mid Fe$

25. a. $2 H^+ + 2 e^- \rightarrow H_2$ $E° = 0.00$ V; $Cu \rightarrow Cu^{2+} + 2 e^-$ $-E° = -0.34$ V

$E°_{cell} = -0.34$ V; no, H^+ cannot oxidize Cu to Cu^{2+} at standard conditions ($E°_{cell} < 0$).

b. $Fe^{3+} + e^- \rightarrow Fe^{2+}$ $E° = 0.77$ V; $2 I^- \rightarrow I_2 + 2 e^-$ $-E° = -0.54$ V

$E°_{cell} = 0.77 - 0.54 = 0.23$ V; yes, Fe^{3+} can oxidize I^- to I_2.

c. $H_2 \rightarrow 2 H^+ + 2 e^-$ $-E° = 0.00$ V; $Ag^+ + e^- \rightarrow Ag$ $E° = 0.80$ V

$E°_{cell} = 0.80$ V; yes, H_2 can reduce Ag^+ to Ag at standard conditions ($E°_{cell} > 0$).

d. $Fe^{2+} \rightarrow Fe^{3+} + e^-$ $-E° = -0.77$ V; $Cr^{3+} + e^- \rightarrow Cr^{2+}$ $E° = -0.50$ V

$E°_{cell} = -0.50 - 0.77 = -1.27$ V;   no, Fe^{2+} cannot reduce Cr^{3+} to Cr^{2+} at standard conditions.

27. Good reducing agents are easily oxidized. The reducing agents are on the right side of the reduction half-reactions listed in Table 11.1. The best reducing agents have the most negative standard reduction potentials (E°); i.e., the best reducing agents have the most positive $-E°$ value. The ordering from worst to best reducing agents is:

	F^-	<	H_2O	<	I_2	<	Cu^+	<	H^-	<	K
$-E°$(V)	-2.87		-1.23		-1.20		-0.16		2.23		2.92

29. a. $2 Br^- \rightarrow Br_2 + 2 e^-$ $-E°_a = -1.09$ V; $2 Cl^- \rightarrow Cl_2 + 2 e^-$ $-E°_a = -1.36$ V; $E°_c > 1.09$ V to oxidize Br^-; $E°_c < 1.36$ V to not oxidize Cl^-; $Cr_2O_7^{2-}$, O_2, MnO_2, and IO_3^- are all possible because when all these oxidizing agents are coupled with Br^-, they give $E°_{cell} > 0$, and when coupled with Cl^-, they give $E°_{cell} < 0$ (assuming standard conditions).

b. $Mn \rightarrow Mn^{2+} + 2 e^-$ $-E°_a = 1.18$ V; $Ni \rightarrow Ni^{2+} + 2 e^-$ $-E°_a = 0.23$ V; any oxidizing agent with -0.23 V $> E°_c > -1.18$ V will work. $PbSO_4$, Cd^{2+}, Fe^{2+}, Cr^{3+}, Zn^{2+}, and H_2O will be able to oxidize Mn but not oxidize Ni (assuming standard conditions).

31. $ClO^- + H_2O + 2 e^- \rightarrow 2 OH^- + Cl^-$ $E°_c = 0.90$ V

$2 NH_3 + 2 OH^- \rightarrow N_2H_4 + 2 H_2O + 2 e^-$ $-E°_a = 0.10$ V

$ClO^-(aq) + 2 NH_3(aq) \rightarrow Cl^-(aq) + N_2H_4(aq) + H_2O(l)$ $E°_{cell} = 1.00$ V

Because $E°_{cell}$ is positive for this reaction, at standard conditions ClO^- can spontaneously oxidize NH_3 to the somewhat toxic N_2H_4.

Cell Potential, Free Energy, and Equilibrium

33. An extensive property is one that depends directly on the amount of substance. The free energy change for a reaction depends on whether 1 mole of product is produced or 2 moles of product are produced or 1 million moles of product are produced. This is not the case for cell potentials, which do not depend on the amount of substance. The equation that relates ΔG to E is $\Delta G = -nFE$. It is the n term that converts the intensive property E into the extensive property ΔG. n is the number of moles of electrons transferred in the balanced reaction that ΔG is associated with.

35. $2\,H_2O + 2\,e^- \rightarrow H_2 + 2\,OH^-$ $\Delta G° = \Sigma n_p \Delta G°_{f,\,products} - \Sigma n_r \Delta G°_{f,\,reactants}$

$$= 2(-157) - [2(-237)] = 160.\ kJ$$

$$\Delta G° = -nFE°,\ \ E° = \frac{-\Delta G°}{nF} = \frac{-1.60 \times 10^5\ J}{(2\ mol\ e^-)(96{,}485\ C/mol\ e^-)} = -0.829\ J/C = -0.829\ V$$

The two values agree to two significant figures (−0.83 V in Table 11.1).

37. Because the cells are at standard conditions, $w_{max} = \Delta G = \Delta G° = -nFE°_{cell}$. See Exercise 11.22 for the balanced overall equations and $E°_{cell}$.

22a. $w_{max} = -(3\ mol\ e^-)(96{,}485\ C/mol\ e^-)(1.34\ J/C) = -3.88 \times 10^5\ J = -388\ kJ$

22b. $w_{max} = -(2\ mol\ e^-)(96{,}485\ C/mol\ e^-)(1.40\ J/C) = -2.70 \times 10^5\ J = -270.\ kJ$

39. Reference Exercise 11.21 for the balanced reactions and standard cell potentials.   The balanced reactions are necessary to determine n, the moles of electrons transferred.

21a. $Cl_2(aq) + 2\,Br^-(aq) \rightarrow Br_2(aq) + 2\,Cl^-(aq)$ $E°_{cell} = 0.27\ V = 0.27\ J/C,\ n = 2\ mol\ e^-$

$$\Delta G° = -nFE°_{cell} = -(2\ mol\ e^-)(96{,}485\ C/mol\ e^-)(0.27\ J/C) = -5.2 \times 10^4\ J = -52\ kJ$$

$$E°_{cell} = \frac{0.0591}{n} \log K,\ \ \log K = \frac{nE°}{0.0591} = \frac{2(0.27)}{0.0591} = 9.14,\ \ K = 10^{9.14} = 1.4 \times 10^9$$

Note: When determining exponents, we will round off to the correct number of significant figures after the calculation is complete in order to help eliminate excessive round-off errors.

21b. $\Delta G° = -(10\ mol\ e^-)(96{,}485\ C/mol\ e^-)(0.09\ J/C) = -9 \times 10^4\ J = -90\ kJ$

$$\log K = \frac{10(0.09)}{0.0591} = 15.2,\ \ K = 10^{15.2} = 2 \times 10^{15}$$

21c. $\Delta G° = -(2\ mol\ e^-)(96{,}485\ C/mol\ e^-)(1.10\ J/C) = -2.12 \times 10^5\ J = -212\ kJ$

$$\log K = \frac{2(1.10)}{0.0591} = 37.225,\ \ K = 1.68 \times 10^{37}$$

21d. $\Delta G° = -(6 \text{ mol e}^-)(96{,}485 \text{ C/mol e}^-)(1.14 \text{ J/C}) = -6.60 \times 10^5 \text{ J} = -660. \text{ kJ}$

$\log K = \dfrac{6(1.14)}{0.0591} = 115.736, \ K = 5.45 \times 10^{115}$

41.   a.

$(4 \text{ H}^+ + \text{NO}_3^- + 3 \text{ e}^- \rightarrow \text{NO} + 2 \text{ H}_2\text{O}) \times 2$ $E_c° = 0.96 \text{ V}$

$(\text{Mn} \rightarrow \text{Mn}^{2+} + 2 \text{ e}^-) \times 3$ $-E_a° = 1.18 \text{ V}$

_____

$3 \text{ Mn(s)} + 8 \text{ H}^+(aq) + 2 \text{ NO}_3^-(aq) \rightarrow 2 \text{ NO(g)} + 4 \text{ H}_2\text{O(l)} + 3 \text{ Mn}^{2+}(aq)$ $E_{cell}° = 2.14 \text{ V}$

$5 \times (2 \text{ e}^- + 2 \text{ H}^+ + \text{IO}_4^- \rightarrow \text{IO}_3^- + \text{H}_2\text{O})$ $E_c° = 1.60 \text{ V}$

$2 \times (\text{Mn}^{2+} + 4 \text{ H}_2\text{O} \rightarrow \text{MnO}_4^- + 8 \text{ H}^+ + 5 \text{ e}^-)$ $-E_a° = -1.51 \text{ V}$

_____

$5 \text{ IO}_4^-(aq) + 2 \text{ Mn}^{2+}(aq) + 3 \text{ H}_2\text{O(l)} \rightarrow 5 \text{ IO}_3^-(aq) + 2 \text{ MnO}_4^-(aq) + 6 \text{ H}^+(aq)$

                                                                     $E_{cell}° = 0.09 \text{ V}$

b.   Nitric acid oxidation (see part a for $E_{cell}°$):

$\Delta G° = -nFE_{cell}° = -(6 \text{ mol e}^-)(96{,}485 \text{ C/mol e}^-)(2.14 \text{ J/C}) = -1.24 \times 10^6 \text{ J} =$

                                                      -1240 kJ

$\log K = \dfrac{nE°}{0.0591} = \dfrac{6(2.14)}{0.0591} = 217, \ K \approx 10^{217}$

Periodate oxidation (see part a for $E_{cell}°$):

$\Delta G° = -(10 \text{ mol e}^-)(96{,}485 \text{ C/mol e}^-)(0.09 \text{ J/C})(1 \text{ kJ}/1000 \text{ J}) = -90 \text{ kJ}$

$\log K = \dfrac{10(0.09)}{0.0591} = 15.2, \ K = 10^{15.2} = 2 \times 10^{15}$

43. $\Delta G° = -nFE° = \Delta H° - T\Delta S°, \ E° = \dfrac{T\Delta S°}{nF} - \dfrac{\Delta H°}{nF}$

If we graph E° versus T, we should get a straight line ($y = mx + b$). The slope (m) of the line is equal to $\Delta S°/nF$, and the y intercept is equal to $-\Delta H°/nF$. From the equation above, E° will have a small temperature dependence when $\Delta S°$ is close to zero.

45.   a. $\text{Cu}^+ + \text{e}^- \rightarrow \text{Cu}$ $E_c° = 0.52 \text{ V}$

           $\text{Cu}^+ \rightarrow \text{Cu}^{2+} + \text{e}^-$ $-E_a° = -0.16 \text{ V}$

_____

$2 \text{ Cu}^+(aq) \rightarrow \text{Cu}^{2+}(aq) + \text{Cu(s)}$ $E_{cell}° = 0.36 \text{ V}; \ \text{spontaneous}$

$\Delta G° = -nFE_{cell}° = -(1 \text{ mol e}^-)(96{,}485 \text{ C/mol e}^-)(0.36 \text{ J/C}) = -34{,}700 \text{ J} = -35 \text{ kJ}$

$$E^\circ_{cell} = \frac{0.0591}{n} \log K, \quad \log K = \frac{nE^\circ}{0.0591} = \frac{1(0.36)}{0.0591} = 6.09, \quad K = 10^{6.09} = 1.2 \times 10^6$$

b. $Fe^{2+} + 2\,e^- \rightarrow Fe$ $E^\circ_c = -0.44$ V

   $(Fe^{2+} \rightarrow Fe^{3+} + e^-) \times 2$ $-E^\circ_a = -0.77$ V

———————————————————————————————————————

  $3\,Fe^{2+}(aq) \rightarrow 2\,Fe^{3+}(aq) + Fe(s)$ $E^\circ_{cell} = -1.21$ V; not spontaneous

c. $HClO_2 + 2\,H^+ + 2\,e^- \rightarrow HClO + H_2O$ $E^\circ_c = 1.65$ V

   $HClO_2 + H_2O \rightarrow ClO_3^- + 3\,H^+ + 2\,e^-$ $-E^\circ_a = -1.21$ V

———————————————————————————————————————

   $2\,HClO_2(aq) \rightarrow ClO_3^-(aq) + H^+(aq) + HClO(aq)$ $E^\circ_{cell} = 0.44$ V; spontaneous

$$\Delta G^\circ = -nFE^\circ_{cell} = -(2 \text{ mol } e^-)(96{,}485 \text{ C/mol } e^-)(0.44 \text{ J/C}) = -84{,}900 \text{ J} = -85 \text{ kJ}$$

$$\log K = \frac{nE^\circ}{0.0591} = \frac{2(0.44)}{0.0591} = 14.89, \quad K = 7.8 \times 10^{14}$$

47. $Al^{3+} + 3\,e^- \rightarrow Al$ $E^\circ_c = -1.66$ V

     $Al + 6\,F^- \rightarrow AlF_6^{3-} + 3\,e^-$ $-E^\circ_a = 2.07$ V

———————————————————————————————————————

  $Al^{3+}(aq) + 6\,F^-(aq) \rightarrow AlF_6^{3-}(aq)$ $E^\circ_{cell} = 0.41$ V $K = ?$

$$\log K = \frac{nE^\circ}{0.0591} = \frac{3(0.41)}{0.0591} = 20.81, \quad K = 10^{20.81} = 6.5 \times 10^{20}$$

49. $CuI + e^- \rightarrow Cu + I^-$ $E^\circ_{CuI} = ?$

    $Cu \rightarrow Cu^+ + e^-$ $-E^\circ_a = -0.52$ V

———————————————————————————————————————

   $CuI(s) \rightarrow Cu^+(aq) + I^-(aq)$ $E^\circ_{cell} = E^\circ_{CuI} - 0.52$ V

For this overall reaction, $K = K_{sp} = 1.1 \times 10^{-12}$:

$$E^\circ_{cell} = \frac{0.0591}{n} \log K_{sp} = \frac{0.0591}{1} \log(1.1 \times 10^{-12}) = -0.71 \text{ V}$$

$$E^\circ_{cell} = -0.71 \text{ V} = E^\circ_{CuI} - 0.52, \quad E^\circ_{CuI} = -0.19 \text{ V}$$

Galvanic Cells: Concentration Dependence

51. $Cr_2O_7^{2-} + 14\,H^+ + 6\,e^- \rightarrow 2\,Cr^{3+} + 7\,H_2O$ $E^\circ_c = 1.33$ V

      $(Al \rightarrow Al^{3+} + 3\,e^-) \times 2$ $-E^\circ_a = 1.66$ V

———————————————————————————————————————

  $Cr_2O_7^{2-} + 14\,H^+ + 2\,Al \rightarrow 2\,Al^{3+} + 2\,Cr^{3+} + 7\,H_2O$ $E^\circ_{cell} = 2.99$ V

$$E = E° - \frac{0.0591}{n} \log Q, \quad E = 2.99 \text{ V} - \frac{0.0591}{6} \log \frac{[Al^{3+}]^2[Cr^{3+}]^2}{[Cr_2O_7^{2-}][H^+]^{14}}$$

$$3.01 = 2.99 - \frac{0.0591}{n} \log \frac{(0.30)^2(0.15)^2}{(0.55)[H^+]^{14}}, \quad \frac{-6(0.02)}{0.0591} = \log\left(\frac{3.7 \times 10^{-3}}{[H^+]^{14}}\right)$$

$$\frac{3.7 \times 10^{-3}}{[H^+]^{14}} = 10^{-2.0} = 0.01, \quad [H^+]^{14} = 0.37, \quad [H^+] = 0.93 = 0.9 \ M, \quad pH = -\log(0.9) = 0.05$$

53. a.
$$Au^{3+} + 3 \ e^- \rightarrow Au \qquad\qquad E_c° = 1.50 \text{ V}$$
$$(Tl \rightarrow Tl^+ + e^-) \times 3 \qquad -E_a° = 0.34 \text{ V}$$
$$\overline{\rule{8cm}{0.4pt}}$$
$$Au^{3+}(aq) + 3 \ Tl(s) \rightarrow Au(s) + 3 \ Tl^+(aq) \qquad E_{cell}° = 1.84 \text{ V}$$

b. $\Delta G° = -nFE_{cell}° = -(3 \text{ mol } e^-)(96,485 \text{ C/mol } e^-)(1.84 \text{ J/C}) = -5.33 \times 10^5 \text{ J} = -533 \text{ kJ}$

$$\log K = \frac{nE°}{0.0591} = \frac{3(1.84)}{0.0591} = 93.401, \quad K = 10^{93.401} = 2.52 \times 10^{93}$$

c. At 25°C, $E_{cell} = E_{cell}° - \dfrac{0.0591}{n} \log Q$, where $Q = \dfrac{[Tl^+]^3}{[Au^{3+}]}$.

$$E_{cell} = 1.84 \text{ V} - \frac{0.0591}{3} \log \frac{[Tl^+]^3}{[Au^{3+}]} = 1.84 - \frac{0.0591}{3} \log \frac{(1.0 \times 10^{-4})^3}{1.0 \times 10^{-2}}$$

$$E_{cell} = 1.84 - (-0.20) = 2.04 \text{ V}$$

55.
$$(Pb^{2+} + 2 \ e^- \rightarrow Pb) \times 3 \qquad\qquad E_c° = -0.13 \text{ V}$$
$$(Al \rightarrow Al^{3+} + 3 \ e^-) \times 2 \qquad -E_a° = 1.66 \text{ V}$$
$$\overline{\rule{8cm}{0.4pt}}$$
$$3 \ Pb^{2+}(aq) + 2 \ Al(s) \rightarrow 3 \ Pb(s) + 2 \ Al^{3+}(aq) \qquad E_{cell}° = 1.53 \text{ V}$$

From the balanced reaction, when the $[Al^{3+}]$ has increased by 0.60 mol/L (Al^{3+} is a product in the spontaneous reaction), then the Pb^{2+} concentration has decreased by 3/2 (0.60 mol/L) = 0.90 M.

$$E_{cell} = 1.53 \text{ V} - \frac{0.0591}{6} \log \frac{[Al^{3+}]^2}{[Pb^{2+}]^3} = 1.53 - \frac{0.0591}{6} \log \frac{(1.60)^2}{(0.10)^3}$$

$$E_{cell} = 1.53 \text{ V} - 0.034 \text{ V} = 1.50 \text{ V}$$

57. a. n = 2 for this reaction (lead goes from $Pb \rightarrow Pb^{2+}$ in $PbSO_4$).

$$E = E_{cell}° - \frac{0.0591}{2} \log\left(\frac{1}{[H^+]^2[HSO_4^-]^2}\right) = 2.04 \text{ V} - \frac{0.0591}{2} \log \frac{1}{(4.5)^2(4.5)^2}$$

2.04 V − (−0.077 V) = 2.12 V

b. We can calculate $\Delta G°$ from $\Delta G° = \Delta H° - T\Delta S°$ and then E° from $\Delta G° = -nFE°$, or we can use the equation derived in the answer to Exercise 11.43.

$$E°_{-20} = \frac{T\Delta S° - \Delta H°}{nF} = \frac{(253\ K)(263.5\ J/K) + 315.9 \times 10^3\ J}{(2\ mol\ e^-)(96{,}485\ C/mol\ e^-)} = 1.98\ J/C = 1.98\ V$$

c. $$E_{-20} = E°_{-20} - \frac{RT}{nF}\ \ln Q = 1.98\ V - \frac{RT}{nF}\ \ln\frac{1}{[H^+]^2[HSO_4^-]^2}$$

$$E_{-20} = 1.98\ V - \frac{(8.3145\ J\ K^{-1}\ mol^{-1})(253\ K)}{(2\ mol\ e^-)(96{,}485\ C/mol\ e^-)}\ \ln\frac{1}{(4.5)^2(4.5)^2} = 1.98\ V - (-0.066\ V)$$
$$= 2.05\ V$$

d. As the temperature decreases, the cell potential decreases.   Also, oil becomes more viscous at lower temperatures, which adds to the difficulty of starting an engine on a cold day. The combination of these two factors results in batteries failing more often on cold days than on warm days.

59. Concentration cell: a galvanic cell in which both compartments contain the same components but at different concentrations. All concentration cells have $E°_{cell} = 0$ because both compartments contain the same contents. The driving force for the cell is the different ion concentrations at the anode and cathode. The cell produces a voltage as long as the ion concentrations are different. Equilibrium for a concentration cell is reached (E = 0) when the ion concentrations in the two compartments are equal.

The net reaction in a concentration cell is:

$$M^{a+}(cathode,\ x\ M) \ \rightarrow\ M^{a+}(anode,\ y\ M)\quad E°_{cell} = 0$$

and the Nernst equation is:

$$E = E° - \frac{0.0591}{n}\ \log Q = -\frac{0.0591}{a}\ \log\frac{[M^{a+}(anode)]}{[M^{a+}(cathode)]},\quad \text{where } a \text{ is the number of electrons transferred.}$$

To register a potential (E > 0), the log Q term must be a negative value. This occurs when $M^{a+}(cathode) > M^{a+}(anode)$. The higher ion concentration is always at the cathode, and the lower ion concentration is always at the anode. The magnitude of the cell potential depends on the magnitude of the differences in ion concentrations between the anode and cathode. The larger the difference in ion concentrations, the more negative is the log Q term, and the more positive is the cell potential. Thus, as the difference in ion concentrations between the anode and cathode compartments increases, the cell potential increases. This can be accomplished by decreasing the ion concentration at the anode and/or by increasing the ion concentration at the cathode.

61.   As is the case for all concentration cells, $E^\circ_{cell} = 0$, and the smaller ion concentration is always in the anode compartment. The general Nernst equation for the Ni | Ni^{2+} (x M) || Ni^{2+}(y M) | Ni concentration cell is:

$$E_{cell} = E^\circ_{cell} - \frac{0.0591}{n} \log Q = \frac{0.0591}{2} \log \frac{[Ni^{2+}]_{anode}}{[Ni^{2+}]_{cathode}}$$

a.   Because both compartments are at standard conditions ([Ni^{2+}] = 1.0 M), $E_{cell} = E^\circ_{cell} = 0$ V. No electron flow occurs.

b.   Cathode = 2.0 M Ni^{2+}; anode = 1.0 M Ni^{2+}; electron flow is always from the anode to the cathode, so electrons flow to the right in the diagram.

$$E_{cell} = \frac{-0.0591}{2} \log \frac{[Ni^{2+}]_{anode}}{[Ni^{2+}]_{cathode}} = \frac{-0.0591}{2} \log \frac{1.0}{2.0} = 8.9 \times 10^{-3} \text{ V}$$

c.   Cathode = 1.0 M Ni^{2+}; anode = 0.10 M Ni^{2+}; electrons flow to the left in the diagram.

$$E_{cell} = \frac{-0.0591}{2} \log \frac{0.10}{1.0} = 0.030 \text{ V}$$

d.   Cathode = 1.0 M Ni^{2+}; anode = 4.0 $\times$ 10^{-5} M Ni^{2+}; electrons flow to the left in the diagram.

$$E_{cell} = \frac{-0.0591}{2} \log \frac{4.0 \times 10^{-5}}{1.0} = 0.13 \text{ V}$$

e.   Because both concentrations are equal, log(2.5/2.5) = log(1.0) = 0, and E_{cell} = 0. No electron flow occurs.

63. $Cu^{2+}(aq) + H_2(g) \rightarrow 2 H^+(aq) + Cu(s)$ E°_{cell} = 0.34 V − 0.00V = 0.34 V and n = 2

Since P_{H_2} = 1.0 atm and [H$^+$] = 1.0 M: $E_{cell} = E^\circ_{cell} - \frac{0.0591}{2} \log \frac{1}{[Cu^{2+}]}$

a. $E_{cell} = 0.34 \text{ V} - \frac{0.0591}{2} \log \frac{1}{2.5 \times 10^{-4}} = 0.34 \text{ V} - 0.11 \text{ V} = 0.23 \text{ V}$

b.   Use the K$_{sp}$ expression to calculate the Cu^{2+} concentration in the cell.

$Cu(OH)_2(s) \rightleftharpoons Cu^{2+}(aq) + 2 OH^-(aq)$ $K_{sp} = 1.6 \times 10^{-19} = [Cu^{2+}][OH^-]^2$

From the problem, [OH$^-$] = 0.10 M, so: $[Cu^{2+}] = \frac{1.6 \times 10^{-19}}{(0.10)^2} = 1.6 \times 10^{-17} M$

$$E_{cell} = E^\circ_{cell} - \frac{0.0591}{2} \log \frac{1}{[Cu^{2+}]} = 0.34 - \frac{0.0591}{2} \log \frac{1}{1.6 \times 10^{-17}} = 0.34 - 0.50$$
$$= -0.16 \text{ V}$$

Because $E_{cell} < 0$, the forward reaction is not spontaneous; but the reverse reaction is spontaneous. The Cu electrode becomes the anode, and $E_{cell} = 0.16$ V for the reverse reaction. The cell reaction is $2\ H^+(aq) + Cu(s) \rightarrow Cu^{2+}(aq) + H_2(g)$.

c. $0.195\ V = 0.34\ V - \dfrac{0.0591}{2}\ \log\dfrac{1}{[Cu^{2+}]}, \quad \log\dfrac{1}{[Cu^{2+}]} = 4.91, \quad [Cu^{2+}] = 10^{-4.91}$
$$= 1.2 \times 10^{-5}\ M$$

Note: When determining exponents, we will carry extra significant figures.

d. $E_{cell} = E^\circ_{cell} - (0.0591/2)\ \log\ (1/[Cu^{2+}]) = E^\circ_{cell} + 0.0296\ \log\ [Cu^{2+}]$; this equation is in the form of a straight-line equation, $y = mx + b$. A graph of E_{cell} versus $\log[Cu^{2+}]$ will yield a straight line with slope equal to 0.0296 V or 29.6 mV.

65.   a. $Ag^+(x\ M,\ anode) \rightarrow Ag^+(0.10\ M,\ cathode)$; for the silver concentration cell, $E^\circ = 0.00$ (as is always the case for concentration cells) and $n = 1$.

$$E = 0.76\ V = 0.00 - \dfrac{0.0591}{1}\ \log\dfrac{[Ag^+]_{anode}}{[Ag^+]_{cathode}}$$

$$0.76 = -0.0591\ \log\dfrac{[Ag^+]_{anode}}{0.10}, \quad \dfrac{[Ag^+]_{anode}}{0.10} = 10^{-12.86}, \quad [Ag^+]_{anode} = 1.4 \times 10^{-14}\ M$$

b. $Ag^+(aq) + 2\ S_2O_3^{2-}(aq) \rightleftharpoons Ag(S_2O_3)_2^{3-}(aq)$

$$K = \dfrac{[Ag(S_2O_3)_2^{3-}]}{[Ag^+][S_2O_3^{2-}]^2} = \dfrac{1.0 \times 10^{-3}}{1.4 \times 10^{-14}(0.050)^2} = 2.9 \times 10^{13}$$

67. $2\ Ag^+(aq) + Cu(s) \rightarrow Cu^{2+}(aq) + 2\ Ag(s)$ $E^\circ_{cell} = 0.80 - 0.34 = 0.46$ V and $n = 2$

Because $[Ag^+] = 1.0\ M$, $E_{cell} = 0.46\ V - \dfrac{0.0591}{2}\ \log\ [Cu^{2+}]$.

Use the equilibrium reaction to calculate the Cu^{2+} concentration in the cell.

$Cu^{2+}(aq) + 4\ NH_3(aq) \rightleftharpoons Cu(NH_3)_4^{2+}(aq)$ $K = \dfrac{[Cu(NH_3)_4^{2+}]}{[Cu^{2+}][NH_3]^4} = 1.0 \times 10^{13}$

From the problem, $[NH_3] = 5.0\ M$ and $[Cu(NH_3)_4^{2+}] = 0.010\ M$:

$$1.0 \times 10^{13} = \dfrac{0.010}{[Cu^{2+}](5.0)^4}, \quad [Cu^{2+}] = 1.6 \times 10^{-18}\ M$$

$$E_{cell} = 0.46 - \dfrac{0.0591}{2}\ \log\ (1.6 \times 10^{-18}) = 0.46 - (-0.53) = 0.99\ V$$

Electrolysis

69. $15 \text{ A} = \dfrac{15 \text{ C}}{\text{s}} \times \dfrac{60 \text{ s}}{\text{min}} \times \dfrac{60 \text{ min}}{\text{mol}} = 5.4 \times 10^4$ C of charge passed in 1 hour

a. $5.4 \times 10^4 \text{ C} \times \dfrac{1 \text{ mol e}^-}{96{,}485 \text{ C}} \times \dfrac{1 \text{ mol Co}}{2 \text{ mol e}^-} \times \dfrac{58.9 \text{ g}}{\text{mol}} = 16$ g Co

b. $5.4 \times 10^4 \text{ C} \times \dfrac{1 \text{ mol e}^-}{96{,}485 \text{ C}} \times \dfrac{1 \text{ mol Hf}}{4 \text{ mol e}^-} \times \dfrac{178.5 \text{ g}}{\text{mol}} = 25$ g Hf

c. $2 \text{ I}^- \rightarrow \text{I}_2 + 2\text{e}^-$; $5.4 \times 10^4 \text{ C} \times \dfrac{1 \text{ mol e}^-}{96{,}485 \text{ C}} \times \dfrac{1 \text{ mol I}_2}{2 \text{ mol e}^-} \times \dfrac{253.8 \text{ g I}_2}{\text{mol I}_2} = 71$ g I_2

d.   Cr is in the +6 oxidation state in CrO_3. Six moles of e^- are needed to produce 1 mol Cr from molten CrO_3.

$5.4 \times 10^4 \text{ C} \times \dfrac{1 \text{ mol e}^-}{96{,}485 \text{ C}} \times \dfrac{1 \text{ mol Cr}}{6 \text{ mol e}^-} \times \dfrac{52.0 \text{ g Cr}}{\text{mol Cr}} = 4.9$ g Cr

71.   The oxidation state of bismuth in BiO^+ is +3 because oxygen has a !2 oxidation state in this ion. Therefore, 3 moles of electrons are required to reduce the bismuth in BiO^+ to Bi(s).

$10.0 \text{ g Bi} \times \dfrac{1 \text{ mol Bi}}{209.0 \text{ g Bi}} \times \dfrac{3 \text{ mol e}^-}{\text{mol Bi}} \times \dfrac{96{,}485 \text{ C}}{\text{mol e}^-} \times \dfrac{1 \text{ s}}{25.0 \text{ C}} = 554 \text{ s} = 9.23$ min

73.   First determine the species present, and then reference Table 11.1 to help you identify each species as a possible oxidizing agent (species reduced) or as a possible reducing agent (species oxidized). Of all the possible oxidizing agents, the species that will be reduced at the cathode will have the most positive E_c° value; the species that will be oxidized at the anode will be the reducing agent with the most positive $-E_a^{\circ}$ value.

a.   Species present: Ni^{2+} and Br^-; Ni^{2+} can be reduced to Ni, and Br^- can be oxidized to Br_2 (from Table 11.1). The reactions are:

Cathode: $Ni^{2+} + 2e^- \rightarrow Ni$ $E_c^{\circ} = -0.23$ V

Anode: $2 \text{ Br}^- \rightarrow \text{Br}_2 + 2 \text{ e}^-$ $-E_a^{\circ} = -1.09$ V

b.   Species present: Al^{3+} and F^-; Al^{3+} can be reduced, and F^- can be oxidized. The reactions are:

Cathode: $Al^{3+} + 3 \text{ e}^- \rightarrow Al$ $E_c^{\circ} = -1.66$ V

Anode: $2 \text{ F}^- \rightarrow \text{F}_2 + 2 \text{ e}^-$ $-E_a^{\circ} = -2.87$ V

c. Species present: Mn^{2+} and I^-; Mn^{2+} can be reduced, and I^- can be oxidized. The reactions are:

Cathode: $Mn^{2+} + 2\ e^- \rightarrow Mn$ $\qquad$ $E^\circ_c = -1.18$ V

Anode: $\quad 2\ I^- \rightarrow I_2 + 2\ e^-$ $\qquad$ $-E^\circ_a = -0.54$ V

d. For aqueous solutions, we must now consider H_2O as a possible oxidizing agent and a possible reducing agent. Species present: Ni^{2+}, Br^-, and H_2O. Possible cathode reactions are:

$Ni^{2+} + 2e^- \rightarrow Ni$ $\qquad\qquad\qquad$ $E^\circ_c = -0.23$ V

$2\ H_2O + 2\ e^- \rightarrow H_2 + 2\ OH^-$ $\qquad$ $E^\circ_c = -0.83$ V

Because it is easier to reduce Ni^{2+} than H_2O (assuming standard conditions), Ni^{2+} will be reduced by the above cathode reaction.

Possible anode reactions are:

$2\ Br^- \rightarrow Br_2 + 2\ e^-$ $\qquad\qquad$ $-E^\circ_a = -1.09$ V

$2\ H_2O \rightarrow O_2 + 4\ H^+ + 4\ e^-$ $\qquad$ $-E^\circ_a = -1.23$ V

Because Br^- is easier to oxidize than H_2O (assuming standard conditions), Br^- will be oxidized by the above anode reaction.

e. Species present: Al^{3+}, F^-, and H_2O; Al^{3+} and H_2O can be reduced. The reduction potentials are $E^\circ_c = -1.66$ V for Al^{3+} and $E^\circ_c = -0.83$ V for H_2O (assuming standard conditions). H_2O should be reduced at the cathode ($2\ H_2O + 2\ e^- \rightarrow H_2 + 2\ OH^-$).

F^- and H_2O can be oxidized. The oxidation potentials are $-E^\circ_a = -2.87$ V for F^- and $-E^\circ_a = -1.23$ V for H_2O (assuming standard conditions). From the potentials, we would predict H_2O to be oxidized at the anode ($2\ H_2O \rightarrow O_2 + 4\ H^+ + 4\ e^-$).

f. Species present: Mn^{2+}, I^-, and H_2O; Mn^{2+} and H_2O can be reduced. The possible cathode reactions are:

$Mn^{2+} + 2\ e^- \rightarrow Mn$ $\qquad\qquad$ $E^\circ_c = -1.18$ V

$2\ H_2O + 2\ e^- \rightarrow H_2 + 2\ OH^-$ $\qquad$ $E^\circ_c = -0.83$ V

Reduction of H_2O should occur at the cathode because E°_c for H_2O is most positive.

I^- and H_2O can be oxidized. The possible anode reactions are:

$2\ I^- \rightarrow I_2 + 2\ e^-$ $\qquad\qquad$ $-E^\circ_a = -0.54$ V

$2\ H_2O \rightarrow O_2 + 4\ H^+ + 4\ e^-$ $\qquad$ $-E^\circ_a = -1.23$ V

Oxidation of I^- will occur at the anode because $-E^\circ_a$ for I^- is most positive.

75. $Au^{3+} + 3 e^- \rightarrow Au$ $E° = 1.50$ V $Ni^{2+} + 2 e^- \rightarrow Ni$ $E° = -0.23$ V
       $Ag^+ + e^- \rightarrow Ag$ $E° = 0.80$ V $Cd^{2+} + 2 e^- \rightarrow Cd$ $E° = -0.40$ V

       $2 H_2O + 2e^- \rightarrow H_2 + 2 OH^-$ $E° = -0.83$ V

Au(s) will plate out first since it has the most positive reduction potential, followed by Ag(s), which is followed by Ni(s), and finally, Cd(s) will plate out last since it has the most negative reduction potential of the metals listed.

77.   To begin plating out Pd: $E_c = 0.62$ V $- \dfrac{0.0591}{2} \log \dfrac{[Cl^-]^4}{[PdCl_4{}^{2-}]} = 0.62 - \dfrac{0.0591}{2} \log \dfrac{(1.0)^4}{0.020}$

       $E_c = 0.62$ V $- 0.050$ V $= 0.57$ V

When 99% of Pd has plated out, $[PdCl_4^-] = \dfrac{1}{100} (0.020) = 0.00020$ M.

       $E_c = 0.62 - \dfrac{0.0591}{2} \log \dfrac{(1.0)^4}{2.0 \times 10^{-4}} = 0.62$ V $- 0.11$V $= 0.51$ V

To begin Pt plating: $E_c = 0.73$ V $- \dfrac{0.0591}{2} \log \dfrac{(1.0)^4}{0.020} = 0.73 - 0.050 = 0.68$ V

When 99% of Pt is plated: $E_c = 0.73 - \dfrac{0.0591}{2} \log \dfrac{(1.0)^4}{2.0 \times 10^{-4}} = 0.73 - 0.11 = 0.62$ V

To begin Ir plating: $E_c = 0.77$ V $- \dfrac{0.0591}{3} \log \dfrac{(1.0)^4}{0.020} = 0.77 - 0.033 = 0.74$ V

When 99% of Ir is plated: $E_c = 0.77 - \dfrac{0.0591}{3} \log \dfrac{(1.0)^4}{2.0 \times 10^{-4}} = 0.77 - 0.073 = 0.70$ V

Yes, since the range of potentials for plating out each metal do not overlap, we should be able to separate the three metals. The exact potential to apply depends on the oxidation reaction. The order of plating will be Ir(s) first, followed by Pt(s), and finally, Pd(s) as the potential is gradually increased.

79.   Alkaline earth metals form +2 ions, so 2 mol of e^- are transferred to form the metal M.

       Mol M $= 748$ s $\times \dfrac{5.00\,\text{C}}{\text{s}} \times \dfrac{1\,\text{mole}^-}{96,485\,\text{C}} \times \dfrac{1\,\text{mol M}}{2\,\text{mole}^-} = 1.94 \times 10^{-2}$ mol M

       Molar mass of M $= \dfrac{0.471\,\text{g M}}{1.94 \times 10^{-2}\,\text{mol M}} = 24.3$ g/mol; $MgCl_2$ was electrolyzed.

81. F_2 is produced at the anode: $2\ F^- \rightarrow F_2 + 2\ e^-$

$$2.00\ h \times \frac{60\ \text{min}}{h} \times \frac{60\ s}{\text{min}} \times \frac{10.0\ C}{s} \times \frac{1\ \text{mol}\ e^-}{96,485\ C} = 0.746\ \text{mol}\ e^-$$

$$0.746\ \text{mol}\ e^- \times \frac{1\ \text{mol}\ F_2}{2\ \text{mol}\ e^-} = 0.373\ \text{mol}\ F_2;\ \ PV = nRT,\ \ V = \frac{nRT}{P}$$

$$V = \frac{(0.373\ \text{mol})(0.08206\ L\ \text{atm}\ K^{-1}\ \text{mol}^{-1})(298\ K)}{1.00\ \text{atm}} = 9.12\ L\ F_2$$

K is produced at the cathode: $K^+ + e^- \rightarrow K$

$$0.746\ \text{mol}\ e^- \times \frac{1\ \text{mol}\ K}{\text{mol}\ e^-} \times \frac{39.10\ g\ K}{\text{mol}\ K} = 29.2\ g\ K$$

83.   In the electrolysis of aqueous sodium chloride, H_2O is reduced in preference to Na^+, and Cl^- is oxidized in preference to H_2O. The anode reaction is $2\ Cl^- \rightarrow Cl_2 + 2\ e^-$, and the cathode reaction is $2\ H_2O + 2\ e^- \rightarrow H_2 + 2\ OH^-$. The overall reaction is $2\ H_2O(l) + 2\ Cl^-\ (aq) \rightarrow Cl_2(g) + H_2(g) + 2\ OH^-\ (aq)$.

From the 1 : 1 mole ratio between Cl_2 and H_2 in the overall balanced reaction, if 6.00 L of $H_2(g)$ is produced, then 6.00 L of $Cl_2(g)$ also will be produced since moles and volume of gas are directly proportional at constant T and P (see Chapter 5 of text).

85.   a.   The spoon is where Cu^{2+} is reduced to Cu, so the spoon will be the cathode. The anode will be the copper strip where Cu is oxidized to Cu^{2+}.

b.   Cathode reaction: $Cu^{2+} + 2\ e^- \rightarrow Cu$; anode reaction: $Cu \rightarrow Cu^{2+} + 2\ e^-$

Additional Exercises

87. $(CO + O^{2-} \rightarrow CO_2 + 2\ e^-) \times 2$
$O_2 + 4\ e^- \rightarrow 2\ O^{2-}$

_____

$2\ CO(g) + O_2(g) \rightarrow 2\ CO_2(g)$

$$\Delta G = -nFE,\ \ E = \frac{-\Delta G}{nF} = \frac{-(-380 \times 10^{-3}\ J)}{(4\ \text{mol}\ e^-)(96,485\ C/\text{mol}\ e^-)} = 0.98\ V$$

89.   For C_2H_5OH, H has a +1 oxidation state, and O has a −2 oxidiation state. This dictates a −2 oxidation state for C. For CO_2, O has a −2 oxidiation state, so carbon has a +4 oxidiation state. Six moles of electrons are transferred per mole of carbon oxidized (C goes from −2 → +4). Two moles of carbon are in the balanced reaction, so n = 12.

$$w_{max} = -1320 \text{ kJ} = \Delta G = -nFE, \; -1320 \times 10^3 \text{ J} = -nFE = -(12 \text{ mol e}^-)(96,485 \text{ C/mol e}^-)E$$

$$E = 1.14 \text{ J/C} = 1.14 \text{ V}$$

91. Cadmium goes from the zero oxidation state to the +2 oxidation state in $Cd(OH)_2$. Because one mole of Cd appears in the balanced reaction, $n = 2$ mol electrons transferred. At standard conditions:

$$w_{max} = \Delta G° = -nFE°, \; w_{max} = -(2 \text{ mol e}^-)(96,485 \text{ C/mol e}^-)(1.10 \text{ J/C}) = -2.12 \times 10^5 \text{ J}$$
$$= -212 \text{ kJ}$$

93. a. Paint: covers the metal surface so that no contact occurs between the metal and air. This only works as long as the painted surface is not scratched.

   b. Durable oxide coatings: covers the metal surface so that no contact occurs between the metal and air.

   c. Galvanizing: coating steel with zinc; Zn forms an effective oxide coating over steel; also, zinc is more easily oxidized than the iron in the steel.

   d. Sacrificial metal: attaching a more easily oxidized metal to an iron surface; the more active metal is preferentially oxidized instead of iron.

   e. Alloying: adding chromium and nickel to steel; the added Cr and Ni form oxide coatings on the steel surface.

   f. Cathodic protection: a more easily oxidized metal is placed in electrical contact with the metal we are trying to protect. It is oxidized in preference to the protected metal. The protected metal becomes the cathode electrode, thus cathodic protection.

95. $Zn \rightarrow Zn^{2+} + 2 e^- \;\; -E_a° = 0.76 \text{ V}; \; Fe \rightarrow Fe^{2+} + 2 e^- \;\; -E_a° = 0.44 \text{ V}$

   It is easier to oxidize Zn than Fe, so the Zn will be oxidized, protecting the iron of the *Monitor's* hull.

97. The potential oxidizing agents are NO_3^- and H^+. Hydrogen ion cannot oxidize Pt under either condition. Nitrate cannot oxidize Pt unless there is Cl^- in the solution. Aqua regia has both Cl^- and NO_3^-. The overall reaction is:

$$(NO_3^- + 4 H^+ + 3 e^- \rightarrow NO + 2 H_2O) \times 2 \qquad E_c° = 0.96 \text{ V}$$
$$(4 Cl^- + Pt \rightarrow PtCl_4^{2-} + 2 e^-) \times 3 \qquad -E_a° = -0.755 \text{ V}$$

$$12 Cl^-(aq) + 3 Pt(s) + 2 NO_3^-(aq) + 8 H^+(aq) \rightarrow 3 PtCl_4^{2-}(aq) + 2 NO(g) + 4 H_2O(l) \qquad E_{cell}° = 0.21 \text{ V}$$

99.   a. $E_{cell} = E_{ref} + 0.05916\ pH$, $0.480\ V = 0.250\ V + 0.05916\ pH$

$$pH = \frac{0.480 - 0.250}{0.05916} = 3.888; \quad \text{uncertainty} = \pm 1\ mV = \pm 0.001\ V$$

$$pH_{max} = \frac{0.481 - 0.250}{0.05916} = 3.905; \quad pH_{min} = \frac{0.479 - 0.250}{0.05916} = 3.871$$

Thus if the uncertainty in potential is $\pm 0.001\ V$, then the uncertainty in pH is ± 0.017, or about ± 0.02 pH units. For this measurement, $[H^+] = 10^{-3.888} = 1.29 \times 10^{-4}\ M$. For an error of $+1\ mV$, $[H^+] = 10^{-3.905} = 1.24 \times 10^{-4}\ M$. For an error of $-1\ mV$, $[H^+] = 10^{-3.871} = 1.35 \times 10^{-4}\ M$. So the uncertainty in $[H^+]$ is $\pm 0.06 \times 10^{-4}\ M = \pm 6 \times 10^{-6}\ M$.

b.   From part a, we will be within ± 0.02 pH units if we measure the potential to the nearest $\pm 0.001\ V$ ($\pm 1\ mV$).

101. $2\ Ag^+(aq) + Cu(s) \rightarrow Cu^{2+}(aq) + 2\ Ag(s)$ $E^\circ_{cell} = 0.80 - 0.34\ V = 0.46\ V$; A galvanic cell produces a voltage as the forward reaction occurs. Any stress that increases the tendency of the forward reaction to occur will increase the cell potential, whereas a stress that decreases the tendency of the forward reaction to occur will decrease the cell potential.

a.   Added Cu^{2+} (a product ion) will decrease the tendency of the forward reaction to occur, which will decrease the cell potential.

b.   Added NH_3 removes Cu^{2+} in the form of $Cu(NH_3)_4^{2+}$. Because a product ion is removed, this will increase the tendency of the forward reaction to occur, which will increase the cell potential.

c.   Added Cl^- removes Ag^+ in the form of $AgCl(s)$. Because a reactant ion is removed, this will decrease the tendency of the forward reaction to occur, which will decrease the cell potential.

d. $Q_1 = \dfrac{[Cu^{2+}]_0}{[Ag^+]_0^2}$; as the volume of solution is doubled, each concentration is halved.

$$Q_2 = \frac{1/2\ [Cu^{2+}]_0}{(1/2\ [Ag^+]_0)^2} = \frac{2[Cu^{2+}]_0}{[Ag^+]_0^2} = 2Q_1$$

The reaction quotient is doubled because the concentrations are halved. Because reactions are spontaneous when $Q < K$, and because Q increases when the solution volume doubles, the reaction is closer to equilibrium, which will decrease the cell potential.

e.   Because $Ag(s)$ is not a reactant in this spontaneous reaction, and because solids do not appear in the reaction quotient expressions, replacing the silver electrode with a platinum electrode will have no effect on the cell potential.

Challenge Problems

103.   Chromium(III) nitrate $[Cr(NO_3)_3]$ has chromium in the +3 oxidation state.

$$1.15 \text{ g Cr} \times \frac{1 \text{ mol Cr}}{52.00 \text{ g}} \times \frac{3 \text{ mol e}^-}{\text{mol Cr}} \times \frac{96,485 \text{ C}}{\text{mol e}^-} = 6.40 \times 10^3 \text{ C of charge}$$

For the Os cell, 6.40×10^3 C of charge also was passed.

$$3.15 \text{ g Os} \times \frac{1 \text{ mol Os}}{190.2 \text{ g}} = 0.0166 \text{ mol Os}; \quad 6.40 \times 10^3 \text{ C} \times \frac{1 \text{ mol e}^-}{96,485 \text{ C}} = 0.0663 \text{ mol e}^-$$

$$\frac{\text{Mol e}^-}{\text{Mol Os}} = \frac{0.0663}{0.0166} = 3.99 \approx 4$$

This salt is composed of Os^{4+} and NO_3^- ions. The compound is $Os(NO_3)_4$, osmium(IV) nitrate.

For the third cell, identify X by determining its molar mass. Two moles of electrons are transferred when X^{2+} is reduced to X.

$$\text{Molar mass} = \frac{2.11 \text{ g X}}{6.40 \times 10^3 \text{ C} \times \dfrac{1 \text{ mol e}^-}{96,485 \text{ C}} \times \dfrac{1 \text{ mol X}}{2 \text{ mol e}^-}} = 63.6 \text{ g/mol}; \quad \text{this is copper (Cu).}$$

105.   a. $Zn(s) + Cu^{2+}(aq) \rightarrow Zn^{2+}(aq) + Cu(s) \quad E^\circ_{cell} = 1.10 \text{ V}$

$$E_{cell} = 1.10 \text{ V} - \frac{0.0591}{2} \log \frac{[Zn^{2+}]}{[Cu^{2+}]}$$

$$E_{cell} = 1.10 \text{ V} - \frac{0.0591}{2} \log \frac{0.10}{2.50} = 1.10 \text{ V} - (-0.041 \text{ V}) = 1.14 \text{ V}$$

b. $10.0 \text{ h} \times \dfrac{60 \text{ min}}{\text{h}} \times \dfrac{60 \text{ s}}{\text{min}} \times \dfrac{10.0 \text{ C}}{\text{s}} \times \dfrac{1 \text{ mol e}^-}{96,485 \text{ C}} \times \dfrac{1 \text{ mol Cu}}{2 \text{ mol e}^-} = 1.87 \text{ mol Cu produced}$

The Cu^{2+} concentration decreases by 1.87 mol/L, and the Zn^{2+} concentration will increase by 1.87 mol/L.

$[Cu^{2+}] = 2.50 - 1.87 = 0.63 \, M; \quad [Zn^{2+}] = 0.10 + 1.87 = 1.97 \, M$

$$E_{cell} = 1.10 \text{ V} - \frac{0.0591}{2} \log \frac{1.97}{0.63} = 1.10 \text{ V} - 0.015 \text{ V} = 1.09 \text{ V}$$

c. $1.87 \text{ mol Zn consumed} \times \dfrac{65.38 \text{ g Zn}}{\text{mol Zn}} = 122 \text{ g Zn}$

Mass of electrode = 200. − 122 = 78 g Zn

$1.87 \text{ mol Cu formed} \times \dfrac{63.55 \text{ g Cu}}{\text{mol Cu}} = 119 \text{ g Cu}$

Mass of electrode = 200. + 119 = 319 g Cu

d. Three things could possibly cause this battery to go dead:

1. All the Zn is consumed.
2. All the Cu^{2+} is consumed.
3. Equilibrium is reached ($E_{cell} = 0$).

We began with 2.50 mol Cu^{2+} and 200. g Zn × 1 mol Zn/65.38 g Zn = 3.06 mol Zn. Cu^{2+} is the limiting reagent and will run out first. To react all the Cu^{2+} requires:

$$2.50 \text{ mol Cu}^{2+} \times \frac{2 \text{ mol e}^-}{\text{mol Cu}^{2+}} \times \frac{96{,}485 \text{ C}}{\text{mol e}^-} \times \frac{1 \text{ s}}{10.0 \text{ C}} \times \frac{1 \text{ h}}{3600 \text{ s}} = 13.4 \text{ h}$$

For equilibrium to be reached: $E = 0 = 1.10 \text{ V} - \dfrac{0.0591}{2} \log \dfrac{[Zn^{2+}]}{[Cu^{2+}]}$

$$\frac{[Zn^{2+}]}{[Cu^{2+}]} = K = 10^{2(1.10)/0.0591} = 1.68 \times 10^{37}$$

This is such a large equilibrium constant that virtually all the Cu^{2+} must react to reach equilibrium. So the battery will go dead in 13.4 hours.

107. a. $3 \times (e^- + 2 H^+ + NO_3^- \rightarrow NO_2 + H_2O)$ $E_c^° = 0.775 \text{ V}$

$2 H_2O + NO \rightarrow NO_3^- + 4 H^+ + 3 e^-$ $-E_a^° = -0.957 \text{ V}$

$2 H^+(aq) + 2 NO_3^-(aq) + NO(g) \rightarrow 3 NO_2(g) + H_2O(l)$ $E_{cell}^° = -0.182 \text{ V}\quad K = ?$

$\log K = \dfrac{E^°}{0.0591} = \dfrac{3(-0.182)}{0.0591} = -9.239,\ K = 10^{-9.239} = 5.77 \times 10^{-10}$

b. Let C = concentration of $HNO_3 = [H^+] = [NO_3^-]$.

$$5.77 \times 10^{-10} = \frac{P_{NO_2}^3}{P_{NO} \times [H^+]^2 \times [NO_3^-]^2} = \frac{P_{NO_2}^3}{P_{NO} \times C^4}$$

If 0.20% NO_2 by moles and $P_{total} = 1.00$ atm:

$$P_{NO_2} = \frac{0.20 \text{ mol } NO_2}{100. \text{ mol total}} \times 1.00 \text{ atm} = 2.0 \times 10^{-3} \text{ atm}; \quad P_{NO} = 1.00 - 0.0020 = 1.00 \text{ atm}$$

$$5.77 \times 10^{-10} = \frac{(2.0 \times 10^{-3})^3}{(1.00)C^4}, \quad C = 1.9 \ M \ HNO_3$$

109. $\quad 2 H^+ + 2 e^- \rightarrow H_2 \qquad\qquad E_c^o = 0.000 \text{ V}$

$\qquad\qquad Fe \rightarrow Fe^{2+} + 2e^- \qquad\qquad -E_a^o = -(-0.440V)$

$2 H^+(aq) + Fe(s) \rightarrow H_2(g) + Fe^{3+}(aq) \qquad E_{cell}^o = 0.440 \text{ V}$

$$E_{cell} = E_{cell}^o - \frac{0.0591}{n} \log Q, \text{ where } n = 2 \text{ and } Q = \frac{P_{H_2} \times [Fe^{3+}]}{[H^+]^2}$$

To determine K_a for the weak acid, first use the electrochemical data to determine the H^+ concentration in the half-cell containing the weak acid.

$$0.333 \text{ V} = 0.440 \text{ V} - \frac{0.0591}{2} \log \frac{1.00 \text{ atm}(1.00 \times 10^{-3} \ M)}{[H^+]^2}$$

$$\frac{0.107(2)}{0.0591} = \log \frac{1.0 \times 10^{-3}}{[H^+]^2}, \quad \frac{1.0 \times 10^{-3}}{[H^+]^2} = 10^{3.621} = 4.18 \times 10^3, \quad [H^+] = 4.89 \times 10^{-4} \ M$$

Now we can solve for the K_a value of the weak acid HA through the normal setup for a weak acid problem.

	HA(aq)	$\rightleftharpoons$	H^+(aq)	+	A^-(aq)	$K_a = \dfrac{[H^+][A^-]}{[HA]}$
Initial	1.00 M		~0		0	
Equil.	1.00 − x		x		x	

$$K_a = \frac{x^2}{1.00 - x}, \text{ where } x = [H^+] = 4.89 \times 10^{-4} \ M, \quad K_a = \frac{(4.89 \times 10^{-4})^2}{1.00 - 4.89 \times 10^{-4}} = 2.39 \times 10^{-7}$$

111.   a. $\qquad (Ag^+ + e^- \rightarrow Ag) \times 2 \qquad\qquad E_c^o = 0.80 \text{ V}$

$\qquad\qquad\qquad Cu \rightarrow Cu^{2+} + 2 e^- \qquad\qquad -E_a^o = -0.34 \text{ V}$

$2 Ag^+(aq) + Cu(s) \rightarrow 2 Ag(s) + Cu^{2+}(aq) \qquad E_{cell}^o = 0.46 \text{ V}$

$$E_{cell} = E_{cell}^o - \frac{0.0591}{n} \log Q, \text{ where } n = 2 \text{ and } Q = \frac{[Cu^{2+}]}{[Ag^+]^2}.$$

To calculate E_{cell}, we need to use the K_{sp} data to determine $[Ag^+]$.

$$AgCl(s) \rightleftharpoons Ag^+(aq) + Cl^-(aq) \quad K_{sp} = 1.6 \times 10^{-10} = [Ag^+][Cl^-]$$

Initial s = solubility (mol/L)   0   0
Equil. s s

$$K_{sp} = 1.6 \times 10^{-10} = s^2, \quad s = [Ag^+] = 1.3 \times 10^{-5} \text{ mol/L}$$

$$E_{cell} = 0.46 \text{ V} - \frac{0.0591}{2} \log \frac{2.0}{(1.3 \times 10^{-5})^2} = 0.46 \text{ V} - 0.30 = 0.16 \text{ V}$$

b. $Cu^{2+}(aq) + 4 NH_3(aq) \rightleftharpoons Cu(NH_4)_4^{2+}(aq) \quad K = 1.0 \times 10^{13} = \dfrac{[Cu(NH_3)_4^{2+}]}{[Cu^{2+}][NH_3]^4}$

Because K is very large for the formation of $Cu(NH_3)_4^{2+}$, the forward reaction is dominant. At equilibrium, essentially all the 2.0 M Cu^{2+} will react to form 2.0 M $Cu(NH_3)_4^{2+}$. This reaction requires 8.0 M NH_3 to react with all the Cu^{2+} in the balanced equation. Therefore, the moles of NH_3 added to 1.0-L solution will be larger than 8.0 mol since some NH_3 must be present at equilibrium. In order to calculate how much NH_3 is present at equilibrium, we need to use the electrochemical data to determine the Cu^{2+} concentration.

$$E_{cell} = E^\circ_{cell} - \frac{0.0591}{n} \log Q, \quad 0.52 \text{ V} = 0.46 \text{ V} - \frac{0.0591}{2} \log \frac{[Cu^{2+}]}{(1.3 \times 10^{-5})^2}$$

$$\log \frac{[Cu^{2+}]}{(1.3 \times 10^{-5})^2} = \frac{-0.06(2)}{0.0591} = -2.03, \quad \frac{[Cu^{2+}]}{(1.3 \times 10^{-5})^2} = 10^{-2.03} = 9.3 \times 10^{-3}$$

$$[Cu^{2+}] = 1.6 \times 10^{-12} = 2 \times 10^{-12} \text{ } M$$

(We carried extra significant figures in the calculation.)

Note: Our assumption that the 2.0 M Cu^{2+} essentially reacts to completion is excellent as only 2×10^{-12} M Cu^{2+} remains after this reaction. Now we can solve for the equilibrium $[NH_3]$.

$$K = 1.0 \times 10^{13} = \frac{[Cu(NH_3)_4^{2+}]}{[Cu^{2+}][NH_3]^4} = \frac{(2.0)}{(2 \times 10^{-12})[NH_3]^4}, \quad [NH_3] = 0.6 \text{ } M$$

Because 1.0 L of solution is present, 0.6 mol NH_3 remains at equilibrium. The total moles of NH_3 added is 0.6 mol plus the 8.0 mol NH_3 necessary to form 2.0 M $Cu(NH_3)_4^{2+}$. Therefore, $8.0 + 0.6 = 8.6$ mol NH_3 was added.

113.

$$(Ag^+ + e^- \rightarrow Ag) \times 2 \qquad E_c^o \quad = 0.80 \text{ V}$$
$$Cd \rightarrow Cd^{2+} + 2 e^- \qquad -E_a^o = 0.40 \text{ V}$$

$$\overline{2 \text{ Ag}^+(aq) + Cd(s) \rightarrow Cd^{2+}(aq) + 2 \text{ Ag}(s) \qquad E_{cell}^o = 1.20 \text{ V}}$$

Overall complex ion reaction:

$$Ag^+(aq) + 2 \text{ NH}_3(aq) \rightarrow Ag(NH_3)_2^+(aq) \quad K = K_1 K_2 = 2.1 \times 10^3 (8.2 \times 10^3) = 1.7 \times 10^7$$

Because K is large, we will let the reaction go to completion, and then solve the back-equilibrium problem.

$$Ag^+ \quad + \quad 2 \text{ NH}_3 \quad \rightleftharpoons \quad Ag(NH_3)_2^+ \qquad K = 1.7 \times 10^7$$

Before	1.00 M	15.0 M	0	
After	0	13.0	1.00	New initial
Change	x	$+2x$	$\leftarrow \quad -x$	
Equil.	x	$13.0 + 2x$	$1.00 - x$	

$$K = \frac{[Ag(NH_3)_2^+]}{[Ag^+][NH_3]^2}; \quad 1.7 \times 10^7 = \frac{1.00 - x}{x(13.0 + 2x)^2} \approx \frac{1.00}{x(13.0)^2}$$

Solving: $x = 3.5 \times 10^{-10} M = [Ag^+]$; assumptions good.

$$E = E^o - \frac{0.0591}{2} \log \frac{[Cd^{2+}]}{[Ag^+]^2} = 1.20 \text{ V} - \frac{0.0591}{2} \log \left[\frac{1.0}{(3.5 \times 10^{-10})^2} \right]$$

$$E = 1.20 - 0.56 = 0.64 \text{ V}$$

115.   a.   From Table 11.1: $2 \text{ H}_2\text{O} + 2 e^- \rightarrow \text{H}_2 + 2 \text{ OH}^-$ $E^o = -0.83 \text{ V}$

$$E_{cell}^o = E_{H_2O}^o - E_{Zr}^o = -0.83 \text{ V} + 2.36 \text{ V} = 1.53 \text{ V}$$

Yes, the reduction of H_2O to H_2 by Zr is spontaneous at standard conditions since $E_{cell}^o > 0$.

b.

$$(2 \text{ H}_2\text{O} + 2 e^- \rightarrow \text{H}_2 + 2 \text{ OH}^-) \times 2$$
$$Zr + 4 \text{ OH}^- \rightarrow ZrO_2 \bullet H_2O + H_2O + 4 e^-$$

$$\overline{3 \text{ H}_2\text{O}(l) + Zr(s) \rightarrow 2 \text{ H}_2(g) + ZrO_2 \bullet H_2O(s)}$$

c. $\Delta G^o = -nFE^o = -(4 \text{ mol } e^-)(96,485 \text{ C/mol } e^-)(1.53 \text{ J/C}) = -5.90 \times 10^5 \text{ J} = -590. \text{ kJ}$

$$E = E^o - \frac{0.0591}{n} \log Q; \text{ at equilibrium, E} = 0 \text{ and Q} = \text{K}.$$

$$E° = \frac{0.0591}{n} \log K, \ \log K = \frac{4(1.53)}{0.0591} = 104, \ K \approx 10^{104}$$

d. $1.00 \times 10^3 \text{ kg Zr} \times \dfrac{1000 \text{ g}}{\text{kg}} \times \dfrac{1 \text{ mol Zr}}{91.22 \text{ g Zr}} \times \dfrac{2 \text{ mol H}_2}{\text{mol Zr}} = 2.19 \times 10^4 \text{ mol H}_2$

$2.19 \times 10^4 \text{ mol H}_2 \times \dfrac{2.016 \text{ g H}_2}{\text{mol H}_2} = 4.42 \times 10^4 \text{ g H}_2$

$V = \dfrac{nRT}{P} = \dfrac{(2.19 \times 10^4 \text{ mol})(0.08206 \text{ L atm K}^{-1} \text{ mol}^{-1})(1273 \text{ K})}{1.0 \text{ atm}} = 2.3 \times 10^6 \text{ L H}_2$

e. Probably yes; less radioactivity overall was released by venting the H_2 than what would have been released if the H_2 had exploded inside the reactor (as happened at Chernobyl). Neither alternative is pleasant, but venting the radioactive hydrogen is the less unpleasant of the two alternatives.

CHAPTER 12

QUANTUM MECHANICS AND ATOMIC THEORY

Light and Matter

21. $\nu = \dfrac{c}{\lambda} = \dfrac{3.00 \times 10^8 \text{ m/s}}{1.0 \times 10^{-2} \text{ m}} = 3.0 \times 10^{10} \text{ s}^{-1}$

$E = h\nu = 6.63 \times 10^{-34} \text{ J s} \times 3.0 \times 10^{10} \text{ s}^{-1} = 2.0 \times 10^{-23} \text{ J/photon}$

$\dfrac{2.0 \times 10^{-23} \text{ J}}{\text{photon}} = \dfrac{6.02 \times 10^{23} \text{ photons}}{\text{mol}} = 12 \text{ J/mol}$

23. Referencing Figure 12.3 of the text, 2.12×10^{-10} m electromagnetic radiation is X rays.

$\lambda = \dfrac{c}{\nu} = \dfrac{2.9979 \times 10^8 \text{ m/s}}{107.1 \times 10^6 \text{ s}^{-1}} = 2.799 \text{ m}$

From the wavelength calculated above, 107.1 MHz electromagnetic radiation is FM radio-waves.

$\lambda = \dfrac{hc}{E} = \dfrac{6.626 \times 10^{-34} \text{ J s} \times 2.998 \times 10^8 \text{ m/s}}{3.97 \times 10^{-19} \text{ J}} = 5.00 \times 10^{-7} \text{ m}$

The 3.97×10^{-19} J/photon electromagnetic radiation is visible (green) light.

The photon energy and frequency order will be the exact opposite of the wavelength ordering because E and ν are both inversely related to λ. From the previously calculated wavelengths, the order of photon energy and frequency is:

FM radiowaves   <   visible (green) light   <   X rays
longest λ shortest λ
lowest ν highest ν
smallest E   largest E

243

25.   a. $\lambda = \dfrac{c}{v} = \dfrac{3.00 \times 10^8 \text{ m/s}}{6.0 \times 10^{13} \text{ s}^{-1}} = 5.0 \times 10^{-6} \text{ m}$

     b.   From Figure 12.3, this is infrared EMR.

     c. $E = hv = 6.63 \times 10^{-34} \text{ J s} \times 6.0 \times 10^{13} \text{ s}^{-1} = 4.0 \times 10^{-20} \text{ J/photon}$

        $\dfrac{4.0 \times 10^{-20} \text{ J}}{\text{photon}} \times \dfrac{6.022 \times 10^{23} \text{ photons}}{\text{mol}} = 2.4 \times 10^4 \text{ J/mol}$

     d.   Frequency and photon energy are directly related ($E = hv$). Because $5.4 \times 10^{13} \text{ s}^{-1}$ EMR has a lower frequency than $6.0 \times 10^{13} \text{ s}^{-1}$ EMR, the $5.4 \times 10^{13} \text{ s}^{-1}$ EMR will have less energetic photons.

27.   The energy needed to remove a single electron is:

        $\dfrac{279.7 \text{ kJ}}{\text{mol}} \times \dfrac{1 \text{ mol}}{6.0221 \times 10^{23}} = 4.645 \times 10^{-22} \text{ kJ} = 4.645 \times 10^{-19} \text{ J}$

        $E = \dfrac{hc}{\lambda}, \ \lambda = \dfrac{hc}{E} = \dfrac{6.6261 \times 10^{-34} \text{ J s} \times 2.9979 \times 10^8 \text{ m/s}}{4.645 \times 10^{-19} \text{ J}} = 4.277 \times 10^{-7} \text{ m} = 427.7 \text{ nm}$

29.   The energy to remove a single electron is:

        $\dfrac{208.4 \text{ kJ}}{\text{mol}} \times \dfrac{1 \text{ mol}}{6.022 \times 10^{23}} = 3.461 \times 10^{-22} \text{ kJ} = 3.461 \times 10^{-19} \text{ J} = E_w$

     Energy of 254-nm light is:

        $E = \dfrac{hc}{\lambda} = \dfrac{(6.626 \times 10^{-34} \text{ J s})(2.998 \times 10^8 \text{ m/s})}{254 \times 10^{-9} \text{ m}} = 7.82 \times 10^{-19} \text{ J}$

        $E_{photon} = E_K + E_w, \ E_K = 7.82 \times 10^{-19} \text{ J} - 3.461 \times 10^{-19} \text{ J} = 4.36 \times 10^{-19} \text{ J} = \text{maximum KE}$

31.   The photoelectric effect refers to the phenomenon in which electrons are emitted from the surface of a metal when light strikes it. The light must have a certain minimum frequency (energy) in order to remove electrons from the surface of a metal. Light having a frequency below the minimum results in no electrons being emitted, whereas light at or higher than the minimum frequency does cause electrons to be emitted. For light having a frequency higher than the minimum frequency, the excess energy is transferred into kinetic energy for the emitted electron. Albert Einstein explained the photoelectric effect by applying quantum theory.

33.   a. $\lambda = \dfrac{h}{mv} = \dfrac{6.626 \times 10^{-34} \text{ J s}}{1.675 \times 10^{-27} \text{ kg} \times (0.0100 \times 2.998 \times 10^8 \text{ m/s})} = 1.32 \times 10^{-13} \text{ m}$

b. $\lambda = \dfrac{h}{mv}$, $v = \dfrac{h}{\lambda m} = \dfrac{6.626 \times 10^{-34} \text{ J s}}{75 \times 10^{-12} \text{ m} \times 1.675 \times 10^{-27} \text{ kg}} = 5.3 \times 10^3 \text{ m/s}$

35. $m = \dfrac{h}{\lambda v} = \dfrac{6.626 \times 10^{-34} \text{ kg m}^2/\text{s}}{3.31 \times 10^{-15} \text{ m} \times (0.0100 \times 2.998 \times 10^8 \text{ m/s})} = 6.68 \times 10^{-26} \text{ kg/atom}$

$\dfrac{6.68 \times 10^{-26} \text{ kg}}{\text{atom}} \times \dfrac{6.022 \times 10^{23} \text{ atoms}}{\text{mol}} \times \dfrac{1000 \text{ g}}{\text{kg}} = 40.2 \text{ g/mol}$

The element is calcium (Ca).

Hydrogen Atom: The Bohr Model

37. a. For hydrogen ($Z = 1$), the energy levels in units of joules are given by the equation $E_n = -2.178 \times 10^{-18}(1/n^2)$. As n increases, the differences between $1/n^2$ for consecutive energy levels becomes smaller and smaller. Consider the difference between $1/n^2$ values for $n = 1$ and $n = 2$ as compared to $n = 3$ and $n = 4$.

For $n = 1$ and $n = 2$:   For $n = 3$ and $n = 4$:

$\dfrac{1}{1^2} - \dfrac{1}{2^2} = 1 - 0.25 = 0.75$ $\dfrac{1}{3^2} - \dfrac{1}{4^2} = 0.1111 - 0.0625 = 0.0486$

Because the differences between $1/n^2$ values for consecutive energy levels decrease as n increases, the energy levels get closer together as n increases.

b. For a spectral transition for hydrogen, $\Delta E = E_f - E_i$:

$$\Delta E = -2.178 \times 10^{-18} \text{ J} \left(\dfrac{1}{n_f^2} - \dfrac{1}{n_i^2} \right)$$

where n_i and n_f are the levels of the initial and final states, respectively. A positive value of ΔE always corresponds to an absorption of light, and a negative value of ΔE always corresponds to an emission of light.

In the diagram, the red line is for the $n_i = 3$ to $n_f = 2$ transition.

$$\Delta E = -2.178 \times 10^{-18} \text{ J} \left(\dfrac{1}{2^2} - \dfrac{1}{3^2} \right) = -2.178 \times 10^{-18} \text{ J} \left(\dfrac{1}{4} - \dfrac{1}{9} \right)$$

$$\Delta E = -2.178 \times 10^{-18} \text{ J} \times (0.2500 - 0.1111) = -3.025 \times 10^{-19} \text{ J}$$

The photon of light must have precisely this energy (3.025×10^{-19} J).

$$|\Delta E| = E_{photon} = h\nu = \dfrac{hc}{\lambda}$$

$$\lambda = \frac{hc}{|\Delta E|} = \frac{6.6261 \times 10^{-34} \text{ J s} \times 2.9979 \times 10^8 \text{ m/s}}{3.025 \times 10^{-19} \text{ J}} = 6.567 \times 10^{-7} \text{ m} = 656.7 \text{ nm}$$

From Figure 12.3, $\lambda = 656.7$ nm is red light so the diagram is correct for the red line.

In the diagram, the green line is for the $n_i = 4$ to $n_f = 2$ transition.

$$\Delta E = -2.178 \times 10^{-18} \text{ J} \left(\frac{1}{2^2} - \frac{1}{4^2} \right) = -4.084 \times 10^{-19} \text{ J}$$

$$\lambda = \frac{hc}{|\Delta E|} = \frac{6.6261 \times 10^{-34} \text{ J s} \times 2.9979 \times 10^8 \text{ m/s}}{4.084 \times 10^{-19} \text{ J}} = 4.864 \times 10^{-7} \text{ m} = 486.4 \text{ nm}$$

From Figure 12.3, $\lambda = 486.4$ nm is green-blue light. The diagram is consistent with this line.

In the diagram, the blue line is for the $n_i = 5$ to $n_f = 2$ transition.

$$\Delta E = -2.178 \times 10^{-18} \text{ J} \left(\frac{1}{2^2} - \frac{1}{5^2} \right) = -4.574 \times 10^{-19} \text{ J}$$

$$\lambda = \frac{hc}{|\Delta E|} = \frac{6.6261 \times 10^{-34} \text{ J s} \times 2.9979 \times 10^8 \text{ m/s}}{4.574 \times 10^{-19} \text{ J}} = 4.343 \times 10^{-7} \text{ m} = 434.3 \text{ nm}$$

From Figure 12.3, $\lambda = 434.3$ nm is blue or blue-violet light. The diagram is consistent with this line also.

39.   There are 4 possible transitions for an electron in the $n = 5$ level ($5 \rightarrow 4$, $5 \rightarrow 3$, $5 \rightarrow 2$, and $5 \rightarrow 1$). If an electron initially drops to the $n = 4$ level, three additional transitions can occur ($4 \rightarrow 3$, $4 \rightarrow 2$, and $4 \rightarrow 1$). Similarly, there are two more transitions from the $n = 3$ level ($3 \rightarrow 2$, $3 \rightarrow 1$) and one more transition for the $n = 2$ level ($2 \rightarrow 1$). There are a total of 10 possible transitions for an electron in the $n = 5$ level for a possible total of 10 different wavelength emissions.

41.   a.   False;   it takes less energy to ionize an electron from $n = 3$ than from the ground state.

      b.   True

      c.   False;   the energy difference between $n = 3$ and $n = 2$ is smaller than the energy difference to $n = 2$ electronic transition than for the $n = 3$ to $n = 1$ transition. E and λ are inversely proportional to each other ($E = hc/\lambda$).

      d.   True

      e.   False;   the ground state in hydrogen is $n = 1$ and all other allowed energy states are called excited states; $n = 2$ is the first excited state, and $n = 3$ is the second excited state.

43. $|\Delta E| = E_{photon} = \dfrac{hc}{\lambda} = \dfrac{6.6261 \times 10^{-34} \text{ J s} \times 2.9979 \times 10^8 \text{ m/s}}{397.2 \times 10^{-9} \text{ m}} = 5.001 \times 10^{-19} \text{ J}$

$\Delta E = -5.001 \times 10^{-19}$ J because we have an emission.

$-5.001 \times 10^{-19} \text{ J} = E_2 - E_n = -2.178 \times 10^{-18} \text{ J} \left(\dfrac{1}{2^2} - \dfrac{1}{n^2} \right)$

$0.2296 = \dfrac{1}{4} - \dfrac{1}{n^2}, \quad \dfrac{1}{n^2} = 0.0204, \quad n = 7$

45. $\Delta E = E_4 - E_n = -E_n = 2.178 \times 10^{-18} \text{ J} \left(\dfrac{1}{n^2} \right)$

$E_{photon} = \dfrac{hc}{\lambda} = \dfrac{6.626 \times 10^{-34} \text{ J s} \times 2.9979 \times 10^8 \text{ m/s}}{1460 \times 10^{-9} \text{ m}} = 1.36 \times 10^{-19} \text{ J}$

$E_{photon} = \Delta E = 1.36 \times 10^{-19} \text{ J} = 2.178 \times 10^{-18} \left(\dfrac{1}{n^2} \right), \quad n^2 = 16.0, \quad n = 4$

47. $E_{photon} = \dfrac{hc}{\lambda} = \dfrac{6.6261 \times 10^{-34} \text{ J s} \times 2.9979 \times 10^8 \text{ m/s}}{253.4 \times 10^{-9} \text{ m}} = 7.839 \times 10^{-19} \text{ J}$

$\Delta E = -7.839 \times 10^{-19}$ J because we have an emission.

The general energy equation for one-electron ions is $E_n = -2.178 \times 10^{-18}$ J $(Z^2)/n^2$, where Z = atomic number.

$\Delta E = -2.178 \times 10^{-18} \text{ J} (Z)^2 \left(\dfrac{1}{n_f^2} - \dfrac{1}{n_i^2} \right), \quad Z = 4 \text{ for Be}^{3+}$

$\Delta E = -7.839 \times 10^{-19} \text{ J} = -2.178 \times 10^{-18} (4)^2 \left(\dfrac{1}{n_f^2} - \dfrac{1}{5^2} \right)$

$\dfrac{7.839 \times 10^{-19}}{2.178 \times 10^{-18} \times 16} + \dfrac{1}{25} = \dfrac{1}{n_f^2}, \quad \dfrac{1}{n_f^2} = 0.06249, \quad n_f = 4$

This emission line corresponds to the $n = 5 \rightarrow n = 4$ electronic transition.

Wave Mechanics and Particle in a Box

49.   a. $\Delta p = m\Delta v = 9.11 \times 10^{-31} \text{ kg} \times 0.100 \text{ m/s} = \dfrac{9.11 \times 10^{-32} \text{ kg m}}{\text{s}}$

$\Delta p \Delta x \geq \dfrac{h}{4\pi}, \quad \Delta x = \dfrac{h}{4\pi \Delta p} = \dfrac{6.626 \times 10^{-34} \text{ J s}}{4 \times 3.142 \times (9.11 \times 10^{-32} \text{ kg m/s})} = 5.79 \times 10^{-4} \text{ m}$

b. $\Delta x = \dfrac{h}{4\pi\,\Delta p} = \dfrac{6.626 \times 10^{-34}\ \text{J s}}{4 \times 3.142 \times 0.145\ \text{kg} \times 0.100\ \text{m/s}} = 3.64 \times 10^{-33}\ \text{m}$

The diameter of an H atom is roughly 2×10^{-8} cm. The uncertainty in the position of the electron is much larger than the size of the atom, whereas, the uncertainty in the position of the baseball is insignificant as compared to the size of a baseball.

51.   At $x = 0$, the value of the square of the wave function must be zero. The particle must be inside the box. For $\psi = A\,\cos(Lx)$, at $x = 0$, $\cos(0) = 1$ and $\psi^2 = A^2$. This violates the boundary condition for a particle in a one-dimensional box.

53. $E_n = \dfrac{n^2 h^2}{8mL^2};\quad \Delta E = E_3 - E_2 = \dfrac{9h^2}{8mL^2} - \dfrac{4h^2}{8mL^2} = \dfrac{5h^2}{8mL^2}$

$\Delta E = \dfrac{hc}{\lambda} = \dfrac{(6.626 \times 10^{-34}\ \text{J s})(2.998 \times 10^8\ \text{m/s})}{8080 \times 10^{-9}\ \text{m}} = 2.46 \times 10^{-20}\ \text{J}$

$\Delta E = 2.46 \times 10^{-20}\ \text{J} = \dfrac{5h^2}{8mL^2} = \dfrac{5(6.626 \times 10^{-34}\ \text{J s})^2}{8(9.109 \times 10^{-31}\ \text{kg})\,L^2},\quad L = 3.50 \times 10^{-9}\ \text{m} = 3.50\ \text{nm}$

55. $E_n = \dfrac{n^2 h^2}{8mL^2};\quad$ as L increases, E_n will decrease, and the spacing between energy levels will also decrease.

57. $E_n = \dfrac{n^2 h^2}{8mL^2},\ n = 1$ for ground state; from equation, as L increases, E_n decreases.

Using numbers: 10^{-6} m box: $E_1 = \dfrac{h^2}{8m}(1 \times 10^{12}\ \text{m}^{-2})$; 10^{-10} m box: $E_1 = \dfrac{h^2}{8m}(1 \times 10^{20}\ \text{m}^{-2})$

As expected, the electron in the 1×10^{-6} m box has the lowest ground state energy.

Orbitals and Quantum Numbers

59.   The 2p orbitals differ from each other in the direction in which they point in space. The 2p and 3p orbitals differ from each other in their size, energy, and number of nodes. A nodal surface in an atomic orbital is a surface in which the probability of finding an electron is zero.

61.   1p: $n = 1$, $\ell = 1$ is not possible; 3f: $n = 3$, $\ell = 3$ is not possible; 2d: $n = 2$, $\ell = 2$ is not possible; in all three incorrect cases, $n = \ell$. The maximum value ℓ can have is $n - 1$, not n.

63.   a.   For $n = 3$, $\ell = 3$ is not possible.

d. m_s cannot equal -1.

     e. ℓ cannot be a negative number.   f.   For $\ell = 1$, m_ℓ cannot equal 2.

The quantum numbers in parts b and c are allowed.

65.   1p, 0 electrons ($\ell \neq 1$ when $n = 1$); $6d_{x^2-y^2}$, 2 electrons (specifies one atomic orbital); 4f, 14 electrons (7 orbitals have 4f designation); $7p_y$, 2 electrons (specifies one atomic orbital); 2s, 2 electrons (specifies one atomic orbital); $n = 3$, 18 electrons (3s, 3p, and 3d orbitals are possible; there are one 3s orbital, three 3p orbitals, and five 3d orbitals).

67.   The diagrams of the orbitals in the text give only 90% probabilities of where the electron may reside. We can never be 100% certain of the location of the electrons due to Heisenberg's uncertainty principle.

69.   For $r = a_o$ and $\theta = 0°$ ($Z = 1$ for H):

$$\psi_{2p_z} = \frac{1}{4(2\pi)^{1/2}}\left(\frac{1}{5.29 \times 10^{-11}}\right)^{3/2} (1)\, e^{-1/2} \cos 0 = 1.57 \times 10^{14}; \quad \psi^2 = 2.46 \times 10^{28}$$

For $r = a_o$ and $\theta = 90°$: $\psi_{2p_z} = 0$ because $\cos 90° = 0$; $\psi^2 = 0$; the xy plane is a node for the $2p_z$ atomic orbital.

Polyelectronic Atoms

71.   He: $1s^2$; Ne: $1s^2 2s^2 2p^6$; Ar: $1s^2 2s^2 2p^6 3s^2 3p^6$; each peak in the diagram corresponds to a subshell with different values of n. Corresponding subshells are closer to the nucleus for heavier elements because of the increased nuclear charge.

73.   Valence electrons are the electrons in the outermost principal quantum level of an atom (those electrons in the highest n value orbitals). The electrons in the lower n value orbitals are all inner core or just core electrons. The key is that the outermost electrons are the valence electrons. When atoms interact with each other, it will be the outermost electrons that are involved in these interactions. In addition, how tightly the nucleus holds these outermost electrons determines atomic size, ionization energy, and other properties of atoms. Elements in the same group have similar valence electron configurations and, as a result, have similar chemical properties.

75.   a. $n = 4$: ℓ can be 0, 1, 2, or 3. Thus we have s (2 e⁻), p (6 e⁻), d (10 e⁻) and f (14 e⁻) orbitals present. Total number of electrons to fill these orbitals is 32.

        b. $n = 5$, $m_\ell = +1$: for $n = 5$, $\ell = 0, 1, 2, 3, 4$; for $\ell = 1, 2, 3, 4$, all can have $m_\ell = +1$. Four distinct orbitals which can hold a maximum of 8 electrons.

        c. $n = 5$, $m_s = +1/2$: for $n = 5$, $\ell = 0, 1, 2, 3, 4$. Number of orbitals = 1, 3, 5, 7, 9 for each value of ℓ, respectively. There are 25 orbitals with $n = 5$. They can hold 50 electrons, and 25 of these electrons can have $m_s = +1/2$.

d. $n = 3$, $\ell = 2$: these quantum numbers define a set of 3d orbitals. There are 5 degenerate 3d orbitals that can hold a total of 10 electrons.

e. $n = 2$, $\ell = 1$: these define a set of 2p orbitals. There are 3 degenerate 2p orbitals that can hold a total of 6 electrons.

f. It is impossible for $n = 0$. Thus no electrons can have this set of quantum numbers.

g. The four quantum numbers completely specify a single electron.

h. $n = 3$: 3s, 3p, and 3d orbitals all have $n = 3$. These orbitals can hold 18 electrons, and 9 of these electrons can have $m_s = +1/2$.

i. $n = 2$, $\ell = 2$: this combination is not possible ($\ell \neq 2$ for $n = 2$). Zero electrons in an atom can have these quantum numbers.

j. $n = 1$, $\ell = 0$, $m_\ell = 0$: these define a 1s orbital that can hold 2 electrons.

77. The following are complete electron configurations. Noble gas shorthand notation could also be used.

Sc: $1s^2 2s^2 2p^6 3s^2 3p^6 4s^2 3d^1$; Fe: $1s^2 2s^2 2p^6 3s^2 3p^6 4s^2 3d^6$

P: $1s^2 2s^2 2p^6 3s^2 3p^3$; Cs: $1s^2 2s^2 2p^6 3s^2 3p^6 4s^2 3d^{10} 4p^6 5s^2 4d^{10} 5p^6 6s^1$

Eu: $1s^2 2s^2 2p^6 3s^2 3p^6 4s^2 3d^{10} 4p^6 5s^2 4d^{10} 5p^6 6s^2 4f^6 5d^1$*

Pt: $1s^2 2s^2 2p^6 3s^2 3p^6 4s^2 3d^{10} 4p^6 5s^2 4d^{10} 5p^6 6s^2 4f^{14} 5d^8$*

Xe: $1s^2 2s^2 2p^6 3s^2 3p^6 4s^2 3d^{10} 4p^6 5s^2 4d^{10} 5p^6$; Br: $1s^2 2s^2 2p^6 3s^2 3p^6 4s^2 3d^{10} 4p^5$

*Note: These electron configurations were written down using only the periodic table. The actual electron configurations are: Eu: $[Xe]6s^2 4f^7$ and Pt: $[Xe]6s^1 4f^{14} 5d^9$

79. Exceptions: Cr, Cu, Nb, Mo, Tc, Ru, Rh, Pd, Ag, Pt, and Au; Tc, Ru, Rh, Pd, and Pt do not correspond to the supposed extra stability of half-filled and filled subshells.

81. The two exceptions are Cr and Cu. Cr: $1s^2 2s^2 2p^6 3s^2 3p^6 4s^1 3p^5$; Cr has 6 unpaired electrons.

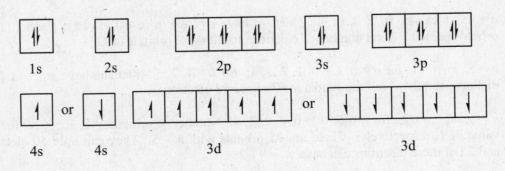

Cu: $1s^2 2s^2 2p^6 3s^2 3p^6 4s^1 3d^{10}$; Cu has 1 unpaired electron.

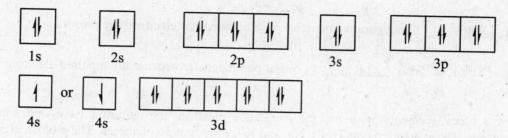

83.   a. 2 valence electrons; $4s^2$ b. 6 valence electrons; $2s^2 2p^4$

c. 7 valence electrons; $7s^2 7p^5$ d. 3 valence electrons; $5s^2 5p^1$

e. 8 valence electrons; $3s^2 3p^6$ f. 5 valence electrons; $6s^2 6p^3$

85.   Element 115, Uup, is in Group 5A under Bi (bismuth):

Uup: $1s^2 2s^2 2p^6 3s^2 3p^6 4s^2 3d^{10} 4p^6 5s^2 4d^{10} 5p^6 6s^2 4f^{14} 5d^{10} 6p^6 7s^2 5f^{14} 6d^{10} 7p^3$

a. $5s^2$, $5p^6$, $5d^{10}$, and $5f^{14}$; 32 electrons have $n = 5$ as one of their quantum numbers

b. $\ell = 3$ are f orbitals. $4f^{14}$ and $5f^{14}$ are the f orbitals used. They are all filled so 28 electrons have $\ell = 3$.

c. p, d, and f orbitals all have one of the degenerate orbitals with $m_\ell = 1$. There are 6 orbitals with $m_\ell = 1$ for the various p orbitals used; there are 4 orbitals with $m_\ell = 1$ for the various d orbitals used; and there are 2 orbitals with $m_\ell = 1$ for the various f orbitals used. We have a total of $6 + 4 + 2 = 12$ orbitals with $m_\ell = 1$. Eleven of these orbitals are filled with 2 electrons, and the 7p orbitals are only half-filled. The number of electrons with $m_\ell = 1$ is $11 \times (2\ e^-) + 1 \times (1\ e^-) = 23$ electrons.

d. The first 112 electrons are all paired; one-half of these electrons (56 e^-) will have $m_s = -1/2$. The 3 electrons in the 7p orbitals singly occupy each of the three degenerate 7p orbitals; the three electrons are spin parallel, so the 7p electrons either have $m_s = +1/2$ or $m_s = -1/2$. Therefore, either 56 electrons have $m_s = -1/2$ or 59 electrons have $m_s = -1/2$.

87.   We get the number of unpaired electrons by examining the incompletely filled subshells. The paramagnetic substances have unpaired electrons, and the ones with no unpaired electrons are not paramagnetic (they are called diamagnetic).

Li: $1s^2 2s^1$ ↑ ; paramagnetic with 1 unpaired electron.
       2s

N: $1s^2 2s^2 2p^3$ ↑ ↑ ↑ ; paramagnetic with 3 unpaired electrons.
         2p

Ni: $[Ar]4s^2 3d^8$ ↑↓ ↑↓ ↑↓ ↑ ↑ ; paramagnetic with 2 unpaired electrons.
          3d

Te: $[Kr]5s^2 4d^{10} 5p^4$ $\uparrow\downarrow$ $\uparrow$ $\uparrow$; paramagnetic with 2 unpaired electrons.
    5p

Ba: $[Xe]6s^2$ $\uparrow\downarrow$; not paramagnetic because no unpaired electrons are present.
    6s

Hg: $[Xe]6s^2 4f^{14} 5d^{10}$ $\uparrow\downarrow$ $\uparrow\downarrow$ $\uparrow\downarrow$ $\uparrow\downarrow$ $\uparrow\downarrow$; not paramagnetic because no unpaired electrons.
    5d

89.   The s block elements with ns^1 for a valence electron configuration have one unpaired electrons. These are elements H, Li, Na, and K for the first 36 elements. The p block elements with $ns^2 np^1$ or $ns^2 np^5$ valence electron configurations have one unpaired electron. These are elements B, Al, and Ga ($ns^2 np^1$) and elements F, Cl, and Br ($ns^2 np^5$) for the first 36 elements. In the d block, Sc ($[Ar]4s^2 3d^1$) and Cu ($[Ar]4s^1 3d^{10}$) each have one unpaired electron. A total of 12 elements from the first 36 elements have one unpaired electron in the ground state.

91.   We get the number of unpaired electrons by examining the incompletely filled subshells.

O: $[He]2s^2 2p^4$ $2p^4$: $\uparrow\downarrow$ $\uparrow$ $\uparrow$ Two unpaired e^-

O^+: $[He]2s^2 2p^3$ $2p^3$: $\uparrow$ $\uparrow$ $\uparrow$ Three unpaired e^-

O^-: $[He]2s^2 2p^5$ $2p^5$: $\uparrow\downarrow$ $\uparrow\downarrow$ $\uparrow$ One unpaired e^-

Os: $[Xe]6s^2 4f^{14} 5d^6$ $5d^6$: $\uparrow\downarrow$ $\uparrow$ $\uparrow$ $\uparrow$ $\uparrow$ Four unpaired e^-

Zr: $[Kr]5s^2 4d^2$ $4d^2$: $\uparrow$ $\uparrow$ __ __ __   Two unpaired e^-

S: $[Ne]3s^2 3p^4$ $3p^4$: $\uparrow\downarrow$ $\uparrow$ $\uparrow$ Two unpaired e^-

F: $[He]2s^2 2p^5$ $2p^5$: $\uparrow\downarrow$ $\uparrow\downarrow$ $\uparrow$ One unpaired e^-

Ar: $[Ne]3s^2 3p^6$ $3p^6$: $\uparrow\downarrow$ $\uparrow\downarrow$ $\uparrow\downarrow$ Zero unpaired e^-

The Periodic Table and Periodic Properties

93.   Ionization energy: $P(g) \rightarrow P^+(g) + e^-$; electron affinity: $P(g) + e^- \rightarrow P^-(g)$

95.   As successive electrons are removed, the net positive charge on the resulting ion increases. This increase in positive charge binds the remaining electrons more firmly, and the ionization energy increases.

The electron configuration for Si is $1s^2 2s^2 2p^6 3s^2 3p^2$. There is a large jump in ionization energy when going from the removal of valence electrons to the removal of core electrons. For silicon, this occurs when the fifth electron is removed since we go from the valence electrons in $n = 3$ to the core electrons in $n = 2$. There should be another big jump when the thirteenth electron is removed, i.e., when a 1s electron is removed.

97.   Size (radius) decreases left to right across the periodic table, and size increases from top to bottom of the periodic table.

   a. S < Se < Te   b. Br < Ni < K   c. F < Si < Ba

   d. Be < Na < Rb   e. Ne < Se < Sr   f. O < P < Fe

   All follow the general radius trend.

99.   a. Ba   b. K

   c. O; in general, Group 6A elements have a lower ionization energy than neighboring Group 5A elements. This is an exception to the general ionization energy trend across the periodic table.

   d. S^{2-}; this ion has the most electrons compared to the other sulfur species present. S^{2-} has the largest number of electron-electron repulsions, which leads to S^{2-} having the largest size and smallest ionization energy.

   e. Cs; this follows the general ionization energy trend.

101. As: $[Ar]4s^2 3d^{10} 4p^3$; Se: $[Ar]4s^2 3d^{10} 4p^4$; the general ionization energy trend predicts that Se should have a higher ionization energy than As. Se is an exception to the general ionization energy trend. There are extra electron-electron repulsions in Se because two electrons are in the same 4p orbital, resulting in a lower ionization energy for Se than predicted.

103. Size also decreases going across a period. Sc and Ti along with Y and Zr are adjacent elements. There are 14 elements (the lanthanides) between La and Hf, making Hf considerably smaller.

105. a. Uus will have 117 electrons. $[Rn]7s^2 5f^{14} 6d^{10} 7p^5$

   b. It will be in the halogen family and will be most similar to astatine (At).

   c. Like the other halogens: NaUus, $Mg(Uus)_2$, $C(Uus)_4$, $O(Uus)_2$

   d. Like the other halogens: $UusO^-$, $UusO_2^-$, $UusO_3^-$, $UusO_4^-$

107. Electron affinity refers to the energy associated with the process of adding an electron to a gaseous substance. Be, N, and Ne all have positive (unfavorable) electron affinity values. In order to add an electron to Be, N, or Ne, energy must be added. Another way of saying this is that Be, N, and Ne become less stable (have a higher energy) when an electron is added to each. To rationalize why those three atoms have positive (unfavorable) electron affinity values, let's see what happens to the electron configuration as an electron is added.

$$Be(g) \; + \; e^- \; \rightarrow \; Be^-(g) \qquad\qquad N(g) \; + \; e^- \; \rightarrow \; N^-(g)$$
$$[He]2s^2 \qquad\qquad [He]2s^2 2p^1 \qquad [He]2s^2 2p^3 \qquad [He]2s^2 2p^4$$

$$Ne(g) \; + \; e^- \; \rightarrow \; Ne^-(g)$$
$$[He]2s^22p^6 \qquad\quad [He]2s^22p^63s^1$$

In each case something energetically unfavorable occurs when an electron is added. For Be, the added electron must go into a higher-energy 2p atomic orbital because the 2s orbital is full. In N, the added electron must pair up with another electron in one of the 2p atomic orbitals; this adds electron-electron repulsions. In Ne, the added electron must be added to a much higher 3s atomic orbital because the n = 2 orbitals are full.

109.   Electron-electron repulsions are much greater in O^- than in S^- because the electron goes into a smaller 2p orbital versus the larger 3p orbital in sulfur. This results in a more favorable (more exothermic) EA for sulfur.

111.   a.   The electron affinity of Mg^{2+} is ΔH for $Mg^{2+}(g) + e^- \rightarrow Mg^+(g)$; this is just the reverse of the second ionization energy for Mg. $EA(Mg^{2+}) = -IE_2(Mg) = -1445$ kJ/mol (Table 12.6)

       b.   EA of Al^+ is ΔH for $Al^+(g) + e^- \rightarrow Al(g)$; $EA(Al^+) = -IE_1(Al) =$
$$-580 \text{ kJ/mol (Table 12.6)}$$

       c.   IE of Cl^- is ΔH for $Cl^-(g) \rightarrow Cl(g) + e^-$; $IE(Cl^-) = -EA(Cl) = +348.7$ kJ/mol
$$\text{(Table 12.8)}$$

       d. $Cl(g) \rightarrow Cl^+(g) + e^- \quad \Delta H = IE_1(Cl) = 1255$ kJ/mol (Table 12.6)

       e. $Cl^+(g) + e^- \rightarrow Cl(g) \quad \Delta H = -IE_1(Cl) = -1255$ kJ/mol $= EA(Cl^+)$

The Alkali Metals

113.   Yes; the ionization energy general trend is to decrease down a group, and the atomic radius trend is to increase down a group. The data in Table 12.9 confirm both of these general trends.

115.   a. $6 Li(s) + N_2(g) \rightarrow 2 Li_3N(s)$ b. $2 Rb(s) + S(s) \rightarrow Rb_2S(s)$

       c. $2 Cs(s) + 2 H_2O(l) \rightarrow 2 CsOH(aq) + H_2(g)$ d. $2 Na(s) + Cl_2(g) \rightarrow 2 NaCl(s)$

117.   For 589.0 nm: $\nu = \dfrac{c}{\lambda} = \dfrac{2.9979 \times 10^8 \text{ m/s}}{589.0 \times 10^{-9} \text{ m}} = 5.090 \times 10^{14}\,s^{-1}$

$$E = h\nu = 6.6261 \times 10^{-34} \text{ J s} \times 5.090 \times 10^{14}\,s^{-1} = 3.373 \times 10^{-19} \text{ J}$$

For 589.6 nm: $\nu = c/\lambda = 5.085 \times 10^{14}\,s^{-1}$; $E = h\nu = 3.369 \times 10^{-19} \text{ J}$

The energies in kJ/mol are:

$$3.373 \times 10^{-19} \text{ J} \times \frac{1 \text{ kJ}}{1000 \text{ J}} \times \frac{6.0221 \times 10^{23}}{\text{mol}} = 203.1 \text{ kJ/mol}$$

$$3.369 \times 10^{-19} \text{ J} \times \frac{1 \text{ kJ}}{1000 \text{ J}} \times \frac{6.0221 \times 10^{23}}{\text{mol}} = 202.9 \text{ kJ/mol}$$

119.   It should be element 119 with ground state electron configuration: $[\text{Rn}]7s^2 5f^{14} 6d^{10} 7p^6 8s^1$

Additional Exercises

121.   a. n b. n and ℓ

123.   Size decreases from left to right and increases going down the periodic table. Thus going one element right and one element down would result in a similar size for the two elements diagonal to each other. The ionization energies will be similar for the diagonal elements since the periodic trends also oppose each other. Electron affinities are harder to predict, but atoms with similar sizes and ionization energies should also have similar electron affinities.

125. $$60 \times 10^6 \text{ km} \times \frac{1000 \text{ m}}{\text{km}} \times \frac{1 \text{ s}}{3.00 \times 10^8 \text{ m}} = 200 \text{ s} \text{ (about 3 minutes)}$$

127. $$\lambda = \frac{hc}{E} = \frac{6.626 \times 10^{-34} \text{ J s} \times 2.998 \times 10^8 \text{ m/s}}{3.59 \times 10^{-19} \text{ J}} = 5.53 \times 10^{-7} \text{ m} \times \frac{100 \text{ cm}}{\text{m}}$$
$$= 5.53 \times 10^{-5} \text{ cm}$$

From the spectrum, $\lambda = 5.53 \times 10^{-5}$ cm is greenish yellow light.

129.   When the p and d orbital functions are evaluated at various points in space, the results sometimes have positive values and sometimes have negative values. The term phase is often associated with the + and − signs. For example, a sine wave has alternating positive and negative phases. This is analogous to the positive and negative values (phases) in the p and d orbitals.

131.   a.

$$Na(g) \rightarrow Na^+(g) + e^- \qquad \qquad IE_1 = 495 \text{ kJ}$$
$$Cl(g) + e^- \rightarrow Cl^-(g) \qquad \qquad EA = -348.7 \text{ kJ}$$
$$\overline{Na(g) + Cl(g) \rightarrow Na^+(g) + Cl^-(g)} \qquad \Delta H = 146 \text{ kJ}$$

b.

$$Mg(g) \rightarrow Mg^+(g) + e^- \qquad \qquad IE_1 = 735 \text{ kJ}$$
$$F(g) + e^- \rightarrow F^-(g) \qquad \qquad EA = -327.8 \text{ kJ}$$
$$\overline{Mg(g) + F(g) \rightarrow Mg^+(g) + F^-(g)} \qquad \Delta H = 407 \text{ kJ}$$

c.

$$Mg^+(g) \rightarrow Mg^{2+}(g) + e^- \qquad \qquad IE_2 = 1445 \text{ kJ}$$
$$F(g) + e^- \rightarrow F^-(g) \qquad \qquad EA = -327.8 \text{ kJ}$$
$$\overline{Mg^+(g) + F(g) \rightarrow Mg^{2+}(g) + F^-(g)} \qquad \Delta H = 1117 \text{ kJ}$$

d. From parts b and c, we get:

$$Mg(g) + F(g) \rightarrow Mg^+(g) + F^-(g) \qquad \Delta H = 407 \text{ kJ}$$
$$Mg^+(g) + F(g) \rightarrow Mg^{2+}(g) + F^-(g) \qquad \Delta H = 1117 \text{ kJ}$$

$$Mg(g) + 2 F(g) \rightarrow Mg^{2+}(g) + 2 F^-(g) \qquad \Delta H = 1524 \text{ kJ}$$

133. Valence electrons are easier to remove than inner core electrons. The large difference in energy between I_2 and I_3 indicates that this element has two valence electrons. This element is most likely an alkaline earth metal since alkaline earth metal elements all have two valence electrons.

135. a. The 4+ ion contains 20 electrons. Thus the electrically neutral atom will contain 24 electrons. The atomic number is 24 which identifies it as chromium.

b. The ground state electron configuration of the ion must be: $1s^2 2s^2 2p^6 3s^2 3p^6 4s^0 3d^2$; there are 6 electrons in s orbitals.

c. 12   d. 2

e. This is the isotope $^{50}_{24}Cr$. There are 26 neutrons in the nucleus.

f. $3.01 \times 10^{23} \text{ atoms} \times \dfrac{1 \text{ mol}}{6.022 \times 10^{23} \text{ atoms}} \times \dfrac{49.9 \text{ g}}{\text{mol}} = 24.9 \text{ g}$

g. $1s^2 2s^2 2p^6 3s^2 3p^6 4s^1 3d^5$ is the ground state electron configuration for Cr. Cr is an exception to the normal filling order.

137. a. Each orbital could hold 4 electrons.

b. The first period corresponds to $n = 1$, which can only have 1s orbitals. The 1s orbital could hold 4 electrons; hence the first period would have four elements. The second period corresponds to $n = 2$, which has 2s and 2p orbitals. These four orbitals can each hold four electrons. A total of 16 elements would be in the second period.

c. 20   d. 28

139. S-type cone receptors: $\lambda = \dfrac{c}{v} = \dfrac{2.998 \times 10^8 \text{ m/s}}{6.00 \times 10^{14} \text{ s}^{-1}} = 5.00 \times 10^{-7} \text{ m} = 500. \text{ nm}$

$$\lambda = \dfrac{2.998 \times 10^8 \text{ m/s}}{7.49 \times 10^{14} \text{ s}^{-1}} = 4.00 \times 10^{-7} \text{ m} = 400. \text{ nm}$$

S-type cone receptors detect 400-500 nm light. From Figure 12.3 in the text, this is violet to green light, respectively.

M-type cone receptors: $\lambda = \dfrac{2.998 \times 10^8 \text{ m/s}}{4.76 \times 10^{14} \text{ s}^{-1}} = 6.30 \times 10^{-7}$ m = 630. nm

$$\lambda = \dfrac{2.998 \times 10^8 \text{ m/s}}{6.62 \times 10^{14} \text{ s}^{-1}} = 4.53 \times 10^{-7} \text{ m} = 453 \text{ nm}$$

M-type cone receptors detect 450-630 nm light. From Figure 12.3 in the text, this is blue to orange light, respectively.

L-type cone receptors: $\lambda = \dfrac{2.998 \times 10^8 \text{ m/s}}{4.28 \times 10^{14} \text{ s}^{-1}} = 7.00 \times 10^{-7}$ m = 700. nm

$$\lambda = \dfrac{2.998 \times 10^8 \text{ m/s}}{6.00 \times 10^{14} \text{ s}^{-1}} = 5.00 \times 10^{-7} \text{ m} = 500. \text{ nm}$$

L-type cone receptors detect 500-700 nm light. This represents green to red light, respectively.

141.   a.   Because wavelength is inversely proportional to energy, the spectral line to the right of B (at a larger wavelength) represents the lowest possible energy transition; this is $n = 4$ to $n = 3$. The B line represents the next lowest energy transition, which is $n = 5$ to $n = 3$, and the A line corresponds to the $n = 6$ to $n = 3$ electronic transition.

b.   Because this spectrum is for a one-electron ion, $E_n = -2.178 \times 10^{-18}$ J (Z^2/n^2). To determine ΔE and, in turn, the wavelength of spectral line A, we must determine Z, the atomic number of the one electron species. Use spectral line B data to determine Z.

$$\Delta E_{5 \to 3} = -2.178 \times 10^{-18} \text{ J} \left(\frac{Z^2}{3^2} - \frac{Z^2}{5^2} \right) = -2.178 \times 10^{-18} \left(\frac{16Z^2}{9 \times 25} \right)$$

$$E = \frac{hc}{\lambda} = \frac{6.6261 \times 10^{-34} \text{ J s}(2.9979 \times 10^8 \text{ m/s})}{142.5 \times 10^{-9} \text{ m}} = 1.394 \times 10^{-18} \text{ J}$$

Because an emission occurs, $\Delta E_{5 \to 3} = -1.394 \times 10^{-18}$ J.

$$\Delta E = -1.394 \times 10^{-18} \text{ J} = -2.178 \times 10^{-18} \text{ J} \left(\frac{16 Z^2}{9 \times 25} \right), \; Z^2 = 9.001, \; Z = 3; \text{ the ion is Li}^{2+}.$$

Solving for the wavelength of line A:

$$\Delta E_{6 \to 3} = -2.178 \times 10^{-18}(3)^2 \left(\frac{1}{3^2} - \frac{1}{6^2} \right) = -1.634 \times 10^{-18} \text{ J}$$

$$\lambda = \frac{hc}{|\Delta E|} = \frac{6.6261 \times 10^{-34} \text{ J s}(2.9979 \times 10^8 \text{ m/s})}{1.634 \times 10^{-18} \text{ J}} = 1.216 \times 10^{-7} \text{ m} = 121.6 \text{ nm}$$

Challenge Problems

143.   For one-electron species, $E_n = -R_H Z^2/n^2$. IE is for the $n = 1 \rightarrow n = \infty$ transition. So:

$$IE = E_\infty - E_1 = -E_1 = R_H Z^2/n^2 = R_H Z^2$$

$$\frac{4.72 \times 10^4 \text{ kJ}}{\text{mol}} \times \frac{1 \text{ mol}}{6.022 \times 10^{23}} \times \frac{1000 \text{ J}}{\text{kJ}} = 2.178 \times 10^{-18} \text{ J } (Z^2); \text{ solving: } Z = 6$$

Element 6 is carbon (X = carbon), and the charge for a one-electron carbon ion is 5+ ($m = 5$). The one-electron ion is C^{5+}.

145.   a.   Because the energy levels E_{xy} are inversely proportional to L^2, the $n_x = 2$, $n_y = 1$ energy level will be lower in energy than the $n_x = 1$, $n_y = 2$ energy level since $L_x > L_y$. The first three energy levels E_{xy} in order of increasing energy are:

$$E_{11} < E_{21} < E_{12}$$

The quantum numbers are:

Ground state (E_{11}) $\rightarrow$ $n_x = 1, n_y = 1$
First excited state (E_{21}) $\rightarrow$ $n_x = 2, n_y = 1$
Second excited state (E_{12}) $\rightarrow$ $n_x = 1, n_y = 2$

b. $E_{21} \rightarrow E_{12}$ is the transition. $E_{xy} = \dfrac{h^2}{8m}\left(\dfrac{n_x^2}{L_x^2} + \dfrac{n_y^2}{L_y^2}\right)$

$$E_{12} = \frac{h^2}{8m}\left[\frac{1^2}{(8.00 \times 10^{-9} \text{ m})^2} + \frac{2^2}{(5.00 \times 10^{-9} \text{ m})^2}\right] = \frac{1.76 \times 10^{17} \, h^2}{8m}$$

$$E_{21} = \frac{h^2}{8m}\left[\frac{2^2}{(8.00 \times 10^{-9} \text{ m})^2} + \frac{1^2}{(5.00 \times 10^{-9} \text{ m})^2}\right] = \frac{1.03 \times 10^{17} \, h^2}{8m}$$

$$\Delta E = E_{12} - E_{21} = \frac{1.76 \times 10^{17} \, h^2}{8m} - \frac{1.03 \times 10^{17} \, h^2}{8m} = \frac{7.3 \times 10^{16} \, h^2}{8m}$$

$$\Delta E = \frac{(7.3 \times 10^{16} \text{ m}^{-2})(6.626 \times 10^{-34} \text{ J s})^2}{8(9.11 \times 10^{-31} \text{ kg})} = 4.4 \times 10^{-21} \text{ J}$$

$$\lambda = \frac{hc}{\Delta E} = \frac{6.626 \times 10^{-34} \text{ J s}(2.998 \times 10^8 \text{ m/s})}{4.4 \times 10^{-21} \text{ J}} = 4.5 \times 10^{-5} \text{ m}$$

147. $\psi_{1s} = \dfrac{1}{\sqrt{\pi}}\left(\dfrac{Z}{a_0}\right)^{3/2} e^{-\sigma}$; $Z = 1$ for H, $\sigma = \dfrac{Zr}{a_0} = \dfrac{-r}{a_0}$, $a_0 = 5.29 \times 10^{-11}$ m

$\psi_{1s} = \dfrac{1}{\sqrt{\pi}}\left(\dfrac{1}{a_0}\right)^{3/2} \exp\left(\dfrac{-r}{a_0}\right)$

Probability is proportional to ψ^2: $\psi_{1s}^2 = \dfrac{1}{\pi}\left(\dfrac{1}{a_0}\right)^3 \exp\left(\dfrac{-2r}{a_0}\right)$ (units of $\psi^2 = $ m^{-3})

a. ψ_{1s}^2 (at nucleus) $= \dfrac{1}{\pi}\left(\dfrac{1}{a_0}\right)^3 \exp\left[\dfrac{-2(0)}{a_0}\right] = 2.15 \times 10^{30}$ m^{-3}

If we assume this probability is constant throughout the 1×10^{-3} pm^3 volume, then the total probability p is $\psi_{1s}^2 \times V$.

1.0×10^{-3} pm$^3 = (1.0 \times 10^{-3}$ pm$) \times (1 \times 10^{-12}$ m/pm$)^3 = 1.0 \times 10^{-39}$ m^3

Total probability $= p = (2.15 \times 10^{30}$ m$^{-3}) \times (1.0 \times 10^{-39}$ m$^3) = 2.2 \times 10^{-9}$

b.   For an electron that is 1.0×10^{-11} m from the nucleus:

$\psi_{1s}^2 = \dfrac{1}{\pi}\left(\dfrac{1}{5.29 \times 10^{-11}}\right)^3 \exp\left[\dfrac{-2(1.0 \times 10^{-11})}{(5.29 \times 10^{-11})}\right] = 1.5 \times 10^{30}$ m^{-3}

$V = 1.0 \times 10^{-39}$ m^3; $p = \psi_{1s}^2 \times V = 1.5 \times 10^{-9}$

c. $\psi_{1s}^2 = 2.15 \times 10^{30}$ m^{-3} $\exp\left[\dfrac{-2(53 \times 10^{-12})}{(5.29 \times 10^{-11})}\right] = 2.9 \times 10^{29}$; $V = 1.0 \times 10^{-39}$ m^3

$p = \psi_{1s}^2 \times V = 2.9 \times 10^{-10}$

d. $V = \dfrac{4}{3}\pi[(10.05 \times 10^{-12}$ m$)^3 - (9.95 \times 10^{-12}$ m$)^3] = 1.3 \times 10^{-34}$ m^3

We shall evaluate ψ_{1s}^2 at the middle of the shell, r = 10.00 pm, and assume ψ_{1s}^2 is constant from r = 9.95 to 10.05 pm. The concentric spheres are assumed centered about the nucleus.

$\psi_{1s}^2 = 2.15 \times 10^{30}$ m^{-3} $\exp\left[\dfrac{-2(10.0 \times 10^{-12}$ m$)}{(5.29 \times 10^{-11}$ m$)}\right] = 1.47 \times 10^{30}$ m^{-3}

$p = (1.47 \times 10^{30}$ m$^{-3})(1.3 \times 10^{-34}$ m$^3) = 1.9 \times 10^{-4}$

e. $V = \dfrac{4}{3}\pi[(52.95 \times 10^{-12}\text{ m})^3 - (52.85 \times 10^{-12}\text{ m})^3] = 4 \times 10^{-33}\text{ m}^3$

Evaluate ψ_{1s}^2 at $r = 52.90$ pm: $\psi_{1s}^2 = 2.15 \times 10^{30}\text{ m}^{-3}\,(e^{-2}) = 2.91 \times 10^{29}\text{ m}^{-3}$; $p = 1 \times 10^{-3}$

149. $E_{xyz} = \dfrac{h^2(n_x^2 + n_y^2 + n_z^2)}{8mL^2}$, where $L = L_x = L_y = L_z$.

The first four energy levels will be filled with the 8 electrons. The first four energy levels are:

$$E_{111} = \frac{h^2(1^2 + 1^2 + 1^2)}{8mL^2} = \frac{3h^2}{8mL^2}$$

$$E_{211} = E_{121} = E_{112} = \frac{6h^2}{8mL^2} \quad \text{(These three energy levels are degenerate.)}$$

The next energy levels correspond to the first excited state. The energy for these levels are:

$$E_{221} = E_{212} = E_{122} = \frac{9h^2}{8mL^2} \quad \text{(These three energy levels are degenerate.)}$$

The electronic transition in question is from one of the degenerate E_{211}, E_{121}, or E_{112} levels to one of the degenerate E_{221}, E_{212}, or E_{122} levels.

$$\Delta E = \frac{9h^2}{8mL^2} - \frac{6h^2}{8mL^2} = \frac{3h^2}{8mL^2}$$

$$\Delta E = \frac{3(6.626 \times 10^{-34}\text{ J s})^2}{8(9.109 \times 10^{-31}\text{ kg})(1.50 \times 10^{-9}\text{ m})^2} = 8.03 \times 10^{-20}\text{ J}$$

$$\lambda = \frac{hc}{\Delta E} = \frac{(6.626 \times 10^{-34}\text{ J s})(2.998 \times 10^8\text{ m/s})}{8.03 \times 10^{-20}\text{ J}} = 2.47 \times 10^{-6}\text{ m} = 2470\text{ nm}$$

151.   a.   Assuming the Bohr model applies to the 1s electron, $E_{1s} = -R_H Z^2/n^2 = -R_H Z^2_{\text{eff}}$, where $n = 1$.

$$IE = E_\infty - E_{1s} = 0 - E_{1s} = R_H Z^2_{\text{eff}}$$

$$\frac{2.462 \times 10^6\text{ kJ}}{\text{mol}} \times \frac{1\text{ mol}}{6.0221 \times 10^{23}} \times \frac{1000\text{ J}}{\text{kJ}} = 2.178 \times 10^{-18}\text{ J }(Z_{\text{eff}})^2, \;\; Z_{\text{eff}} = 43.33$$

b.   Silver is element 47, so $Z = 47$ for silver. Our calculated Z_{eff} value is slightly less than 47. Electrons in other orbitals can penetrate the 1s orbital. Thus a 1s electron can be slightly shielded from the nucleus, giving a Z_{eff} close to but less than Z.

CHAPTER 13

BONDING: GENERAL CONCEPTS

Chemical Bonds and Electronegativity

11. Electronegativity is the ability of an atom in a molecule to attract electrons to itself. Electronegativity is a bonding term. Electron affinity is the energy change when an electron is added to a substance. Electron affinity deals with isolated atoms in the gas phase.

A covalent bond is a sharing of electron pair(s) in a bond between two atoms. An ionic bond is a complete transfer of electrons from one atom to another to form ions. The electrostatic attraction of the oppositely charged ions is the ionic bond.

A pure covalent bond is an equal sharing of shared electron pair(s) in a bond. A polar covalent bond is an unequal sharing.

Ionic bonds form when there is a large difference in electronegativity between the two atoms bonding together. This usually occurs when a metal with a small electronegativity is bonded to a nonmetal having a large electronegativity. A pure covalent bond forms between atoms having identical or nearly identical eletronegativities. A polar covalent bond forms when there is an intermediate electronegativity difference. In general, nonmetals bond together by forming covalent bonds, either pure covalent or polar covalent.

Ionic bonds form due to the strong electrostatic attraction between two oppositely charged ions. Covalent bonds form because the shared electrons in the bond are attracted to two different nuclei, unlike the isolated atoms where electrons are only attracted to one nuclei. The attraction to another nuclei overrides the added electron-electron repulsions.

13.

	(IE – EA)	(IE – EA)/502	EN (text)	2006/502 = 4.0
F	2006 kJ/mol	4.0	4.0	
Cl	1604 kJ/mol	3.2	3.0	
Br	1463 kJ/mol	2.9	2.8	
I	1302 kJ/mol	2.6	2.5	

The values calculated from IE and EA show the same trend as (and agree fairly closely) with the values given in the text.

261

15.   Using the periodic table, we expect the general trend for electronegativity to be:

   1.   Increase as we go from left to right across a period

   2.   Decrease as we go down a group

   a. $C < N < O$ b. $Se < S < Cl$ c. $Sn < Ge < Si$

   d. $Tl < Ge < S$ e. $Rb < K < Na$ f. $Ga < B < O$

17.   The general trends in electronegativity used in Exercises 13.15 and 13.16 are only rules of thumb. In this exercise we use experimental values of electronegativities and can begin to see several exceptions. The order of EN using Figure 13.3 is:

   a. $C (2.5) < N (3.0) < O (3.5)$ same as predicted

   b. $Se (2.4) < S (2.5) < Cl (3.0)$ same

   c. $Si (1.8) = Ge (1.8) = Sn (1.8)$ different   d. $Tl (1.8) = Ge (1.8) < S (2.5)$ different

   e. $Rb (0.8) = K (0.8) < Na (0.9)$ different   f. $Ga (1.6) < B (2.0) < O (3.5)$ same

Most polar bonds using actual EN values:

   a.   Si–F and Ge–F (Ge–F predicted)   b.   P–Cl (same as predicted)

   c.   S–F (same as predicted)   d.   Ti–Cl (same as predicted)

   e.   Si–H and Sn–H (Sn–H predicted)   f.   Al–Br (Tl–Br predicted)

19.   Ionic character is proportional to the difference in electronegativity values between the two elements forming the bond. Using the trend in electronegativity, the order will be:

   Br–Br < N–O < C–F < Ca–O < K–F
   least   most
   ionic character   ionic character

Note that Br–Br, N–O and C–F bonds are all covalent bonds since the elements are all nonmetals. The Ca–O and K–F bonds are ionic, as is generally the case when a metal forms a bond with a nonmetal.

21.   Of the compounds listed, P_2O_5 is the only compound containing only covalent bonds. $(NH_4)_2SO_4$, $Ca_3(PO_4)_2$, K_2O, and KCl are all compounds composed of ions, so they exhibit ionic bonding. The polyatomic ions in $(NH_4)_2SO_4$ are NH_4^+ and SO_4^{2-}. Covalent bonds exist between the N and H atoms in NH_4^+ and between the S and O atoms in SO_4^{2-}. Therefore, $(NH_4)_2SO_4$ contains both ionic and covalent bonds. The same is true for $Ca_3(PO_4)_2$. The bonding is ionic between the Ca^{2+} and PO_4^{3-} ions and covalent between the P and O atoms in PO_4^{3-}. Therefore, $(NH_4)_2SO_4$ and $Ca_3(PO_4)_2$ are the compounds with both ionic and covalent bonds.

Ions and Ionic Compounds

23.   Anions are larger than the neutral atom, and cations are smaller than the neutral atom. For anions, the added electrons increase the electron-electron repulsions. To counteract this, the size of the electron cloud increases, placing the electrons further apart from one another. For cations, as electrons are removed, there are fewer electron-electron repulsions, and the electron cloud can be pulled closer to the nucleus.

Isoelectronic: same number of electrons. Two variables, the number of protons and the number of electrons, determine the size of an ion. Keeping the number of electrons constant, we only have to consider the number of protons to predict trends in size. The ion with the most protons attracts the same number of electrons most strongly, resulting in a smaller size.

25.   a. $Cu > Cu^+ > Cu^{2+}$ b. $Pt^{2+} > Pd^{2+} > Ni^{2+}$ c. $O^{2-} > O^- > O$

d. $La^{3+} > Eu^{3+} > Gd^{3+} > Yb^{3+}$ e. $Te^{2-} > I^- > Cs^+ > Ba^{2+} > La^{3+}$

For answer a, as electrons are removed from an atom, size decreases. Answers b and d follow the radius trend. For answer c, as electrons are added to an atom, size increases. Answer e follows the trend for an isoelectronic series, i.e., the smallest ion has the most protons.

27. Rb^+: $[Ar]4s^2 3d^{10} 4p^6$; Ba^{2+}: $[Kr]5s^2 4d^{10} 5p^6$; Se^{2-}: $[Ar]4s^2 3d^{10} 4p^6$

I^-: $[Kr]5s^2 4d^{10} 5p^6$

29.   a. Cs_2S is composed of Cs^+ and S^{2-}. Cs^+ has the same electron configuration as Xe, and S^{2-} has the same configuration as Ar.

b. SrF_2; Sr^{2+} has the Kr electron configuration, and F^- has the Ne configuration.

c. Ca_3N_2; Ca^{2+} has the Ar electron configuration, and N^{3-} has the Ne configuration.

d. $AlBr_3$; Al^{3+} has the Ne electron configuration, and Br^- has the Kr configuration.

31. Se^{2-}, Br^-, Rb^+, Sr^{2+}, Y^{3+}, and Zr^{4+} are some ions that are isoelectronic with Kr (36 electrons). In terms of size, the ion with the most protons will hold the electrons tightest and will be the smallest. The size trend is:

$$Zr^{4+} < Y^{3+} < Sr^{2+} < Rb^+ < Br^- < Se^{2-}$$
smallest   largest

33.   a. Al^{3+} and S^{2-} are the expected ions. The formula of the compound would be Al_2S_3 (aluminum sulfide).

b. K^+ and N^{3-}; K_3N, potassium nitride

c. Mg^{2+} and Cl^-; $MgCl_2$, magnesium chloride

d. Cs^+ and Br^-; CsBr, cesium bromide

35.

$$K(s) \rightarrow K(g) \qquad \Delta H = 90. \text{ kJ} \quad \text{(sublimation)}$$
$$K(g) \rightarrow K^+(g) + e^- \qquad \Delta H = 419 \text{ kJ} \quad \text{(ionization energy)}$$
$$1/2 \, Cl_2(g) \rightarrow Cl(g) \qquad \Delta H = 239/2 \text{ kJ (bond energy)}$$
$$Cl(g) + e^- \rightarrow Cl^-(g) \qquad \Delta H = -349 \text{ kJ} \quad \text{(electron affinity)}$$
$$K^+(g) + Cl^-(g) \rightarrow KCl(s) \qquad \Delta H = -690. \text{ kJ} \quad \text{(lattice energy)}$$

$$K(s) + 1/2 \, Cl_2(g) \rightarrow KCl(s) \qquad \Delta H_f^o = -411 \text{ kJ/mol}$$

37.   Use Figure 13.11 as a template for this problem.

$$Li(s) \rightarrow Li(g) \qquad \Delta H_{sub} = ?$$
$$Li(g) \rightarrow Li^+(g) + e^- \qquad \Delta H = 520. \text{ kJ}$$
$$1/2 \, I_2(g) \rightarrow I(g) \qquad \Delta H = 151/2 \text{ kJ}$$
$$I(g) + e^- \rightarrow I^-(g) \qquad \Delta H = -295 \text{ kJ}$$
$$Li^+(g) + I^-(g) \rightarrow LiI(s) \qquad \Delta H = -753 \text{ kJ}$$

$$Li(s) + 1/2 \, I_2(g) \rightarrow LiI(s) \qquad \Delta H = -292 \text{ kJ}$$

$$\Delta H_{sub} + 520. + 151/2 - 295 - 753 = -292, \ \Delta H_{sub} = 161 \text{ kJ}$$

39.   a.   From the data given, less energy is required to produce $Mg^+(g) + O^-(g)$ than to produce $Mg^{2+}(g) + O^{2-}(g)$. However, the lattice energy for $Mg^{2+}O^{2-}$ will be much more exothermic than for Mg^+O^- (due to the greater charges in $Mg^{2+}O^{2-}$). The favorable lattice energy term will dominate and $Mg^{2+}O^{2-}$ forms.

b. Mg^+ and O^- both have unpaired electrons. In Mg^{2+} and O^{2-} there are no unpaired electrons. Hence Mg^+O^- would be paramagnetic; $Mg^{2+}O^{2-}$ would be diamagnetic. Paramagnetism can be detected by measuring the mass of a sample in the presence and absence of a magnetic field. The apparent mass of a paramagnetic substance will be larger in a magnetic field because of the force between the unpaired electrons and the field.

41. Ca^{2+} has a greater charge than Na^+, and Se^{2-} is smaller than Te^{2-}. The effect of charge on the lattice energy is greater than the effect of size. We expect the trend from most exothermic to least exothermic to be:

$$CaSe > CaTe > Na_2Se > Na_2Te$$

(−2862)   (−2721)   (−2130)   (−2095 kJ/mol)   This is what we observe.

Bond Energies

43.   a. 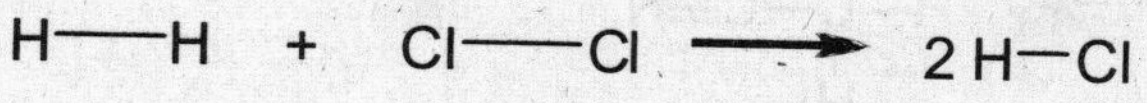   H——H + Cl——Cl ⟶ 2 H–Cl

Bonds broken: Bonds formed:

1 H–H (432 kJ/mol)   2 H–Cl (427 kJ/mol)
1 Cl–Cl (239 kJ/mol)

$\Delta H = \Sigma D_{broken} - \Sigma D_{formed}$, $\Delta H = 432 \text{ kJ} + 239 \text{ kJ} - 2(427) \text{ kJ} = -183 \text{ kJ}$

b. 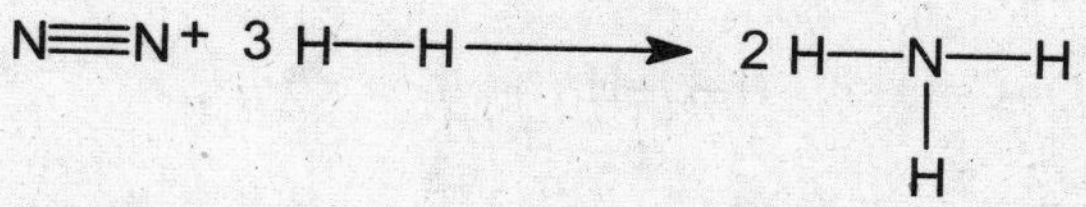   N≡N + 3 H——H ⟶ 2 H——N——H
                                                             |
                                                             H

Bonds broken: Bonds formed:

1 N ≡ N (941 kJ/mol)   6 N–H (391 kJ/mol)
3 H–H (432 kJ/mol)

$\Delta H = 941 \text{ kJ} + 3(432) \text{ kJ} - 6(391) \text{ kJ} = -109 \text{ kJ}$

c.   Sometimes some of the bonds remain the same between reactants and products. To save time, only break and form bonds that are involved in the reaction.

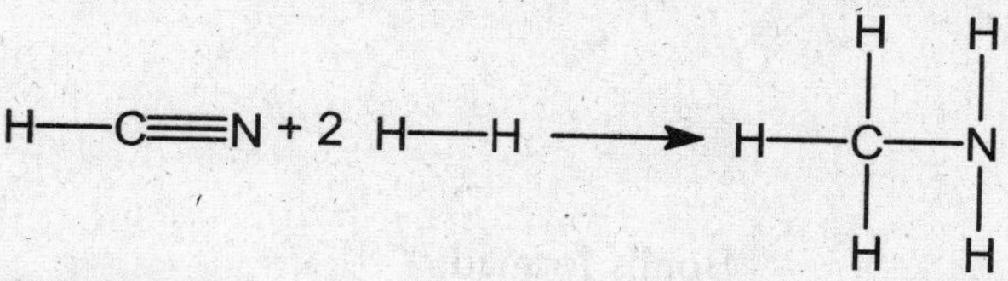

H——C≡N + 2 H——H ⟶ H——C——N

Bonds broken: Bonds formed:

1 C≡N (891 kJ/mol)   1 C–N (305 kJ/mol)
2 H–H (432 kJ/mol)   2 C–H (413 kJ/mol)
                                                         2 N–H (391 kJ/mol)

$\Delta H = 891 \text{ kJ} + 2(432 \text{ kJ}) - [305 \text{ kJ} + 2(413 \text{ kJ}) + 2(391 \text{ kJ})] = -158 \text{ kJ}$

d. 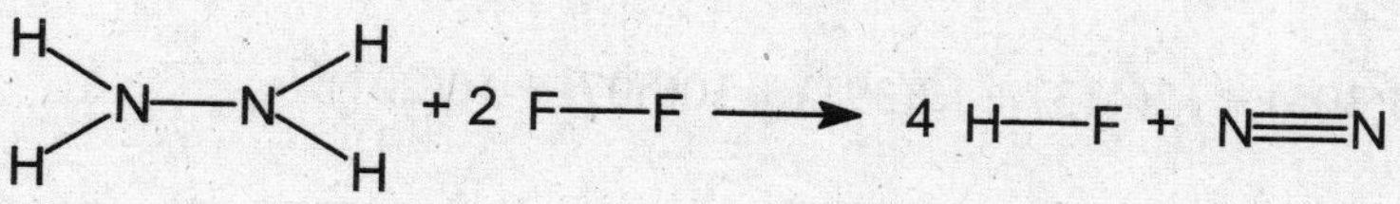   H₂N——NH₂ + 2 F——F ⟶ 4 H——F + N≡N

Bonds broken: Bonds formed:

1 N–N (160. kJ/mol)   4 H–F (565 kJ/mol)
4 N–H (391 kJ/mol)   1 N≡N (941 kJ/mol)
2 F–F (154 kJ/mol)

$\Delta H = 160. \text{ kJ} + 4(391 \text{ kJ}) + 2(154 \text{ kJ}) - [4(565 \text{ kJ}) + 941 \text{ kJ}] = -1169 \text{ kJ}$

45.

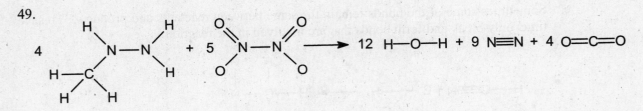

Bonds broken: 1 C–N (305 kJ/mol)   Bonds formed: 1 C–C (347 kJ/mol)

$\Delta H = \Sigma D_{broken} - \Sigma D_{formed}$, $\Delta H = 305 - 347 = -42$ kJ

Note: Sometimes some of the bonds remain the same between reactants and products. To save time, only break and form bonds that are involved in the reaction.

47. $H-C\equiv C-H + 5/2\ O=O \rightarrow 2\ O=C=O + H-O-H$

Bonds broken: Bonds formed:

   2 C–H (413 kJ/mol)   2 × 2 C=O (799 kJ/mol)
   1 C≡C (839 kJ/mol)   2 O–H (467 kJ/mol)
   5/2 O=O (495 kJ/mol)

$\Delta H = 2(413\ kJ) + 839\ kJ + 5/2\ (495\ kJ) - [4(799\ kJ) + 2(467\ kJ)] = -1228$ kJ

49.

Bonds broken: Bonds formed:

   9 N–N (160. kJ/mol)   24 O–H (467 kJ/mol)
   4 N–C (305 kJ/mol)   9 N≡N (941 kJ/mol)
   12 C–H (413 kJ/mol)   8 C=O (799 kJ/mol)
   12 N–H (391 kJ/mol)
   10 N=O (607 kJ/mol)
   10 N–O (201 kJ/mol)

$\Delta H = 9(160.) + 4(305) + 12(413) + 12(391) + 10(607) + 10(201)$

$- [24(467) + 9(941) + 8(799)]$

$\Delta H = 20{,}388\ kJ - 26{,}069\ kJ = -5681$ kJ

51.   Because both reactions are highly exothermic, the high temperature is not needed to provide energy. It must be necessary for some other reason. The reason is to increase the speed of the reaction. This will be discussed in Chapter 15 on kinetics.

53.   a. $HF(g) \rightarrow H(g) + F(g)$ $\Delta H = 565$ kJ
         $H(g) \rightarrow H^+(g) + e^-$ $\Delta H = 1312$ kJ
         $F(g) + e^- \rightarrow F^-(g)$ $\Delta H = -327.8$ kJ

         $HF(g) \rightarrow H^+(g) + F^-(g)$ $\Delta H = 1549$ kJ

    b. $HCl(g) \rightarrow H(g) + Cl(g)$ $\Delta H = 427$ kJ
         $H(g) \rightarrow H^+(g) + e^-$ $\Delta H = 1312$ kJ
      $Cl(g) + e^- \rightarrow Cl^-(g)$ $\Delta H = -348.7$ kJ

         $HCl(g) \rightarrow H^+(g) + Cl^-(g)$ $\Delta H = 1390.$ kJ

    c. $HI(g) \rightarrow H(g) + I(g)$ $\Delta H = 295$ kJ
         $H(g) \rightarrow H^+(g) + e^-$ $\Delta H = 1312$ kJ
     $I(g) + e^- \rightarrow I^-(g)$ $\Delta H = -295.2$ kJ

         $HI(g) \rightarrow H^+(g) + I^-(g)$ $\Delta H = 1312$ kJ

    d. $H_2O(g) \rightarrow OH(g) + H(g)$ $\Delta H = 467$ kJ
         $H(g) \rightarrow H^+(g) + e^-$ $\Delta H = 1312$ kJ
    $OH(g) + e^- \rightarrow OH^-(g)$ $\Delta H = -180.$ kJ

         $H_2O(g) \rightarrow H^+(g) + OH^-(g)$ $\Delta H = 1599$ kJ

55. $NH_3(g) \rightarrow N(g) + 3 H(g)$

$\Delta H° = 3D_{NH} = 472.7 \text{ kJ} + 3(216.0 \text{ kJ}) - (-46.1 \text{ kJ}) = 1166.8 \text{ kJ}$

$$D_{NH} = \frac{1166.8 \text{ kJ}}{3 \text{ mol NH bonds}} = 388.93 \text{ kJ/mol}$$

$D_{calc} = 389$ kJ/mol compared with 391 kJ/mol in the table. There is good agreement.

Lewis Structures and Resonance

57.   Drawing Lewis structures is mostly trial and error. However, the first two steps are always the same. These steps are (1) count the valence electrons available in the molecule/ion, and (2) attach all atoms to each other with single bonds (called the skeletal structure). Unless noted otherwise, the atom listed first is assumed to be the atom in the middle, called the central atom, and all other atoms in the formula are attached to this atom. The most notable exceptions to the rule are formulas that begin with H, e.g., H_2O, H_2CO, etc. Hydrogen can never be a central atom since this would require H to have more than two electrons. In these compounds, the atom listed second is assumed to be the central atom.

After counting valence electrons and drawing the skeletal structure, the rest is trial and error. We place the remaining electrons around the various atoms in an attempt to satisfy the octet rule (or duet rule for H). Keep in mind that practice makes perfect. After practicing, you can (and will) become very adept at drawing Lewis structures.

a.   HCN has $1 + 4 + 5 = 10$ valence electrons.

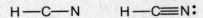

Skeletal   Lewis
structure   structure

Skeletal structure uses 4 e⁻; 6 e⁻ remain

b. PH_3 has $5 + 3(1) = 8$ valence electrons.

H—P—H   H—P̈—H
    | |
    H   H

Skeletal   Lewis
structure   structure

Skeletal structures uses 6 e⁻; 2 e⁻ remain

c. $CHCl_3$ has $4 + 1 + 3(7) = 26$ valence electrons.

Skeletal   Lewis
structure   structure

d. NH_4^+ has $5 + 4(1) ! 1 = 8$ valence electrons.

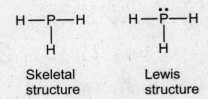

Lewis
structure

Note: Subtract valence electrons for positive charged ions.

e. H_2CO has $2(1) + 4 + 6 = 12$ valence electrons.

f. SeF_2 has $6 + 2(7) = 20$ valence electrons.

g. CO_2 has $4 + 2(6) = 16$ valence electrons.

h. O_2 has $2(6) = 12$ valence electrons.

i.   HBr has $1 + 7 = 8$ valence electrons.

59.   Molecules/ions that have the same number of valence electrons and the same number of atoms will have similar Lewis structures.

61.   Ozone: O_3 has $3(6) = 18$ valence electrons. Two resonance structures can be drawn.

Sulfur dioxide: SO_2 has $6 + 2(6) = 18$ valence electrons. Two resonance structures are possible.

Sulfur trioxide: SO_3 has $6 + 3(6) = 24$ valence electrons. Three resonance structures are possible.

63. CH_3NCO has $4 + 3(1) + 5 + 4 + 6 = 22$ valence electrons. The order of the elements in the formula give the skeletal structure.

65.   Benzene has $6(4) + 6(1) = 30$ valence electrons. Two resonance structures can be drawn for benzene. The actual structure of benzene is an average of these two resonance structures; i.e., all carbon-carbon bonds are equivalent with a bond length and bond strength somewhere between a single and a double bond.

67.   Borazine ($B_3N_3H_6$) has $3(3) + 3(5) + 6(1) = 30$ valence electrons. The possible resonance
       structures are similar to those of benzene in Exercise 13.65.

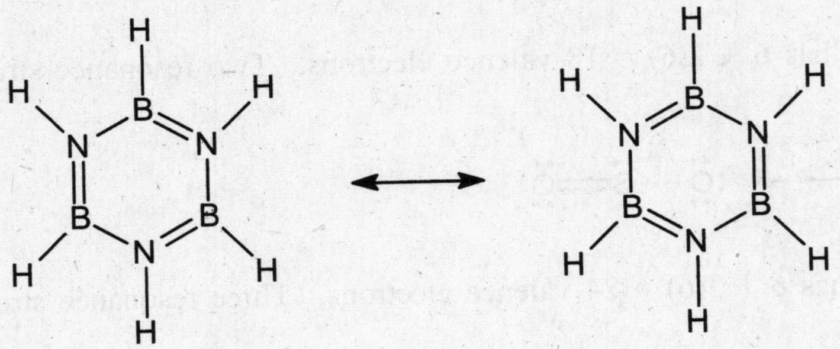

69.   Statements a and c are true. For statement a, XeF_2 has 22 valence electrons and it is
       impossible to satisfy the octet rule for all atoms with this number of electrons. The best Lewis
       structure is:

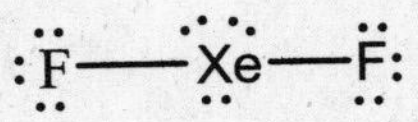

       For statement c, NO^+ has 10 valence electrons, whereas NO^- has 12 valence electrons. The
       Lewis structures are:

       Because a triple bond is stronger than a double bond, NO^+ has a stronger bond.

       For statement b, SF_4 has five electron pairs around the sulfur in the best Lewis structure; it is
       an exception to the octet rule. Because OF_4 has the same number of valence electrons as SF_4,
       OF_4 would also have to be an exception to the octet rule. However, Row 2 elements such as
       O never have more than 8 electrons around them, so OF_4 does not exist. For statement d, two
       resonance structures can be drawn for ozone:

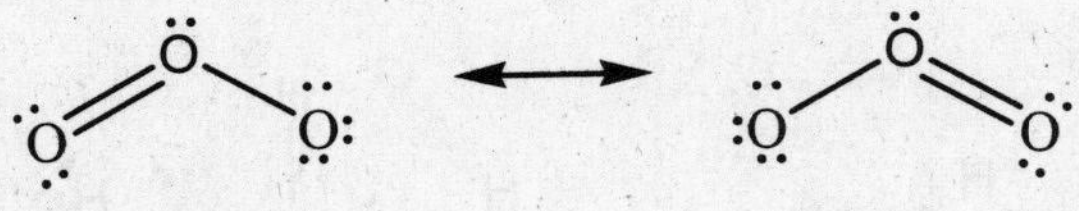

       When resonance structures can be drawn, the actual bond lengths and strengths are all equal
       to each other. Even though each Lewis structure implies the two O–O bonds are different,
       this is not the case in real life. In real life, both of the O–O bonds are equivalent. When
       resonance structures can be drawn, you can think of the bonding as an average of all of the
       resonance structures.

71.   PF$_5$, 5 + 5(7) = 40 valence electrons   SF$_4$, 6 + 4(7) = 34 e$^-$

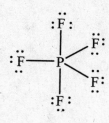

ClF$_3$, 7 + 3(7) = 28 e$^-$ Br$_3^-$, 3(7) + 1 = 22 e$^-$

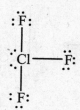

Row 3 and heavier nonmetals can have more than 8 electrons around them when they have to. Row 3 and heavier elements have empty d orbitals that are close in energy to valence s and p orbitals. These empty d orbitals can accept extra electrons.

For example, P in PF$_5$ has its five valence electrons in the 3s and 3p orbitals. These s and p orbitals have room for three more electrons, and if it has to, P can use the empty 3d orbitals for any electrons above 8.

73.   CO$_3^{2-}$ has 4 + 3(6) + 2 = 24 valence electrons.

Three resonance structures can be drawn for CO$_3^{2-}$. The actual structure for CO$_3^{2-}$ is an average of these three resonance structures. That is, the three C–O bond lengths are all equivalent, with a length somewhere between a single and a double bond. The actual bond length of 136 pm is consistent with this resonance view of CO$_3^{2-}$.

75.   H$_2$NOH (14 e$^-$):   H—N—O—H   Single bond between N and O

N_2O (16 e⁻): $\overset{..}{N}=N=\overset{..}{\underset{..}{O}}$ ⟷ $:N\equiv N-\overset{..}{\underset{..}{O}}:$ ⟷ $:\overset{..}{\underset{..}{N}}-N\equiv O:$

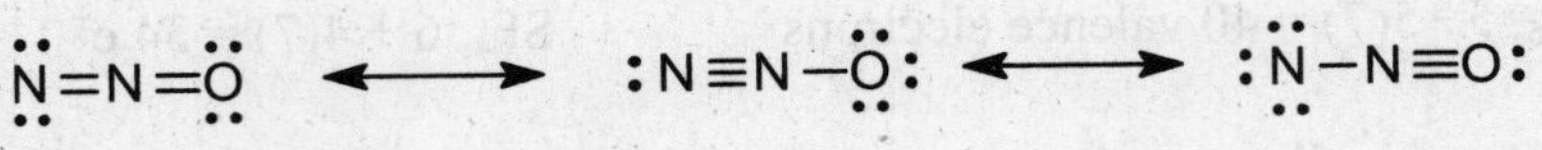

Average of a double bond between N and O

NO^+ (10 e⁻): $\left[:N\equiv O:\right]^+$ Triple bond between N and O

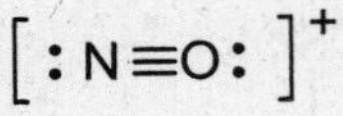

NO_2^- (18 e⁻): $\left[\overset{..}{O}=\overset{..}{N}-\overset{..}{\underset{..}{O}}:\right]^-$ ⟷ $\left[:\overset{..}{\underset{..}{O}}-\overset{..}{N}=\overset{..}{O}\right]^-$

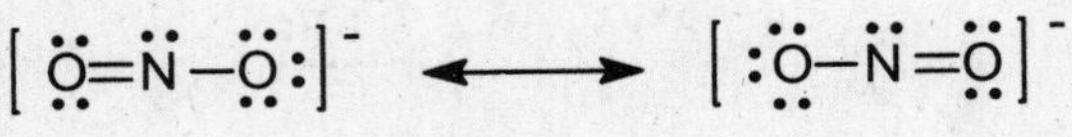

Average of 1 1/2 bond between N and O

NO_3^- (24 e⁻): $\left[\begin{array}{c}:\overset{..}{O}:\\N\\:\overset{..}{\underset{..}{O}}\quad\overset{..}{\underset{..}{O}}:\end{array}\right]^-$ ⟷ $\left[\begin{array}{c}:\overset{..}{O}:\\N\\\overset{..}{\underset{..}{O}}\quad:\overset{..}{\underset{..}{O}}:\end{array}\right]^-$ ⟷ $\left[\begin{array}{c}:\overset{..}{\underset{..}{O}}:\\N\\:\overset{..}{\underset{..}{O}}\quad\overset{..}{\underset{..}{O}}:\end{array}\right]^-$

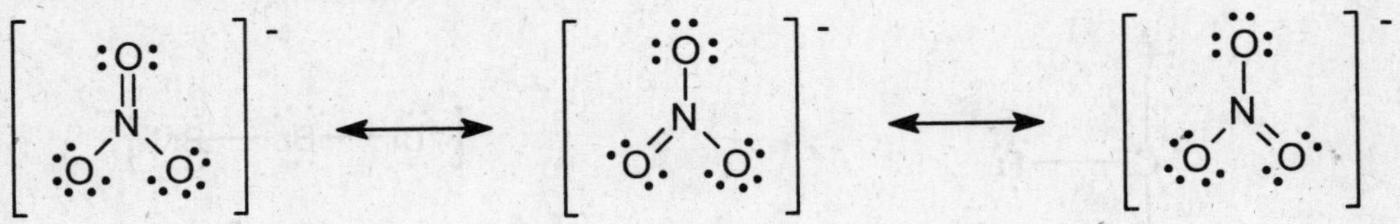

Average of 1 1/3 bond between N and O

From the Lewis structures, the order from shortest → longest N–O bond is:

$$NO^+ < N_2O < NO_2^- < NO_3^- < H_2NOH$$

Formal Charge

77.  $:C\equiv O:$ Carbon: FC = 4 – 2 – 1/2(6) = –1; oxygen: FC = 6 – 2 – 1/2(6) = +1

Electronegativity predicts the opposite polarization. The two opposing effects seem to partially cancel to give a much less polar molecule than expected.

79.   See Exercise 13.58a for the Lewis structures of $POCl_3$, SO_4^{2-}, ClO_4^- and PO_4^{3-}. All of these compounds/ions have similar Lewis structures to those of SO_2Cl_2 and XeO_4 shown below.

a. $POCl_3$: P, FC = 5 – 1/2(8) = +1   b. SO_4^{2-}: S, FC = 6 – 1/2(8) = +2

c. ClO_4^-: Cl, FC = 7 – 1/2(8) = +3   d. PO_4^{3-}: P, FC = 5 – 1/2(8) = +1

e. SO_2Cl_2, 6 + 2(6) + 2(7) = 32 e⁻   f. XeO_4, 8 + 4(6) = 32 e⁻

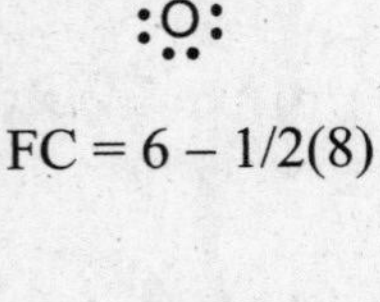

$$\begin{array}{c}:\overset{..}{O}:\\:\overset{..}{\underset{..}{Cl}}-S-\overset{..}{\underset{..}{Cl}}:\\:\overset{..}{\underset{..}{O}}:\end{array}\qquad\qquad\begin{array}{c}:\overset{..}{O}:\\:\overset{..}{\underset{..}{O}}-Xe-\overset{..}{\underset{..}{O}}:\\:\overset{..}{\underset{..}{O}}:\end{array}$$

S, FC = 6 – 1/2(8) = +2 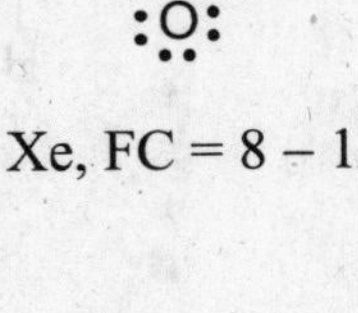   Xe, FC = 8 – 1/2(8) = +4

g. ClO_3^-, $7 + 3(6) + 1 = 26$ e⁻   h. NO_4^{3-}, $5 + 4(6) + 3 = 32$ e⁻

$$\left[\begin{array}{c} :\ddot{O}\!-\!\ddot{C}l\!-\!\ddot{O}: \\ \ddot{\underset{\cdot\cdot}{O}}: \end{array} \right]^{-}$$

$$\left[\begin{array}{c} :\ddot{O}: \\ :\ddot{O}\!-\!N\!-\!\ddot{O}: \\ :\ddot{O}: \end{array} \right]^{3-}$$

Cl, FC = 7 − 2 − 1/2(6) = +2   N, FC = 5 − 1/2(8) = +1

81.   SCl, $6 + 7 = 13$; the formula could be SCl (13 valence electrons), S_2Cl_2 (26 valence electrons), S_3Cl_3 (39 valence electrons), etc. For a formal charge of zero on S, we will need each sulfur in the Lewis structure to have two bonds to it and two lone pairs [FC = 6 − 4 − 1/2(4) = 0]. Cl will need one bond and three lone pairs for a formal charge of zero [FC = 7 − 6 − 1/2(2) = 0]. Since chlorine wants only one bond to it, it will not be a central atom here. With this in mind, only S_2Cl_2 can have a Lewis structure with a formal charge of zero on all atoms. The structure is:

$$:\ddot{C}l\!-\!\!-\!\ddot{S}\!-\!\!-\!\ddot{S}\!-\!\!-\!\ddot{C}l:$$

83. O_2F_2 has $2(6) + 2(7) = 26$ valence e⁻. The formal charge and oxidation number (state) of each atom is below the Lewis structure of O_2F_2.

$$:\ddot{F}\!-\!\!-\!\ddot{O}\!-\!\!-\!\ddot{O}\!-\!\!-\!\ddot{F}:$$

Formal Charge   0   0   0   0

Oxid. Number   -1   +1   +1   -1

Oxidation states are more useful when accounting for the reactivity of O_2F_2. We are forced to assign +1 as the oxidation state for oxygen due to the bonding to fluorine. Oxygen is very electronegative, and +1 is not a stable oxidation state for this element.

Molecular Structure and Polarity

85.   a. V-shaped or bent   b. see-saw   c. trigonal pyramid

       d. trigonal bipyramid   e. tetrahedral

87.   The first step always is to draw a valid Lewis structure when predicting molecular structure. When resonance is possible, only one of the possible resonance structures is necessary to predict the correct structure because all resonance structures give the same structure. The Lewis structures are in Exercises 13.57, 13.58 and 13.60. The structures and bond angles for each follow.

       13.57 a. HCN: linear, 180°   b. PH₃: trigonal pyramid, <109.5°

   c. $CHCl_3$: tetrahedral, 109.5°   d. NH_4^+: tetrahedral, 109.5°

   e. H_2CO: trigonal planar, 120°   f. SeF_2: V-shaped or bent, <109.5°

   g. CO_2: linear, 180°   h and i. O_2 and HBr are both linear, but there is no bond angle in either.

Note: PH_3 and SeF_2 both have lone pairs of electrons on the central atom, which result in bond angles that are something less than predicted from a tetrahedral arrangement (109.5°). However, we cannot predict the exact number. For these cases we will just insert a less than sign to indicate this phenomenon.

13.58 a. All are tetrahedral; 109.5°

      b. All are trigonal pyramid; <109.5°

      c. All are V-shaped; <109.5°

13.60 a. NO_2^-: V-shaped, ≈120°; NO_3^-: trigonal planar, 120°

         N_2O_4: trigonal planar, 120° about both N atoms

      b. OCN^-, SCN^-, and N_3^- are all linear with 180° bond angles.

89.   From the Lewis structures (see Exercise 13.71), Br_3^- would have a linear molecular structure, ClF_3 would have a T-shaped molecular structure, and SF_4 would have a see-saw molecular structure. For example, consider ClF_3 (28 valence electrons):

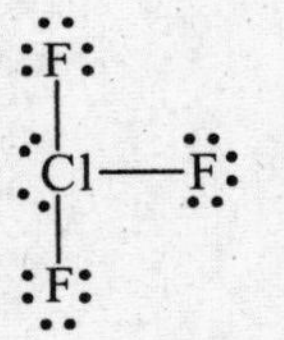

The central Cl atom is surrounded by five electron pairs, which requires a trigonal bipyramid geometry. Since there are three bonded atoms and two lone pairs of electrons about Cl, we describe the molecular structure of ClF_3 as T-shaped with predicted bond angles of about 90°. The actual bond angles will be slightly less than 90° due to the stronger repulsive effect of the lone pair electrons as compared to the bonding electrons.

91.   a. $XeCl_2$ has $8 + 2(7) = 22$ valence electrons.

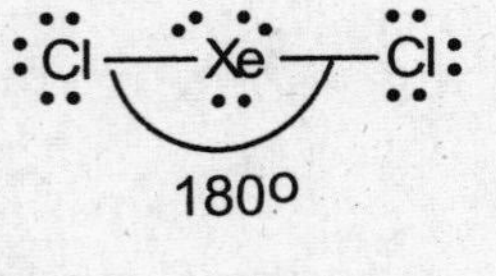

There are five pairs of electrons about the central Xe atom. The structure will be based on a trigonal bipyramid geometry. The most stable arrangement of the atoms in $XeCl_2$ is a linear molecular structure with a 180° bond angle.

    b. ICl_3 has $7 + 3(7) = 28$ valence electrons.

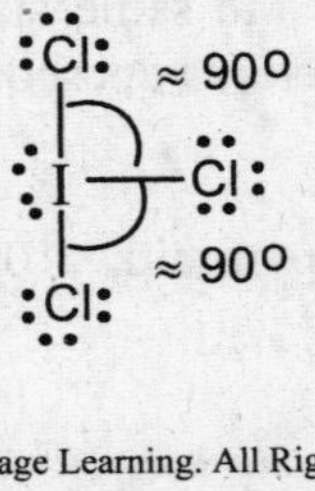

T-shaped; The ClICl angles are ≈ 90°. Since the lone pairs will take up more space, the ClICl bond angles will probably be slightly less than 90°.

c. TeF₄ has 6 + 4(7) = 34
    valence electrons.

d. PCl₅ has 5 + 5(7) = 40
    valence electrons.

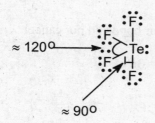

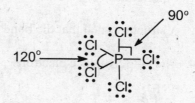

See-saw or teeter-totter
or distorted tetrahedron

Trigonal bipyramid

All the species in this exercise have five pairs of electrons around the central atom. All the structures are based on a trigonal bipyramid geometry, but only in PCl₅ are all the pairs bonding pairs. Thus PCl₅ is the only one we describe the molecular structure as trigonal bipyramid. Still, we had to begin with the trigonal bipyramid geometry to get to the structures (and bond angles) of the others.

93.   Let us consider the molecules with three pairs of electrons around the central atom first; these molecules are SeO₃ and SeO₂, and both have a trigonal planar arrangement of electron pairs. Both these molecules have polar bonds, but only SeO₂ has an overall net dipole moment. The net effect of the three bond dipoles from the three polar Se–O bonds in SeO₃ will be to cancel each other out when summed together. Hence SeO₃ is nonpolar since the overall molecule has no resulting dipole moment. In SeO₂, the two Se–O bond dipoles do not cancel when summed together; hence SeO₂ has a net dipole moment (is polar). Since O is more electronegative than Se, the negative end of the dipole moment is between the two O atoms, and the positive end is around the Se atom. The arrow in the following illustration represents the overall dipole moment in SeO₂. Note that to predict polarity for SeO₂, either of the two resonance structures can be used.

The other molecules in Exercise 13.88 (PCl₃, SCl₂, and SiF₄) have a tetrahedral arrangement of electron pairs. All have polar bonds; in SiF₄ the individual bond dipoles cancel when summed together, and in PCl₃ and SCl₂ the individual bond dipoles do not cancel. Therefore, SiF₄ has no net dipole moment (is nonpolar), and PCl₃ and SCl₂ have net dipole moments (are polar). For PCl₃, the negative end of the dipole moment is between the more electronegative chlorine atoms, and the positive end is around P. For SCl₂, the negative end is between the more electronegative Cl atoms, and the positive end of the dipole moment is around S.

95.   The two general requirements for a polar molecule are:

1. Polar bonds

2. A structure such that the bond dipoles of the polar bonds do not cancel

CF_4, 4 + 4(7) = 32 valence electrons XeF_4, 8 + 4(7) = 36 e⁻

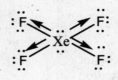

Tetrahedral, 109.5°   Square planar, 90°

SF_4, 6 + 4(7) = 34 e⁻

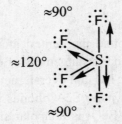

See-saw, ≈ 90°, ≈ 120°

The arrows indicate the individual bond dipoles in the three molecules (the arrows point to the more electronegative atom in the bond, which will be the partial negative end of the bond dipole). All three of these molecules have polar bonds. To determine the polarity of the overall molecule, we sum the effect of all of the individual bond dipoles. In CF_4, the fluorines are symmetrically arranged about the central carbon atom. The net result is for all the individual C–F bond dipoles to cancel each other out, giving a nonpolar molecule. In XeF_4, the 4 Xe–F bond dipoles are also symmetrically arranged, and XeF_4 is also nonpolar. The individual bond dipoles cancel out when summed together. In SF_4, we also have four polar bonds. But in SF_4 the bond dipoles are not symmetrically arranged, and they do not cancel each other out. SF_4 is polar. It is the positioning of the lone pair that disrupts the symmetry in SF_4.

CO_2, 4 + 2(6) = 16 e⁻   COS, 4 + 6 + 6 = 16 e⁻

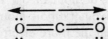

CO_2 and COS both have a linear molecular structure with a 180° bond angle. CO_2 is nonpolar because the individual bond dipoles cancel each other out, but COS is polar. By replacing an O with a less electronegative S atom, the molecule is not symmetric any more. The individual bond dipoles do not cancel because the C–S bond dipole is smaller than the C–O bond dipole resulting in a polar molecule.

97.   Only statement c is true. The bond dipoles in CF_4 and KrF_4 are arranged in a manner that they all cancel each other out, making them nonpolar molecules (CF_4 has a tetrahedral molecular structure, whereas KrF_4 has a square planar molecular structure). In SeF_4 the bond dipoles in this see-saw molecule do not cancel each other out, so SeF_4 is polar. For statement a, all the molecules have either a trigonal planar geometry or a trigonal bipyramid geometry, both of which have 120° bond angles. However, $XeCl_2$ has three lone pairs and two bonded chlorine atoms around it. $XeCl_2$ has a linear molecular structure with a 180° bond angle. With three lone pairs, we no longer have a 120° bond angle in $XeCl_2$. For statement b, SO_2 has a V-shaped molecular structure with a bond angle of about 120°. CS_2 is linear with a 180° bond angle and SCl_2 is V-shaped but with an approximate 109.5° bond angle. The three compounds do not have the same bond angle. For statement d, central atoms adopt a geometry to minimize electron repulsions, not maximize them.

99.   The formula is EF_2O^{2-}, and the Lewis structure has 28 valence electrons.

   $28 = x + 2(7) + 6 + 2$, $x = 6$ valence electrons for element E

   Element E must belong to the Group 6A elements since E has six valence electrons. E must also be a Row 3 or heavier element since this ion has more than eight electrons around the central E atom (Row 2 elements never have more than eight electrons around them). Some possible identities for E are S, Se and Te. The ion has a T-shaped molecular structure (see Exercise 13.89) with bond angles of $\approx 90°$.

101.   Molecules that have an overall dipole moment are called polar molecules, and molecules that do not have an overall dipole moment are called nonpolar molecules.

   a. OCl_2, $6 + 2(7) = 20$ e⁻ KrF_2, $8 + 2(7) = 22$ e⁻

   V-shaped, polar; OCl_2 is polar because   Linear, nonpolar; The molecule is
   the two O–Cl bond dipoles don't cancel   nonpolar because the two Kr–F
   each other. The resulting dipole moment   bond dipoles cancel each other.
   is shown in the drawing.

   BeH_2, $2 + 2(1) = 4$ e⁻ SO_2, $6 + 2(6) = 18$ e⁻

   Linear, nonpolar; Be–H bond dipoles   V-shaped, polar; The S–O bond
   are equal and point in opposite directions.   dipoles do not cancel, so SO_2 is polar
   They cancel each other. BeH_2 is nonpolar.   (has a net dipole moment). Only one
                                                      resonance structure is shown.

Note: All four species contain three atoms. They have different structures because the number of lone pairs of electrons around the central atom are different in each case.

b. SO_3, $6 + 3(6) = 24 \text{ e}^-$

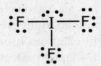

NF$_3$, $5 + 3(7) = 26 \text{ e}^-$

Trigonal planar, nonpolar;
bond dipoles cancel. Only one
resonance structure is shown.

Trigonal pyramid, polar;
bond dipoles do not cancel.

IF$_3$ has $7 + 3(7) = 28$ valence electrons.

T-shaped, polar; bond dipoles do not cancel.

Note: Each molecule has the same number of atoms but different structures because of differing numbers of lone pairs around each central atom.

c. CF$_4$, $4 + 4(7) = 32 \text{ e}^-$

SeF$_4$, $6 + 4(7) = 34 \text{ e}^-$

Tetrahedral, nonpolar;
bond dipoles cancel.

See-saw, polar;
bond dipoles do not cancel.

KrF$_4$, $8 + 4(7) = 36$ valence electrons

Square planar, nonpolar;
bond dipoles cancel.

Note: Again, each molecule has the same number of atoms but different structures because of differing numbers of lone pairs around the central atom.

d. IF_5, $7 + 5(7) = 42 \ e^-$ AsF_5, $5 + 5(7) = 40 \ e^-$

Square pyramid, polar;   Trigonal bipyramid, nonpolar;
bond dipoles do not cancel.   bond dipoles cancel.

Note: Yet again, the molecules have the same number of atoms but different structures because of the presence of differing numbers of lone pairs.

103.   All these molecules have polar bonds that are symmetrically arranged about the central atoms. In each molecule the individual bond dipoles cancel to give no net overall dipole moment. All these molecules are nonpolar even though they all contain polar bonds.

Additional Exercises

105. CO_3^{2-} has $4 + 3(6) + 2 = 24$ valence electrons.

HCO_3^- has $1 + 4 + 3(6) + 1 = 24$ valence electrons.

H_2CO_3 has $2(1) + 4 + 3(6) = 24$ valence electrons.

The Lewis structures for the reactants and products are:

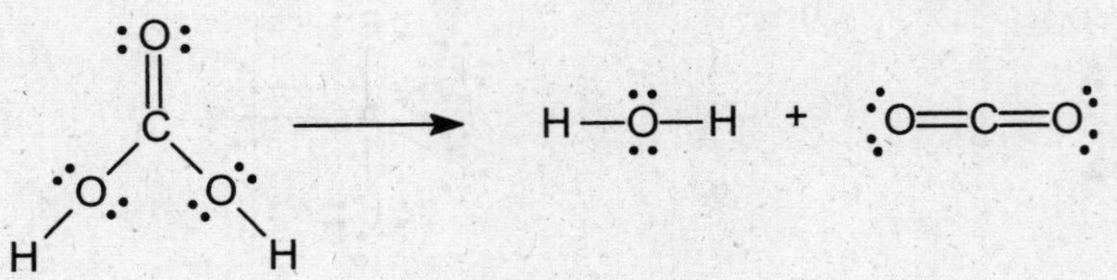

Bonds broken:

2 C–O (358 kJ/mol)

1 O–H (467 kJ/mol)

Bonds formed:

1 C=O (799 kJ/mol)

1 O–H (467 kJ/mol)

$\Delta H = 2(358) + 467 - (799 + 467) = -83$ kJ; the carbon-oxygen double bond is stronger than two carbon-oxygen single bonds; hence CO_2 and H_2O are more stable than H_2CO_3.

107.   As the halogen atoms get larger, it becomes more difficult to fit three halogen atoms around the small nitrogen atom, and the NX_3 molecule becomes less stable.

109.   The stable species are:

a.   NaBr: In $NaBr_2$, the sodium ion would have a 2+ charge, assuming that each bromine has a 1– charge. Sodium doesn't form stable Na^{2+} compounds.

b. ClO_4^-: ClO_4 has 31 valence electrons, so it is impossible to satisfy the octet rule for all atoms in ClO_4. The extra electron from the 1– charge in ClO_4^- allows for complete octets for all atoms.

c. XeO_4: We can't draw a Lewis structure that obeys the octet rule for SO_4 (30 electrons), unlike with XeO_4 (32 electrons).

d. SeF_4: Both compounds require the central atom to expand its octet. O is too small and doesn't have low-energy d orbitals to expand its octet (which is true for all Row 2 elements).

111.   a.   Radius: $N^+ < N < N^-$; IE: $N^- < N < N^+$

N^+ has the fewest electrons held by the seven protons in the nucleus whereas N^- has the most electrons held by the seven protons. The seven protons in the nucleus will hold the electrons most tightly in N^+ and least tightly in N^-. Therefore, N^+ has the smallest radius with the largest ionization energy (IE), and N^- is the largest species with the smallest IE.

b.   Radius: $Cl^+ < Cl < Se < Se^-$; IE: $Se^- < Se < Cl < Cl^+$

The general trends tell us that Cl has a smaller radius than Se and a larger IE than Se. Cl^+, with fewer electron-electron repulsions than Cl, will be smaller than Cl and have a larger IE. Se^-, with more electron-electron repulsions than Se, will be larger than Se and have a smaller IE.

c.   Radius: $Sr^{2+} < Rb^+ < Br^-$; IE: $Br^- < Rb^+ < Sr^{2+}$

These ions are isoelectronic. The species with the most protons (Sr^{2+}) will hold the electrons most tightly and will have the smallest radius and largest IE.   The ion with the fewest protons (Br^-) will hold the electrons least tightly and will have the largest radius and smallest IE.

113.   Assuming 100.00 g of compound: $42.81 \text{ g F} \times \dfrac{1 \text{ mol F}}{19.00 \text{ g F}} = 2.253 \text{ mol F}$

The number of moles of X in XF_5 is: $2.253 \text{ mol F} \times \dfrac{1 \text{ mol X}}{5 \text{ mol F}} = 0.4506 \text{ mol X}$

This number of moles of X has a mass of 57.19 g (= 100.00 g – 42.81 g). The molar mass of X is:

$$\dfrac{57.19 \text{ g X}}{0.4506 \text{ mol X}} = 126.9 \text{ g/mol}; \text{ This is element I.}$$

IF_5, $7 + 5(7) = 42 \text{ e}^-$

The molecular structure is square pyramid.

115.   Yes, each structure has the same number of effective pairs around the central atom, giving the same predicted molecular structure for each compound/ion. (A multiple bond is counted as a single group of electrons.)

Challenge Problems

117. KrF_2, $8 + 2(7) = 22 \text{ e}^-$; from the Lewis structure, we have a trigonal bipyramid arrangement of electron pairs with a linear molecular structure.

Hyperconjugation assumes that the overall bonding in KrF_2 is a combination of covalent and ionic contributions (see Section 13.12 of the text for discussion of hyperconjugation). Using hyperconjugation, two resonance structures are possible that keep the linear structure.

119.

$$2 \text{ Li}^+(g) + 2 \text{ Cl}^-(g) \rightarrow 2 \text{ LiCl}(s) \qquad \Delta H = 2(-829 \text{ kJ})$$
$$2 \text{ Li}(g) \rightarrow 2 \text{ Li}^+(g) + 2 \text{ e}^- \qquad \Delta H = 2(520.\text{ kJ})$$
$$2 \text{ Li}(s) \rightarrow 2 \text{ Li}(g) \qquad \Delta H = 2(166 \text{ kJ})$$
$$2 \text{ HCl}(g) \rightarrow 2 \text{ H}(g) + 2 \text{ Cl}(g) \qquad \Delta H = 2(427 \text{ kJ})$$
$$2 \text{ Cl}(g) + 2 \text{ e}^- \rightarrow 2 \text{ Cl}^-(g) \qquad \Delta H = 2(-349 \text{ kJ})$$
$$2 \text{ H}(g) \rightarrow \text{ H}_2(g) \qquad \Delta H = -(432 \text{ kJ})$$

$$2 \text{ Li}(s) + 2 \text{ HCl}(g) \rightarrow 2 \text{ LiCl}(s) + \text{H}_2(g) \qquad \Delta H = -562 \text{ kJ}$$

121. a. $N(NO_2)_2^-$ contains $5 + 2(5) + 4(6) + 1 = 40$ valence electrons.

The most likely structures are:

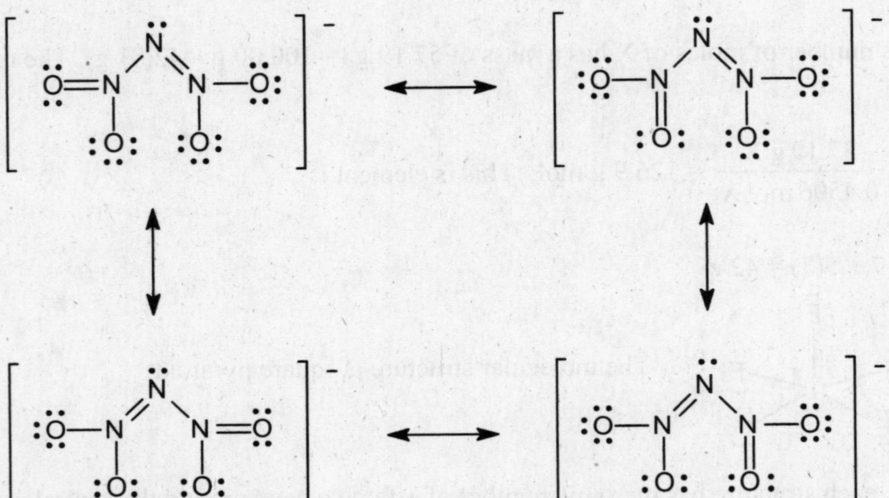

There are other possible resonance structures, but these are most likely.

b. The NNN and all ONN and ONO bond angles should be about 120°.

c. $NH_4N(NO_2)_2(s) \rightarrow 2 \text{ N}_2(g) + 2 \text{ H}_2O(g) + O_2(g)$; break and form all bonds.

Bonds broken: Bonds formed:

4 N–H (391 kJ/mol)   2 N≡N (941 kJ/mol)
1 N–N (160. kJ/mol)   4 H–O (467 kJ/mol)
1 N=N (418 kJ/mol)   1 O=O (495 kJ/mol)
3 N–O (201 kJ/mol)   _____
1 N=O (607 kJ/mol) ΣD_{formed} = 4245 kJ
_____

ΣD_{broken} = 3352 kJ

$\Delta H = \Sigma D_{broken} - \Sigma D_{formed} = 3352 \text{ kJ} - 4245 \text{ kJ} = -893 \text{ kJ}$

d. To estimate ΔH, we completely ignored the ionic interactions between NH_4^+ and $N(NO_2)_2^-$ in the solid phase. In addition, we assumed the bond energies in Table 13.6 applied to the $N(NO_2)^-$ bonds in any one of the resonance structures above. This is a bad assumption since molecules that exhibit resonance generally have stronger overall bonds than predicted. All these assumptions give an estimated ΔH value which is too negative.

123.   a.   i. $C_6H_6N_{12}O_{12} \rightarrow 6\ CO + 6\ N_2 + 3\ H_2O + 3/2\ O_2$

The NO_2 groups have one N–O single bond and one N=O double bond, and each carbon atom has one C–H single bond. We must break and form all bonds.

Bonds broken:	Bonds formed:
3 C–C (347 kJ/mol)	6 C≡O (1072 kJ/mol)
6 C–H (413 kJ/mol)	6 N≡N (941 kJ/mol)
12 C–N (305 kJ/mol)	6 H–O (467 kJ/mol)
6 N–N (160. kJ/mol)	3/2 O=O (495 kJ/mol)
6 N–O (201 kJ/mol)	
6 N=O (607 kJ/mol)	ΣD_{formed} = 15,623 kJ

ΣD_{broken} = 12,987 kJ

$\Delta H = \Sigma D_{broken} - \Sigma D_{formed} = 12{,}987\ kJ - 15{,}623\ kJ = -2636\ kJ$

ii. $C_6H_6N_{12}O_{12} \rightarrow 3\ CO + 3\ CO_2 + 6\ N_2 + 3\ H_2O$

Note: The bonds broken will be the same for all three reactions.

Bonds formed:

3 C≡O (1072 kJ/mol)
6 C=O (799 kJ/mol)
6 N≡N (941 kJ/mol)
6 H–O (467 kJ/mol)

ΣD_{formed} = 16,458 kJ

$\Delta H = 12{,}987\ kJ - 16{,}458\ kJ = -3471\ kJ$

iii. $C_6H_6N_{12}O_{12} \rightarrow 6\ CO_2 + 6\ N_2 + 3\ H_2$

Bonds formed:

12 C=O (799 kJ/mol)
6 N≡N (941 kJ/mol)
3 H–H (432 kJ/mol)

ΣD_{formed} = 16,530. kJ

$\Delta H = 12{,}987\ kJ - 16{,}530.\ kJ = -3543\ kJ$

b. Reaction iii yields the most energy per mole of CL-20, so it will yield the most energy per kilogram.

$$\frac{-3543 \text{ kJ}}{\text{mol}} \times \frac{1 \text{ mol}}{438.23 \text{ g}} \times \frac{1000 \text{ g}}{\text{kg}} = -8085 \text{ kJ/kg}$$

125.   The reaction is: $1/2 \ I_2(s) + 1/2 \ Cl_2(g) \rightarrow ICl(g)$ $\Delta H_f^{\circ} = ?$

Using Hess's law:

$1/2 \ I_2(s) \rightarrow 1/2 \ I_2(g)$	$\Delta H = 1/2 \ (62 \text{ kJ})$	(Appendix 4)
$1/2 \ I_2(g) \rightarrow I(g)$	$\Delta H = 1/2 \ (149 \text{ kJ})$	(Table 13.6)
$1/2 \ Cl_2(g) \rightarrow Cl(g)$	$\Delta H = 1/2 \ (239 \text{ kJ})$	(Table 13.6)
$I(g) + Cl(g) \rightarrow ICl(g)$	$\Delta H = -208 \text{ kJ}$	(Table 13.6)

$1/2 \ I_2(s) + 1/2 \ Cl_2(g) \rightarrow ICl(g)$ $\Delta H = 17 \text{ kJ}$ so $\Delta H_f^{\circ} = 17 \text{ kJ/mol}$

CHAPTER 14

COVALENT BONDING: ORBITALS

The Localized Electron Model and Hybrid Orbitals

9.   The valence orbitals of the nonmetals are the s and p orbitals. The lobes of the p orbitals are 90° and 180° apart from each other. If the p orbitals were used to form bonds, then all bond angles should be 90° or 180°. This is not the case. In order to explain the observed geometry (bond angles) that molecules exhibit, we need to make up (hybridize) orbitals that point to where the bonded atoms and lone pairs are located. We know the geometry; we hybridize orbitals to explain the geometry.

Sigma bonds have shared electrons in the area centered on a line joining the atoms. The orbitals that overlap to form the sigma bonds must overlap head to head or end to end. The hybrid orbitals about a central atom always are directed at the bonded atoms. Hybrid orbitals will always overlap head to head to form sigma bonds.

11.   We use d orbitals when we have to, i.e., we use d orbitals when the central atom on a molecule has more than eight electrons around it. The d orbitals are necessary to accommodate the electrons over eight. Row 2 elements never have more than eight electrons around them, so they never hybridize d orbitals. We rationalize this by saying there are no d orbitals close in energy to the valence 2s and 2p orbitals (2d orbitals are forbidden energy levels). However, for Row 3 and heavier elements, there are 3d, 4d, 5d, etc. orbitals that will be close in energy to the valence s and p orbitals. It is Row 3 and heavier nonmetals that hybridize d orbitals when they have to.

For sulfur, the valence electrons are in 3s and 3p orbitals. Therefore, 3d orbitals are closest in energy and are available for hybridization. Arsenic would hybridize 4d orbitals to go with the valence 4s and 4p orbitals, whereas iodine would hybridize 5d orbitals since the valence electrons are in $n = 5$.

13. H_2O has $2(1) + 6 = 8$ valence electrons.

H_2O has a tetrahedral arrangement of the electron pairs about the O atom that requires sp^3 hybridization. Two of the four sp^3 hybrid orbitals are used to form bonds to the two hydrogen atoms, and the other two sp^3 hybrid orbitals hold the two lone pairs on oxygen. The two O–H bonds are formed from overlap of the sp^3 hybrid orbitals from oxygen with the 1s atomic orbitals from the hydrogen atoms. Each O–H covalent bond is called a sigma (σ) bond since the shared electron pair in each bond is centered in an area on a line running between the two atoms.

15.   Ethane, C_2H_6, has $2(4) + 6(1) = 14$ valence electrons. The Lewis structure is:

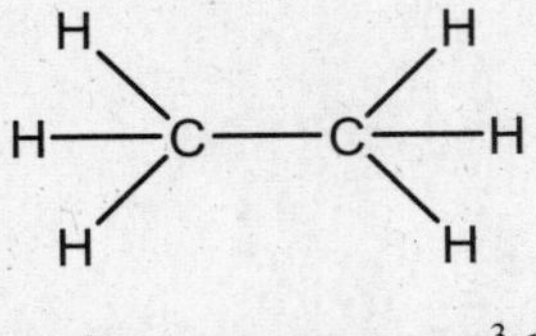

The carbon atoms are sp^3 hybridized. The six C–H sigma bonds are formed from overlap of the sp^3 hybrid orbitals on C with the 1s atomic orbitals from the hydrogen atoms. The carbon-carbon sigma bond is formed from overlap of an sp^3 hybrid orbital on each C atom.

Ethanol, C_2H_6O has $2(4) + 6(1) + 6 = 20$ e$^-$

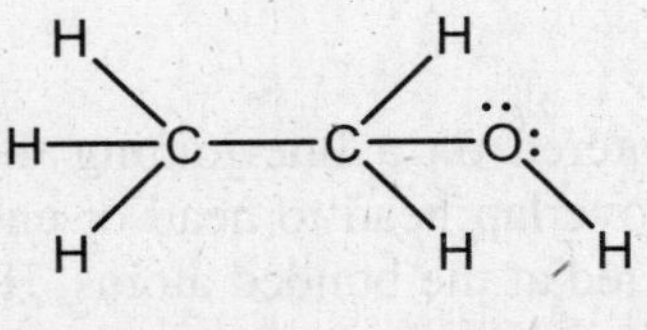

The two C atoms and the O atom are sp^3 hybridized. All bonds are formed from overlap with these sp^3 hybrid orbitals. The C–H and O–H sigma bonds are formed from overlap of sp^3 hybrid orbitals with hydrogen 1s atomic orbitals. The C–C and C–O sigma bonds are formed from overlap of the sp^3 hybrid orbitals on each atom.

17.   See Exercises 13.57, 13.58, and 13.60 for the Lewis structures. To predict the hybridization, first determine the arrangement of electron pairs about each central atom using the VSEPR model, then utilize the information in Figure 14.24 of the text to deduce the hybridization required for that arrangement of electron pairs.

13.57 a. HCN; C is sp hybridized.   b. PH_3; P is sp^3 hybridized.

       c. $CHCl_3$; C is sp^3 hybridized.   d. NH_4^+; N is sp^3 hybridized.

       e. H_2CO; C is sp^2 hybridized.   f. SeF_2; Se is sp^3 hybridized.

       g. CO_2; C is sp hybridized.   h. O_2; Each O atom is sp^2 hybridized.

       i. HBr; Br is sp^3 hybridized.

13.58 a. All the central atoms are sp^3 hybridized.

       b. All the central atoms are sp^3 hybridized.

       c. All the central atoms are sp^3 hybridized.

13.60 a. In NO_2^- and NO_3^-, N is sp^2 hybridized. In N_2O_4, both central N atoms are also sp^2 hybridized.

   b. In OCN^- and SCN^-, the central carbon atoms in each ion are sp hybridized, and in N_3^-, the central N atom is also sp hybridized.

19.   The two nitrogen atoms in urea both have a tetrahedral arrangement of electron pairs, so both of these atoms are sp^3 hybridized. The carbon atom has a trigonal planar arrangement of electron pairs, so C is sp^2 hybridized. O is also sp^2 hybridized because it also has a trigonal planar arrangement of electron pairs.

   Each of the four N–H sigma bonds are formed from overlap of an sp^3 hybrid orbital from nitrogen with a 1s orbital from hydrogen. Each of the two N–C sigma bonds are formed from an sp^3 hybrid orbital from N with an sp^2 hybrid orbital from carbon. The double bond between carbon and oxygen consists of one σ and one π bond. The σ bond in the double bond is formed from overlap of a carbon sp^2 hybrid orbital with an oxygen sp^2 hybrid orbital. The π bond in the double bond is formed from overlap of the unhybridized p atomic orbitals. Carbon and oxygen each have one unhybridized p atomic orbital, and they are assumed to be parallel to each other. When two parallel p atomic orbitals overlap side to side, a π bond results.

21.   a.

   tetrahedral sp^3
   109.5°   nonpolar

   b.

   trigonal pyramid sp^3
   < 109.5°   polar

   The angles in NF_3 should be slightly less than 109.5° because the lone pair requires more space than the bonding pairs.

   c.

   V-shaped sp^3
   < 109.5°   polar

   d.

   trigonal planar sp^2
   120°   nonpolar

e.

H—Be—H

linear   sp
180°   nonpolar

f.

see-saw   dsp^3
a) ≈120°, b) ≈90°   polar

g.

trigonal bipyramid   dsp^3
a) 90°,   b) 120°   nonpolar

h.

:F—Kr—F:

linear   dsp^3
180°   nonpolar

i.

square planar   d^{2}sp^3
90°   nonpolar

j.

octahedral   d^{2}sp^3
90°   nonpolar

k.

square pyramid   d^{2}sp^3
≈90°   polar

l.

T-shaped   dsp^3
≈90°   polar

23.

For the p orbitals to properly line up to form the π bond, all six atoms are forced into the same plane. If the atoms are not in the same plane, then the π bond could not form since the p orbitals would no longer be parallel to each other.

25.   To complete the Lewis structures, just add lone pairs of electrons to satisfy the octet rule for the atoms with fewer than eight electrons.

Biacetyl ($C_4H_6O_2$) has $4(4) + 6(1) + 2(6) = 34$ valence electrons.

All CCO angles are 120°. The six atoms are not forced to lie in the same plane because of free rotation about the carbon-carbon single (sigma) bonds. There are 11 σ and 2 π bonds in biacetyl.

Acetoin ($C_4H_8O_2$) has $4(4) + 8(1) + 2(6) = 36$ valence electrons.

The carbon with the doubly bonded O is sp² hybridized. The other 3 C atoms are sp³ hybridized. Angle a = 120° and angle b = 109.5°. There are 13 σ and 1 π bonds in acetoin.

Note: All single bonds are σ bonds, all double bonds are one σ and one π bond, and all triple bonds are one σ and two π bonds.

27.   To complete the Lewis structure, just add lone pairs of electrons to satisfy the octet rule for the atoms that have fewer than eight electrons.

a. 6   b. 4   c. The center N in −N=N=N group

d. 33 σ   e. 5 π   f. 180°   g. ≈109.5°   h. sp^3

29.   a. Add lone pairs to complete octets for each O and N.

Azodicarbonamide   methyl cyanoacrylate

Note: NH$_2$, CH$_2$ (H$_2$C), and CH$_3$ are shorthand for nitrogen or carbon atoms singly bonded to hydrogen atoms.

b. In azodicarbonamide, the two carbon atoms are sp^2 hybridized, the two nitrogen atoms with hydrogens attached are sp^3 hybridized, and the other two nitrogens are sp^2 hybridized. In methyl cyanoacrylate, the CH$_3$ carbon is sp^3 hybridized, the carbon with the triple bond is sp hybridized, and the other three carbons are sp^2 hybridized.

c. Azodicarbonamide contains three π bonds and methyl cyanoacrylate contains four π bonds.

d. a) ≈109.5°   b) 120°   c) ≈120°   d) 120°   e) 180°

f) 120°   g) ≈109.5°   h) 120°

31.   CO, 4 + 6 = 10 e⁻;   CO₂, 4 + 2(6) = 16 e⁻;   C₃O₂, 3(4) + 2(6) = 24 e⁻

There is no molecular structure for the diatomic CO molecule. The carbon in CO is sp hybrid-ized. CO_2 is a linear molecule, and the central carbon atom is sp hybridized. C_3O_2 is a linear molecule with all the central carbon atoms exhibiting sp hybridization.

The Molecular Orbital (MO) Model

33.   a.

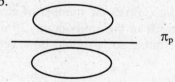

When p orbitals are combined head-to-head and the phases are the same sign (the orbital lobes have the same sign), a sigma bonding molecular orbital is formed.

b.

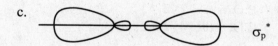

When parallel p orbitals are combined in-phase (the signs match up), a pi bonding molecular orbital is formed.

c.

σ_p^*

When p orbitals are combined head-to-head and the phases are opposite (the orbital lobes have opposite signs), a sigma antibonding molecular orbital is formed.

d.

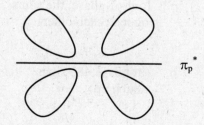

When parallel p orbitals are combined out-of-phase (the orbital lobes have opposite signs), a pi antibonding molecular orbital is formed.

35. From experiment, B_2 is paramagnetic. If the σ_{2p} MO is lower in energy than the two degenerate π_{2p} MOs, the electron configuration for B_2 would have all electrons paired. Experiment tells us we must have unpaired electrons. Therefore, the MO diagram is modified to have the π_{2p} orbitals lower in energy than the σ_{2p} orbitals. This gives two unpaired electrons in the electron configuration for B_2, which explains the paramagnetic properties of B_2. The model allowed for s and p orbitals to mix, which shifted the energy of the σ_{2p} orbital to above that of the π_{2p} orbitals.

37. a. H_2 has two valence electrons to put in the MO diagram, whereas He_2 has four valence electrons.

H_2: $(\sigma_{1s})^2$ Bond order = B.O. = $(2-0)/2 = 1$
He_2: $(\sigma_{1s})^2(\sigma_{1s}*)^2$ B.O. = $(2-2)/2 = 0$

H_2 has a nonzero bond order, so MO theory predicts it will exist. The H_2 molecule is stable with respect to the two free H atoms. He_2 has a bond order of zero, so it should not form. The He_2 molecule is not more stable than the two free He atoms.

b. See Figure 14.41 for the MO energy-level diagrams of B_2, C_2, N_2, O_2, and F_2. B_2 and O_2 have unpaired electrons in their electron configuration, so they are predicted to be paramagnetic. C_2, N_2, and F_2 have no unpaired electrons in the MO diagrams; they are all diamagnetic.

c. From the MO energy diagram in Figure 14.41, N_2 maximizes the number of electrons in the lower-energy bonding orbitals and has no electrons in the antibonding 2p molecular orbitals. N_2 has the highest possible bond order of three, so it should be a very strong (stable) bond.

d. NO^+ has $5 + 6 - 1 = 10$ valence electrons to place in the MO diagram, and NO^- has $5 + 6 + 1 = 12$ valence electrons. The MO diagram for these two ions is assumed to be the same as that used for N_2.

NO^+: $(\sigma_{2s})^2(\sigma_{2s}*)^2(\pi_{2p})^4(\sigma_{2p})^2$ B.O. = $(8-2)/2 = 3$
NO^-: $(\sigma_{2s})^2(\sigma_{2s}*)^2(\pi_{2p})^4(\sigma_{2p})^2(\pi_{2p}*)^2$ B.O. = $(8-4)/2 = 2$

NO^+ has a larger bond order than NO^-, so NO^+ should be more stable than NO^-.

39.

These molecular orbitals are sigma MOs because the electron density is cylindrically symmetric about the internuclear axis.

41. N_2: $(\sigma_{2s})^2(\sigma_{2s}*)^2(\pi_{2p})^4(\sigma_{2p})^2$ B.O. $= (8 - 2)/2 = 3$

We need to decrease the bond order from 3 to 2.5. There are two ways to do this. One is to add an electron to form N_2^-. This added electron goes into one of the π_{2p}^* orbitals, giving a bond order of $(8 - 3)/2 = 2.5$. We could also remove a bonding electron to form N_2^+. The bond order for N_2^+ is also 2.5 $[= (7 - 2)/2]$.

43.   CN: $(\sigma_{2s})^2(\sigma_{2s}*)^2(\pi_{2p})^4(\sigma_{2p})^1$

NO: $(\sigma_{2s})^2(\sigma_{2s}*)^2(\pi_{2p})^4(\sigma_{2p})^2(\pi_{2p}*)^1$

O_2^{2+}: $(\sigma_{2s})^2(\sigma_{2s}*)^2(\sigma_{2p})^2(\pi_{2p})^4$

N_2^{2+}: $(\sigma_{2s})^2(\sigma_{2s}*)^2(\pi_{2p})^4$

If the added electron goes into a bonding orbital, the bond order would increase, making the species more stable and more likely to form. Between CN and NO, CN would most likely form CN^- since the bond order increases (unlike NO^-, where the added electron goes into an antibonding orbital). Between O_2^{2+} and N_2^{2+}, N_2^+ would most likely form since the bond order increases (unlike O_2^{2+} going to O_2^+).

45.   The π bonds between S atoms and between C and S atoms are not as strong. The atomic orbitals do not overlap with each other as well as the smaller atomic orbitals of C and O overlap.

47.   Side-to-side overlap of these d orbitals would produce a π molecular orbital. There would be no probability of finding an electron on the axis joining the two nuclei, which is characteristic of π MOs.

49.   a.   The electron density would be closer to F on average. The F atom is more electronegative than the H atom, and the 2p orbital of F is lower in energy than the 1s orbital of H.

b.   The bonding MO would have more fluorine 2p character since it is closer in energy to the fluorine 2p atomic orbital.

c.   The antibonding MO would place more electron density closer to H and would have a greater contribution from the higher-energy hydrogen 1s atomic orbital.

51.   Molecules that exhibit resonance have delocalized π bonding. This is a fancy way of saying that the π electrons are not permanently stationed between two specific atoms but instead can roam about over the surface of a molecule. We use the concept of delocalized π electrons to explain why molecules that exhibit resonance have equal bonds in terms of strength. Because the π electrons can roam about over the entire surface of the molecule, the π electrons are shared by all the atoms in the molecule, giving rise to equal bond strengths.

The classic example of delocalized π electrons is benzene (C_6H_6). Figure 14.50 shows the π molecular orbital system for benzene. Each carbon in benzene is sp^2 hybridized, leaving one unhybridized p atomic orbital. All six of the carbon atoms in benzene have an unhybridized p orbital pointing above and below the planar surface of the molecule. Instead of just two unhybridized p orbitals overlapping, we say all six of the unhybridized p orbitals overlap, resulting in delocalized π electrons roaming about above and below the entire surface of the benzene molecule.

SO_2, $6 + 2(6) = 18$ e$^-$

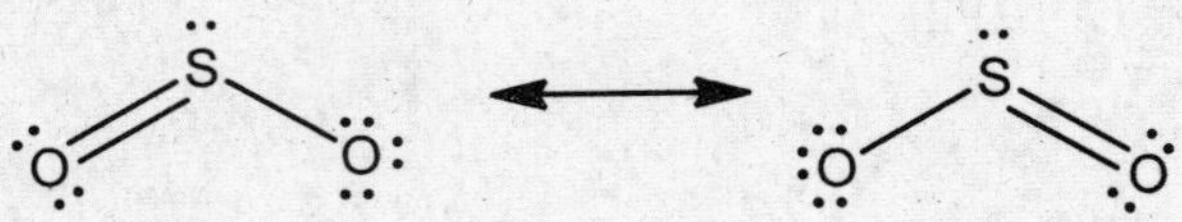

In SO_2 the central sulfur atom is sp^2 hybridized. The unhybridized p atomic orbital on the central sulfur atom will overlap with parallel p orbitals on each adjacent O atom. All three of these p orbitals overlap together, resulting in the π electrons moving about above and below the surface of the SO_2 molecule. With the delocalized π electrons, the S–O bond lengths in SO_2 are equal (and not different, as each individual Lewis structure indicates).

53.   The Lewis structures for CO_3^{2-} are (24 e$^-$):

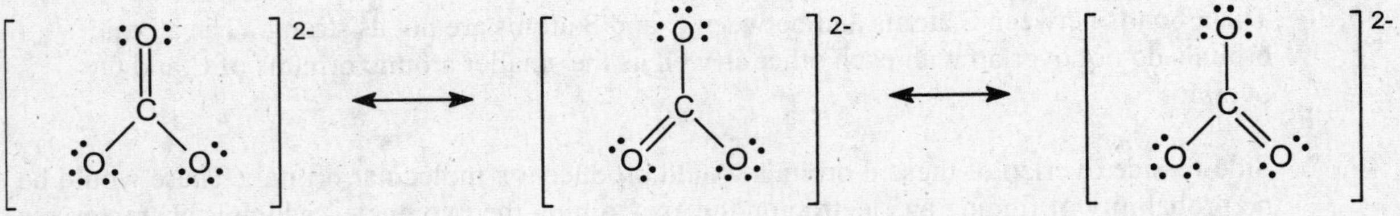

In the localized electron view, the central carbon atom is sp^2 hybridized; the sp^2 hybrid orbitals are used to form the three sigma bonds in CO_3^{2-}. The central C atom also has one unhybridized p atomic orbital that overlaps with another p atomic orbital from one of the oxygen atoms to form the π bond in each resonance structure. This localized π bond moves (resonates) from one position to another. In the molecular orbital model for CO_3^{2-}, all four atoms in CO_3^{2-} have a p atomic orbital that is perpendicular to the plane of the ion. All four of these p orbitals overlap at the same time to form a delocalized π bonding system where the π electrons can roam above and below the entire surface of the ion. The π molecular orbital system for CO_3^{2-} is analogous to that for NO_3^- which is shown in Figure 14.51 of the text.

Spectroscopy

55.   reduced mass $= \mu = \dfrac{m_1 m_2}{m_1 + m_2} = \dfrac{(1.0078)(78.918)}{1.0078 + 78.918}$ amu $\times \dfrac{1.66054 \times 10^{-27} \text{ kg}}{\text{amu}}$

$$= 1.6524 \times 10^{-27} \text{ kg}$$

$\nu_o = \dfrac{1}{2\pi} \sqrt{\dfrac{k}{\mu}} = \dfrac{c}{\lambda} = 2.9979 \times 10^{10}$ cm s$^{-1} \times 2650.$ cm$^{-1} = 7.944 \times 10^{13}$ s^{-1}

$$7.944 \times 10^{13} \text{ s}^{-1} = \frac{1}{2\pi} \sqrt{\frac{k}{\mu}} = \frac{1}{2\pi} \sqrt{\frac{k}{1.6524 \times 10^{-27} \text{ kg}}}$$

Solving for k, the force constant: $k = 411.7 \text{ kg s}^{-2} = 411.7 \text{ N m}^{-1}$

Note: one newton = $1 \text{ N} = 1 \text{ kg m s}^{-2}$.

57.   a. $\Delta E = 2hB(J_i + 1) = h\nu = \dfrac{hc}{\lambda}, \quad \dfrac{c}{\lambda} = 2B(J_i + 1)$

$$\frac{c}{\lambda} = 2B(0 + 1) = 2B = \frac{2.998 \times 10^8 \text{ m s}^{-1}}{2.60 \times 10^{-3} \text{ m}} = 1.15 \times 10^{11} \text{ s}^{-1}$$

$$B = \frac{1.15 \times 10^{-11} \text{ s}^{-1}}{2} = 5.75 \times 10^{10} \text{ s}^{-1}$$

$$I = \frac{h}{8\pi^2 B} = \frac{6.626 \times 10^{-34} \text{ J s}}{8\pi^2 (5.75 \times 10^{10} \text{ s}^{-1})} = 1.46 \times 10^{-46} \text{ kg m}^2$$

$$I = \mu R_e^2, \quad \mu = \frac{m_1 m_2}{m_1 + m_2} = \frac{(12.000)(15.995)}{12.000 + 15.995} \text{ amu} \times \frac{1.66054 \times 10^{-27} \text{ kg}}{\text{amu}}$$
$$= 1.1385 \times 10^{-26} \text{ kg}$$

$$R_e^2 = \frac{I}{\mu} = \frac{1.46 \times 10^{-46} \text{ kg m}^2}{1.1385 \times 10^{-26} \text{ kg}} = 1.28 \times 10^{-20} \text{ m}^2, \quad R_e = \text{bond length} = 1.13 \times 10^{-10} \text{ m}$$
$$= 113 \text{ pm}$$

b. $\nu = \dfrac{\Delta E}{h} = 2B(J_i + 1) = 2B(2 + 1) = 6B$

From part a, $B = 5.75 \times 10^{10} \text{ s}^{-1}$, so: $\nu = 6(5.75 \times 10^{10} \text{ s}^{-1}) = 3.45 \times 10^{11} \text{ s}^{-1}$

59.   a.   The $-CH_2$ group neighbors a $-CH_3$ group, so a quartet of peaks should result (iv).

b.   The $-CH_3$ H atoms are separated by more than three sigma bonds from other H atoms, so no spin-spin coupling should occur. A singlet peak should result (i).

c.   The $-CH_2$ H atoms neighbor two H atoms, so a triplet peak should result (iii).

d.   The two H atoms in the $-CH_2F$ group neighbor one H atom, so a doublet peak should result (ii).

Additional Exercises

61.   a. XeO_3, $8 + 3(6) = 26 \ e^-$ b. XeO_4, $8 + 4(6) = 32 \ e^-$

trigonal pyramid; sp^3 tetrahedral; sp^3

c. $XeOF_4$, $8 + 6 + 4(7) = 42 \ e^-$ d. $XeOF_2$, $8 + 6 + 2(7) = 28 \ e^-$

square pyramid; d^2sp^3 T-shaped; dsp^3

e. XeO_3F_2 has $8 + 3(6) + 2(7) = 40$ valence electrons.

trigonal bipyramid; dsp^3

63.   a.   No, some atoms are in different places. Thus these are not resonance structures; they are different compounds.

b.   For the first Lewis structure, all nitrogens are sp^3 hybridized and all carbons are sp^2 hybridized. In the second Lewis structure, all nitrogens and carbons are sp^2 hybridized.

c.   For the reaction:

Bonds broken: Bonds formed:

3 C=O (745 kJ/mol)   3 C≡N (615 kJ/mol)
3 C–N (305 kJ/mol)   3 C–O (358 kJ/mol)
3 N–H (391 kJ/mol)   3 O–H (467 kJ/mol)

$\Delta H = 3(745) + 3(305) + 3(391) - [3(615) + 3(358) + 3(467)]$

$\Delta H = 4323 \text{ kJ} - 4320 \text{ kJ} = 3 \text{ kJ}$

The bonds are slightly stronger in the first structure with the carbon-oxygen double bonds since ΔH for the reaction is positive. However, the value of ΔH is so small that the best conclusion is that the bond strengths are comparable in the two structures.

65.

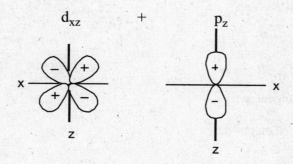

The two orbitals will overlap side-to-side, so when the orbitals are in-phase, a π bonding molecular orbital would form.

67.   Molecule A has a tetrahedral arrangement of electron pairs because it is sp^3 hybridized. Molecule B has 6 electron pairs about the central atom, so it is d^2sp^3 hybridized. Molecule C has two σ and two π bonds to the central atom, so it either has two double bonds to the central atom (as in CO_2) or one triple bond and one single bond (as in HCN). Molecule C is consistent with a linear arrangement of electron pairs exhibiting sp hybridization. There are many correct possibilities for each molecule; an example of each is:

Molecule A: CH_4 Molecule B: XeF_4 Molecule C: CO_2 or HCN

tetrahedral; 109.5°; sp^3 square planar; 90°; d^2sp^3 linear; 180°; sp

69.   a. BH_3 has $3 + 3(1) = 6$ valence electrons.

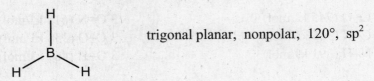

trigonal planar, nonpolar, 120°, sp^2

b. N_2F_2 has $2(5) + 2(7) = 24$ valence electrons.

Can also be:

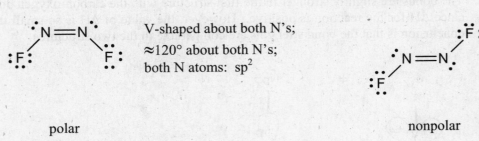

V-shaped about both N's;
≈120° about both N's;
both N atoms: sp^2

polar   nonpolar

These are distinctly different molecules.

c. C_4H_6 has $4(4) + 6(1) = 22$ valence electrons.

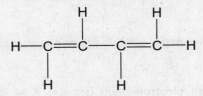

All C atoms are trigonal planar with 120° bond angles and sp^2 hybridization. Because C and H have about equal electronegativities, the C–H bonds are essentially nonpolar, so the molecule is nonpolar. All neutral compounds composed of only C and H atoms are nonpolar.

d. ICl_3 has $7 + 3(7) = 28$ valence electrons.

:Cl: ≈ 90°

:I—Cl:

:Cl: ≈ 90°

T-shaped, polar
≈90°, dsp^3

71.   For carbon, nitrogen, and oxygen atoms to have formal charge values of zero, each C atom will form four bonds to other atoms and have no lone pairs of electrons, each N atom will form three bonds to other atoms and have one lone pair of electrons, and each O atom will form two bonds to other atoms and have two lone pairs of electrons. The Lewis structure for caffeine that has a formal charge of zero for all atoms is:

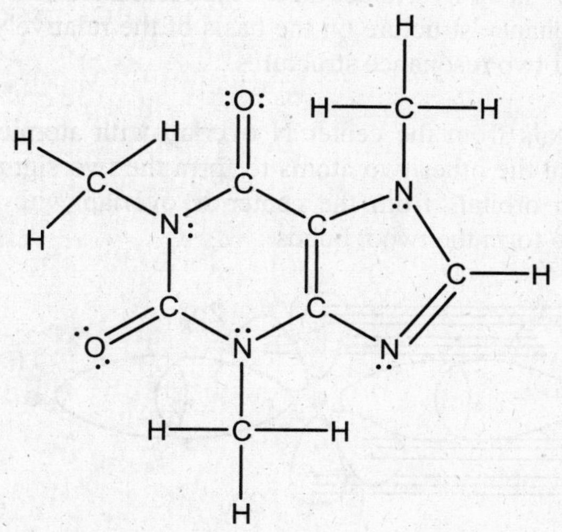

The three C atoms each bonded to three H atoms are sp^3 hybridized (tetrahedral geometry); the other five C atoms with trigonal planar geometry are sp^2 hybridized. The one N atom with the double bond is sp^2 hybridized, and the other three N atoms are sp^3 hybridized. The answers to the questions are:

- 6 total C and N atoms are sp^2 hybridized
- 6 total C and N atoms are sp^3 hybridized
- 0 C and N atoms are sp hybridized (linear geometry)
- 25 σ bonds and 4 π bonds

73.   a.   The Lewis structures for NNO and NON are:

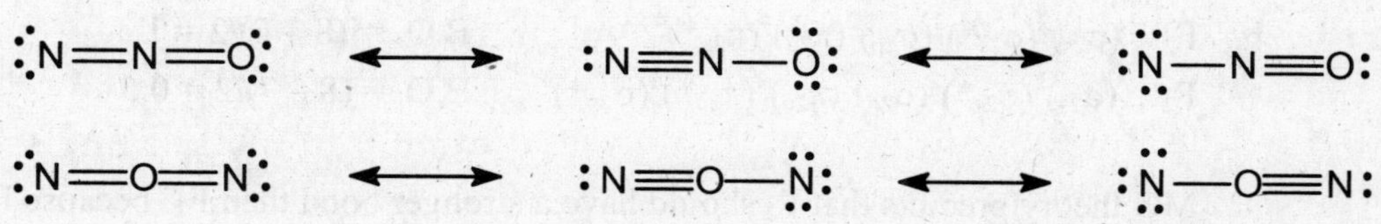

The NNO structure is correct. From the Lewis structures we would predict both NNO and NON to be linear. However, we would predict NNO to be polar and NON to be nonpolar. Since experiments show N_2O to be polar, then NNO is the correct structure.

        b.   Formal charge = number of valence electrons of atoms − [(number of lone pair electrons) + 1/2 (number of shared electrons)].

$$:N=N=O: \longleftrightarrow \quad :N\equiv N-\ddot{\underset{..}{O}}: \longleftrightarrow \quad :\ddot{\underset{..}{N}}-N\equiv O:$$

　　-1　　+1　　0　　　　　　　　　0　　+1　　-1　　　　　　　　-2　　+1　　+1

The formal charges for the atoms in the various resonance structures are below each atom. The central N is sp hybridized in all the resonance structures. We can probably ignore the third resonance structure on the basis of the relatively large formal charges as compared to the first two resonance structures.

c. The sp hybrid orbitals from the center N overlap with atomic orbitals (or appropriate hybrid orbitals) from the other two atoms to form the two sigma bonds. The remaining two unhybridized p orbitals from the center N overlap with two p orbitals from the peripheral N atom to form the two π bonds.

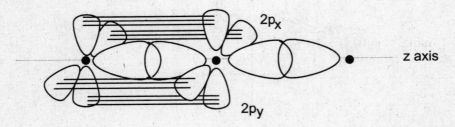

Challenge Problems

75.　　a.　$F_2^-(g) \rightarrow F(g) + F^-(g)$　$\Delta H = F_2^-$ bond energy; using Hess's law:

$$\begin{array}{ll} F_2^-(g) \rightarrow F_2(g) + e^- & \Delta H = 290.\ kJ\ (IE\ for\ F_2^-) \\ F_2(g) \rightarrow 2\ F(g) & \Delta H = 154\ kJ\ (BE\ for\ F_2\ from\ Table\ 13.6) \\ F(g) + e^- \rightarrow F^-(g) & \Delta H = -327.8\ kJ\ (EA\ for\ F\ from\ Table\ 12.8) \end{array}$$

$$F_2^-(g) \rightarrow F(g) + F^-(g) \quad \Delta H = 116\ kJ;\ BE\ for\ F_2^- = 116\ kJ/mol$$

Note that F_2^- has a smaller bond energy than F_2.

b.　F_2:　$(\sigma_{2s})^2(\sigma_{2s}*)^2(\sigma_{2p})^2(\pi_{2p})^4(\pi_{2p}*)^4$　　　　B.O. = (8 − 6)/2 = 1

　　F_2^-:　$(\sigma_{2s})^2(\sigma_{2s}*)^2(\sigma_{2p})^2(\pi_{2p})^4(\pi_{2p}*)^4(\sigma_{2p}*)^1$　B.O. = (8 − 7)/2 = 0.5

MO theory predicts that F_2 should have a stronger bond than F_2^- because F_2 has the larger bond order. As determined in part a, F_2 indeed has a stronger bond because the F_2 bond energy (154 kJ/mol) is greater than the F_2^- bond energy (116 kJ/mol).

77.　　a.　$E = \dfrac{hc}{\lambda} = \dfrac{(6.626 \times 10^{-34}\ J\ s)(2.998 \times 10^8\ m/s)}{25 \times 10^{-9}\ m} = 7.9 \times 10^{-18}\ J$

$$7.9 \times 10^{-18}\ J \times \frac{6.022 \times 10^{23}}{mol} \times \frac{1\ kJ}{1000\ J} = 4800\ kJ/mol$$

Using ΔH values from the various reactions, 25-nm light has sufficient energy to ionize N_2 and N and to break the triple bond. Thus N_2, N_2^+, N, and N^+ will all be present, assuming excess N_2.

b. To produce atomic nitrogen but no ions, the range of energies of the light must be from 941 kJ/mol to just below 1402 kJ/mol.

$$\frac{941 \text{ kJ}}{\text{mol}} \times \frac{1 \text{ mol}}{6.022 \times 10^{23}} \times \frac{1000 \text{ J}}{1 \text{ kJ}} = 1.56 \times 10^{-18} \text{ J/photon}$$

$$\lambda = \frac{hc}{E} = \frac{(6.6261 \times 10^{-34} \text{ J s})(2.998 \times 10^8 \text{ m/s})}{1.56 \times 10^{-18} \text{ J}} = 1.27 \times 10^{-7} \text{ m} = 127 \text{ nm}$$

$$\frac{1402 \text{ kJ}}{\text{mol}} \times \frac{1 \text{ mol}}{6.0221 \times 10^{23}} \times \frac{1000 \text{ J}}{\text{kJ}} = 2.328 \times 10^{-18} \text{ J/photon}$$

$$\lambda = \frac{hc}{E} = \frac{(6.6261 \times 10^{-34} \text{ J s})(2.9979 \times 10^8 \text{ m/s})}{2.328 \times 10^{-18} \text{ J}} = 8.533 \times 10^{-8} \text{ m} = 85.33 \text{ nm}$$

Light with wavelengths in the range of 85.33 nm $< \lambda \leq$ 127 nm will produce N but no ions.

c. N_2: $(\sigma_{2s})^2(\sigma_{2s}*)^2(\pi_{2p})^4(\sigma_{2p})^2$; the electron removed from N_2 is in the σ_{2p} molecular orbital, which is lower in energy than the 2p atomic orbital from which the electron in atomic nitrogen is removed. Because the electron removed from N_2 is lower in energy than the electron removed from N, the ionization energy of N_2 is greater than that for N.

79. The complete Lewis structure follows. All but two of the carbon atoms are sp^3 hybridized. The two carbon atoms that contain the double bond are sp^2 hybridized (see *).

No; most of the carbons are not in the same plane since a majority of carbon atoms exhibit the non-planar tetrahedral structure. *Note*: CH, CH$_2$, H$_2$C, and CH$_3$ are shorthand for carbon atoms singly bonded to hydrogen atoms.

81. a. The CO bond is polar with the negative end around the more electronegative oxygen atom. We would expect metal cations to be attracted to and to bond to the oxygen end of CO on the basis of electronegativity.

    b.

$$:C\equiv O:\qquad FC\ (carbon) = 4 - 2 - 1/2(6) = -1$$

$$FC\ (oxygen) = 6 - 2 - 1/2(6) = +1$$

From formal charge, we would expect metal cations to bond to the carbon (with the negative formal charge).

    c. In molecular orbital theory, only orbitals with proper symmetry overlap to form bonding orbitals. The metals that form bonds to CO are usually transition metals, all of which have outer electrons in the d orbitals. The only molecular orbitals of CO that have proper symmetry to overlap with d orbitals are the π_{2p}^* orbitals, whose shape is similar to the d orbitals (see Figure 14.36). Since the antibonding molecular orbitals have more carbon character, one would expect the bond to form through carbon.

83.   One of the resonance structures for benzene is:

To break C$_6$H$_6$(g) into C(g) and H(g) requires the breaking of 6 C–H bonds, 3 C=C bonds and 3 C–C bonds:

$$C_6H_6(g) \rightarrow 6\ C(g) + 6\ H(g) \qquad \Delta H = 6D_{C-H} + 3D_{C=C} + 3D_{C-C}$$

$$\Delta H = 6(413\ kJ) + 3(614\ kJ) + 3(347\ kJ) = 5361\ kJ$$

The question asks for ΔH_f° for C$_6$H$_6$(g), which is ΔH for the reaction:

$$6\ C(s) + 3\ H_2(g) \rightarrow C_6H_6(g) \qquad \Delta H = \Delta H_{f,\ C_6H_6(g)}^{\circ}$$

To calculate ΔH for this reaction, we will use Hess's law along with the value ΔH_f° for C(g) and the bond energy value for H$_2$ (D_{H_2} = 432 kJ/mol).

$$6\ C(g) + 6\ H(g) \rightarrow C_6H_6(g) \qquad \Delta H_1 = -5361\ kJ$$

$$6\ C(s) \rightarrow 6\ C(g) \qquad \Delta H_2 = \ 6(717\ kJ)$$

$$3\ H_2(g) \rightarrow 6\ H(g) \qquad \Delta H_3 = \ 3(432\ kJ)$$

$$6\ C(s) + 3\ H_2(g) \rightarrow C_6H_6(g) \qquad \Delta H = \Delta H_1 + \Delta H_2 + \Delta H_3 = 237\ kJ;\ \Delta H^\circ_{f,\ C_6H_6(g)} = 237\ kJ/mol$$

The experimental ΔH°_f for $C_6H_6(g)$ is more stable (lower in energy) by 154 kJ than the ΔH°_f calculated from bond energies ($83 - 237 = -154$ kJ). This extra stability is related to benzene's ability to exhibit resonance. Two equivalent Lewis structures can be drawn for benzene. The π bonding system implied by each Lewis structure consists of three localized π bonds. This is not correct because all C–C bonds in benzene are equivalent. We say the π electrons in benzene are delocalized over the entire surface of C_6H_6 (see Section 14.5 of the text). The large discrepancy between ΔH°_f values is due to the delocalized π electrons, whose effect was not accounted for in the calculated ΔH°_f value. The extra stability associated with benzene can be called resonance stabilization. In general, molecules that exhibit resonance are usually more stable than predicted using bond energies.

85.   The two isomers having the formula C_2H_6O are:

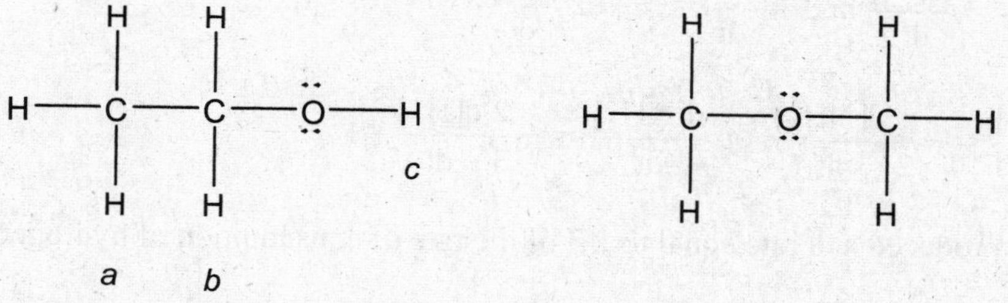

The first structure has three types of hydrogens (a, b, and c), so the signals should be seen in three different regions of the NMR spectra. The overall relative areas of the three signals should be in a 3:2:1 ratio due to the number of a, b, and c hydrogens. The signal for the a hydrogens will be split into a triplet signal due to the two neighboring b hydrogens. The signal for the b hydrogens should be split into a quintet signal due to the three neighboring a hydrogens plus the neighboring c hydrogen (four total protons). The c hydrogen signal should be split into a triplet signal due to the two neighboring b hydrogens. In practice, however, the c hydrogen bonded to the oxygen does not behave "normally". This O–H hydrogen generally behaves as if it was more than three sigma bonds apart from the b hydrogens. Therefore, the spectrum will most likely have a triplet signal for the a hydrogens, a quartet signal for the b hydrogens, and a singlet signal for the c hydrogen.

In the second structure, all six hydrogens are equivalent, so only one signal will appear in the NMR spectrum. This signal will be a singlet signal because the hydrogens in the two –CH_3 groups are more than three sigma bonds apart from each other (no splitting occurs).

CHAPTER 15

CHEMICAL KINETICS

Reaction Rates

11. 0.0120/0.0080 = 1.5; reactant B is used up 1.5 times faster than reactant A. This corresponds to a 3 to 2 mole ratio between B and A in the balanced equation. 0.0160/0.0080 = 2; product C is produced twice as fast as reactant A is used up. So the coefficient for C is twice the coefficient for A. A possible balanced equation is $2A + 3B \rightarrow 4C$.

13. Using the coefficients in the balanced equation to relate the rates:

$$\frac{d[H_2]}{dt} = 3\frac{d[N_2]}{dt} \text{ and } \frac{d[NH_3]}{dt} = -2\frac{d[N_2]}{dt}$$

So: $\dfrac{1}{3}\dfrac{d[H_2]}{dt} = -\dfrac{1}{2}\dfrac{d[NH_3]}{dt}$ or $\dfrac{d[NH_3]}{dt} = -\dfrac{2}{3}\dfrac{d[H_2]}{dt}$

Ammonia is produced at a rate equal to 2/3 of the rate of consumption of hydrogen.

15. Rate = $k[Cl]^{1/2}[CHCl_3]$, $\dfrac{mol}{L\,s} = k\left(\dfrac{mol}{L}\right)^{1/2}\left(\dfrac{mol}{L}\right)$; k must have units of $L^{1/2}\,mol^{-1/2}\,s^{-1}$.

Rate Laws from Experimental Data: Initial Rates Method

17. a. In the first two experiments, [NO] is held constant and [Cl$_2$] is doubled. The rate also doubled. Thus the reaction is first order with respect to Cl$_2$. Or mathematically: Rate = $k[NO]^x[Cl_2]^y$.

$$\frac{0.36}{0.18} = \frac{k(0.10)^x(0.20)^y}{k(0.10)^x(0.10)^y} = \frac{(0.20)^y}{(0.10)^y}, \ 2.0 = 2.0^y, \ y = 1$$

We can get the dependence on NO from the second and third experiments. Here, as the NO concentration doubles (Cl$_2$ concentration is constant), the rate increases by a factor of four. Thus, the reaction is second order with respect to NO. Or mathematically:

304

$$\frac{1.45}{0.36} = \frac{k(0.20)^x(0.20)}{k(0.10)^x(0.20)} = \frac{(0.20)^x}{(0.10)^x}, \quad 4.0 = 2.0^x, \quad x = 2; \quad \text{so,} \quad \text{Rate} = k[NO]^2[Cl_2].$$

Try to examine experiments where only one concentration changes at a time. The more variables that change, the harder it is to determine the orders. Also, these types of problems can usually be solved by inspection. In general, we will solve using a mathematical approach, but keep in mind that you probably can solve for the orders by simple inspection of the data.

b. The rate constant k can be determined from the experiments. From experiment 1:

$$\frac{0.18\ \text{mol}}{\text{L min}} = k\left(\frac{0.10\ \text{mol}}{\text{L}}\right)^2\left(\frac{0.10\ \text{mol}}{\text{L}}\right), \quad k = 180\ \text{L}^2\ \text{mol}^{-2}\ \text{min}^{-1}$$

From the other experiments:

$$k = 180\ \text{L}^2\ \text{mol}^{-2}\ \text{min}^{-1}\ \text{(second exp.);} \quad k = 180\ \text{L}^2\ \text{mol}^{-2}\ \text{min}^{-1}\ \text{(third exp.)}$$

The average rate constant is $k_{mean} = 1.8 \times 10^2\ \text{L}^2\ \text{mol}^{-2}\ \text{min}^{-1}$.

19. a. Rate = $k[NOCl]^n$; using experiments two and three:

$$\frac{2.66 \times 10^4}{6.64 \times 10^3} = \frac{k(2.0 \times 10^{16})^n}{k(1.0 \times 10^{16})^n}, \quad 4.01 = 2.0^n, \quad n = 2; \quad \text{Rate} = k[NOCl]^2$$

b. $$\frac{5.98 \times 10^4\ \text{molecules}}{\text{cm}^3\ \text{s}} = k\left(\frac{3.0 \times 10^{16}\ \text{molecules}}{\text{cm}^3}\right)^2, \quad k = 6.6 \times 10^{-29}\ \text{cm}^3\ \text{molecules}^{-1}\ \text{s}^{-1}$$

The other three experiments give (6.7, 6.6, and 6.6) $\times 10^{-29}\ \text{cm}^3\ \text{molecules}^{-1}\ \text{s}^{-1}$, respectively. The mean value for k is $6.6 \times 10^{-29}\ \text{cm}^3\ \text{molecules}^{-1}\ \text{s}^{-1}$.

c. $$\frac{6.6 \times 10^{-29}\ \text{cm}^3}{\text{molecules s}} \times \frac{1\ \text{L}}{1000\ \text{cm}^3} \times \frac{6.022 \times 10^{23}\ \text{molecules}}{\text{mol}} = \frac{4.0 \times 10^{-8}\ \text{L}}{\text{mol s}}$$

21. a. Rate = $k[ClO_2]^x[OH^-]^y$; From the first two experiments:

$$2.30 \times 10^{-1} = k(0.100)^x(0.100)^y \quad \text{and} \quad 5.75 \times 10^{-2} = k(0.0500)^x(0.100)^y$$

Dividing the two rate laws: $4.00 = \dfrac{(0.100)^x}{(0.0500)^x} = 2.00^x, \quad x = 2$

Comparing the second and third experiments:

$$2.30 \times 10^{-1} = k(0.100)(0.100)^y \quad \text{and} \quad 1.15 \times 10^{-1} = k(0.100)(0.0500)^y$$

Dividing: $2.00 = \dfrac{(0.100)^y}{(0.050)^y} = 2.0^y,\ y = 1$

The rate law is: Rate = $k[ClO_2]^2[OH^-]$

2.30×10^{-1} mol L^{-1} s^{-1} = $k(0.100\ \text{mol/L})^2(0.100\ \text{mol/L})$, $k = 2.30 \times 10^2$ L^2 mol^{-2} s^{-1}
$\hspace{11cm} = k_{mean}$

b.   Rate = $\dfrac{2.30 \times 10^2\ \text{L}^2}{\text{mol}^2\ \text{s}} \times \left(\dfrac{0.175\ \text{mol}}{\text{L}}\right)^2 \times \dfrac{0.0844\ \text{mol}}{\text{L}} = 0.594$ mol L^{-1} s^{-1}

23.   Rate = $k[H_2SeO_3]^x[H^+]^y[I^-]^z$; comparing the first and second experiments:

$\dfrac{3.33 \times 10^{-7}}{1.66 \times 10^{-7}} = \dfrac{k(2.0 \times 10^{-4})^x(2.0 \times 10^{-2})^y(2.0 \times 10^{-2})^z}{k(1.0 \times 10^{-4})^x(2.0 \times 10^{-2})^y(2.0 \times 10^{-2})^z}$, $2.01 = 2.0^x,\ x = 1$

Comparing the first and fourth experiments:

$\dfrac{6.66 \times 10^{-7}}{1.66 \times 10^{-7}} = \dfrac{k(1.0 \times 10^{-4})(4.0 \times 10^{-2})^y(2.0 \times 10^{-2})^z}{k(1.0 \times 10^{-4})(2.0 \times 10^{-2})^y(2.0 \times 10^{-2})^z}$, $4.01 = 2.0^y,\ y = 2$

Comparing the first and sixth experiments:

$\dfrac{13.2 \times 10^{-7}}{1.66 \times 10^{-7}} = \dfrac{k(1.0 \times 10^{-4})(2.0 \times 10^{-2})^2(4.0 \times 10^{-2})^z}{k(1.0 \times 10^{-4})(2.0 \times 10^{-2})^2(2.0 \times 10^{-2})^z}$

$7.95 = 2.0^z,\ \log(7.95) = z \log(2.0),\ z = \dfrac{\log(7.95)}{\log(2.0)} = 2.99 \approx 3$

Rate = $k[H_2SeO_3][H^+]^2[I^-]^3$

Experiment 1:

$\dfrac{1.66 \times 10^{-7}\ \text{mol}}{\text{L s}} = k\left(\dfrac{1.0 \times 10^{-4}\ \text{mol}}{\text{L}}\right)\left(\dfrac{2.0 \times 10^{-2}\ \text{mol}}{\text{L}}\right)^2\left(\dfrac{2.0 \times 10^{-2}\ \text{mol}}{\text{L}}\right)^3$

$k = 5.19 \times 10^5$ L^5 mol^{-5} s^{-1} = 5.2×10^5 L^5 mol^{-5} s^{-1} = k_{mean}

25.   Rate = $k[I^-]^x[OCl^-]^y[OH^-]^z$; Comparing the first and second experiments:

$\dfrac{18.7 \times 10^{-3}}{9.4 \times 10^{-3}} = \dfrac{k(0.0026)^x(0.012)^y(0.10)^z}{k(0.0013)^x(0.012)^y(0.10)^z}$, $2.0 = 2.0^x,\ x = 1$

Comparing the first and third experiments:

$$\frac{9.4 \times 10^{-3}}{4.7 \times 10^{-3}} = \frac{k(0.0013)(0.012)^y(0.10)^z}{k(0.0013)(0.0060)^y(0.10)^z}, \quad 2.0 = 2.0^y, \quad y = 1$$

Comparing the first and sixth experiments:

$$\frac{4.8 \times 10^{-3}}{9.4 \times 10^{-3}} = \frac{k(0.0013)(0.012)(0.20)^z}{k(0.0013)(0.012)(0.10)^z}, \quad 1/2 = 2.0^z, \quad z = -1$$

$\text{Rate} = \dfrac{k[I^-][OCl^-]}{[OH^-]}$; the presence of OH^- decreases the rate of the reaction.

For the first experiment:

$$\frac{9.4 \times 10^{-3} \text{ mol}}{L \text{ s}} = k \frac{(0.0013 \text{ mol}/L)(0.012 \text{ mol}/L)}{(0.10 \text{ mol}/L)}, \quad k = 60.3 \text{ s}^{-1} = 60. \text{ s}^{-1}$$

For all experiments, $k_{mean} = 60. \text{ s}^{-1}$.

Integrated Rate Laws

27.   Zero order: $t_{1/2} = \dfrac{[A]_0}{2k}$;   first order: $t_{1/2} = \dfrac{\ln 2}{k}$;   second order: $t_{1/2} = \dfrac{1}{k[A]_0}$

For a first-order reaction, if the first half-life equals 20. s, the second half-life will also be 20. s because the half-life for a first-order reaction is concentration-independent. The second half-life for a zero-order reaction will be 1/2(20.) = 10. s. This is because the half-life for a zero-order reaction has a direct relationship with concentration (as the concentration decreases by a factor of 2, the half-life decreases by a factor of 2). Because a second-order reaction has an inverse relationship between $t_{1/2}$ and $[A]_0$, the second half-life will be 40. s (twice the first half-life value).

29.   a.   Because the $\ln[A]$ versus time plot was linear, the reaction is first order in A. The slope of the $\ln[A]$ versus time plot equals $-k$. Therefore, the rate law, the integrated rate law, and the rate constant value are:

$$\text{Rate} = k[A]; \quad \ln[A] = -kt + \ln[A]_0; \quad k = 2.97 \times 10^{-2} \text{ min}^{-1}$$

b.   The half-life expression for a first order rate law is:

$$t_{1/2} = \frac{\ln 2}{k} = \frac{0.6931}{k}, \quad t_{1/2} = \frac{0.6931}{2.97 \times 10^{-2} \text{ min}^{-1}} = 23.3 \text{ min}$$

c. 2.50×10^{-3} M is 1/8 of the original amount of A present initially, so the reaction is 87.5% complete. When a first-order reaction is 87.5% complete (or 12.5% remains), then the reaction has gone through 3 half-lives:

$$100\% \;\rightarrow\; 50.0\% \;\rightarrow\; 25.0\% \;\rightarrow\; 12.5\%; \quad t = 3 \times t_{1/2} = 3 \times 23.3 \text{ min} = 69.9 \text{ min}$$
$$ t_{1/2} t_{1/2} t_{1/2}$$

Or we can use the integrated rate law:

$$\ln\!\left(\frac{[A]}{[A]_0}\right) = -kt, \quad \ln\!\left(\frac{2.50 \times 10^{-3} \ M}{2.00 \times 10^{-2} \ M}\right) = -(2.97 \times 10^{-2} \text{ min}^{-1})t$$

$$t = \frac{\ln(0.125)}{-2.97 \times 10^{-2} \text{ min}^{-1}} = 70.0 \text{ min}$$

31.   The first assumption to make is that the reaction is first order. For a first order reaction, a graph of $\ln[H_2O_2]$ versus time will yield a straight line. If this plot is not linear, then the reaction is not first order, and we make another assumption.

Time (s)	$[H_2O_2]$ (mol/L)	$\ln[H_2O_2]$
0	1.00	0.000
120.	0.91	−0.094
300.	0.78	−0.25
600.	0.59	−0.53
1200.	0.37	−0.99
1800.	0.22	−1.51
2400.	0.13	−2.04
3000.	0.082	−2.50
3600.	0.050	−3.00

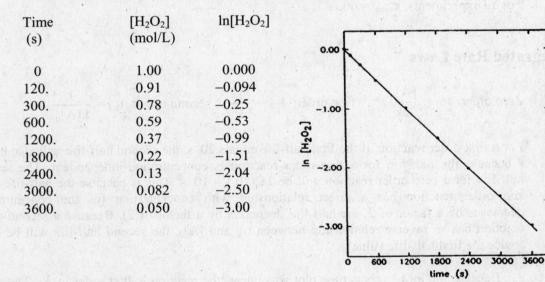

Note: We carried extra significant figures in some of the natural log values in order to reduce round-off error. For the plots, we will do this most of the time when the natural log function is involved.

The plot of $\ln[H_2O_2]$ versus time is linear. Thus the reaction is first order. The differential rate law and integrated rate law are Rate $= \dfrac{-d[H_2O_2]}{dt} = k[H_2O_2]$ and $\ln[H_2O_2] = -kt + \ln[H_2O_2]_0$.

We determine the rate constant k by determining the slope of the $\ln[H_2O_2]$ versus time plot (slope $= -k$). Using two points on the curve gives:

$$\text{slope} = -k = \frac{\Delta y}{\Delta x} = \frac{0 - (3.00)}{0 - 3600.} = -8.3 \times 10^{-4} \text{ s}^{-1}, \ k = 8.3 \times 10^{-4} \text{ s}^{-1}$$

To determine $[H_2O_2]$ at 4000. s, use the integrated rate law, where $[H_2O_2]_0 = 1.00 \ M$.

$$\ln[H_2O_2] = -kt + \ln[H_2O_2]_0 \quad \text{or} \quad \ln\left(\frac{[H_2O_2]}{[H_2O_2]_0}\right) = -kt$$

$$\ln\left(\frac{[H_2O_2]}{1.00}\right) = -8.3 \times 10^{-4} \text{ s}^{-1} \times 4000. \text{ s}, \quad \ln[H_2O_2] = -3.3, \ [H_2O_2] = e^{-3.3} = 0.037 \ M$$

33.   Assume the reaction is first order and see if the plot of $\ln[NO_2]$ versus time is linear. If this isn't linear, try the second order plot of $1/[NO_2]$ versus time. The data and plots follow.

Time (s)	$[NO_2]$ (M)	$\ln[NO_2]$	$1/[NO_2]$ (M^{-1})
0	0.500	−0.693	2.00
1.20×10^3	0.444	−0.812	2.25
3.00×10^3	0.381	−0.965	2.62
4.50×10^3	0.340	−1.079	2.94
9.00×10^3	0.250	−1.386	4.00
1.80×10^4	0.174	−1.749	5.75

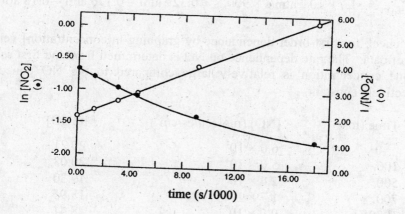

The plot of $1/[NO_2]$ versus time is linear. The reaction is second order in NO_2. The differential rate law and integrated rate law are $\text{Rate} = k[NO_2]^2$ and $\dfrac{1}{[NO_2]} = kt + \dfrac{1}{[NO_2]_0}$.

The slope of the plot $1/[NO_2]$ versus time gives the value of k. Using a couple of points on the plot:

$$\text{slope} = k = \frac{\Delta y}{\Delta x} = \frac{(5.75 - 2.00) \ M^{-1}}{(1.80 \times 10^4 - 0) \text{ s}} = 2.08 \times 10^{-4} \text{ L mol}^{-1} \text{ s}^{-1}$$

To determine $[NO_2]$ at 2.70×10^4 s, use the integrated rate law, where $1/[NO_2]_0 = 1/0.500\ M = 2.00\ M^{-1}$.

$$\frac{1}{[NO_2]} = kt + \frac{1}{[NO_2]_0}, \quad \frac{1}{[NO_2]} = \frac{2.08 \times 10^{-4}\ L}{mol\ s} \times 2.70 \times 10^4\ s + 2.00\ M^{-1}$$

$$\frac{1}{[NO_2]} = 7.62, \quad [NO_2] = 0.131\ M$$

35.   From the data, the pressure of C_2H_5OH decreases at a constant rate of 13 torr for every 100. s. Since the rate of disappearance of C_2H_5OH is not dependent on concentration, the reaction is zero order in C_2H_5OH.

$$k = \frac{13\ torr}{100.\ s} \times \frac{1\ atm}{760\ torr} = 1.7 \times 10^{-4}\ atm/s$$

The rate law and integrated rate law are:

$$Rate = k = 1.7 \times 10^{-4}\ atm/s; \quad P_{C_2H_5OH} = -kt + 250.\ torr\left(\frac{1\ atm}{760\ torr}\right) = -kt + 0.329\ atm$$

At 900. s:

$$P_{C_2H_5OH} = -1.7 \times 10^{-4}\ atm/s \times 900.\ s + 0.329\ atm = 0.176\ atm = 0.18\ atm = 130\ torr$$

37.   a.   We check for first-order dependence by graphing ln[concentration] versus time for each set of data. The rate dependence on NO is determined from the first set of data since the ozone concentration is relatively large compared to the NO concentration, so it is effectively constant.

Time (ms)	[NO] (molecules/cm^3)	ln[NO]
0	6.0×10^8	20.21
100.	5.0×10^8	20.03
500.	2.4×10^8	19.30
700.	1.7×10^8	18.95
1000.	9.9×10^7	18.41

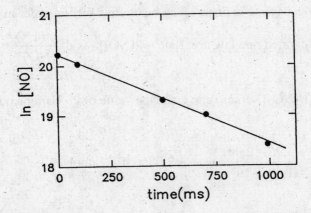

Because ln[NO] versus t is linear, the reaction is first order with respect to NO.

We follow the same procedure for ozone using the second set of data. The data and plot are:

Time (ms)	$[O_3]$ (molecules/cm^3)	ln$[O_3]$
0	1.0×10^{10}	23.03
50.	8.4×10^{9}	22.85
100.	7.0×10^{9}	22.67
200.	4.9×10^{9}	22.31
300.	3.4×10^{9}	21.95

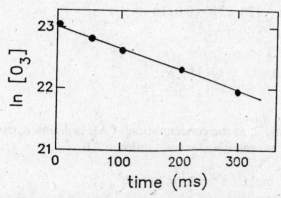

The plot of ln$[O_3]$ versus t is linear. Hence the reaction is first order with respect to ozone.

b. Rate = k[NO]$[O_3]$ is the overall rate law.

c. For NO experiment, Rate = k'[NO] and k' = –(slope from graph of ln[NO] versus t).

$$k' = -\text{slope} = -\frac{18.41 - 20.21}{(1000. - 0) \times 10^{-3} \text{ s}} = 1.8 \text{ s}^{-1}$$

For ozone experiment, Rate = k''$[O_3]$ and k'' = –(slope from ln$[O_3]$ versus t).

$$k'' = -\text{slope} = -\frac{(21.95 - 23.03)}{(300. - 0) \times 10^{-3} \text{ s}} = 3.6 \text{ s}^{-1}$$

d. From NO experiment, Rate = k[NO]$[O_3]$ = k'[NO] where k' = k$[O_3]$.

$k' = 1.8 \text{ s}^{-1}$ = k(1.0×10^{14} molecules/cm^3), k = 1.8×10^{-14} cm^3 molecules^{-1} s^{-1}

We can check this from the ozone data. Rate = k''$[O_3]$ = k[NO]$[O_3]$, where k'' = k[NO].

$k'' = 3.6 \text{ s}^{-1}$ = k(2.0×10^{14} molecules/cm^3), k = 1.8×10^{-14} cm^3 molecules^{-1} s^{-1}

Both values of k agree.

39.   Because $[V]_0 >> [AV]_0$, the concentration of V is essentially constant in this experiment. We have a pseudo-first-order reaction in AV:

$$Rate = k[AV][V] = k'[AV], \text{ where } k' = k[V]_0$$

The slope of the ln[AV] versus time plot is equal to $-k'$.

$$k' = -slope = 0.32 \text{ s}^{-1}; \quad k = \frac{k'}{[V]_0} = \frac{0.32 \text{ s}^{-1}}{0.20 \text{ mol/L}} = 1.6 \text{ L mol}^{-1} \text{ s}^{-1}$$

41.   For a first-order reaction, the integrated rate law is $\ln([A]/[A]_0) = -kt$. Solving for k:

$$\ln\left(\frac{0.250 \text{ mol/L}}{1.00 \text{ mol/L}}\right) = -k \times 120. \text{ s}, \quad k = 0.0116 \text{ s}^{-1}$$

$$\ln\left(\frac{0.350 \text{ mol/L}}{2.00 \text{ mol/L}}\right) = -0.0116 \text{ s}^{-1} \times t, \quad t = 150. \text{ s}$$

43.   Comparing experiments 1 and 2, as the concentration of AB is doubled, the initial rate increases by a factor of 4. The reaction is second order in AB.

$$Rate = k[AB]^2, \quad 3.20 \times 10^{-3} \text{ mol L}^{-1} \text{ s}^{-1} = k_1(0.200 \, M)^2$$

$$k = 8.00 \times 10^{-2} \text{ L mol}^{-1} \text{ s}^{-1} = k_{mean}$$

For a second-order reaction:

$$t_{1/2} = \frac{1}{k[AB]_0} = \frac{1}{8.00 \times 10^{-2} \text{ L mol}^{-1} \text{ s}^{-1} \times 1.00 \text{ mol/L}} = 12.5 \text{ s}$$

45.   a.   When a reaction is 75.0% complete (25.0% of reactant remains), this represents two half-lives (100% → 50% → 25%). The first-order half-life expression is $t_{1/2} = (\ln 2)/k$. Because there is no concentration dependence for a first-order half-life, 320. s = two half-lives, $t_{1/2} = 320./2 = 160.$ s. This is both the first half-life, the second half-life, etc.

b. $t_{1/2} = \dfrac{\ln 2}{k}, \quad k = \dfrac{\ln 2}{t_{1/2}} = \dfrac{\ln 2}{160. \text{ s}} = 4.33 \times 10^{-3} \text{ s}^{-1}$

At 90.0% complete, 10.0% of the original amount of the reactant remains, so $[A] = 0.100[A]_0$.

$$\ln\left(\frac{[A]}{[A]_0}\right) = -kt, \quad \ln\frac{0.100[A]_0}{[A]_0} = -(4.33 \times 10^{-3} \text{ s}^{-1})t, \quad t = \frac{\ln(0.100)}{-4.33 \times 10^{-3} \text{ s}^{-1}} = 532 \text{ s}$$

47.   a.   The integrated rate law for this zero-order reaction is $[HI] = -kt + [HI]_0$.

$$[HI] = -kt + [HI]_0, \quad [HI] = -\left(\frac{1.20 \times 10^{-4} \text{ mol}}{\text{L s}}\right) \times \left(25 \text{ min} \times \frac{60 \text{ s}}{\text{min}}\right) + \frac{0.250 \text{ mol}}{\text{L}}$$

$$[HI] = -0.18 \text{ mol/L} + 0.250 \text{ mol/L} = 0.07 \text{ } M$$

b. $[HI] = 0 = -kt + [HI]_0, \quad kt = [HI]_0, \quad t = \dfrac{[HI]_0}{k}$

$$t = \frac{0.250 \text{ mol/L}}{1.20 \times 10^{-4} \text{ mol L}^{-1} \text{ s}^{-1}} = 2080 \text{ s} = 34.7 \text{ min}$$

49.   The consecutive half-life values of 24 hours, then 12 hours, show a direct relationship with concentration; as the concentration decreases, the half-life decreases. Assuming the drug reaction is either zero, first, or second order, only a zero order reaction shows this direct relationship between half-life and concentration. Therefore, assume the reaction is zero order in the drug.

$$t_{1/2} = \frac{[A]_0}{2k}, \quad k = \frac{[A]_0}{2t_{1/2}} = \frac{2.0 \times 10^{-3} \text{ mol/L}}{2(24 \text{ h})} = 4.2 \times 10^{-5} \text{ mol L}^{-1} \text{ h}^{-1}$$

51.   a.   Because $[A]_0 << [B]_0$ or $[C]_0$, the B and C concentrations remain constant at $1.00 \text{ } M$ for this experiment. Thus, rate $= k[A]^2[B][C] = k'[A]^2$ where $k' = k[B][C]$.

For this pseudo-second-order reaction:

$$\frac{1}{[A]} = k't + \frac{1}{[A]_0}, \quad \frac{1}{3.26 \times 10^{-5} \text{ } M} = k'(3.00 \text{ min}) + \frac{1}{1.00 \times 10^{-4} \text{ } M}$$

$$k' = 6890 \text{ L mol}^{-1} \text{ min}^{-1} = 115 \text{ L mol}^{-1} \text{ s}^{-1}$$

$$k' = k[B][C], \quad k = \frac{k'}{[B][C]}, \quad k = \frac{115 \text{ L mol}^{-1} \text{ s}^{-1}}{(1.00 \text{ } M)(1.00 \text{ } M)} = 115 \text{ L}^3 \text{ mol}^{-3} \text{ s}^{-1}$$

b.   For this pseudo-second-order reaction:

$$\text{Rate} = k'[A]^2, \quad t_{1/2} = \frac{1}{k'[A]_0} = \frac{1}{115 \text{ L mol}^{-1} \text{ s}^{-1}(1.00 \times 10^{-4} \text{ mol/L})} = 87.0 \text{ s}$$

c. $\dfrac{1}{[A]} = k't + \dfrac{1}{[A]_0} = 115 \text{ L mol}^{-1} \text{ s}^{-1} \times 600. \text{ s} + \dfrac{1}{1.00 \times 10^{-4} \text{ mol/L}} = 7.90 \times 10^4 \text{ L/mol}$

$$[A] = 1/7.90 \times 10^4 \text{ L/mol} = 1.27 \times 10^{-5} \text{ mol/L}$$

From the stoichiometry in the balanced reaction, 1 mol of B reacts with every 3 mol of A.

Amount A reacted = $1.00 \times 10^{-4} \, M - 1.27 \times 10^{-5} \, M = 8.7 \times 10^{-5} \, M$

Amount B reacted = $8.7 \times 10^{-5} \, \text{mol/L} \times \dfrac{1 \, \text{mol B}}{3 \, \text{mol A}} = 2.9 \times 10^{-5} \, M$

[B] = $1.00 \, M - 2.9 \times 10^{-5} \, M = 1.00 \, M$

As we mentioned in part a, the concentration of B (and C) remain constant because the A concentration is so small compared to the B (or C) concentration.

Reaction Mechanisms

53.   In a unimolecular reaction, a single reactant molecule decomposes to products. In a bimolecular reaction, two molecules collide to give products. The probability of the simultaneous collision of three molecules with enough energy and the proper orientation is very small, making termolecular steps very unlikely.

55.   For elementary reactions, the rate law can be written using the coefficients in the balanced equation to determine the orders.

    a. Rate = $k[CH_3NC]$ b. Rate = $k[O_3][NO]$

    c. Rate = $k[O_3]$ d. Rate = $k[O_3][O]$

    e. Rate = $k\left[{}^{14}_{6}C\right]$ or Rate = kN, where N = the number of ${}^{14}_{6}C$ atoms (convention)

57.   A mechanism consists of a series of elementary reactions in which the rate law for each step can be determined using the coefficients in the balanced equations. For a plausible mechanism, the rate law derived from a mechanism must agree with the rate law determined from experiment. To derive the rate law from the mechanism, the rate of the reaction is assumed to equal the rate of the slowest step in the mechanism.

Because step 1 is the rate determining step, the rate law for this mechanism is Rate = $k[C_4H_9Br]$. To get the overall reaction, we sum all the individual steps of the mechanism. Summing all steps gives:

$$C_4H_9Br \rightarrow C_4H_9{}^{+} + Br^{-}$$
$$C_4H_9{}^{+} + H_2O \rightarrow C_4H_9OH_2{}^{+}$$
$$C_4H_9OH_2{}^{+} + H_2O \rightarrow C_4H_9OH + H_3O^{+}$$

$$C_4H_9Br + 2 \, H_2O \rightarrow C_4H_9OH + Br^{-} + H_3O^{+}$$

Intermediates in a mechanism are species that are neither reactants nor products but that are formed and consumed during the reaction sequence in the mechanism. The intermediates for this mechanism are $C_4H_9{}^{+}$ and $C_4H_9OH_2{}^{+}$.

59.   Let's determine the rate law for each mechanism. If the rate law derived from the mechanism is the same as the experimental rate law, then the mechanism is possible (assuming the sum of all the steps in the mechanism gives the overall balanced equation). When deriving rate laws from a mechanism, we must substitute for all intermediate concentrations.

a.   Rate = $k_1[NO][O_2]$;   <u>not possible</u>

b.   Rate = $k_2[NO_3][NO]$ and $k_1[NO][O_2] = k_{-1}[NO_3]$ or $[NO_3] = \dfrac{k_1}{k_{-1}}[NO][O_2]$

Rate = $\dfrac{k_2k_1}{k_{-1}}[NO]^2[O_2]$;   <u>possible</u>

c.   Rate = $k_1[NO]^2$;   <u>not possible</u>   d.   Rate = $k_2[N_2O_2]$ and $[N_2O_2] = \dfrac{k_1}{k_{-1}}[NO]^2$

Rate = $\dfrac{k_2k_1}{k_{-1}}[NO]^2$;   <u>not possible</u>

Only the mechanism in b is consistent, so only mechanism b is a possible mechanism for this reaction.

61.   Rate = $k_3[Br^-][H_2BrO_3^+]$;   we must substitute for the intermediate concentration. Because steps 1 and 2 are fast-equilibrium steps, rate forward reaction = rate reverse reaction.

$k_2[HBrO_3][H^+] = k_{-2}[H_2BrO_3^+]$; $k_1[BrO_3^-][H^+] = k_{-1}[HBrO_3]$

$[HBrO_3] = \dfrac{k_1}{k_{-1}}[BrO_3^-][H^+]$; $[H_2BrO_3^+] = \dfrac{k_2}{k_{-2}}[HBrO_3][H^+] = \dfrac{k_2k_1}{k_{-2}k_{-1}}[BrO_3^-][H^+]^2$

Rate = $\dfrac{k_3k_2k_1}{k_{-2}k_{-1}}[Br^-][BrO_3^-][H^+]^2 = k[Br^-][BrO_3^-][H^+]^2$

63.   Rate = $k_2[I^-][HOCl]$;   from the fast-equilibrium first step:

$k_1[OCl^-] = k_{-1}[HOCl][OH^-]$, $[HOCl] = \dfrac{k_1[OCl^-]}{k_{-1}[OH^-]}$;   substituting into the rate equation:

Rate = $\dfrac{k_2k_1[I^-][OCl^-]}{k_{-1}[OH^-]} = \dfrac{k[I^-][OCl^-]}{[OH^-]}$

65.   a. $MoCl_5^-$

b.   Rate = $\dfrac{d[NO_2^-]}{dt} = k_2[NO_3^-][MoCl_5^-]$ (Only the last step contains NO_2^-.)

We use the steady-state assumption to substitute for the intermediate concentration in the rate law. The steady-state approximation assumes that the concentration of an intermediate remains constant; i.e., d[intermediate]/dt = 0. To apply the steady-state assumption, we write rate laws for all steps where the intermediate is produced and equate the sum of these rate laws to the sum of the rate laws where the intermediate is consumed. Applying the steady-state approximation to $MoCl_5^-$:

$$\frac{d[MoCl_5^-]}{dt} = 0, \text{ so } k_1[MoCl_6^{2-}] = k_{-1}[MoCl_5^-][Cl^-] + k_2[NO_3^-][MoCl_5^-]$$

$$[MoCl_5^-] = \frac{k_1[MoCl_6^{2-}]}{k_{-1}[Cl^-] + k_2[NO_3^-]}; \text{ Rate} = \frac{d[NO_2^-]}{dt} = \frac{k_1k_2[NO_3^-][MoCl_6^{2-}]}{k_{-1}[Cl^-] + k_2[NO_3^-]}$$

67.   a.   Rate $= \dfrac{d[E]}{dt} = k_2[B^*]$; assume $\dfrac{d[B^*]}{dt} = 0$, then $k_1[B]^2 = k_{-1}[B][B^*] + k_2[B^*]$.

$$[B^*] = \frac{k_1[B]^2}{k_{-1}[B] + k_2}; \text{ the rate law is: Rate} = \frac{d[E]}{dt} = \frac{k_1k_2[B]^2}{k_{-1}[B] + k_2}$$

   b.   When $k_2 \ll k_{-1}[B]$, then Rate $= \dfrac{d[E]}{dt} = \dfrac{k_1k_2[B]^2}{k_{-1}[B]} = \dfrac{k_1k_2}{k_{-1}}[B]$.

The reaction is first order when the rate of the second step is very slow (when k_2 is very small).

   c.   Collisions between B molecules only transfer energy from one B to another. This occurs at a much faster rate than the decomposition of an energetic B molecule (B^*).

Temperature Dependence of Rate Constants and the Collision Model

69.   Two reasons are:

   1) The collision must involve enough energy to produce the reaction; i.e., the collision energy must be equal to or exceed the activation energy.

   2) The relative orientation of the reactants when they collide must allow formation of any new bonds necessary to produce products.

71.   a. $T_2 > T_1$; as temperature increases, the distribution of collision energies shifts to the right. That is, as temperature increases, there are fewer collision energies with small energies and more collisions with large energies.

   b.   As temperature increases, more of the collisions have the required activation energy necessary to convert reactants into products. Hence, the rate of the reaction increases with increasing temperature.

73. $k = A \exp(-E_a/RT)$ or $\ln k = \dfrac{-E_a}{RT} + \ln A$ (the Arrhenius equation)

For two conditions: $\ln\left(\dfrac{k_2}{k_1}\right) = \dfrac{E_a}{R}\left(\dfrac{1}{T_1} - \dfrac{1}{T_2}\right)$ (Assuming A is temperature independent.)

Let $k_1 = 3.52 \times 10^{-7}$ L mol^{-1}s^{-1}, $T_1 = 555$ K; $k_2 = ?$, $T_2 = 645$ K; $E_a = 186 \times 10^3$ J/mol

$$\ln\left(\frac{k_2}{3.52 \times 10^{-7}}\right) = \frac{1.86 \times 10^5 \text{ J/mol}}{8.3145 \text{ J K}^{-1}\text{ mol}^{-1}}\left(\frac{1}{555 \text{ K}} - \frac{1}{645 \text{ K}}\right) = 5.6$$

$$\frac{k_2}{3.52 \times 10^{-7}} = e^{5.6} = 270, \quad k_2 = 270(3.52 \times 10^{-7}) = 9.5 \times 10^{-5} \text{ L mol}^{-1}\text{ s}^{-1}$$

75. $\ln\left(\dfrac{k_2}{k_1}\right) = \dfrac{E_a}{R}\left(\dfrac{1}{T_1} - \dfrac{1}{T_2}\right); \quad \dfrac{k_2}{k_1} = 7.00, T_1 = 295$ K, $E_a = 54.0 \times 10^3$ J/mol

$$\ln(7.00) = \frac{54.0 \times 10^3 \text{ J/mol}}{8.3145 \text{ J K}^{-1}\text{ mol}^{-1}}\left(\frac{1}{295 \text{ K}} - \frac{1}{T_2}\right), \quad \frac{1}{295 \text{ K}} - \frac{1}{T_2} = 3.00 \times 10^{-4}$$

$$\frac{1}{T_2} = 3.09 \times 10^{-3}, \quad T_2 = 324 \text{ K} = 51°\text{C}$$

77.   From the Arrhenius equation in logarithmic form ($\ln k = -E_a/RT + \ln A$), a graph of $\ln k$ versus. $1/T$ should yield a straight line with a slope equal to $-E_a/R$ and a y intercept equal to $\ln A$.

   a.   Slope $= -E_a/R$, $E_a = 1.10 \times 10^4$ K $\times \dfrac{8.3145 \text{ J}}{\text{K mol}} = 9.15 \times 10^4$ J/mol $= 91.5$ kJ/mol

   b. The units for A are the same as the units for k (s^{-1}).

       y intercept $= \ln A$, $A = e^{33.5} = 3.54 \times 10^{14}$ s^{-1}

   c. $\ln k = -E_a/RT + \ln A$ or $k = A \exp(-E_a/RT)$

       $$k = 3.54 \times 10^{14} \text{ s}^{-1} \times \exp\left(\frac{-9.15 \times 10^{-4} \text{ J/mol}}{8.3145 \text{ J K}^{-1}\text{ mol}^{-1} \times 298 \text{ K}}\right) = 3.24 \times 10^{-2} \text{ s}^{-1}$$

79.   The Arrhenius equation is $k = A\exp(-E_a/RT)$ or, in logarithmic form, $\ln k = -E_a/RT + \ln A$. Hence a graph of $\ln k$ versus $1/T$ should yield a straight line with a slope equal to $-E_a/R$ since the logarithmic form of the Arrhenius equation is in the form of a straight-line equation, $y = mx + b$. *Note*: We carried one extra significant figure in the following $\ln k$ values in order to reduce round-off error.

T (K)	1/T (K^{-1})	k (L mol^{-1} s^{-1})	ln k
195	5.13×10^{-3}	1.08×10^9	20.80
230.	4.35×10^{-3}	2.95×10^9	21.81
260.	3.85×10^{-3}	5.42×10^9	22.41
298	3.36×10^{-3}	12.0×10^9	23.21
369	2.71×10^{-3}	35.5×10^9	24.29

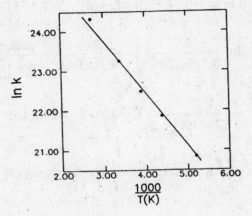

Using a couple of points on the plot:

$$\text{slope} = \frac{20.95 - 23.65}{5.00 \times 10^{-3} - 3.00 \times 10^{-3}} = \frac{-2.70}{2.00 \times 10^{-3}} = -1.35 \times 10^3 \text{ K} = \frac{-E_a}{R}$$

$E_a = 1.35 \times 10^3 \text{ K} \times 8.3145 \text{ J K}^{-1} \text{ mol}^{-1} = 1.12 \times 10^4 \text{ J/mol} = 11.2 \text{ kJ/mol}$

From the best straight line (by calculator): slope $= -1.43 \times 10^3$ K and $E_a = 11.9$ kJ/mol

81.   In the following reaction profiles R = reactants, P = products, E_a = activation energy, ΔE = overall energy change for the reaction, and RC = reaction coordinate, which is the same as reaction progress.

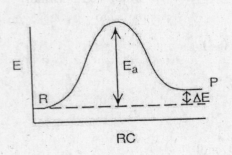
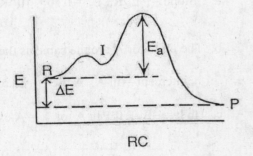

The second reaction profile represents a two-step reaction since an intermediate plateau appears between the reactants and the products. This plateau (see I in plot) represents the energy of the intermediate. The general reaction mechanism for this reaction is:

$$R \rightarrow I$$
$$\underline{I \rightarrow P}$$
$$R \rightarrow P$$

In a mechanism, the rate of the slowest step determines the rate of the reaction.   The activation energy for the slowest step will be the largest energy barrier that the reaction must overcome.   Since the second hump in the diagram is at the highest energy, the second step has the largest activation energy and will be the rate determining step (the slow step).

83.

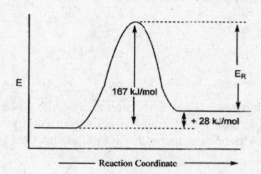

The activation energy for the reverse reaction is E_R in the diagram.

$E_R = 167 - 28 = 139$ kJ/mol

Catalysis

85.   a.   The blue plot is the catalyzed pathway. The catalyzed pathway has the lower activation. This is why the catalyzed pathway is faster.

   b. ΔE_1 represents the activation energy for the uncatalyzed pathway.

   c. ΔE_2 represents the energy difference between the reactants and products. Note that ΔE_2 is the same for both the catalyzed and the uncatalyzed pathways. It is the activation energy that is different for a catalyzed pathway versus an uncatalyzed pathway.

   d.   Because the products have a higher total energy as compared to reactants, this is an endothermic reaction.

87.   a.   W because it has a lower activation energy than the Os catalyst.

   b. $k_w = A_w \exp[-E_a(W)/RT]$; $k_{uncat} = A_{uncat} \exp[-E_a(uncat)/RT]$; assume $A_w = A_{uncat}$.

$$\frac{k_w}{k_{uncat}} = \exp\left(\frac{-E_a(W)}{RT} + \frac{E_a(uncat)}{RT}\right)$$

$$\frac{k_w}{k_{uncat}} = \exp\left(\frac{-163,000 \text{ J}/\text{mol} + 335,000 \text{ J}/\text{mol}}{8.3145 \text{ J K}^{-1} \text{ mol}^{-1} \times 298 \text{ K}}\right) = 1.41 \times 10^{30}$$

The W-catalyzed reaction is approximately 10^{30} times faster than the uncatalyzed reaction.

c. Because $[H_2]$ is in the denominator of the rate law, the presence of H_2 decreases the rate of the reaction. For the decomposition to occur, NH_3 molecules must be adsorbed on the surface of the catalyst. If H_2 is also adsorbed on the catalyst surface, then there are fewer sites for NH_3 molecules to be adsorbed, and the rate decreases.

89.   The mechanism for the chlorine catalyzed destruction of ozone is:

$$O_3 + Cl \rightarrow O_2 + ClO \quad \text{(slow)}$$
$$ClO + O \rightarrow O_2 + Cl \quad \text{(fast)}$$
$$\overline{}$$
$$O_3 + O \rightarrow 2\,O_2$$

Because the chlorine atom-catalyzed reaction has a lower activation energy, the Cl-catalyzed rate is faster. Hence Cl is a more effective catalyst. Using the activation energy, we can estimate the efficiency that Cl atoms destroy ozone compared to NO molecules (see Exercise 15.88c).

At 25°C: $\dfrac{k_{Cl}}{k_{NO}} = \exp\left[\dfrac{-E_a(Cl)}{RT} + \dfrac{E_a(NO)}{RT}\right] = \exp\left[\dfrac{(-2100 + 11{,}900)\,J/mol}{(8.3145 \times 298)\,J/mol}\right] = e^{3.96} = 52$

At 25°C, the Cl catalyzed reaction is roughly 52 times faster than the NO-catalyzed reaction, assuming the frequency factor A is the same for each reaction.

91.   At high [S], the enzyme is completely saturated with substrate. Once the enzyme is completely saturated, the rate of decomposition of ES can no longer increase, and the overall rate remains constant.

93. $\text{Rate} = \dfrac{-d[A]}{dt} = k[A]^x$

Assuming the catalyzed and uncatalyzed reaction have the same form and orders and because concentrations are assumed equal, rate $\propto 1/\Delta t$, where $\Delta t = \Delta\text{time}$.

$$\dfrac{\text{Rate}_{cat}}{\text{Rate}_{un}} = \dfrac{\Delta t_{un}}{\Delta t_{cat}} = \dfrac{2400\,\text{yr}}{\Delta t_{cat}} \quad \text{and} \quad \dfrac{\text{rate}_{cat}}{\text{rate}_{un}} = \dfrac{k_{cat}}{k_{un}}$$

$$\dfrac{\text{Rate}_{cat}}{\text{Rate}_{un}} = \dfrac{k_{cat}}{k_{un}} = \dfrac{A\exp[-E_a(cat)/RT]}{A\exp[-E_a(un)/RT]} = \exp\left[\dfrac{-E_a(cat) + E_a(un)}{RT}\right]$$

$$\dfrac{k_{cat}}{k_{un}} = \exp\left(\dfrac{-5.90 \times 10^4\,J/mol + 1.84 \times 10^5\,J/mol}{8.3145\,J\,K^{-1}\,mol^{-1} \times 600.\,K}\right) = 7.62 \times 10^{10}$$

$$\dfrac{\Delta t_{un}}{\Delta t_{cat}} = \dfrac{\text{rate}_{cat}}{\text{rate}_{un}} = \dfrac{k_{cat}}{k_{un}}, \quad \dfrac{2400\,\text{yr}}{\Delta t_{cat}} = 7.62 \times 10^{10}, \quad \Delta t_{cat} = 3.15 \times 10^{-8}\,\text{yr} \approx 1\,\text{s}$$

Additional Exercises

95. The most common method to experimentally determine the differential rate law is the method of initial rates. Once the differential rate law is determined experimentally, the integrated rate law can be derived. However, sometimes it is more convenient and more accurate to collect concentration versus time data for a reactant. When this is the case, then we do "proof" plots to determine the integrated rate law. Once the integrated rate law is determined, the differential rate law can be determined. Either experimental procedure allows determination of both the integrated and the differential rate law; and which rate law is determined by experiment and which is derived is usually decided by which data are easiest and most accurately collected.

97. Rate $= k[DNA]^x[CH_3I]^y$; comparing the second and third experiments:

$$\frac{1.28 \times 10^{-3}}{6.40 \times 10^{-4}} = \frac{k(0.200)^x(0.200)^y}{k(0.100)^x(0.200)^y}, \quad 2.00 = 2.00^x, \quad x = 1$$

Comparing the first and second experiments:

$$\frac{6.40 \times 10^{-4}}{3.20 \times 10^{-4}} = \frac{k(0.100)(0.200)^y}{k(0.100)(0.100)^y}, \quad 2.00 = 2.00^y, \quad y = 1$$

The rate law is Rate $= k[DNA][CH_3I]$.

Mechanism I is possible because the derived rate law from the mechanism (Rate $= k[DNA][CH_3I]$) agrees with the experimentally determined rate law. The derived rate law for Mechanism II will equal the rate of the slowest step. This is step 1 in the mechanism giving a derived rate law that is Rate $= k[CH_3I]$. Because this rate law does not agree with experiment, Mechanism II would not be a possible mechanism for the reaction.

99.

Heating Time	Untreated		Deacidifying		Antioxidant	
(days)	s	ln s	s	ln s	s	ln s
0.00	100.0	4.605	100.1	4.606	114.6	4.741
1.00	67.9	4.218	60.8	4.108	65.2	4.177
2.00	38.9	3.661	26.8	3.288	28.1	3.336
3.00	16.1	2.779	–	–	11.3	2.425
6.00	6.8	1.92	–	–	–	–

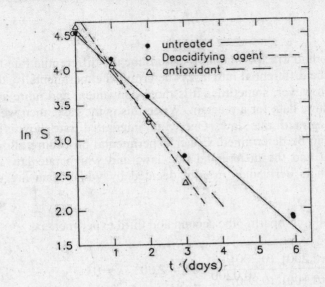

a. We used a calculator to fit the data by least squares. The results follow.

Untreated: ln s = –(0.465)t + 4.55, k = 0.465 d^{-1}

Deacidifying agent: ln s = –(0.659)t + 4.66, k = 0.659 d^{-1}

Antioxidant: ln s = –(0.779)t + 4.84, k = 0.779 d^{-1}

b. No, the silk degrades more rapidly with the additives since k increases.

c. $t_{1/2}$ = (ln 2)/k; untreated: $t_{1/2}$ = 1.49 day; deacidifying agent: $t_{1/2}$ = 1.05 day;

antioxidant: $t_{1/2}$ = 0.890 day.

101. Carbon cannot form the fifth bond necessary for the transition state because of the small atomic size of carbon and because carbon doesn't have low-energy d orbitals available to expand the octet.

103. a.

t (s)	$[C_4H_6]$ (M)	$\ln[C_4H_6]$	$1/[C_4H_6]$ (M^{-1})
0	0.01000	–4.6052	1.000×10^2
1000.	0.00629	–5.069	1.59×10^2
2000.	0.00459	–5.384	2.18×10^2
3000.	0.00361	–5.624	2.77×10^2

The plot of $1/[C_4H_6]$ versus t is linear, thus the reaction is second order in butadiene. From the plot (not included), the integrated rate law is:

$$\frac{1}{[C_4H_6]} = (5.90 \times 10^{-2} \text{ L mol}^{-1} \text{ s}^{-1})t + 100.0 \ M^{-1}$$

b. When dimerization is 1.0% complete, 99.0% of C_4H_6 is left.

$[C_4H_6] = 0.990(0.01000) = 0.00990\ M$; $\dfrac{1}{0.00990} = (5.90 \times 10^{-2})t + 100.0$

$t = 17.1\ s \approx 20\ s$

c. 2.0% complete, $[C_4H_6] = 0.00980\ M$; $\dfrac{1}{0.00980} = (5.90 \times 10^{-2})t + 100.0$,

$t = 34.6\ s \approx 30\ s$

d. $\dfrac{1}{[C_4H_6]} = kt + \dfrac{1}{[C_4H_6]_0}$; $[C_4H_6]_0 = 0.0200\ M$; at $t = t_{1/2}$, $[C_4H_6] = 0.0100\ M$.

$\dfrac{1}{0.0100} = (5.90 \times 10^{-2})t_{1/2} + \dfrac{1}{0.0200}$, $t_{1/2} = 847\ s = 850\ s$

Or: $t_{1/2} = \dfrac{1}{k[A]_0} = \dfrac{1}{(5.90 \times 10^{-2}\ L\ mol^{-1}\ s^{-1})(2.00 \times 10^{-2}\ M)} = 847\ s$

e. From Exercise 15.32, $k = 1.4 \times 10^{-2}\ L\ mol^{-1}\ s^{-1}$ at 500. K. From this problem, $k = 5.90 \times 10^{-2}\ L\ mol^{-1}\ s^{-1}$ at 620. K.

$\ln\left(\dfrac{k_2}{k_1}\right) = \dfrac{E_a}{R}\left(\dfrac{1}{T_1} - \dfrac{1}{T_2}\right)$, $\ln\left(\dfrac{5.90 \times 10^{-2}}{1.4 \times 10^{-2}}\right) = \dfrac{E_a}{8.3145\ J\ K^{-1}\ mol^{-1}}\left(\dfrac{1}{500.\ K} - \dfrac{1}{620.\ K}\right)$

$12 = E_a(3.9 \times 10^{-4})$, $E_a = 3.1 \times 10^4\ J/mol = 31\ kJ/mol$

105. $k = A\ exp(-E_a/RT)$; $\dfrac{k_{cat}}{k_{uncat}} = \dfrac{A_{cat}\ exp(-E_{a,\ cat}/RT)}{A_{uncat}\ exp(-E_{a,\ uncat}/RT)} = exp\left(\dfrac{-E_{a,\ cat} + E_{a,\ uncat}}{RT}\right)$

$2.50 \times 10^3 = \dfrac{k_{cat}}{k_{uncat}} = exp\left(\dfrac{-E_{a,\ cat} + 5.00 \times 10^4\ J/mol}{8.3145\ J\ K^{-1}\ mol^{-1} \times 310.\ K}\right)$

$\ln(2.50 \times 10^3) \times 2.58 \times 10^3\ J/mol = -E_{a,\ cat} + 5.00 \times 10^4\ J/mol$

$E_{a,\ cat} = 5.00 \times 10^4\ J/mol - 2.02 \times 10^4\ J/mol = 2.98 \times 10^4\ J/mol = 29.8\ kJ/mol$

107. a.

T (K)	1/T (K^{-1})	k (min^{-1})	ln k
298.2	3.353×10^{-3}	178	5.182
293.5	3.407×10^{-3}	126	4.836
290.5	3.442×10^{-3}	100.	4.605

A plot of ln k versus 1/T gives a straight line (plot not included). The equation for the straight line is:

$$\ln k = -6.48 \times 10^3 (1/T) + 26.9$$

For the $\ln k$ versus $1/T$ plot, slope $= -E_a/R = -6.48 \times 10^3$ K.

$$-6.48 \times 10^3 \text{ K} = -E_a/8.3145 \text{ J K}^{-1} \text{ mol}^{-1}, \quad E_a = 5.39 \times 10^4 \text{ J/mol} = 53.9 \text{ kJ/mol}$$

b. $\ln k = -6.48 \times 10^3 (1/288.2) + 26.9 = 4.42, \quad k = e^{4.42} = 83 \text{ min}^{-1}$

About 83 chirps per minute per insect. Note: We carried extra significant figures.

c. k gives the number of chirps per minute. The number or chirps in 15 s is k/4.

T (°C)	T (°F)	k (min^{-1})	42 + 0.80(k/4)
25.0	77.0	178	78° F
20.3	68.5	126	67°F
17.3	63.1	100.	62°F
15.0	59.0	83	59°F

The rule of thumb appears to be fairly accurate, almost ±1°F.

109. $\text{Rate} = \dfrac{d[Cl_2]}{dt} = k_2[NO_2Cl][Cl]; \quad$ Assume $\dfrac{d[Cl]}{dt} = 0$, then:

$$k_1[NO_2Cl] = k_{-1}[NO_2][Cl] + k_2[NO_2Cl][Cl], \quad [Cl] = \frac{k_1[NO_2Cl]}{k_{-1}[NO_2] + k_2[NO_2Cl]}$$

$$\text{Rate} = \frac{d[Cl_2]}{dt} = \frac{k_1 k_2 [NO_2Cl]^2}{k_{-1}[NO_2] + k_2[NO_2Cl]}$$

Challenge Problems

111. a. $\text{Rate} = k[CH_3X]^x[Y]^y$; for experiment 1, [Y] is in large excess, so its concentration will be constant. $\text{Rate} = k'[CH_3X]^x$, where $k' = k(3.0 \text{ M})^y$.

A plot (not included) of $\ln[CH_3X]$ versus t is linear ($x = 1$). The integrated rate law is:

$$\ln[CH_3X] = -(0.93)t - 3.99; \quad k' = 0.93 \text{ h}^{-1}$$

For experiment 2, [Y] is again constant, with $\text{Rate} = k''[CH_3X]^x$, where $k'' = k(4.5 \text{ M})^y$.
The natural log plot is linear again with an integrated rate law:

$$\ln[CH_3X] = -(0.93)t - 5.40; \quad k'' = 0.93 \text{ h}^{-1}$$

Dividing the rate-constant values: $\dfrac{k'}{k''} = \dfrac{0.93}{0.93} = \dfrac{k(3.0)^y}{k(4.5)^y}, \quad 1.0 = (0.67)^y, \quad y = 0$

Reaction is first order in CH_3X and zero order in Y. The overall rate law is:

$$Rate = k[CH_3X], \text{ where } k = 0.93 \text{ h}^{-1} \text{ at } 25°C$$

b. $t_{1/2} = (\ln 2)/k = 0.6931/(7.88 \times 10^8 \text{ h}^{-1}) = 8.80 \times 10^{-10} \text{ hour}$

c. $\ln\left(\dfrac{k_2}{k_1}\right) = \dfrac{E_a}{R}\left(\dfrac{1}{T_1} - \dfrac{1}{T_2}\right)$, $\ln\left(\dfrac{7.88 \times 10^8}{0.93}\right) = \dfrac{E_a}{8.3145 \text{ J K}^{-1} \text{ mol}^{-1}}\left(\dfrac{1}{298 \text{ K}} - \dfrac{1}{358 \text{ K}}\right)$

$E_a = 3.0 \times 10^5 \text{ J/mol} = 3.0 \times 10^2 \text{ kJ/mol}$

d. From part a, the reaction is first order in CH_3X and zero order in Y. From part c, the activation energy is close to the C-X bond energy. A plausible mechanism that explains the results in parts a and c is:

$$CH_3X \rightarrow CH_3 + X \quad \text{(slow)}$$

$$CH_3 + Y \rightarrow CH_3Y \quad \text{(fast)}$$

Note: This is a possible mechanism because the derived rate law is the same as the experimental rate law (and the sum of the steps gives the overall balanced equation).

113. $\dfrac{-d[A]}{dt} = k[A]^3$, $\displaystyle\int_{[A]_0}^{[A]_t} \dfrac{d[A]}{[A]^3} = -\int_0^t k \, dt$

$\displaystyle\int x^n \, dx = \dfrac{x^{n+1}}{n+1}$; so: $-\dfrac{1}{2[A]^2}\Bigg|_{[A]_0}^{[A]_t} = -kt$, $-\dfrac{1}{2[A]_t^2} + \dfrac{1}{2[A]_0^2} = -kt$

For the half-life equation, $[A]_t = 1/2[A]_0$:

$$-\dfrac{1}{2\left(\dfrac{1}{2}[A]_0\right)^2} + \dfrac{1}{2[A]_0^2} = -kt_{1/2}, -\dfrac{4}{2[A]_0^2} + \dfrac{1}{2[A]_0^2} = -kt_{1/2}$$

$$-\dfrac{3}{2[A]_0^2} = -kt_{1/2}, t_{1/2} = \dfrac{3}{2[A]_0^2 k}$$

The first half-life is $t_{1/2} = 40.$ s and corresponds to going from $[A]_0$ to $1/2[A]_0$. The second half-life corresponds to going from $1/2 [A]_0$ to $1/4 [A]_0$.

First half-life $= \dfrac{3}{2[A]_0^2 k}$; second half-life $= \dfrac{3}{2\left(\dfrac{1}{2}[A]_0\right)^2 k} = \dfrac{6}{[A]_0^2 k}$

$$\frac{\text{First half} - \text{life}}{\text{Second half} - \text{life}} = \frac{\dfrac{3}{2[A]_0^2\, k}}{\dfrac{6}{[A]_0^2\, k}} = 3/12 = 1/4$$

Because the first half-life is 40. s, the second half-life will be four times this, or 160 s.

115.   Rate = $k[A]^x[B]^y[C]^z$; during the course of experiment 1, [A] and [C] are essentially constant, and Rate = $k'[B]^y$, where $k' = k[A]_0^x[C]_0^z$.

[B] (M)	time (s)	ln[B]	1/[B] (M^{-1})
1.0×10^{-3}	0	−6.91	1.0×10^3
2.7×10^{-4}	1.0×10^5	−8.22	3.7×10^3
1.6×10^{-4}	2.0×10^5	−8.74	6.3×10^3
1.1×10^{-4}	3.0×10^5	−9.12	9.1×10^3
8.5×10^{-5}	4.0×10^5	−9.37	12×10^3
6.9×10^{-5}	5.0×10^5	−9.58	14×10^3
5.8×10^{-5}	6.0×10^5	−9.76	17×10^3

A plot of 1/[B] versus t is linear (plot not included). The reaction is second order in B, and the integrated rate equation is:

$$1/[B] = (2.7 \times 10^{-2}\ \text{L mol}^{-1}\ \text{s}^{-1})t + 1.0 \times 10^3\ M^{-1};\ \ k' = 2.7 \times 10^{-2}\ \text{L mol}^{-1}\ \text{s}^{-1}$$

For experiment 2, [B] and [C] are essentially constant, and Rate = $k''[A]^x$, where $k'' = k[B]_0^y[C]_0^z = k[B]_0^2[C]_0^z$.

[A] (M)	time (s)	ln[A]	1/[A] (M^{-1})
1.0×10^{-2}	0	−4.61	1.0×10^2
8.9×10^{-3}	1.0	−4.72	110
7.1×10^{-3}	3.0	−4.95	140
5.5×10^{-3}	5.0	−5.20	180
3.8×10^{-3}	8.0	−5.57	260
2.9×10^{-3}	10.0	−5.84	340
2.0×10^{-3}	13.0	−6.21	5.0×10^2

A plot of ln[A] versus t is linear. The reaction is first order in A, and the integrated rate law is:

$$\ln[A] = -(0.123\ \text{s}^{-1})t - 4.61;\ \ k'' = 0.123\ \text{s}^{-1}$$

Note: We will carry an extra significant figure in k''.

Experiment 3:   [A] and [B] are constant; Rate = $k'''[C]^z$

The plot of [C] versus t is linear. Thus $z = 0$.

The overall rate law is Rate = $k[A][B]^2$.

From experiment 1 (to determine k):

$$k' = 2.7 \times 10^{-2} \text{ L mol}^{-1}\text{ s}^{-1} = k[A]_0^x[C]_0^z = k[A]_0 = k(2.0\ M),\ \ k = 1.4 \times 10^{-2} \text{ L}^2 \text{ mol}^{-2} \text{ s}^{-1}$$

From experiment 2: $k'' = 0.123 \text{ s}^{-1} = k[B]_0^2,\ \ k = \dfrac{0.123 \text{ s}^{-1}}{(3.0\ M)^2} = 1.4 \times 10^{-2} \text{ L}^2 \text{ mol}^{-2} \text{ s}^{-1}$

Thus Rate = $k[A][B]^2$ and $k = 1.4 \times 10^{-2} \text{ L}^2 \text{ mol}^{-2} \text{ s}^{-1}$.

117.   Rate $= \dfrac{-d[N_2O_5]}{dt} = k_1[M][N_2O_5] - k_{-1}[NO_3][NO_2][M]$

Assume $d[NO_3]/dt = 0$, so $k_1[N_2O_5][M] = k_{-1}[NO_3][NO_2][M] + k_2[NO_3][NO_2] + $
$$k_3[NO_3][NO].$$

$$[NO_3] = \frac{k_1[N_2O_5][M]}{k_{-1}[NO_2][M] + k_2[NO_2] + k_3[NO]}$$

Assume $\dfrac{d[NO]}{dt} = 0$, so $k_2[NO_3][NO_2] = k_3[NO_3][NO]$, $[NO] = \dfrac{k_2}{k_3}[NO_2]$.

Substituting: $[NO_3] = \dfrac{k_1[N_2O_5][M]}{k_{-1}[NO_2][M] + k_2[NO_2] + \dfrac{k_2 k_3}{k_3}[NO_2]} = \dfrac{k_1[N_2O_5][M]}{[NO_2](k_{-1}[M] + 2k_2)}$

Solving for the rate law:

$$\text{Rate} = \frac{-d[N_2O_5]}{dt} = k_1[N_2O_5][M] - \frac{k_{-1}k_1[NO_2][N_2O_5][M]^2}{[NO_2](k_{-1}[M] + 2k_2)} = k_1[N_2O_5][M]$$

$$- \frac{k_{-1}k_1[M]^2[N_2O_5]}{k_{-1}[M] + 2k_2}$$

$$\text{Rate} = \frac{-d[N_2O_5]}{dt} = \left(k_1 - \frac{k_{-1}k_1[M]}{k_{-1}[M] + 2k_2} \right)[N_2O_5][M];\ \ \text{simplifying:}$$

$$\text{Rate} = \frac{-d[N_2O_5]}{dt} = \frac{2k_1 k_2[M][N_2O_5]}{k_{-1}[M] + 2k_2}$$

119.   a.   Rate $= (k_1 + k_2[H^+])[I^-]^m[H_2O_2]^n$

In all the experiments the concentration of H_2O_2 is small compared to the concentrations of I^- and H^+. Therefore, the concentrations of I^- and H^+ are effectively constant, and the rate law reduces to:

Rate $= k_{obs}[H_2O_2]^n$, where $k_{obs} = (k_1 + k_2[H^+])[I^-]^m$

Because all plots of $\ln[H_2O_2]$ versus time are linear, the reaction is first order with respect to H_2O_2 ($n = 1$). The slopes of the $\ln[H_2O_2]$ versus time plots equal $-k_{obs}$, which equals $-(k_1 + k_2[H^+])[I^-]^m$. To determine the order of I^-, compare the slopes of two experiments in which I^- changes and H^+ is constant. Comparing the first two experiments:

$$\frac{\text{slope (exp. 2)}}{\text{slope (exp. 1)}} = \frac{-0.360}{-0.120} = \frac{-[k_1 + k_2(0.0400\ M)](0.3000\ M)^m}{-[k_1 + k_2(0.0400\ M)](0.1000\ M)^m}$$

$$3.00 = \left(\frac{0.3000\ M}{0.1000\ M}\right)^m = (3.000)^m, \ m = 1$$

The reaction is also first order with respect to I^-.

b.   The slope equation has two unknowns, k_1 and k_2. To solve for k_1 and k_2, we must have two equations. We need to take one of the first set of three experiments and one of the second set of three experiments to generate the two equations in k_1 and k_2.

Experiment 1: slope $= -(k_1 + k_2[H^+])[I^-]$

-0.120 min$^{-1} = -[k_1 + k_2(0.0400\ M)](0.1000\ M)$ or $1.20 = k_1 + k_2(0.0400)$

Experiment 4:

-0.0760 min$^{-1} = -[k_1 + k_2(0.0200\ M)](0.0750\ M)$ or $1.01 = k_1 + k_2(0.0200)$

Subtracting 4 from 1:

$1.20 = \ \ k_1 + k_2(0.0400)$
$-1.01 = -k_1 - k_2(0.0200)$
_____
$0.19 = \ \ \ \ \ \ \ \ k_2(0.0200), \ k_2 = 9.5\ L^2\ mol^{-2}\ min^{-1}$

$1.20 = k_1 + 9.5(0.0400), \ k_1 = 0.82\ L\ mol^{-1}\ min^{-1}$

c.   There are two pathways, one involving H^+ with rate $= k_2[H^+][I^-][H_2O_2]$ and another not involving H^+ with rate $= k_1[I^-][H_2O_2]$. The overall rate of reaction depends on which of these two pathways dominates, and this depends on the H^+ concentration.

CHAPTER 16

LIQUIDS AND SOLIDS

Intermolecular Forces and Physical Properties

11.   Intermolecular forces are the relatively weak forces between molecules that hold the molecules together in the solid and liquid phases. Intramolecular forces are the forces within a molecule. These are the covalent bonds in a molecule. Intramolecular forces (covalent bonds) are much stronger than intermolecular forces.

Dipole forces are the forces that act between polar molecules. The electrostatic attraction between the partial positive end of one polar molecule and the partial negative end of another is the dipole force. Dipole forces are generally weaker than hydrogen bonding. Both of these forces are due to dipole moments in molecules. Hydrogen bonding is given a separate name from dipole forces because hydrogen bonding is a particularly strong dipole force. Any neutral molecule that has a hydrogen covalently bonded to N, O, or F exhibits the relatively strong hydrogen bonding intermolecular forces.

London dispersion forces are accidental-induced dipole forces. Like dipole forces, London dispersion forces are electrostatic in nature. Dipole forces are the electrostatic forces between molecules having a permanent dipole. London dispersion forces are the electrostatic forces between molecules having an accidental or induced dipole. All covalent molecules (polar and nonpolar) have London dispersion forces, but only polar molecules (those with permanent dipoles) exhibit dipole forces.

13.   Fusion refers to a solid converting to a liquid, and vaporization refers to a liquid converting to a gas. Only a fraction of the hydrogen bonds are broken in going from the solid phase to the liquid phase. Most of the hydrogen bonds are still present in the liquid phase and must be broken during the liquid to gas phase transition. Thus the enthalpy of vaporization is much larger than the enthalpy of fusion since more intermolecular forces are broken during the vaporization process.

15.   Ionic compounds have ionic forces. Covalent compounds all have London dispersion (LD) forces, whereas polar covalent compounds have dipole forces and/or hydrogen-bonding forces. For hydrogen bonding (H-bonding) forces, the covalent compound must have either a N–H, O–H, or F–H bond in the molecule.

   a. LD only   b. dipole, LD   c. H-bonding, LD

   d. ionic   e. LD only (CH_4 in a nonpolar covalent compound.)

329

f.   dipole, LD   g.   ionic   h.   ionic

i.   LD mostly; C−F bonds are polar, but polymers such as teflon are so large that LD forces are the predominant intermolecular forces.

j.   LD   k.   dipole, LD   l.   H-bonding, LD

m.   dipole, LD   n.   LD only

17.   Boiling points and freezing points are assumed directly related to the strength of the intermolecular forces, whereas vapor pressure is inversely related to the strength of the intermolecular forces.

a.   HBr; HBr is polar, whereas Kr and Cl_2 are nonpolar. HBr has dipole forces unlike Kr and Cl_2. So HBr has the stronger intermolecular forces and the highest boiling point.

b.   NaCl; the ionic forces in NaCl are much stronger than the intermolecular forces for molecular substances, so NaCl has the highest melting point.

c. I_2; all are nonpolar, so the largest molecule (I_2) will have the strongest LD (London Dispersion) forces and the lowest vapor pressure.

d. N_2; nonpolar and smallest, so it has the weakest intermolecular forces.

e. CH_4; smallest, nonpolar molecule, so it has the weakest LD forces.

f.   HF; HF can form relatively strong H-bonding interactions, unlike the others.

g. $CH_3CH_2CH_2OH$; H-bonding, unlike the others, so it has strongest intermolecular forces.

19.   a.   Neopentane is more compact than n-pentane. There is less surface-area contact among neopentane molecules. This leads to weaker LD (London Dispersion) forces and a lower boiling point.

b.   HF is capable of H-bonding; HCl is not.

c.   LiCl is ionic, and HCl is a molecular solid with only dipole forces and LD forces. Ionic forces are much stronger than the forces for molecular solids.

d.   n-Hexane is a larger molecule, so it has stronger LD forces.

21.   Ar exists as individual atoms that are held together in the condensed phases by London dispersion forces. The molecule that will have a boiling point closest to Ar will be a nonpolar substance with about the same molar mass as Ar (39.95 g/mol); this same size nonpolar substance will have about equivalent strength of London dispersion forces. Of the choices, only Cl_2 (70.90 g/mol) and F_2 (38.00 g/mol) are nonpolar. Because F_2 has a molar mass closest to that of Ar, one would expect the boiling point of F_2 to be close to that of Ar.

23.   The electrostatic potential diagrams indicate that ethanol and acetone are polar substances, and that propane is a nonpolar substance. Ethanol, with the O−H covalent bond, will exhibit relatively strong hydrogen bonding intermolecular forces in addition to London dispersion forces. The polar acetone will exhibit dipole forces in addition to London dispersion forces,

and the nonpolar propane will only exhibit London dispersion (LD) forces. Because all three compounds have about the same molar mass, the relative strengths of the LD forces should be about the same. Therefore, ethanol (with the H-bonding capacity) should have the highest boiling point, with polar acetone having the next highest boiling point, and the nonpolar propane, with the weakest intermolecular forces, will have the lowest boiling point.

25.   A single hydrogen bond in H_2O has a strength of 21 kJ/mol. Each H_2O molecule forms two H-bonds. Thus it should take 42 kJ/mol of energy to break all of the H-bonds in water. Consider the phase transitions:

$$\text{Solid} \xrightarrow{\text{6.0 kJ}} \text{liquid} \xrightarrow{\text{40.7 kJ}} \text{vapor} \qquad \Delta H_{sub} = \Delta H_{fus} + \Delta H_{vap}$$

It takes a total of 46.7 kJ/mol to convert solid H_2O to vapor (ΔH_{sub}). This would be the amount of energy necessary to disrupt all of the intermolecular forces in ice. Thus $(42 \div 46.7) \times 100 = 90.\%$ of the attraction in ice can be attributed to H-bonding.

Properties of Liquids

27.   a.   Surface tension: the resistance of a liquid to an increase in its surface area.

        b.   Viscosity: the resistance of a liquid to flow.

        c.   Melting point: the temperature (at constant pressure) where a solid converts entirely to a liquid as long as heat is applied. A more detailed definition is the temperature at which the solid and liquid states have the same vapor pressure under conditions where the total pressure is constant.

        d.   Boiling point: the temperature (at constant pressure) where a liquid converts entirely to a gas as long as heat is applied. The detailed definition is the temperature at which the vapor pressure of the liquid is exactly equal to the external pressure.

        e.   Vapor pressure: the pressure of the vapor over a liquid at equilibrium.

        As the strengths of intermolecular forces increase, surface tension, viscosity, melting point, and boiling point increase, whereas vapor pressure decreases.

29.   Water is a polar substance, and wax is a nonpolar substance; they are not attracted to each other. A molecule at the surface of a drop of water is subject to attractions only by water molecules below it and to each side. The effect of this uneven pull on the surface water molecules tends to draw them into the body of the water and causes the droplet to assume the shape that has the minimum surface area, a sphere.

31.   The structure of H_2O_2 is $H - O - O - H$, which produces greater hydrogen bonding than in water. Thus the intermolecular forces are stronger in H_2O_2 than in H_2O resulting in a higher normal boiling point for H_2O_2 and a lower vapor pressure.

Structures and Properties of Solids

33.   a.   Crystalline solid: Regular, repeating structure

      Amorphous solid: Irregular arrangement of atoms or molecules

      b.   Ionic solid: Made up of ions held together by ionic bonding.

      Molecular solid: Made up of discrete covalently bonded molecules held together in the solid phase by weaker forces (LD, dipole or hydrogen bonds).

      c.   Molecular solid: Discrete, individual molecules

      Covalent network solid: No discrete molecules; a covalent network solid is one large molecule; the interparticle forces are the covalent bonds between atoms.

      d.   Metallic solid: Completely delocalized electrons, conductor of electricity (ions in a sea of electrons)

      Covalent network solid: Localized electrons; insulator or semiconductor

35.   a.   Both forms of carbon are network solids. In diamond, each carbon atom is surrounded by a tetrahedral arrangement of other carbon atoms to form a huge molecule. Each carbon atom is covalently bonded to four other carbon atoms.

      The structure of graphite is based on layers of carbon atoms arranged in fused six-membered rings. Each carbon atom in a particular layer of graphite is surrounded by three other carbons in a trigonal planar arrangement. This requires sp^2 hybridization. Each carbon has an unhybridized p atomic orbital; all of these p orbitals in each six-membered ring overlap with each other to form a delocalized π electron system.

      b.   Silica is a network solid having an empirical formula of SiO_2. The silicon atoms are singly bonded to four oxygens. Each silicon atom is at the center of a tetrahedral arrangement of oxygen atoms that are shared with other silicon atoms. The structure of silica is based on a network of SiO_4 tetrahedra with shared oxygen atoms rather than discrete SiO_2 molecules.

      Silicates closely resemble silica. The structure is based on interconnected SiO_4 tetrahedra. However, in contrast to silica, where the O/Si ratio is 2:1, silicates have O/Si ratios greater than 2:1 and contain silicon-oxygen anions. To form a neutral solid silicate, metal cations are needed to balance the charge. In other words, silicates are salts containing metal cations and polyatomic silicon-oxygen anions.

      When silica is heated above its melting point and cooled rapidly, an amorphous (disordered) solid called glass results. Glass more closely resembles a very viscous solution than it does a crystalline solid. To affect the properties of glass, several different additives are thrown into the mixture. Some of these additives are Na_2CO_3, B_2O_3, and K_2O, with each compound serving a specific purpose relating to the properties of glass.

37.   A crystalline solid will have the simpler diffraction pattern because a regular, repeating arrangement is necessary to produce planes of atoms that will diffract the X rays in regular patterns. An amorphous solid does not have a regular repeating arrangement and will produce a complicated diffraction pattern.

39. $\lambda = \dfrac{2d \sin \theta}{n} = \dfrac{2 \times 1.36 \times 10^{-10} \text{ m} \times \sin 15.0°}{1} = 7.04 \times 10^{-11} \text{ m} = 0.704 \text{ Å}$

41. $n\lambda = 2d \sin \theta$, $d = \dfrac{n\lambda}{2 \sin \theta} = \dfrac{1 \times 1.54 \text{ Å}}{2 \times \sin 14.22°} = 3.13 \text{ Å} = 3.13 \times 10^{-10} \text{ m} = 313 \text{ pm}$

43.   A cubic closest packed structure has a face-centered cubic unit cell. In a face-centered cubic unit, there are:

$$8 \text{ corners} \times \frac{1/8 \text{ atom}}{\text{corner}} + 6 \text{ faces} \times \frac{1/2 \text{ atom}}{\text{face}} = 4 \text{ atoms}$$

The atoms in a face-centered cubic unit cell touch along the face diagonal of the cubic unit cell. Using the Pythagorean formula, where l = length of the face diagonal and r = radius of the atom:

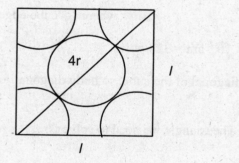

$$l^2 + l^2 = (4r)^2$$

$$2l^2 = 16r^2$$

$$l = r\sqrt{8}$$

$l = r\sqrt{8} = 197 \times 10^{-12} \text{ m} \times \sqrt{8} = 5.57 \times 10^{-10} \text{ m} = 5.57 \times 10^{-8} \text{ cm}$

Volume of a unit cell = $l^3 = (5.57 \times 10^{-8} \text{ cm})^3 = 1.73 \times 10^{-22} \text{ cm}^3$

Mass of a unit cell = 4 Ca atoms $\times \dfrac{1 \text{ mol Ca}}{6.022 \times 10^{23} \text{ atoms}} \times \dfrac{40.08 \text{ g Ca}}{\text{mol Ca}} = 2.662 \times 10^{-22} \text{ g Ca}$

Density = $\dfrac{\text{mass}}{\text{volume}} = \dfrac{2.662 \times 10^{-22} \text{ g}}{1.73 \times 10^{-22} \text{ cm}^3} = 1.54 \text{ g/cm}^3$

45.   There are four Ni atoms in each unit cell. For a unit cell:

$$\text{density} = \frac{\text{mass}}{\text{volume}} = 6.84 \text{ g/cm}^3 = \frac{4 \text{ Ni atoms} \times \dfrac{1 \text{ mol Ni}}{6.022 \times 10^{23} \text{ atoms}} \times \dfrac{58.69 \text{ g Ni}}{\text{mol Ni}}}{l^3}$$

Solving: $l = 3.85 \times 10^{-8}$ cm = cube edge length

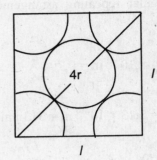

For a face centered cube:

$$(4r)^2 = l^2 + l^2 = 2l^2$$

$$r\sqrt{8} = l, r = l/\sqrt{8}$$

$$r = 3.85 \times 10^{-8} \text{ cm}/\sqrt{8}$$

$$r = 1.36 \times 10^{-8} \text{ cm} = 136 \text{ pm}$$

47.   For a body-centered unit cell: $8 \text{ corners} \times \dfrac{1/8 \text{ Ti}}{\text{corner}} + 1 \text{ Ti at body center} = 2 \text{ Ti atoms}$

All body-centered unit cells have two atoms per unit cell. For a unit cell:

$$\text{density} = 4.50 \text{ g/cm}^3 = \frac{2 \text{ atoms Ti} \times \dfrac{1 \text{ mol Ti}}{6.022 \times 10^{23} \text{ atoms}} \times \dfrac{47.88 \text{ g Ti}}{\text{mol Ti}}}{l^3},$$

where l = cube edge length

Solving: l = edge length of unit cell = 3.28×10^{-8} cm = 328 pm

Assume Ti atoms just touch along the body diagonal of the cube, so body diagonal = 4 × radius of atoms = 4r.

The triangle we need to solve is:

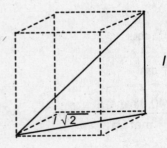

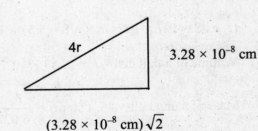

$(4r)^2 = (3.28 \times 10^{-8} \text{ cm})^2 + [(3.28 \times 10^{-8} \text{ cm})\sqrt{2}]^2$,   r = 1.42×10^{-8} cm = 142 pm

For a body-centered cubic unit cell, the radius of the atom is related to the cube edge length by $4r = l\sqrt{3}$ or $l = 4r/\sqrt{3}$.

49.   If gold has a face-centered cubic structure, then there are four atoms per unit cell, and from Exercise 16.43:

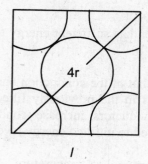

$$2l^2 = 16r^2$$

$$l = r\sqrt{8} = (144 \text{ pm})\sqrt{8} = 407 \text{ pm}$$

$$l = 407 \times 10^{-12} \text{ m} = 4.07 \times 10^{-8} \text{ cm}$$

$$\text{Density} = \frac{4 \text{ atoms Au} \times \dfrac{1 \text{ mol Au}}{6.022 \times 10^{23} \text{ atoms}} \times \dfrac{197.0 \text{ g Au}}{\text{mol Au}}}{(4.07 \times 10^{-8} \text{ cm})^3} = 19.4 \text{ g/cm}^3$$

If gold has a body-centered cubic structure, then there are two atoms per unit cell, and from Exercise 16.47:

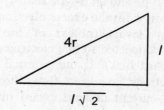

$$16r^2 = l^2 + 2l^2$$

$$l = 4r/\sqrt{3} = 333 \text{ pm} = 333 \times 10^{-12} \text{ m}$$

$$l = 333 \times 10^{-10} \text{ cm} = 3.33 \times 10^{-8} \text{ cm}$$

$$\text{Density} = \frac{2 \text{ atoms Au} \times \dfrac{1 \text{ mol Au}}{6.022 \times 10^{23} \text{ atoms}} \times \dfrac{197.0 \text{ g Au}}{\text{mol Au}}}{(3.33 \times 10^{-8} \text{ cm})^3} = 17.7 \text{ g/cm}^3$$

The measured density of gold is consistent with a face-centered cubic unit cell.

51.   Conductor: The energy difference between the filled and unfilled molecular orbitals is minimal. We call this energy difference the band gap. Because the band gap is minimal, electrons can easily move into the conduction bands (the unfilled molecular orbitals).

       Insulator: Large band gap; electrons do not move from the filled molecular orbitals to the conduction bands since the energy difference is large.

Semiconductor: Small band gap; the energy difference between the filled and unfilled molecular orbitals is smaller than in insulators, so some electrons can jump into the conduction bands. The band gap, however, is not as small as with conductors, so semiconductors have intermediate conductivity.

a.   As the temperature is increased, more electrons in the filled molecular orbitals have sufficient kinetic energy to jump into the conduction bands (the unfilled molecular orbitals).

b.   A photon of light is absorbed by an electron that then has sufficient energy to jump into the conduction bands.

c.   An impurity either adds electrons at an energy near that of the conduction bands (n-type) or creates holes (unfilled energy levels) at energies in the previously filled molecular orbitals (p-type). Both n-type and p-type semiconductors increase conductivity by creating an easier path for electrons to jump from filled to unfilled energy levels.

In conductors, electrical conductivity is inversely proportional to temperature. Increases in temperature increase the motions of the atoms, which gives rise to increased resistance (decreased conductivity). In a semiconductor, electrical conductivity is directly proportional to temperature. An increase in temperature provides more electrons with enough kinetic energy to jump from the filled molecular orbitals to the conduction bands, increasing conductivity.

53.   A rectifier is a device that produces a current that flows in one direction from an alternating current that flows in both directions. In a p-n junction, a p-type and an n-type semi-conductor are connected. The natural flow of electrons in a p-n junction is for the excess electrons in the n-type semiconductor to move to the empty energy levels (holes) of the p-type semiconductor. Only when an external electric potential is connected so that electrons flow in this natural direction will the current flow easily (forward bias). If the external electric potential is connected in reverse of the natural flow of electrons, no current flows through the system (reverse bias). A p-n junction only transmits a current under forward bias, thus converting the alternating current to direct current.

55.   In has fewer valence electrons than Se. Thus Se doped with In would be a p-type semiconductor.

57. $E_{gap} = 2.5 \text{ eV} \times 1.6 \times 10^{-19} \text{ J/eV} = 4.0 \times 10^{-19} \text{ J}$; we want $E_{gap} = E_{light}$.

$$E_{light} = \frac{hc}{\lambda}, \quad \lambda = \frac{hc}{E} = \frac{(6.63 \times 10^{-34} \text{ J s})(3.00 \times 10^8 \text{ m/s})}{4.0 \times 10^{-19} \text{ J}} = 5.0 \times 10^{-7} \text{ m} = 5.0 \times 10^2 \text{ nm}$$

59.   Sodium chloride structure: $8 \text{ corners} \times \dfrac{1/8 \text{ Cl}^-}{\text{corner}} + 6 \text{ faces} \times \dfrac{1/2 \text{ Cl}^-}{\text{face}} = 4 \text{ Cl}^- \text{ ions}$

$12 \text{ edges} \times \dfrac{1/4 \text{ Na}^+}{\text{edge}} + 1 \text{ Na}^+ \text{ at body center} = 4 \text{ Na}^+ \text{ ions; NaCl is the formula.}$

Cesium chloride structure: 1 Cs^+ ion at body center; $8 \text{ corners} \times \dfrac{1/8 \text{ Cl}^-}{\text{corner}} = 1 \text{ Cl}^- \text{ ion; CsCl}$ is the formula.

Zinc sulfide structure: There are four Zn^{2+} ions inside the cube.

$$8 \text{ corners} \times \frac{1/8 \, S^{2-}}{\text{corner}} + 6 \text{ faces} \times \frac{1/2 \, S^{2-}}{\text{face}} = 4 \, S^{2-} \text{ ions}; \quad ZnS \text{ is the formula.}$$

Titanium oxide structure: $8 \text{ corners} \times \dfrac{1/8 \, Ti^{4+}}{\text{corner}} + 1 \, Ti^{4+} \text{ at body center} = 2 \, Ti^{4+} \text{ ions}$

$$4 \text{ faces} \times \frac{1/2 \, O^{2-}}{\text{face}} + 2 \, O^{2-} \text{ inside cube} = 4 \, O^{2-} \text{ ions}; \quad TiO_2 \text{ is the formula.}$$

61. The structures of most binary ionic solids can be explained by the closest packing of spheres. Typically, the larger ions, usually the anions, are packed in one of the closest packing arrangements, and the smaller cations fit into holes among the closest packed anions. There are different types of holes within the closest packed anions that are determined by the number of spheres that form them. Which of the three types of holes are filled usually depends on the relative size of the cation to the anion. Ionic solids will always try to maximize electrostatic attractions among oppositely charged ions and minimize the repulsions among ions with like charges.

The structure of sodium chloride can be described in terms of a cubic closest packed array of Cl^- ions with Na^+ ions in all of the octahedral holes. An octahedral hole is formed between six Cl^- anions. The number of octrahedral holes is the same as the number of packed ions. So in the face-centered unit cell of sodium chloride, there are four net Cl^- ions and four net octahedral holes. Because the stoichiometry dictates a 1:1 ratio between the number of Cl^- anions and Na^+ cations, all of the octahedral holes must be filled with Na^+ ions.

In zinc sulfide, the sulfide anions also occupy the lattice points of a cubic closest packing arrangement. But instead of having the cations in octahedral holes, the Zn^{2+} cations occupy tetrahedral holes. A tetrahedral hole is the empty space created when four spheres are packed together. There are twice as many tetrahedral holes as packed anions in the closest packed structure. Therefore, each face-centered unit cell of sulfide anions contains four net S^{2-} ions and eight net tetrahedral holes. For the 1:1 stoichiometry to work out, only one-half of the tetrahedral holes are filled with Zn^{2+} ions. This gives four S^{2-} ions and four Zn^{2+} ions per unit cell for an empirical formula of ZnS.

63. There is one octahedral hole per closest packed anion in a closest packed structure. If one-half of the octahedral holes are filled, then there is a 2:1 ratio of fluoride ions to cobalt ions in the crystal. The formula is CoF_2.

65. Mn ions at 8 corners: $8(1/8) = 1$ Mn ion; F ions at 12 edges: $12(1/4) = 3$ F ions; the formula is MnF_3. Assuming fluoride is 1− charged, then the charge on Mn is 3+.

67. CsCl is a simple cubic array of Cl^- ions with Cs^+ in the middle of each unit cell. There is one Cs^+ and one Cl^- ion in each unit cell. Cs^+ and Cl^- touch along the body diagonal.

Body diagonal $= 2r_{Cs^+} + 2r_{Cl^-} = \sqrt{3} \, l$, $l =$ length of cube edge

In each unit cell:

$$\text{mass} = 1 \text{ CsCl unit}(1 \text{ mol}/6.022 \times 10^{23} \text{ units})(168.4 \text{ g/mol}) = 2.796 \times 10^{-22} \text{ g}$$

$$\text{volume} = l^3 = \text{mass/density} = 2.796 \times 10^{-22} \text{ g}/3.97 \text{ g cm}^{-3} = 7.04 \times 10^{-23} \text{ cm}^3$$

$$l^3 = 7.04 \times 10^{-23} \text{ cm}^3, \;\; l = 4.13 \times 10^{-8} \text{ cm} = 413 \text{ pm} = \text{length of cube edge}$$

$$2r_{Cs^+} + 2r_{Cl^-} = \sqrt{3}\, l = \sqrt{3}\,(413 \text{ pm}) = 715 \text{ pm}$$

The distance between ion centers $= r_{Cs^+} + r_{Cl^-} = 715 \text{ pm}/2 = 358 \text{ pm}.$

From ionic radius: $r_{Cs^+} = 169 \text{ pm}$ and $r_{Cl^-} = 181 \text{ pm}$; $r_{Cs^+} + r_{Cl^-} = 169 + 181 = 350. \text{ pm}$

The distance calculated from the density is 8 pm (2.3%) greater than that calculated from tables of ionic radii.

69.   For a cubic hole to be filled, the cation to anion radius ratio is between $0.732 < r_+/r_- < 1.00$.

     CsBr: Cs^+ radius = 169 pm, Br^- radius = 195 pm; $r_+/r_- = 169/195 = 0.867$

           From the radius ratio, Cs^+ should occupy cubic holes. The structure should be the CsCl structure. The actual structure is the CsCl structure.

     KF: K^+ radius = 133 pm, F^- radius = 136 pm; $r_+/r_- = 133/136 = 0.978$

           Again, we would predict a structure similar to CsCl, i.e., cations in the middle of a simple cubic array of anions. The actual structure is the NaCl structure.

     The radius ratio rules fail for KF. Exceptions are common for crystal structures.

71.   Al: 8 corners $\times \dfrac{1/8 \text{ Al}}{\text{corner}} = 1$ Al;   Ni: 6 face centers $\times \dfrac{1/2 \text{ Ni}}{\text{face center}} = 3$ Ni

     The composition of the specific phase of the superalloy is $AlNi_3$.

73.   With a cubic closest packed array of oxygen ions, we have 4 O^{2-} ions per unit cell. We need to balance the total 8– charge of the anions with a 8+ charge from the Al^{3+} and Mg^{2+} cations. The only simple combination of ions that gives a 8+ charge is 2 Al^{3+} ions and 1 Mg^{2+} ion. The formula is Al_2MgO_4.

     There are an equal number of octahedral holes as anions (4) in a cubic closest packed array and twice the number of tetrahedral holes as anions in a cubic closest packed array. For the stoichiometry to work out, we need 2 Al^{3+} and 1 Mg^{2+} per unit cell. Hence one-half the octahedral holes are filled with Al^{3+} ions, and one-eighth the tetrahedral holes are filled with Mg^{2+} ions.

75.   a.   Y: 1 Y in center; Ba: 2 Ba in center

Cu: 8 corners × $\dfrac{1/8\,Cu}{corner}$ = 1 Cu, 8 edges × $\dfrac{1/4\,Cu}{edge}$ = 2 Cu, total = 3 Cu atoms

O: 20 edges × $\dfrac{1/4\,O}{edge}$ = 5 oxygen, 8 faces × $\dfrac{1/2\,O}{face}$ = 4 oxygen, total = 9 O atoms

Formula: $YBa_2Cu_3O_9$

b.   The structure of this superconductor material follows the alternative perovskite structure described in Exercise 16.74b. The $YBa_2Cu_3O_9$ structure is three of these cubic perovskite unit cells stacked on top of each other. The oxygen atoms are in the same places, Cu takes the place of Ti, two of the calcium atoms are replaced by two barium atoms, and one Ca is replaced by Y.

c.   Y, Ba, and Cu are the same. Some oxygen atoms are missing.

12 edges × $\dfrac{1/4\,O}{edge}$ = 3 O, 8 faces × $\dfrac{1/2\,O}{face}$ = 4 O, total = 7 O atoms

Superconductor formula is $YBa_2Cu_3O_7$.

Phase Changes and Phase Diagrams

77.   A volatile liquid is one that evaporates relatively easily. Volatile liquids have large vapor pressures because the intermolecular forces that prevent evaporation are relatively weak.

79.   a.   As the intermolecular forces increase, the rate of evaporation decreases.

b.   As temperature increases, the rate of evaporation increases.

c.   As surface area increases, the rate of evaporation increases.

81. $C_2H_5OH(l) \rightarrow C_2H_5OH(g)$ is an endothermic process. Heat is absorbed when liquid ethanol vaporizes; the internal heat from the body provides this heat, which results in the cooling of the body.

83.   The mathematical equation that relates the vapor pressure of a substance to temperature is:

$$\underset{y}{\ln P_{vap}} = \underset{m}{-\dfrac{\Delta H_{vap}}{R}} \underset{x}{\left(\dfrac{1}{T}\right)} \underset{+\,b}{+\,C}$$

As shown above, this equation is in the form of the straight-line equation. If one plots $\ln P_{vap}$ versus $1/T$ with temperature in Kelvin, the slope of the straight line is $-\Delta H_{vap}/R$. Because ΔH_{vap} is always positive, the slope of the straight line will be negative.

85. $\ln\left(\dfrac{P_1}{P_2}\right) = \dfrac{\Delta H_{vap}}{R}\left(\dfrac{1}{T_2} - \dfrac{1}{T_1}\right),\quad \ln\left(\dfrac{836\ \text{torr}}{213\ \text{torr}}\right) = \dfrac{\Delta H_{vap}}{8.3145\ \text{J K}^{-1}\ \text{mol}^{-1}}\left(\dfrac{1}{313\ \text{K}} - \dfrac{1}{353\ \text{K}}\right)$

Solving: $\Delta H_{vap} = 3.1 \times 10^4$ J/mol; for the normal boiling point, P = 1.00 atm = 760. torr.

$\ln\left(\dfrac{760.\ \text{torr}}{213\ \text{torr}}\right) = \dfrac{3.1 \times 10^4\ \text{J/mol}}{8.3145\ \text{J K}^{-1}\ \text{mol}^{-1}}\left(\dfrac{1}{313\ \text{K}} - \dfrac{1}{T_1}\right),\quad \dfrac{1}{313} - \dfrac{1}{T_1} = 3.4 \times 10^{-4}$

$T_1 = 350.\ \text{K} = 77^\circ\text{C}$; the normal boiling point of CCl_4 is 77°C.

87.   If we graph $\ln P_{vap}$ versus $1/T$ with temperature in Kelvin, the slope of the resulting straight line will be $-\Delta H_{vap}/R$.

P_{vap}	$\ln P_{vap}$	T (Li)	1/T	T (Mg)	1/T
1 torr	0	1023 K	$9.775 \times 10^{-4}\ \text{K}^{-1}$	893 K	$11.2 \times 10^{-4}\ \text{K}^{-1}$
10.	2.3	1163	8.598×10^{-4}	1013	9.872×10^{-4}
100.	4.61	1353	7.391×10^{-4}	1173	8.525×10^{-4}
400.	5.99	1513	6.609×10^{-4}	1313	7.616×10^{-4}
760.	6.63	1583	6.317×10^{-4}	1383	7.231×10^{-4}

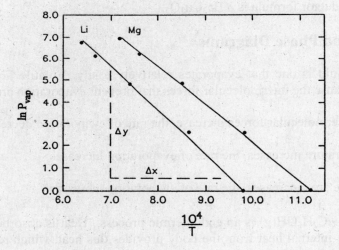

For Li:

We get the slope by taking two points (x, y) that are on the line we draw. For a line, slope = $\Delta y/\Delta x$; or we can determine the straight-line equation using a calculator. The general straight-line equation is $y = mx + b$ where m = slope and b = y intercept.

The equation of the Li line is: $\ln P_{vap} = -1.90 \times 10^4 (1/T) + 18.6$, slope = -1.90×10^4 K

Slope = $-\Delta H_{vap}/R$, $\Delta H_{vap} = -\text{slope} \times R = 1.90 \times 10^4\ \text{K} \times 8.3145\ \text{J K}^{-1}\ \text{mol}^{-1}$

$\Delta H_{vap} = 1.58 \times 10^5$ J/mol = 158 kJ/mol

For Mg:

The equation of the line is: $\ln P_{vap} = -1.67 \times 10^4 (1/T) + 18.7$, slope $= -1.67 \times 10^4$ K

$\Delta H_{vap} = -\text{slope} \times R = 1.67 \times 10^4 \text{ K} \times 8.3145 \text{ J K}^{-1} \text{ mol}^{-1}$

$\Delta H_{vap} = 1.39 \times 10^5 \text{ J/mol} = 139 \text{ kJ/mol}$

The bonding is stronger in Li because ΔH_{vap} is larger for Li.

89.

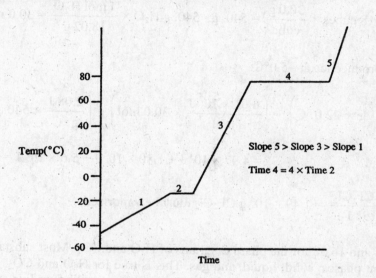

Slope 5 > Slope 3 > Slope 1

Time 4 = 4 × Time 2

91.   To calculate q_{total}, break up the heating process into five steps.

$H_2O(s, -20.°C) \rightarrow H_2O(s, 0°C)$, $\Delta T = 20.°C$

$$q_1 = s_{ice} \times m \times \Delta T = \frac{2.1 \text{ J}}{\text{g }°\text{C}} \times 5.00 \times 10^2 \text{ g} \times 20.°\text{C} = 2.1 \times 10^4 \text{ J} = 21 \text{ kJ}$$

$H_2O(s, 0°C) \rightarrow H_2O(l, 0°C)$, $q_2 = 5.00 \times 10^2 \text{ g } H_2O \times \dfrac{1 \text{ mol}}{18.02 \text{ g}} \times \dfrac{6.01 \text{ kJ}}{\text{mol}} = 167 \text{ kJ}$

$H_2O(l, 0°C) \rightarrow H_2O(l, 100.°C)$, $q_3 = \dfrac{4.2 \text{ J}}{\text{g }°\text{C}} \times 5.00 \times 10^2 \text{ g} \times 100.°\text{C} = 2.1 \times 10^5 \text{ J} = 210 \text{ kJ}$

$H_2O(l, 100.°C) \rightarrow H_2O(g, 100.°C)$, $q_4 = 5.00 \times 10^2 \text{ g} \times \dfrac{1 \text{ mol}}{18.02 \text{ g}} \times \dfrac{40.7 \text{ kJ}}{\text{mol}} = 1130 \text{ kJ}$

$H_2O(g, 100.°C) \rightarrow H_2O(g, 250.°C)$, $q_5 = \dfrac{2.0 \text{ J}}{\text{g }°\text{C}} \times 5.00 \times 10^2 \text{ g} \times 150.°\text{C} = 1.5 \times 10^5 \text{ J}$

$= 150 \text{ kJ}$

$q_{total} = q_1 + q_2 + q_3 + q_4 + q_5 = 21 + 167 + 210 + 1130 + 150 = 1680 \text{ kJ}$

93.   Heat released = $0.250 \text{ g Na} \times \dfrac{1 \text{ mol}}{22.99 \text{ g}} \times \dfrac{368 \text{ kJ}}{2 \text{ mol}} = 2.00 \text{ kJ}$

To melt 50.0 g of ice requires: $50.0 \text{ g ice} \times \dfrac{1 \text{ mol H}_2\text{O}}{18.02 \text{ g}} \times \dfrac{6.01 \text{ kJ}}{\text{mol}} = 16.7 \text{ kJ}$

The reaction doesn't release enough heat to melt all of the ice. The temperature will remain at 0°C.

95.   Total mass H_2O = 18 cubes $\times \dfrac{30.0 \text{ g}}{\text{cube}}$ = 540. g; 540. g $H_2O \times \dfrac{1 \text{ mol H}_2\text{O}}{18.02 \text{ g}}$ = 30.0 mol H_2O

Heat removed to produce ice at –5.0°C:

$$\left(\dfrac{4.18 \text{ J}}{\text{g °C}} \times 540. \text{ g} \times 22.0 \, ^\circ\text{C} \right) + \left(\dfrac{6.01 \times 10^3 \text{ J}}{\text{mol}} \times 30.0 \text{ mol} \right) + \left(\dfrac{2.08 \text{ J}}{\text{g °C}} \times 540. \text{ g} \times 5.0 \, ^\circ\text{C} \right)$$

$$= 4.97 \times 10^4 \text{ J} + 1.80 \times 10^5 \text{ J} + 5.6 \times 10^3 \text{ J} = 2.35 \times 10^5 \text{ J}$$

$2.35 \times 10^5 \text{ J} \times \dfrac{1 \text{ g CF}_2\text{Cl}_2}{158 \text{ J}} = 1.49 \times 10^3 \text{ g CF}_2\text{Cl}_2$ must be vaporized.

97.   See Figures 16.55 and 16.58 for the phase diagrams of H_2O and CO_2. Most substances exhibit only three different phases: solid, liquid, and gas. This is true for H_2O and CO_2. Also typical of phase diagrams is the positive slopes for both the liquid-gas equilibrium line and the solid-gas equilibrium line. This is also true for both H_2O and CO_2. The solid-liquid equilibrium line also generally has a positive slope. This is true for CO_2 but not for H_2O. In the H_2O phase diagram, the slope of the solid-liquid line is negative. The determining factor for the slope of the solid-liquid line is the relative densities of the solid and liquid phases. The solid phase is denser than the liquid phase in most substances; for these substances, the slope of the solid-liquid equilibrium line is positive. For water, the liquid phase is denser than the solid phase, which corresponds to a negative-sloping solid-liquid equilibrium line. Another difference between H_2O and CO_2 is the normal melting points and normal boiling points. The term normal just dictates a pressure of 1 atm. H_2O has a normal melting point (0°C) and a normal boiling point (100°C), but CO_2 does not. At 1 atm pressure, CO_2 only sublimes (goes from the solid phase directly to the gas phase). There are no temperatures at 1 atm for CO_2 where the solid and liquid phases are in equilibrium or where the liquid and gas phases are in equilibrium. There are other differences, but those discussed above are the major ones.

The relationship between melting points and pressure is determined by the slope of the solid-liquid equilibrium line. For most substances (CO_2 included), the positive slope of the solid-liquid line shows a direct relationship between the melting point and pressure. As pressure increases, the melting point increases. Water is just the opposite since the slope of the solid-liquid line in water is negative. Here, the melting point of water is inversely related to the pressure.

For boiling points, the positive slope of the liquid-gas equilibrium line indicates a direct relationship between the boiling point and pressure. This direct relationship is true for all substances, including H_2O and CO_2.

The critical temperature for a substance is defined as the temperature above which the vapor cannot be liquefied no matter what pressure is applied. The critical temperature, like the boiling-point temperature, is directly related to the strength of the intermolecular forces. Since H_2O exhibits relatively strong hydrogen bonding interactions and CO_2 only exhibits London dispersion forces, one would expect a higher critical temperature for H_2O than for CO_2.

99.   The critical temperature is the temperature above which the vapor cannot be liquefied no matter what pressure is applied. Since N_2 has a critical temperature below room temperature (~22°C), it cannot be liquefied at room temperature. NH_3, with a critical temperature above room temperature, can be liquefied at room temperature.

101.   A: solid;   B: liquid;   C: vapor

D: solid + vapor;   E: solid + liquid + vapor

F: liquid + vapor;   G: liquid + vapor;   H: vapor

triple point: E;   critical point: G

Normal freezing point: temperature at which solid-liquid line is at 1.0 atm (see following plot).

Normal boiling point: temperature at which liquid-vapor line is at 1.0 atm (see following plot).

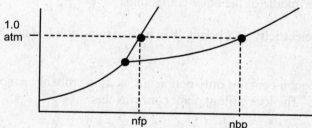

Because the solid-liquid equilibrium line has a positive slope, the solid phase is denser than the liquid phase.

103.   a. two

b. Higher-pressure triple point: graphite, diamond, and liquid; lower-pressure triple point at ~10^7 Pa: graphite, liquid, and vapor

c. It is converted to diamond (the more dense solid form).

d. Diamond is more dense, which is why graphite can be converted to diamond by applying pressure.

105.   Because the density of the liquid phase is greater than the density of the solid phase, the slope of the solid-liquid boundary line is negative (as in H_2O). With a negative slope, the melting points increase with a decrease in pressure, so the normal melting point of X should be greater than 225°C.

107.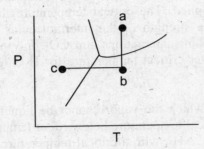

As P is lowered, we go from a to b on the phase diagram. The water boils. The evaporation of the water is endothermic, and the water is cooled (b → c), forming some ice. If the pump is left on, the ice will sublime until none is left. This is the basis of freeze drying.

Additional Exercises

109.   The strength of intermolecular forces determines relative boiling points. The types of intermolecular forces for covalent compounds are London dispersion forces, dipole forces, and hydrogen bonding. Because the three compounds are assumed to have similar molar mass and shape, the strength of the London dispersion forces will be about equal among the three compounds. One of the compounds will be nonpolar, so it only has London dispersion forces. The other two compounds will be polar, so they have additional dipole forces and will boil at a higher temperature than the nonpolar compound. One of the polar compounds probably has an H covalently bonded to either N, O, or F. This gives rise to the strongest type of covalent intermolecular forces, hydrogen bonding. The compound that hydrogen bonds will have the highest boiling point, whereas the polar compound with no hydrogen bonding will boil at a temperature in the middle of the other compounds.

111.   If TiO_2 conducts electricity as a liquid, then it is an ionic solid; if not, then TiO_2 is a network solid.

113. B_2H_6: This compound contains only nonmetals, so it is probably a molecular solid with covalent bonding. The low boiling point confirms this.

        SiO_2: This is the empirical formula for quartz, which is a network solid.

        CsI: This is a metal bonded to a nonmetal, which generally form ionic solids. The electrical conductivity in aqueous solution confirms this.

        W: Tungsten is a metallic solid as the conductivity data confirms.

115.      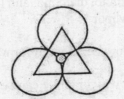   R = radius of the sphere

        r = radius of trigonal hole

For a right-angle triangle (opposite side not drawn in):

$$\cos 30° = \frac{\text{adjacent}}{\text{hypotenuse}} = \frac{R}{R+r}, \quad 0.866 = \frac{R}{R+r}$$

$$(0.866)r = R - (0.866)R, \quad r = \left(\frac{0.134}{0.866}\right) \times R = (0.155)R$$

The cation must have a radius that is 0.155 times the radius of the spheres to just fit into the trigonal hole.

117.   Ar is cubic closest packed. There are four Ar atoms per unit cell and with a face-centered unit cell, the atoms touch along the face diagonal. Let l = length of cube edge.

Face diagonal = $4r = l\sqrt{2}$, $\quad l = 4(190.\ \text{pm})/\sqrt{2} = 537\ \text{pm} = 5.37 \times 10^{-8}\ \text{cm}$

$$\text{Density} = \frac{\text{mass}}{\text{volume}} = \frac{4\ \text{atoms} \times \dfrac{1\ \text{mol}}{6.022 \times 10^{23}\ \text{atoms}} \times \dfrac{39.95\ \text{g}}{\text{mol}}}{(5.37 \times 10^{-8}\ \text{cm})^3} = 1.71\ \text{g/cm}^3$$

119.   Assuming 100.00 g: $28.31\ \text{g O} \times \dfrac{1\ \text{mol}}{15.999\ \text{g}} = 1.769\ \text{mol O}; \quad 71.69\ \text{g Ti} \times \dfrac{1\ \text{mol}}{47.88\ \text{g}}$

$$= 1.497\ \text{mol Ti}$$

Dividing one mole quantity by the other, gives formulas of $\text{TiO}_{1.182}$ or $\text{Ti}_{0.8462}\text{O}$.

For $\text{Ti}_{0.8462}\text{O}$, let $x = \text{Ti}^{2+}$ per mol O^{2-} and $y = \text{Ti}^{3+}$ per mol O^{2-}. Setting up two equations and solving:

$x + y = 0.8462$ and $2x + 3y = 2$; $\quad 2x + 3(0.8462 - x) = 2$

$x = 0.539\ \text{mol Ti}^{2+}/\text{mol O}^{2-}$ and $y = 0.307\ \text{mol Ti}^{3+}/\text{mol O}^{2-}$

$$\frac{0.539}{0.8462} \times 100 = 63.7\% \text{ of the titanium is Ti}^{2+} \text{ and } 36.3\% \text{ is Ti}^{3+}.$$

121.   For a face-centered unit cell, the radius r of an atom is related to the length of a cube edge l by the equation $l = r\sqrt{8}$ (see Exercise 43).

Radius = $r = l/\sqrt{8} = 392 \times 10^{-12}\ \text{m}/\sqrt{8} = 1.39 \times 10^{-10}\ \text{m} = 1.39 \times 10^{-8}\ \text{cm}$

The volume of a unit cell is l^3, so the mass of the unknown metal (X) in a unit cell is:

$$\text{volume} \times \text{density} = (3.92 \times 10^{-8} \text{ cm})^3 \times \frac{21.45 \text{ g X}}{\text{cm}^3} = 1.29 \times 10^{-21} \text{ g X}$$

Because each face-centered unit cell contains four atoms of X:

$$\text{mol X in unit cell} = 4 \text{ atoms X} \times \frac{1 \text{ mol X}}{6.022 \times 10^{23} \text{ atoms X}} = 6.642 \times 10^{-24} \text{ mol X}$$

Therefore, each unit cell contains 1.29×10^{-21} g X, which is equal to 6.642×10^{-24} mol X. The molar mass of X is:

$$\frac{1.29 \times 10^{-21} \text{ g X}}{6.642 \times 10^{-24} \text{ mol X}} = 194 \text{ g/mol}$$

The atomic mass would be 194 amu. From the periodic table, the best choice for the metal is platinum.

123.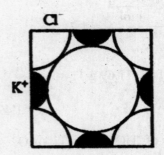

Assuming K^+ and Cl^- just touch along the cube edge l:

$$l = 2(314 \text{ pm}) = 628 \text{ pm} = 6.28 \times 10^{-8} \text{ cm}$$

$$\text{Volume of unit cell} = l^3 = (6.28 \times 10^{-8} \text{ cm})^3$$

The unit cell contains four K^+ and four Cl^- ions. For a unit cell:

$$\text{density} = \frac{4 \text{ KCl formula units} \times \dfrac{1 \text{ mol KCl}}{6.022 \times 10^{23} \text{ formula units}} \times \dfrac{74.55 \text{ g KCl}}{\text{mol KCl}}}{(6.28 \times 10^{-8} \text{ cm})^3}$$

$$= 2.00 \text{ g/cm}^3$$

125. $24.7 \text{ g } C_6H_6 \times \dfrac{1 \text{ mol}}{78.11 \text{ g}} = 0.316 \text{ mol } C_6H_6$

$$P_{C_6H_6} = \frac{nRT}{V} = \frac{0.316 \text{ mol} \times \dfrac{0.08296 \text{ L atm}}{\text{K mol}} \times 293.2 \text{ K}}{100.0 \text{ L}} = 0.0760 \text{ atm, or } 57.8 \text{ torr}$$

127. $w = -P\Delta V;$ assume constant P of 1.00 atm.

$$V_{373} = \frac{nRT}{P} = \frac{1.00 (0.08206)(373)}{1.00} = 30.6 \text{ L for 1 mol of water vapor}$$

Because the density of $H_2O(l)$ is 1.00 g/cm^3, 1.00 mol of $H_2O(l)$ occupies 18.0 cm^3 or 0.0180 L.

$w = -1.00$ atm (30.6 L – 0.0180 L) = –30.6 L atm

$w = -30.6$ L atm × 101.3 J L^{-1} atm^{-1} = -3.10×10^3 J = -3.10 kJ

$\Delta E = q + w = 41.16$ kJ $- 3.10$ kJ $= 38.06$ kJ

$\dfrac{38.06}{41.16} \times 100 = 92.47\%$ of the energy goes to increase the internal energy of the water.

The remainder of the energy (7.53%) goes to do work against the atmosphere.

129. The three sheets of B's form a cube. In this cubic unit cell there are B atoms at every face, B atoms at every edge, B atoms at every corner, and one B atom in the center. A representation of this cube is:

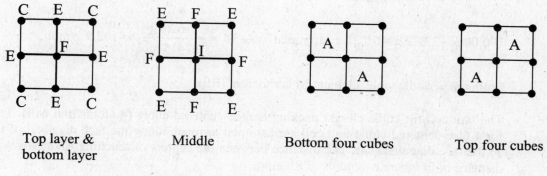

Top layer & bottom layer

Middle

Bottom four cubes

Top four cubes

C = corner;   E = edge;   F = face;   I = inside

Each unit cell of B atoms contains four A atoms. The number of B atoms in each unit cell is:

(8 corners × 1/8) + (6 faces × 1/2) + (12 edges × 1/4) + (1 middle B) = 8 B atoms

The empirical formula is AB_2.

Each A atom is in a cubic hole of B atoms, so eight B atoms surround each A atom. This will also be true in the extended lattice. The structure of B atoms in the unit cell is a cubic arrangement with B atoms at every face, edge, corner, and center of the cube.

Challenge Problems

131. $E = \dfrac{h^2(n_x^2 + n_y^2 + n_z^2)}{8mL^2}$; $E_{111} = \dfrac{3h^2}{8mL^2}$; $E_{112} = \dfrac{6h^2}{8mL^2}$; $\Delta E = E_{112} - E_{111} = \dfrac{3h^2}{8mL^2}$

$\Delta E = \dfrac{hc}{\lambda} = \dfrac{(6.626 \times 10^{-34}\ \text{J s})(2.998 \times 10^8\ \text{m/s})}{9.50 \times 10^{-9}\ \text{m}} = 2.09 \times 10^{-17}$ J

$$L^2 = \frac{3h^2}{8m\Delta E}, \quad L = \left(\frac{3h^2}{8m\Delta E}\right)^{1/2} = \left[\frac{3(6.626 \times 10^{-34} \text{ J s})^2}{8(9.109 \times 10^{-31} \text{ kg})(2.09 \times 10^{-17} \text{ J})}\right]^{1/2}$$

$$L = 9.30 \times 10^{-11} \text{ m} = 93.0 \text{ pm}$$

The sphere that fits in this cube will touch the cube at the center of each face. The diameter of the sphere will equal the length of the cube. So:

$$2r = L, \quad r = 93.0/2 = 46.5 \text{ pm}$$

133.   Assuming 100.00 g of MO_2:

$$23.72 \text{ g O} \times \frac{1 \text{ mol O}}{16.00 \text{ g O}} = 1.483 \text{ mol O}$$

$$1.483 \text{ mol O} \times \frac{1 \text{ mol M}}{2 \text{ mol O}} = 0.7415 \text{ mol M}$$

$$100.00 \text{ g} - 23.72 \text{ g} = 76.28 \text{ g M}; \quad \text{molar mass M} = \frac{76.28 \text{ g}}{0.7415 \text{ mol}} = 102.9 \text{ g/mol}$$

From the periodic table, element M is rhodium (Rh).

The unit cell for cubic closest packing is face-centered cubic (4 atoms/unit cell). The atoms for a face-centered cubic unit cell are assumed to touch along the face diagonal of the cube, so the face diagonal = 4r. The distance between the centers of touching Rh atoms will be the distance of 2r where r = radius of Rh atom.

Face diagonal = $\sqrt{2} \; l$, where l = cube edge.

Face diagonal = 4r = $2 \times 269.0 \times 10^{-12}$ m = 5.380×10^{-10} m

$$\sqrt{2} \; l = 4r = 5.38 \times 10^{-10} \text{ m}, \quad l = \frac{5.38 \times 10^{-10} \text{ m}}{\sqrt{2}} = 3.804 \times 10^{-10} \text{ m} = 3.804 \times 10^{-8} \text{ cm}$$

$$\text{Density} = \frac{4 \text{ atoms Rh} \times \dfrac{1 \text{ mol Rh}}{6.0221 \times 10^{23} \text{ atoms}} \times \dfrac{102.9 \text{ g Rh}}{\text{mol Rh}}}{(3.804 \times 10^{-8} \text{ cm})^3} = 12.42 \text{ g/cm}^3$$

135.   For water vapor at 30.0°C and 31.824 torr:

$$\text{density} = \frac{P(\text{molar mass})}{RT} = \frac{\left(\dfrac{31.824 \text{ atm}}{760}\right)\left(\dfrac{18.015 \text{ g}}{\text{mol}}\right)}{\dfrac{0.08206 \text{ L atm}}{\text{K mol}} \times 303.2 \text{ K}} = 0.03032 \text{ g/L}$$

The volume of one molecule is proportional to d^3, where d is the average distance between molecules. For a large sample of molecules, the volume is still proportional to d^3. So:

$$\frac{V_{gas}}{V_{liq}} = \frac{d^3_{gas}}{d^3_{liq}}$$

If we have 0.99567 g H_2O, then V_{liq}= 1.0000 cm^3 = 1.0000 × 10^{-3} L.

V_{gas} = 0.99567 g × 1 L/0.03032 g = 32.84 L

$$\frac{d^3_{gas}}{d^3_{liq}} = \frac{32.84 \text{ L}}{1.0000 \times 10^{-3} \text{ L}} = 3.284 \times 10^4, \quad \frac{d_{gas}}{d_{liq}} = (3.284 \times 10^4)^{1/3} = 32.02, \quad \frac{d_{liq}}{d_{gas}} = 0.03123$$

137.   a.   Structure a:

Ba: 2 Ba inside unit cell; Tl: 8 corners × $\dfrac{1/8 \text{ Tl}}{\text{corner}}$ = 1 Tl; Cu: 4 edges × $\dfrac{1/4 \text{ Cu}}{\text{edge}}$ = 1 Cu

O: 6 faces × $\dfrac{1/2 \text{ O}}{\text{face}}$ + 8 edges × $\dfrac{1/4 \text{ O}}{\text{edge}}$ = 5 O

Formula = $TlBa_2CuO_5$.

Structure b:

Tl and Ba are the same as in structure a.

Ca: 1 Ca inside unit cell; Cu: 8 edges × $\dfrac{1/4 \text{ Cu}}{\text{edge}}$ = 2 Cu

O: 10 faces × $\dfrac{1/2 \text{ O}}{\text{edge}}$ + 8 edges × $\dfrac{1/4 \text{ O}}{\text{edge}}$ = 7 O

Formula = $TlBa_2CaCu_2O_7$.

Structure c:

Tl and Ba are the same and two Ca atoms are located inside the unit cell.

Cu: 12 edges × $\dfrac{1/4 \text{ Cu}}{\text{edge}}$ = 3 Cu; O: 14 faces × $\dfrac{1/2 \text{ O}}{\text{face}}$ + 8 edges × $\dfrac{1/4 \text{ O}}{\text{edge}}$ = 9 O

Formula = $TlBa_2Ca_2Cu_3O_9$.

Structure d: Following similar calculations, formula = $TlBa_2Ca_3Cu_4O_{11}$.

b. Structure a has one planar sheet of Cu and O atoms and the number increases by one for each of the remaining structures. The order of superconductivity temperature from lowest to highest temperature is a < b < c < d.

c. $TlBa_2CuO_5$: $3 + 2(2) + x + 5(-2) = 0$, $x = +3$
    Only Cu^{3+} is present in each formula unit.

    $TlBa_2CaCu_2O_7$: $3 + 2(2) + 2 + 2(x) + 7(-2) = 0$, $x = +5/2$
    Each formula unit contains 1 Cu^{2+} and 1 Cu^{3+}.

    $TlBa_2Ca_2Cu_3O_9$: $3 + 2(2) + 2(2) + 3(x) + 9(-2) = 0$, $x = +7/3$
    Each formula unit contains 2 Cu^{2+} and 1 Cu^{3+}.

    $TlBa_2Ca_3Cu_4O_{11}$: $3 + 2(2) + 3(2) + 4(x) + 11(-2) = 0$, $x = +9/4$
    Each formula unit contains 3 Cu^{2+} and 1 Cu^{3+}.

d. This superconductor material achieves variable copper oxidation states by varying the numbers of Ca, Cu, and O in each unit cell. The mixtures of copper oxidation states is discussed in part c. The superconductor material in Exercise 16.75 achieves variable copper oxidation states by omitting oxygen at various sites in the lattice.

139.   For an octahedral hole, the geometry is:

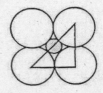

$2R + 2r$
$45°$
$2R$

$R = Cl^-$ radius

$r = Li^+$ radius

From the diagram: $\cos 45° = \dfrac{\text{adjacent}}{\text{hypotenuse}} = \dfrac{2R}{2R + 2r} = \dfrac{R}{R + r}$, $0.707 = \dfrac{R}{R + r}$

$R = 0.707(R + r)$, $r = (0.414)R$

LiCl unit cell:

length of cube edge = $2R + 2r = 514$ pm

$2R + 2(0.414\ R) = 514$ pm, $R = 182$ pm $= Cl^-$ radius

$2(182\ pm) + 2r = 514$ pm, $r = 75$ pm $= Li^+$ radius

From Figure 13.8, the Li^+ radius is 60 pm, and the Cl^- radius is 181 pm. The Li^+ ion is much smaller than calculated. This probably means that the ions are not actually in contact with each other. The octahedral holes are larger than the Li^+ ion.

CHAPTER 17

PROPERTIES OF SOLUTIONS

Solution Composition

13. a. $HNO_3(l) \rightarrow H^+(aq) + NO_3^-(aq)$ b. $Na_2SO_4(s) \rightarrow 2\ Na^+(aq) + SO_4^{2-}(aq)$

   c. $Al(NO_3)_3(s) \rightarrow Al^{3+}(aq) + 3\ NO_3^-(aq)$ d. $SrBr_2(s) \rightarrow Sr^{2+}(aq) + 2\ Br^-(aq)$

   e. $KClO_4(s) \rightarrow K^+(aq) + ClO_4^-(aq)$ f. $NH_4Br(s) \rightarrow NH_4^+(aq) + Br^-(aq)$

   g. $NH_4NO_3(s) \rightarrow NH_4^+(aq) + NO_3^-(aq)$ h. $CuSO_4(s) \rightarrow Cu^{2+}(aq) + SO_4^{2-}(aq)$

   i. $NaOH(s) \rightarrow Na^+(aq) + OH^-(aq)$

15. $\text{Molality} = \dfrac{40.0\text{ g EG}}{60.0\text{ g H}_2\text{O}} \times \dfrac{1000\text{ g}}{\text{kg}} \times \dfrac{1\text{ mol EG}}{62.07\text{ g}} = 10.7\text{ mol/kg}$

   $\text{Molarity} = \dfrac{40.0\text{ g EG}}{100.0\text{ g solution}} \times \dfrac{1.05\text{ g}}{\text{cm}^3} \times \dfrac{1000\text{ cm}^3}{\text{L}} \times \dfrac{1\text{ mol}}{62.07\text{ g}} = 6.77\text{ mol/L}$

   $40.0\text{ g EG} \times \dfrac{1\text{ mol}}{62.07\text{ g}} = 0.644\text{ mol EG}; \quad 60.0\text{ g H}_2\text{O} \times \dfrac{1\text{ mol}}{18.02\text{ g}} = 3.33\text{ mol H}_2\text{O}$

   $\chi_{EG} = \dfrac{0.644}{3.33 + 0.644} = 0.162 = \text{mole fraction ethylene glycol}$

17. $50.0\text{ mL toluene} \times \dfrac{0.867\text{ g}}{\text{mL}} = 43.4\text{ g toluene}; \quad 125\text{ mL benzene} \times \dfrac{0.874\text{ g}}{\text{mL}} = 109\text{ g benzene}$

   $\text{Mass \% toluene} = \dfrac{\text{mass of toluene}}{\text{total mass}} \times 100 = \dfrac{43.4\text{ g}}{43.4\text{ g} + 109\text{ g}} \times 100 = 28.5\%$

   $\text{Molarity} = \dfrac{43.4\text{ g toluene}}{175\text{ mL soln}} \times \dfrac{1000\text{ mL}}{\text{L}} \times \dfrac{1\text{ mol toluene}}{92.13\text{ g toluene}} = 2.69\text{ mol/L}$

351

$$\text{Molality} = \frac{43.4 \text{ g toluene}}{109 \text{ g benzene}} \times \frac{1000 \text{ g}}{\text{kg}} \times \frac{1 \text{ mol toluene}}{92.13 \text{ g toluene}} = 4.32 \text{ mol/kg}$$

$$43.4 \text{ g toluene} \times \frac{1 \text{ mol}}{92.13 \text{ g}} = 0.471 \text{ mol toluene}$$

$$109 \text{ g benzene} \times \frac{1 \text{ mol benzene}}{78.11 \text{ g benzene}} = 1.40 \text{ mol benzene}; \quad \chi_{\text{toluene}} = \frac{0.471}{0.471 + 1.40} = 0.252$$

19.   If we have 1.00 L of solution:

$$1.37 \text{ mol citric acid} \times \frac{192.1 \text{ g}}{\text{mol}} = 263 \text{ g citric acid}$$

$$1.00 \times 10^3 \text{ mL solution} \times \frac{1.10 \text{ g}}{\text{mL}} = 1.10 \times 10^3 \text{ g solution}$$

$$\text{Mass \% of citric acid} = \frac{263 \text{ g}}{1.10 \times 10^3 \text{ g}} \times 100 = 23.9\%$$

In 1.00 L of solution, we have 263 g citric acid and $(1.10 \times 10^3 - 263) = 840$ g of H_2O.

$$\text{Molality} = \frac{1.37 \text{ mol citric acid}}{0.84 \text{ kg } H_2O} = 1.6 \text{ mol/kg}$$

$$840 \text{ g } H_2O \times \frac{1 \text{ mol}}{18.0 \text{ g}} = 47 \text{ mol } H_2O; \quad \chi_{\text{citric acid}} = \frac{1.37}{47 + 1.37} = 0.028$$

21.   Because the density of water is 1.00 g/mL, 100.0 mL of water has a mass of 100. g.

$$\text{Density} = \frac{\text{mass}}{\text{volume}} = \frac{10.0 \text{ g } H_3PO_4 + 100. \text{ g } H_2O}{104 \text{ mL}} = 1.06 \text{ g/mL} = 1.06 \text{ g/cm}^3$$

$$\text{Mol } H_3PO_4 = 10.0 \text{ g} \times \frac{1 \text{ mol}}{97.99 \text{ g}} = 0.102 \text{ mol } H_3PO_4$$

$$\text{Mol } H_2O = 100. \text{ g} \times \frac{1 \text{ mol}}{18.02 \text{ g}} = 5.55 \text{ mol } H_2O$$

$$\text{Mole fraction of } H_3PO_4 = \frac{0.102 \text{ mol } H_3PO_4}{(0.102 + 5.55) \text{ mol}} = 0.0180$$

$$\chi_{H_2O} = 1.0000 - 0.0180 = 0.9820$$

CHAPTER 17   PROPERTIES OF SOLUTIONS

$$\text{Molarity} = \frac{0.102 \text{ mol } H_3PO_4}{0.104 \text{ L soln}} = 0.981 \text{ mol/L}$$

$$\text{Molality} = \frac{0.102 \text{ mol } H_3PO_4}{0.100 \text{ kg solvent}} = 1.02 \text{ mol/kg}$$

Thermodynamics of Solutions and Solubility

23.   "Like dissolves like" refers to the nature of the intermolecular forces. Polar solutes and ionic solutes dissolve in polar solvents because the types of intermolecular forces present in solute and solvent are similar. When they dissolve, the strengths of the intermolecular forces in solution are about the same as in pure solute and pure solvent. The same is true for nonpolar solutes in nonpolar solvents. The strengths of the intermolecular forces (London dispersion forces) are about the same in solution as in pure solute and pure solvent. In all cases of like dissolves like, the magnitude of ΔH_{soln} is either a small positive number (endothermic) or a small negative number (exothermic), with a value close to zero. For polar solutes in nonpolar solvents and vice versa, ΔH_{soln} is a very large, unfavorable value (very endothermic). Because the energetics are so unfavorable, polar solutes do not dissolve in nonpolar solvents, and vice versa.

25.   Although the enthalpy change is endothermic when NaCl dissolves in water, the magnitude of ΔH_{soln} will be a value close to zero ($\Delta H_{soln} \approx 0$). This is typical for soluble ionic compounds.   So energy is not the reason why ionic solids like NaCl are so soluble in water. The answer lies in nature's tendency toward the higher probability of the mixed state; solutions form due to an increase in entropy ($\Delta S_{soln} > 0$). In the mixed state, the Na^+ and Cl^- ions have access to a larger volume and therefore have more positions available to them. When a solution forms, positional probability increases, which translates into an increase in entropy. The positive ΔS_{soln} furnishes the driving force for NaCl to dissolve in water.

27.   Using Hess's law:

$$NaI(s) \rightarrow Na^+(g) + I^-(g) \qquad\qquad \Delta H = -\Delta H_{LE} = -(-686 \text{ kJ/mol})$$
$$Na^+(g) + I^-(g) \rightarrow Na^+(aq) + I^-(aq) \qquad \Delta H = \Delta H_{hyd} = -694 \text{ kJ/mol}$$

$$\overline{NaI(s) \rightarrow Na^+(aq) + I^-(aq) \qquad\qquad\qquad \Delta H_{soln} = -8 \text{ kJ/mol}}$$

ΔH_{soln} refers to the heat released or gained when a solute dissolves in a solvent. Here, an ionic compound dissolves in water.

29.   Both $Al(OH)_3$ and NaOH are ionic compounds. Because the lattice energy is proportional to the charge of the ions, the lattice energy of aluminum hydroxide is greater than that of sodium hydroxide. The attraction of water molecules for Al^{3+} and OH^- cannot overcome the larger lattice energy, and $Al(OH)_3$ is insoluble. For NaOH, the favorable hydration energy is large enough to overcome the smaller lattice energy, and NaOH is soluble.

31.   Water exhibits H-bonding in the pure state and is classified as a polar solvent. Water will dissolve other polar solutes and ionic solutes.

a. NH_3; similar to water, NH_3 is capable of H-bonding, unlike PH_3.

b. CH_3CN; CH_3CN is polar, while CH_3CH_3 is nonpolar.

c. CH_3CO_2H; CH_3CO_2H is capable of H-bonding, unlike the other compound.

33.   As the length of the hydrocarbon chain increases, the solubility decreases. The —OH end of the alcohols can hydrogen-bond with water. The hydrocarbon chain, however, is basically nonpolar and interacts poorly with water. As the hydrocarbon chain gets longer, a greater portion of the molecule cannot interact with the water molecules, and the solubility decreases; i.e., the effect of the —OH group decreases as the alcohols get larger.

35.   Structure effects refer to solute and solvent having similar polarities in order for solution formation to occur. Hydrophobic solutes are mostly nonpolar substances that are "water-fearing." Hydrophilic solutes are mostly polar or ionic substances that are "water-loving." Pressure has little effect on the solubilities of solids or liquids; it does significantly affect the solubility of a gas. Henry's law states that the amount of a gas dissolved in a solution is directly proportional to the pressure of the gas above the solution (C = kP). The equation for Henry's law works best for dilute solutions of gases that do not dissociate in or react with the solvent. $HCl(g)$ does not follow Henry's law because it dissociates into $H^+(aq)$ and $Cl^-(aq)$ in solution (HCl is a strong acid). For O_2 and N_2, Henry's law works well since these gases do not react with the water solvent.

An increase in temperature can either increase or decrease the solubility of a solid solute in water. It is true that a solute dissolves more rapidly with an increase in temperature, but the amount of solid solute that dissolves to form a saturated solution can either decrease or increase with temperature. The temperature effect is difficult to predict for solid solutes. However, the temperature effect for gas solutes is easier to predict because the solubility of a gas typically decreases with increasing temperature.

37. $P_{gas} = kC$, $0.790 \text{ atm} = k \times \dfrac{8.21 \times 10^{-4} \text{ mol}}{L}$, $k = 962$ L atm/mol

$P_{gas} = kC$, $1.10 \text{ atm} = \dfrac{962 \text{ L atm}}{mol} \times C$, $C = 1.14 \times 10^{-3}$ mol/L

39.   As the temperature increases the gas molecules will have a greater average kinetic energy. A greater fraction of the gas molecules in solution will have a kinetic energy greater than the attractive forces between the gas molecules and the solvent molecules. More gas molecules are able to escape to the vapor phase and the solubility of the gas decreases.

Vapor Pressures of Solution

41. $P_{soln} = \chi_{C_2H_5OH} P^o_{C_2H_5OH}$; $\chi_{C_2H_5OH} = \dfrac{\text{mol } C_2H_5OH \text{ in solution}}{\text{total mol in solution}}$

$53.6 \text{ g } C_3H_8O_3 \times \dfrac{1 \text{ mol } C_3H_8O_3}{92.09 \text{ g}} = 0.582 \text{ mol } C_3H_8O_3$

$133.7 \text{ g } C_2H_5OH \times \dfrac{1 \text{ mol } C_2H_5OH}{46.07 \text{ g}} = 2.90 \text{ mol } C_2H_5OH$; total mol $= 0.582 + 2.90 = 3.48$ mol

$113 \text{ torr} = \dfrac{2.90 \text{ mol}}{3.48 \text{ mol}} \times P^o_{C_2H_5OH}$, $P^o_{C_2H_5OH} = 136$ torr

43. $P = \chi P^o$; $710.0 \text{ torr} = \chi(760.0 \text{ torr})$, $\chi = 0.9342 =$ mole fraction of methanol

45. $P_B = \chi_B P^o_B$, $\chi_B = P_B / P^o_B = 0.900 \text{ atm}/0.930 \text{ atm} = 0.968$

$0.968 = \dfrac{\text{mol benzene}}{\text{total mol}}$;   mol benzene $= 78.11 \text{ g } C_6H_6 \times \dfrac{1 \text{ mol}}{78.11 \text{ g}} = 1.000$ mol

Let $x =$ mol solute; then: $\chi_B = 0.968 = \dfrac{1.000 \text{ mol}}{1.000 + x}$, $0.968 + (0.968)x = 1.000$, $x = 0.033$ mol

Molar mass $= \dfrac{10.0 \text{ g}}{0.033 \text{ mol}} = 303 \text{ g/mol} \approx 3.0 \times 10^2 \text{ g/mol}$

47. **a.** $25 \text{ mL } C_5H_{12} \times \dfrac{0.63 \text{ g}}{\text{mL}} \times \dfrac{1 \text{ mol}}{72.15} = 0.22 \text{ mol } C_5H_{12}$

$45 \text{ mL } C_6H_{14} \times \dfrac{0.66 \text{ g}}{\text{mL}} \times \dfrac{1 \text{ mol}}{86.17} = 0.34 \text{ mol } C_6H_{14}$; total mol $= 0.22 + 0.34 = 0.56$ mol

$\chi^L_{pen} = \dfrac{\text{mol pentane in solution}}{\text{total mol in solution}} = \dfrac{0.22 \text{ mol}}{0.56 \text{ mol}} = 0.39$, $\chi^L_{hex} = 1.00 - 0.39 = 0.61$

$P_{pen} = \chi^L_{pen} P^o_{pen} = 0.39(511 \text{ torr}) = 2.0 \times 10^2 \text{ torr}$; $P_{hex} = 0.61(150. \text{ torr}) = 92$ torr

$P_{total} = P_{pen} + P_{hex} = 2.0 \times 10^2 + 92 = 292 \text{ torr} = 290$ torr

b.   From Chapter 5 on gases, the partial pressure of a gas is proportional to the number of moles of gas present. For the vapor phase:

$\chi^V_{pen} = \dfrac{\text{mol pentane in vapor}}{\text{total mol vapor}} = \dfrac{P_{pen}}{P_{total}} = \dfrac{2.0 \times 10^2 \text{ torr}}{290 \text{ torr}} = 0.69$

Note: In the *Solutions Guide*, we added V or L to the mole fraction symbol to emphasize which value we are solving. If the L or V is omitted, then the liquid phase is assumed.

49. $P_{total} = P_{meth} + P_{prop}$, $174 \text{ torr} = \chi^L_{meth}(303 \text{ torr}) + \chi^L_{prop}(44.6 \text{ torr})$; $\chi^L_{prop} = 1.000 - \chi^L_{meth}$

$174 = 303\chi^L_{meth} + (1.000 - \chi^L_{meth})44.6 \text{ torr}$, $\dfrac{129}{258} = \chi^L_{meth} = 0.500$

$\chi^L_{prop} = 1.000 - 0.500 = 0.500$

51. $50.0 \text{ g CH}_3\text{COCH}_3 \times \dfrac{1 \text{ mol}}{58.08 \text{ g}} = 0.861 \text{ mol acetone}$

$50.0 \text{ g CH}_3\text{OH} \times 1 \text{ mol}/32.04 \text{ g} = 1.56 \text{ mol methanol}$

$\chi^L_{acetone} = \dfrac{0.861}{0.861 + 1.56} = 0.356$; $\chi^L_{methanol} = 1.000 - \chi^L_{acetone} = 0.644$

$P_{total} = P_{methanol} + P_{acetone} = 0.644(143 \text{ torr}) + 0.356(271 \text{ torr}) = 92.1 \text{ torr} + 96.5 \text{ torr}$

$= 188.6 \text{ torr}$

Because partial pressures are proportional to the moles of gas present, in the vapor phase:

$\chi^V_{acetone} = \dfrac{P_{acetone}}{P_{total}} = \dfrac{96.5 \text{ torr}}{188.6 \text{ torr}} = 0.512$; $\chi^V_{methanol} = 1.000 - 0.512 = 0.488$

The actual vapor pressure of the solution (161 torr) is less than the calculated pressure assuming ideal behavior (188.6 torr). Therefore, the solution exhibits negative deviations from Raoult's law. This occurs when the solute-solvent interactions are stronger than in pure solute and pure solvent.

53.   No, the solution is not ideal. For an ideal solution, the strengths of intermolecular forces in the solution are the same as in pure solute and pure solvent. This results in $\Delta H_{soln} = 0$ for an ideal solution. ΔH_{soln} for methanol-water is not zero. Because $\Delta H_{soln} < 0$, this solution exhibits negative deviation from Raoult's law.

55.   Solutions of A and B have vapor pressures less than ideal (see Figure 17.11 of the text), so this plot shows negative deviations from Rault's law. Negative deviations occur when the intermolecular forces are stronger in solution than in pure solvent and solute. This results in an exothermic enthalpy of solution. The only statement that is false is e. A substance boils when the vapor pressure equals the external pressure. Because $\chi_B = 0.6$ has a lower vapor pressure at the temperature of the plot than either pure A or pure B, one would expect this solution to require the highest temperature in order for the vapor pressure to reach the external pressure. Therefore, the solution with $\chi_B = 0.6$ will have a higher boiling point than either pure A or pure B. (Note that because $P^o_B > P^o_A$, B is more volatile than A.)

Colligative Properties

57. $\pi = MRT$, $M = \dfrac{\pi}{RT} = \dfrac{15 \text{ atm}}{0.08206 \text{ L atm K}^{-1} \text{ mol}^{-1} \times 295 \text{ K}} = 0.62 \text{ M}$

$$\frac{0.62 \text{ mol}}{L} \times \frac{342.30 \text{ g}}{\text{mol } C_{12}H_{22}O_{11}} = 212 \text{ g/L} \approx 210 \text{ g/L}$$

Dissolve 210 g of sucrose in some water and dilute to 1.0 L in a volumetric flask. To get 0.62 ±0.01 mol/L, we need 212 ±3 g sucrose.

59.   Molality = $m = \dfrac{\text{mol solute}}{\text{kg solvent}} = \dfrac{27.0 \text{ g } N_2H_4CO}{150.0 \text{ g } H_2O} \times \dfrac{1000 \text{ g}}{\text{kg}} \times \dfrac{1 \text{ mol } N_2H_4CO}{60.06 \text{ g } N_2H_4CO} = 3.00$ molal

$$\Delta T_b = K_b m = \frac{0.51 \,^{\circ}\text{C}}{\text{molal}} \times 3.00 \text{ molal} = 1.5\,^{\circ}\text{C}$$

The boiling point is raised from 100.0°C to 101.5°C (assuming P = 1 atm).

61.   Molality = $m = \dfrac{50.0 \text{ g } C_2H_6O_2}{50.0 \text{ g } H_2O} \times \dfrac{1000 \text{ g}}{\text{kg}} \times \dfrac{1 \text{ mol}}{62.07 \text{ g}} = 16.1$ mol/kg

$\Delta T_f = K_f m = 1.86 \,^{\circ}\text{C/molal} \times 16.1 \text{ molal} = 29.9\,^{\circ}\text{C};$ $T_f = 0.0\,^{\circ}\text{C} - 29.9\,^{\circ}\text{C} = -29.9\,^{\circ}\text{C}$

$\Delta T_b = K_b m = 0.51\,^{\circ}\text{C/molal} \times 16.1 \text{ molal} = 8.2\,^{\circ}\text{C};$ $T_b = 100.0\,^{\circ}\text{C} + 8.2\,^{\circ}\text{C} = 108.2\,^{\circ}\text{C}$

63. $m = \dfrac{24.0 \text{ g} \times \dfrac{1 \text{ mol}}{58.0 \text{ g}}}{0.600 \text{ kg}} = 0.690 \text{ mol/kg};$ $\Delta T_b = K_b m = 0.51\,^{\circ}\text{C kg/mol} \times 0.690 \text{ mol/kg} = 0.35\,^{\circ}\text{C}$

$T_b = 99.725\,^{\circ}\text{C} + 0.35\,^{\circ}\text{C} = 100.08\,^{\circ}\text{C}$

65. $\Delta T_b = 77.85\,^{\circ}\text{C} - 76.50\,^{\circ}\text{C} = 1.35\,^{\circ}\text{C};$ $m = \dfrac{\Delta T_b}{K_b} = \dfrac{1.35 \,^{\circ}\text{C}}{5.03 \,^{\circ}\text{C kg/mol}} = 0.268$ mol/kg

Mol biomolecule = 0.0150 kg solvent $\times \dfrac{0.268 \text{ mol hydrocarbon}}{\text{kg solvent}} = 4.02 \times 10^{-3}$ mol

From the problem, 2.00 g biomolecule was used, which must contain 4.02×10^{-3} mol biomolecule. The molar mass of the biomolecule is:

$$\frac{2.00 \text{ g}}{4.02 \times 10^{-3} \text{ mol}} = 498 \text{ g/mol}$$

67. $\Delta T_f = K_f m,$ $m = \dfrac{\Delta T_f}{K_f} = \dfrac{0.300\,^{\circ}\text{C}}{5.12 \,^{\circ}\text{C kg/mol}} = \dfrac{5.86 \times 10^{-2} \text{ mol thyroxine}}{\text{kg benzene}}$

The moles of thyroxine present are:

$$0.0100 \text{ kg benzene} \times \frac{5.86 \times 10^{-2} \text{ mol thyroxine}}{\text{kg benzene}} = 5.86 \times 10^{-4} \text{ mol thyroxine}$$

From the problem, 0.455 g thyroxine was used; this must contain 5.86×10^{-4} mol thyroxine. The molar mass of the thyroxine is:

$$\text{molar mass} = \frac{0.455 \text{ g}}{5.86 \times 10^{-4} \text{ mol}} = 776 \text{ g/mol}$$

69. $M = \dfrac{1.0 \text{ g}}{\text{L}} \times \dfrac{1 \text{ mol}}{9.0 \times 10^4 \text{ g}} = 1.1 \times 10^{-5} \text{ mol/L}; \quad \pi = MRT$

At 298 K: $\pi = \dfrac{1.1 \times 10^{-5} \text{ mol}}{\text{L}} \times \dfrac{0.08206 \text{ L atm}}{\text{K mol}} \times 298 \text{ K} \times \dfrac{760 \text{ torr}}{\text{atm}}, \quad \pi = 0.20 \text{ torr}$

Because $d = 1.0 \text{ g/cm}^3$, 1.0 L of solution has a mass of 1.0 kg. Because only 1.0 g of protein is present per liter solution, 1.0 kg of H_2O is present, and molality equals molarity to the correct number of significant figures.

$$\Delta T_f = K_f m = \frac{1.86 \text{ }^\circ\text{C}}{\text{molal}} \times 1.1 \times 10^{-5} \text{ molal} = 2.0 \times 10^{-5} \text{ }^\circ\text{C}$$

71. $\pi = MRT = \dfrac{0.1 \text{ mol}}{\text{L}} \times \dfrac{0.08206 \text{ L atm}}{\text{K mol}} \times 298 \text{ K} = 2.45 \text{ atm} \approx 2 \text{ atm}$

$\pi = 2 \text{ atm} \times \dfrac{760 \text{ mm Hg}}{\text{atm}} \approx 2000 \text{ mm} \approx 2 \text{ m}$

The osmotic pressure would support a mercury column of approximately 2 m. The height of a fluid column in a tree will be higher because Hg is more dense than the fluid in a tree. If we assume the fluid in a tree is mostly H_2O, then the fluid has a density of 1.0 g/cm^3. The density of Hg is 13.6 g/cm^3.

Height of fluid $\approx$ 2 m $\times$ 13.6 $\approx$ 30 m

73.   With addition of salt or sugar, the osmotic pressure inside the fruit cells (and bacteria) is less than outside the cell. Water will leave the cells, which will dehydrate any bacteria present, causing them to die.

Properties of Electrolyte Solutions

75. $P = \chi P^\circ$; 19.6 torr = χ_{H_2O}(23.8 torr), $\chi_{H_2O} = 0.824$; $\chi_{\text{solute}} = 1.000 - 0.824 = 0.176$

0.176 is the mole fraction of all the solute particles present. Because NaCl dissolves to produce two ions in solution (Na^+ and Cl^-), 0.176 is the mole fraction of Na^+ and Cl^- ions present. The mole fraction of NaCl is 1/2 (0.176) = 0.0880 = χ_{NaCl}.

At 45°C, $P_{\text{soln}} = 0.824(71.9 \text{ torr}) = 59.2 \text{ torr}$.

77. $Na_3PO_4(s) \rightarrow 3\ Na^+(aq) + PO_4^{3-}(aq)$, i = 4.0; $CaBr_2(s) \rightarrow Ca^{2+}(aq) + 2\ Br^-(aq)$, i = 3.0

$KCl(s) \rightarrow K^+(aq) + Cl^-(aq)$, i = 2.0

The effective particle concentrations of the solutions are (assuming complete dissociation):

4.0(0.010 molal) = 0.040 molal for the Na_3PO_4 solution; 3.0(0.020 molal) = 0.060 molal for the $CaBr_2$ solution; 2.0(0.020 molal) = 0.040 molal for the KCl solution; slightly greater than 0.020 molal for the HF solution because HF only partially dissociates in water (it is a weak acid).

a.   The 0.010 *m* Na_3PO_4 solution and the 0.020 *m* KCl solution both have effective particle concentrations of 0.040 *m* (assuming complete dissociation), so both of these solutions should have the same boiling point as the 0.040 *m* $C_6H_{12}O_6$ solution (a nonelectrolyte).

b. $P = \chi P°$; as the solute concentration decreases, the solvent's vapor pressure increases because χ increases. Therefore, the 0.020 *m* HF solution will have the highest vapor pressure because it has the smallest effective particle concentration.

c. $\Delta T = K_f m$; the 0.020 *m* $CaBr_2$ solution has the largest effective particle concentration, so it will have the largest freezing point depression (largest ΔT).

79.   There are six cations and six anions in the illustration which indicates six solute formula units initially. There are a total of 10 solute particles in solution (a combined ion pair counts as one solute particle). So the value for the van't Hoff factor is:

$$i = \frac{\text{moles of particles in solution}}{\text{moles of solute dissolved}} = \frac{10}{6} = 1.67$$

81.   For $CaCl_2$: $i = \dfrac{\Delta T_f}{K_f m} = \dfrac{0.440°C}{1.86\ °C/molal \times 0.091\ molal} = 2.6$

Percent $CaCl_2$ ionized $= \dfrac{2.6 - 1.0}{3.0 - 1.0} \times 100 = 80.\%$; 20.% ion association occurs.

For CsCl: $i = \dfrac{\Delta T_f}{K_f m} = \dfrac{0.320°C}{1.86\ °C/molal \times 0.091\ molal} = 1.9$

Percent CsCl ionized $= \dfrac{1.9 - 1.0}{2.0 - 1.0} \times 100 = 90.\%$; 10% ion association occurs.

The ion association is greater in the $CaCl_2$ solution.

83. $NaCl(s) \rightarrow Na^+(aq) + Cl^-(aq)$, i = 2.0

$$\pi = iMRT = 2.0 \times \frac{0.10\ mol}{L} \times \frac{0.08206\ L\ atm}{K\ mol} \times 293\ K\ = 4.8\ atm$$

A pressure greater than 4.8 atm should be applied to ensure purification by reverse osmosis.

85.   a.   MgCl$_2$, i (observed) = 2.7

$\Delta T_f = iK_f m = 2.7 \times 1.86\ °C/molal \times 0.050\ molal = 0.25°C; \quad T_f = -0.25°C$

$\Delta T_b = iK_b m = 2.7 \times 0.51\ °C/molal \times 0.050\ molal = 0.069°C; \quad T_b = 100.069°C$

b.   FeCl$_3$, i (observed) = 3.4

$\Delta T_f = iK_f m = 3.4 \times 1.86\ °C/molal \times 0.050\ molal = 0.32°C; \quad T_f = -0.32°C$

$\Delta T_b = iK_b m = 3.4 \times 0.51°C/molal \times 0.050\ molal = 0.087°C; \quad T_b = 100.087°C$

87. $\pi = iMRT, \quad M = \dfrac{\pi}{iRT} = \dfrac{2.50\ atm}{2.00 \times \dfrac{0.08206\ L\ atm}{K\ mol} \times 298\ K} = 5.11 \times 10^{-2}\ mol/L$

Molar mass of compound = $\dfrac{0.500\ g}{0.1000\ L \times \dfrac{5.11 \times 10^{-2}\ mol}{L}} = 97.8\ g/mol$

Additional Exercises

89.   Both solutions and colloids have suspended particles in some medium. The major difference between the two is the size of the particles. A colloid is a suspension of relatively large particles compared to a solution. Because of this, colloids will scatter light, whereas solutions will not. The scattering of light by a colloidal suspension is called the Tyndall effect.

91.   Coagulation is the destruction of a colloid by the aggregation of many suspended particles to form a large particle that settles out of solution.

93.   A 92 proof ethanol solution is 46% C$_2$H$_5$OH by volume. Assuming 100.0 mL of solution:

mol ethanol = $46\ mL\ C_2H_5OH \times \dfrac{0.79\ g}{mL} \times \dfrac{1\ mol\ C_2H_5OH}{46.07\ g} = 0.79\ mol\ C_2H_5OH$

molarity = $\dfrac{0.79\ mol}{0.1000\ L} = 7.9\ M$ ethanol

95. $\Delta T = K_f m, \quad m = \dfrac{\Delta T}{K_f} = \dfrac{2.79°C}{1.86\ °C/molal} = 1.50\ molal$

a. $\Delta T = K_b m, \quad \Delta T = 0.51°C/molal \times 1.50\ molal = 0.77°C, \quad T_b = 100.77°C$

b. $P_{soln} = \chi_{water}P°_{water}, \quad \chi_{water} = \dfrac{mol\ H_2O}{mol\ H_2O + mol\ solute}$

Assuming 1.00 kg of water, we have 1.50 mol solute, and:

$$mol\ H_2O = 1.00 \times 10^3\ g\ H_2O \times \frac{1\ mol\ H_2O}{18.02\ g\ H_2O} = 55.5\ mol\ H_2O$$

$$\chi_{water} = \frac{55.5\ mol}{1.50 + 55.5} = 0.974; \quad P_{soln} = (0.974)(23.76\ mm\ Hg) = 23.1\ mm\ Hg$$

c.   We assumed ideal behavior in solution formation, we assumed the solute was nonvolatile, and we assumed i = 1 (no ions formed).

97.   Benzoic acid is capable of hydrogen-bonding, but a significant part of benzoic acid is the nonpolar benzene ring.   In benzene, a hydrogen-bonded dimer forms.

The dimer is relatively nonpolar and thus more soluble in benzene than in water. Because benzoic acid forms dimers in benzene, the effective solute particle concentration will be less than 1.0 molal. Therefore, the freezing-point depression would be less than 5.12°C ($\Delta T_f = K_f m$).

99.   a. $NH_4NO_3(s) \rightarrow NH_4^+(aq) + NO_3^-(aq)$ $\Delta H_{soln} = ?$

Heat gain by dissolution process = heat loss by solution; We will keep all quantities positive in order to avoid sign errors. Because the temperature of the water decreased, the dissolution of NH_4NO_3 is endothermic (ΔH is positive). Mass of solution = 1.60 + 75.0 = 76.6 g

$$Heat\ loss\ by\ solution\ = \frac{4.18\ J}{°C\ g} \times 76.6\ g \times (25.00°C - 23.34°C) = 532\ J$$

$$\Delta H_{soln} = \frac{532\ J}{1.60\ g\ NH_4NO_3} \times \frac{80.05\ g\ NH_4NO_3}{mol\ NH_4NO_3} = 2.66 \times 10^4\ J/mol = 26.6\ kJ/mol$$

b. We will use Hess's law to solve for the lattice energy. The lattice energy equation is:

$NH_4^+(g) + NO_3^-(g) \rightarrow NH_4NO_3(s)$ ΔH = lattice energy

$NH_4^+(g) + NO_3^-(g) \rightarrow NH_4^+(aq) + NO_3^-(aq)$ $\Delta H = \Delta H_{hyd} = -630.\ kJ/mol$
$NH_4^+(aq) + NO_3^-(aq) \rightarrow NH_4NO_3(s)$ $\Delta H = -\Delta H_{soln} = -26.6\ kJ/mol$

$NH_4^+(g) + NO_3^-(g) \rightarrow NH_4NO_3(s)$ $\Delta H = \Delta H_{hyd} - \Delta H_{soln} = -657\ kJ/mol$

101. $\chi_{pen}^{V} = 0.15 = \dfrac{P_{pen}}{P_{total}}$; $P_{pen} = \chi_{pen}^{L} P_{pen}^{o}$; $P_{total} = P_{pen} + P_{hex} = \chi_{pen}^{L}(511) + \chi_{hex}^{L}(150.)$

Because $\chi_{hex}^{L} = 1.000 - \chi_{pen}^{L}$: $P_{total} = \chi_{pen}^{L}(511) + (1.000 - \chi_{pen}^{L})(150.) = 150. + 361\chi_{pen}^{L}$

$\chi_{pen}^{V} = \dfrac{P_{pen}}{P_{total}}$, $0.15 = \dfrac{\chi_{pen}^{L}(511)}{150. + 361\chi_{pen}^{L}}$, $0.15(150. + 361\chi_{pen}^{L}) = 511\chi_{pen}^{L}$

$23 + 54\chi_{pen}^{L} = 511\chi_{pen}^{L}$, $\chi_{pen}^{L} = \dfrac{23}{457} = 0.050$

103. $14.2 \text{ mg } CO_2 \times \dfrac{12.01 \text{ mg C}}{44.01 \text{ mg } CO_2} = 3.88 \text{ mg C}$; $\% \text{ C} = \dfrac{3.88 \text{ mg}}{4.80 \text{ mg}} \times 100 = 80.8\% \text{ C}$

$1.65 \text{ mg } H_2O \times \dfrac{2.016 \text{ mg H}}{18.02 \text{ mg } H_2O} = 0.185 \text{ mg H}$; $\% \text{ H} = \dfrac{0.185 \text{ mg}}{4.80 \text{ mg}} \times 100 = 3.85\% \text{ H}$

Mass % O = $100.00 - (80.8 + 3.85) = 15.4\% \text{ O}$

Out of 100.00 g:

$80.8 \text{ g C} \times \dfrac{1 \text{ mol}}{12.01 \text{ g}} = 6.73 \text{ mol C}$; $\dfrac{6.73}{0.963} = 6.99 \approx 7$

$3.85 \text{ g H} \times \dfrac{1 \text{ mol}}{1.008 \text{ g}} = 3.82 \text{ mol H}$; $\dfrac{3.82}{0.963} = 3.97 \approx 4$

$15.4 \text{ g O} \times \dfrac{1 \text{ mol}}{16.00 \text{ g}} = 0.963 \text{ mol O}$; $\dfrac{0.963}{0.963} = 1.00$

Therefore, the empirical formula is C_7H_4O.

$\Delta T_f = K_f m$, $m = \dfrac{\Delta T_f}{K_f} = \dfrac{22.3 \,^{\circ}C}{40. \,^{\circ}C / molal} = 0.56 \text{ molal}$

Mol anthraquinone = $0.0114 \text{ kg camphor} \times \dfrac{0.56 \text{ mol anthraquinone}}{\text{kg camphor}} = 6.4 \times 10^{-3} \text{ mol}$

Molar mass = $\dfrac{1.32 \text{ g}}{6.4 \times 10^{-3} \text{ mol}} = 210 \text{ g/mol}$

The empirical mass of C_7H_4O is $7(12) + 4(1) + 16 \approx 104$ g/mol. Because the molar mass is twice the empirical mass, the molecular formula is $C_{14}H_8O_2$.

105.   a. $m = \dfrac{\Delta T_f}{K_f} = \dfrac{1.32°C}{5.12\ °C\ kg/mol} = 0.258\ mol/kg$

Mol unknown = $0.01560\ kg \times \dfrac{0.258\ mol\ unknown}{kg} = 4.02 \times 10^{-3}\ mol$

Molar mass of unknown = $\dfrac{1.22\ g}{4.02 \times 10^{-3}\ mol} = 303\ g/mol$

Uncertainty in temperature = $\dfrac{0.04}{1.32} \times 100 = 3\%$

A 3% uncertainty in 303 g/mol = 9 g/mol. So, molar mass = 303 ±9 g/mol.

b.   No, codeine could not be eliminated since its molar mass is in the possible range including the uncertainty.

c.   We would like the uncertainty to be ±1 g/mol. We need the freezing-point depression to be about 10 times what it was in this problem. Two possibilities are:

1.   make the solution 10 times more concentrated (may be solubility problem)

2.   use a solvent with a larger K_f value, e.g., camphor

107. $M_3X_2(s)\ \rightarrow\ 3\ M^{2+}(aq)\ +\ 2\ X^{3-}(aq)$ $K_{sp} = [M^{2+}]^3[X^{3-}]^2$

Initial   s = solubility (mol/L)   0   0
Equil.   3s   2s

$K_{sp} = (3s)^3(2s)^2 = 108s^5$;   total ion concentration = $3s + 2s = 5s$.

$\pi = iMRT,\ iM$ = total ion concentration $= \dfrac{\pi}{RT} = \dfrac{2.64 \times 10^{-2}\ atm}{0.08206\ L\ atm\ K^{-1}\ mol^{-1} \times 298\ K}$

$= 1.08 \times 10^{-3}\ mol/L$

$5s = 1.08 \times 10^{-3}\ mol/L,\ s = 2.16 \times 10^{-4}\ mol/L$

$K_{sp} = 108s^5 = 108(2.16 \times 10^{-4})^5 = 5.08 \times 10^{-17}$

109. $m = \dfrac{0.100\ g \times \dfrac{1\ mol}{100.0\ g}}{0.5000\ kg} = 2.00 \times 10^{-3}\ mol/kg \approx 2.00 \times 10^{-3}\ mol/L$ (dilute solution)

$\Delta T_f = iK_f m,\ 0.0056°C = i \times 1.86\ °C/molal \times 2.00 \times 10^{-3}\ molal,\ i = 1.5$

If i = 1.0, percent dissociation = 0%, and if i = 2.0, percent dissociation = 100%. Because i = 1.5, the weak acid is 50.% dissociated.

$$HA \rightleftharpoons H^+ + A^- \qquad K_a = \frac{[H^+][A^-]}{[HA]}$$

Because the weak acid is 50.% dissociated:

$$[H^+] = [A^-] = [HA]_o \times 0.50 = 2.00 \times 10^{-3} \, M \times 0.50 = 1.0 \times 10^{-3} \, M$$

$$[HA] = [HA]_0 - \text{amount HA reacted} = 2.00 \times 10^{-3} \, M - 1.0 \times 10^{-3} \, M = 1.0 \times 10^{-3} \, M$$

$$K_a = \frac{[H^+][A^-]}{[HA]} = \frac{(1.0 \times 10^{-3})(1.0 \times 10^{-3})}{1.0 \times 10^{-3}} = 1.0 \times 10^{-3}$$

111. $iM = \dfrac{\pi}{RT} = \dfrac{0.3950 \text{ atm}}{0.08206 \text{ L atm K}^{-1} \text{ mol}^{-1}(298.2 \text{ K})} = 0.01614 \text{ mol/L} = \text{total ion concentration}$

$$0.01614 \text{ mol/L} = M_{Mg^{2+}} + M_{Na^+} + M_{Cl^-}; \quad M_{Cl^-} = 2M_{Mg^{2+}} + M_{Na^+} \text{ (charge balance)}$$

Combining: $0.01614 = 3M_{Mg^{2+}} + 2M_{Na^+}$

Let x = mass $MgCl_2$ and y = mass NaCl; then $x + y = 0.5000$ g.

$$M_{Mg^{2+}} = \frac{x}{95.218} \text{ and } M_{Na^+} = \frac{y}{58.443} \quad \text{(Because V = 1.000 L.)}$$

$$\text{Total ion concentration} = \frac{3x}{95.218} + \frac{2y}{58.443} = 0.01614 \text{ mol/L}$$

Rearranging: $3x + (3.2585)y = 1.537$

Solving by simultaneous equations:

$$\begin{aligned} 3x + (3.2585)y &= 1.537 \\ -3(x + y) &= -3(0.5000) \\ \hline (0.2585)y &= 0.037, \qquad y = 0.14 \text{ g NaCl} \end{aligned}$$

Mass $MgCl_2 = 0.5000 \text{ g} - 0.14 \text{ g} = 0.36 \text{ g}$;   mass % $MgCl_2 = \dfrac{0.36 \text{ g}}{0.5000 \text{ g}} \times 100 = 72\%$

Challenge Problems

113.   From the problem, $\chi_{C_6H_6}^L = \chi_{CCl_4}^L = 0.500$. We need the pure vapor pressures (P^o) in order to calculate the vapor pressure of the solution.

$$C_6H_6(l) \rightleftharpoons C_6H_6(g) \quad K = P_{C_6H_6} = P_{C_6H_6}^o \text{ at } 25°C$$

$$\Delta G^{\circ}_{rxn} = \Delta G^{\circ}_{f, C_6H_6(g)} - \Delta G^{\circ}_{f, C_6H_6(l)} = 129.66 \text{ kJ/mol} - 124.50 \text{ kJ/mol} = 5.16 \text{ kJ/mol}$$

$$\Delta G^{\circ} = -RT \ln K, \ \ln K = \frac{-\Delta G^{\circ}}{RT} = \frac{-5.16 \times 10^3 \text{ J/mol}}{(8.3145 \text{ J K}^{-1} \text{ mol}^{-1})(298 \text{ K})} = -2.08$$

$$K = P^{\circ}_{C_6H_6} = e^{-2.08} = 0.125 \text{ atm}$$

For CCl$_4$: $\Delta G^{\circ}_{rxn} = \Delta G^{\circ}_{f, CCl_4(g)} - \Delta G^{\circ}_{f, CCl_4(l)} = -60.59 \text{ kJ/mol} - (-65.21 \text{ kJ/mol})$

$$= 4.62 \text{ kJ/mol}$$

$$K = P^{\circ}_{CCl_4} = \exp\left(\frac{-\Delta G^{\circ}}{RT}\right) = \exp\left(\frac{-4620 \text{ J/mol}}{8.3145 \text{ J K}^{-1} \text{ mol}^{-1} \times 298 \text{ K}}\right) = 0.155 \text{ atm}$$

$$P_{C_6H_6} = \chi^{L}_{C_6H_6} P^{\circ}_{C_6H_6} = 0.500(0.125 \text{ atm}) = 0.0625 \text{ atm}; \ P_{CCl_4} = 0.500(0.155 \text{ atm})$$

$$= 0.0775 \text{ atm}$$

$$\chi^{V}_{C_6H_6} = \frac{P_{C_6H_6}}{P_{total}} = \frac{0.0625 \text{ atm}}{0.0625 \text{ atm} + 0.0775 \text{ atm}} = \frac{0.0625}{0.1400} = 0.446; \ \chi^{V}_{CCl_4} = 1.000 - 0.446$$

$$= 0.554$$

115.   For the second vapor collected, $\chi^{V}_{B,2} = 0.714$ and $\chi^{V}_{T,2} = 0.286$. Let $\chi^{L}_{B,2}$ = mole fraction of benzene in the second solution and $\chi^{L}_{T,2}$ = mole fraction of toluene in the second solution.

$$\chi^{L}_{B,2} + \chi^{L}_{T,2} = 1.000$$

$$\chi^{V}_{B,2} = 0.714 = \frac{P_B}{P_{total}} = \frac{P_B}{P_B + P_T} = \frac{\chi^{L}_{B,2}(750.0 \text{ torr})}{\chi^{L}_{B,2}(750.0 \text{ torr}) + (1.000 - \chi^{L}_{B,2})(300.0 \text{ torr})}$$

Solving: $\chi^{L}_{B,2} = 0.500 = \chi^{L}_{T,2}$

This second solution came from the vapor collected from the first (initial) solution, so, $\chi^{V}_{B,1} = \chi^{V}_{T,1} = 0.500$. Let $\chi^{L}_{B,1}$ = mole fraction benzene in the first solution and $\chi^{L}_{T,1}$ = mole fraction of toluene in first solution. $\chi^{L}_{B,1} + \chi^{L}_{T,1} = 1.000$.

$$\chi^{V}_{B,1} = 0.500 = \frac{P_B}{P_{total}} = \frac{P_B}{P_B + P_T} = \frac{\chi^{L}_{B,1}(750.0 \text{ torr})}{\chi^{L}_{B,1}(750.0 \text{ torr}) + (1.000 - \chi^{L}_{B,1})(300.0 \text{ torr})}$$

Solving: $\chi^{L}_{B,1} = 0.286$; the original solution had $\chi_B = 0.286$ and $\chi_T = 0.714$.

117.   a.   Assuming MgCO$_3$(s) does not dissociate, the solute concentration in water is:

$$\frac{560 \ \mu\text{g MgCO}_3\text{(s)}}{\text{mL}} = \frac{560 \text{ mg}}{\text{L}} = \frac{560 \times 10^{-3} \text{ g}}{\text{L}} \times \frac{1 \text{ mol MgCO}_3}{84.32 \text{ g}}$$

$$= 6.6 \times 10^{-3} \text{ mol MgCO}_3\text{/L}$$

An applied pressure of 8.0 atm will purify water up to a solute concentration of:

$$M = \frac{\pi}{RT} = \frac{8.0 \text{ atm}}{0.08206 \text{ L atm K}^{-1} \text{ mol}^{-1} \times 300. \text{ K}} = \frac{0.32 \text{ mol}}{L}$$

When the concentration of $MgCO_3(s)$ reaches 0.32 mol/L, the reverse osmosis unit can no longer purify the water. Let V = volume (L) of water remaining after purifying 45 L of H_2O. When V + 45 L of water has been processed, the moles of solute particles will equal:

$$6.6 \times 10^{-3} \text{ mol/L} \times (45 \text{ L} + \text{V}) = 0.32 \text{ mol/L} \times \text{V}$$

Solving: $0.30 = (0.32 - 0.0066) \times \text{V}, \text{ V} = 0.96 \text{ L}$

The minimum total volume of water that must be processed is 45 L + 0.96 L = 46 L.

Note: If $MgCO_3$ does dissociate into Mg^{2+} and CO_3^{2-} ions, then the solute concentration increases to $1.3 \times 10^{-2} M$, and at least 47 L of water must be processed.

b. No; a reverse osmosis system that applies 8.0 atm can only purify water with a solute concentration of less than 0.32 mol/L. Salt water has a solute concentration of 2(0.60 M) = 1.2 mol/L ions. The solute concentration of salt water is much too high for this reverse osmosis unit to work.

119. a. Assuming no ion association between $SO_4^{2-}(aq)$ and $Fe^{3+}(aq)$, then i = 5 for $Fe_2(SO_4)_3$.

$$\pi = iMRT = 5 \times 0.0500 \text{ mol/L} \times 0.08206 \text{ L atm K}^{-1} \text{ mol}^{-1} \times 298 \text{ K} = 6.11 \text{ atm}$$

b. $Fe_2(SO_4)_3(aq) \rightarrow 2 \, Fe^{3+}(aq) + 3 \, SO_4^{2-}(aq)$

Under ideal circumstances, 2/5 of π calculated above results from Fe^{3+} and 3/5 results from SO_4^{2-}. The contribution to π from SO_4^{2-} is 3/5 × 6.11 atm = 3.67 atm. Because SO_4^{2-} is assumed unchanged in solution, the SO_4^{2-} contribution in the actual solution will also be 3.67 atm. The contribution to the actual π from the $Fe(H_2O)_6^{3+}$ dissociation reaction is 6.73 − 3.67 = 3.06 atm.

The initial concentration of $Fe(H_2O)_6^{2+}$ is 2(0.0500) = 0.100 M. The setup for the weak acid problem is:

$$Fe(H_2O)_6^{3+} \quad \rightarrow \quad H^+ \quad + \quad Fe(OH)(H_2O)_5^{2+} \qquad K_a = \frac{[H^+][Fe(OH)(H_2O)_5^{2+}]}{[Fe(H_2O)_6^{3+}]}$$

Initial   0.100 M ~0   0

        x mol/L of $Fe(H_2O)_6^{3+}$ reacts to reach equilibrium

Equil.   0.100 − x x x

$$\text{Total ion concentration} = iM = \frac{\pi}{RT} = \frac{3.06 \text{ atm}}{0.08206 \text{ L atm K}^{-1} \text{ mol}^{-1}(298 \text{ K})} = 0.125 \, M$$

$$0.125\ M = 0.100 - x + x + x = 0.100 + x,\ \ x = 0.025\ M$$

$$K_a = \frac{[H^+][Fe(OH)(H_2O)_5^{2+}]}{[Fe(H_2O)_6^{3+}]} = \frac{x^2}{0.100 - x} = \frac{(0.025)^2}{(0.100 - 0.025)} = \frac{(0.025)^2}{0.075} = 8.3 \times 10^{-3}$$

121.   a. $\pi = iMRT,\ \ iM = \dfrac{\pi}{RT} = \dfrac{7.83\ atm}{0.08206\ L\ atm\ K^{-1}\ mol^{-1} \times 298\ K} = 0.320\ mol/L$

Assuming 1.000 L of solution:

total mol solute particles = mol Na^+ + mol Cl^- + mol NaCl = 0.320 mol

mass solution = 1000. mL $\times \dfrac{1.071\ g}{mL}$ = 1071 g solution

mass NaCl in solution = 0.0100 × 1071 g = 10.7 g NaCl

mol NaCl added to solution = 10.7 g $\times \dfrac{1\ mol}{58.44\ g}$ = 0.183 mol NaCl

Some of this NaCl dissociates into Na^+ and Cl^- (two moles of ions per mole of NaCl), and some remains undissociated. Let x = mol undissociated NaCl = mol ion pairs.

Mol solute particles = 0.320 mol = 2(0.183 − x) + x

0.320 = 0.366 − x, x = 0.046 mol ion pairs

Fraction of ion pairs = $\dfrac{0.046}{0.183}$ = 0.25, or 25%

b. $\Delta T = K_f m$, where K_f = 1.86 °C kg/mol; from part a, 1.000 L of solution contains 0.320 mol of solute particles. To calculate the molality of the solution, we need the kilograms of solvent present in 1.000 L of solution.

Mass of 1.000 L solution = 1071 g; mass of NaCl = 10.7 g

Mass of solvent in 1.000 L solution = 1071 g − 10.7 g = 1060. g

$\Delta T = 1.86\ °C\ kg/mol \times \dfrac{0.320\ mol}{1.060\ kg} = 0.562°C$

Assuming water freezes at 0.000°C, then T_f = −0.562°C.

CHAPTER 18

THE REPRESENTATIVE ELEMENTS

Group 1A Elements

1. The gravity of the earth is not strong enough to keep the light H_2 molecules in the atmosphere.

3. a. $\Delta H° = 2(-46 \text{ kJ}) = -92 \text{ kJ}$; $\Delta S° = 2(193 \text{ J/K}) - [3(131 \text{ J/K}) + 192 \text{ J/K}] = -199 \text{ J/K}$

    $\Delta G° = \Delta H° - T\Delta S° = -92 \text{ kJ} - 298 \text{ K}(-0.199 \text{ kJ/K}) = -33 \text{ kJ}$

    b. Because $\Delta G°$ is negative, this reaction is spontaneous at standard conditions.

    c. $\Delta G° = 0$ when $T = \dfrac{\Delta H°}{\Delta S°} = \dfrac{-92 \text{ kJ}}{-0.199 \text{ kJ/K}} = 460 \text{ K}$

    At $T < 460$ K and standard pressures, the favorable $\Delta H°$ term dominates, and the reaction is spontaneous ($\Delta G° < 0$).

5. The first illustration is an example of a covalent hydride like H_2O. Covalent hydrides are just binary covalent compounds formed between hydrogen and some other nonmetal and exists as individual molecules. The middle illustration represents interstitial (or metallic) hydrides. In interstitial hydrides, hydrogen atoms occupy the holes of a transition metal crystal. These hydrides are more like solid solutions than true compounds. The third illustration represents ionic (or salt-like) hydrides like LiH. Ionic hydrides form when hydrogen reacts with a metal from Group 1A or 2A. The metals lose electrons to form cations and the hydrogen atoms gain electrons to form the hydride anions (H^-). These are just ionic compounds formed between a metal and hydrogen.

7. Alkali metals have a ns^1 valence shell electron configuration. Alkali metals lose this valence electron with relative ease to form M^+ cations when in ionic compounds. They all are easily oxidized. Therefore, in order to prepare the pure metals, alkali metals must be produced in the absence of materials (H_2O, O_2) that are capable of oxidizing them. The method of preparation is electrochemical processes, specifically, electrolysis of molten chloride salts and reduction of alkali salts with Mg and H_2. In all production methods, H_2O and O_2 must be absent.

9. Hydrogen forms many compounds in which the oxidation state is +1, as do the Group 1A elements. For example, H_2SO_4 and HCl as compared to Na_2SO_4 and NaCl. On the other hand, hydrogen forms diatomic H_2 molecules and is a nonmetal, whereas the Group 1A elements are metals. Hydrogen also forms compounds with a −1 oxidation state, which is not characteristic of Group 1A metals, e.g., NaH.

368

11. $4 \text{ Li}(s) + O_2(g) \rightarrow 2 \text{ Li}_2O(s)$

$16 \text{ Li}(s) + S_8(s) \rightarrow 8 \text{ Li}_2S(s); \quad 2 \text{ Li}(s) + Cl_2(g) \rightarrow 2 \text{ LiCl}(s)$

$12 \text{ Li}(s) + P_4(s) \rightarrow 4 \text{ Li}_3P(s); \quad 2 \text{ Li}(s) + H_2(g) \rightarrow 2 \text{ LiH}(s)$

$2 \text{ Li}(s) + 2 \text{ H}_2O(l) \rightarrow 2 \text{ LiOH}(aq) + H_2(g); \quad 2 \text{ Li}(s) + 2 \text{ HCl}(aq) \rightarrow 2 \text{ LiCl}(aq) + H_2(g)$

Group 2A Elements

13. $MgCl_2$ is composed of Mg^{2+} ions; Mg^{2+} gains 2 electrons to form magnesium metal in electrolysis.

$$1.00 \times 10^6 \text{ g Mg} \times \frac{1 \text{ mol Mg}}{24.31 \text{ g Mg}} \times \frac{2 \text{ mol e}^-}{1 \text{ mol Mg}} \times \frac{96{,}485 \text{ C}}{1 \text{ mol e}^-} \times \frac{1 \text{ s}}{5.00 \times 10^4 \text{ C}} \times \frac{1 \text{ h}}{3600 \text{ s}}$$
$$= 44.1 \text{ hours}$$

15.   The alkaline earth ions that give water the hard designation are Ca^{2+} and Mg^{2+}. These ions interfere with the action of detergents and form unwanted precipitates with soaps. Large-scale water softeners remove Ca^{2+} by precipitating out the calcium ions as $CaCO_3$. In homes, Ca^{2+} and Mg^{2+} (plus other cations) are removed by ion exchange. See Figure 18.6 for a schematic of a typical cation exchange resin.

17. $CaCO_3(s) + H_2SO_4(aq) \rightarrow CaSO_4(aq) + H_2O(l) + CO_2(g)$

19. $\dfrac{1 \text{ mg F}^-}{L} \times \dfrac{1 \text{ g}}{1000 \text{ mg}} \times \dfrac{1 \text{ mol F}^-}{19.0 \text{ g F}^-} = 5.3 \times 10^{-5} \, M \text{ F}^- = 5 \times 10^{-5} \, M \text{ F}^-$

$CaF_2(s) \rightleftharpoons Ca^{2+}(aq) + 2 \text{ F}^-(aq) \quad K_{sp} = [Ca^{2+}][\text{F}^-]^2 = 4.0 \times 10^{-11}$; precipitation will occur when $Q > K_{sp}$. Let's calculate $[Ca^{2+}]$ so that $Q = K_{sp}$.

$Q = 4.0 \times 10^{-11} = [Ca^{2+}]_0[\text{F}^-]_0^2 = [Ca^{2+}]_0(5 \times 10^{-5})^2, \quad [Ca^{2+}]_0 = 2 \times 10^{-2} \, M$

$CaF_2(s)$ will precipitate when $[Ca^{2+}]_0 > 2 \times 10^{-2} \, M$. Therefore, hard water should have a calcium ion concentration of less than $2 \times 10^{-2} \, M$ to avoid precipitate formation.

Group 3A Elements

21.   The valence electron configuration of Group 3A elements is ns^2np^1. The lightest Group 3A element, boron, is a nonmetal because most of its compounds are covalent. Aluminum, although commonly thought of as a metal, does have some nonmetallic properties because its bonds to other nonmetals have significant covalent character. The other Group 3A elements have typical metal characteristics; their compounds formed with nonmetals are ionic. From this discussion, metallic character increases as the Group 3A elements get larger.

As discussed above, boron is a nonmetal in both properties and compounds formed. However aluminum has physical properties of metals, such as high thermal and electrical conductivities and a lustrous appearance. The compounds of aluminum with other nonmetals, however, do have some nonmetallic properties because the bonds have significant covalent character.

23. $B_2H_6(g) + 3\ O_2(g) \rightarrow 2\ B(OH)_3(s)$

25. $2\ Ga(s) + 3\ F_2(g) \rightarrow 2\ GaF_3(s);\ \ 4\ Ga(s) + 3\ O_2(g) \rightarrow 2\ Ga_2O_3(s)$

      $16\ Ga(s) + 3\ S_8(s) \rightarrow 8\ Ga_2S_3(s)$

      $2\ Ga(s) + 6\ HCl(aq) \rightarrow 2\ GaCl_3(aq) + 3\ H_2(g)$

Group 4A Elements

27.   The valence electron configuration of Group 4A elements is ns^2np^2. The two most important elements on earth are both Group 4A elements. They are carbon, found in all biologically important molecules, and silicon, found in most of the compounds that make up the earth's crust. They are important because they are so prevalent in compounds necessary for life and the geologic world.

      Group 4A shows an increase in metallic character as the elements get heavier. Carbon is a typical nonmetal, silicon and germanium have properties of both metals and nonmetals so they are classified as semimetals, and tin and lead have typical metallic characteristics.

29. 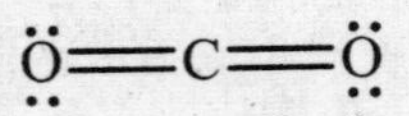 $\ddot{O}\!=\!\!=\!\!C\!=\!\!=\!\ddot{O}$

      The darker green orbitals about carbon are sp hybrid orbitals. The lighter green orbitals about each oxygen are sp^2 hybrid orbitals, and the gold orbitals about all of the atoms are unhybridized p atomic orbitals. In each double bond in CO_2, one sigma and one π bond exists. The two carbon-oxygen sigma bonds are formed from overlap of sp hybrid orbitals from carbon with a sp^2 hybrid orbital from each oxygen. The two carbon-oxygen π bonds are formed from side-to-side overlap of the unhybridized p atomic orbitals from carbon with an unhybridized p atomic orbital from each oxygen. These two π bonds are oriented perpendicular to each other as illustrated in the figure.

31.   White tin is stable at normal temperatures. Gray tin is stable at temperatures below 13.2°C. Thus, for the phase change: $Sn(gray) \rightarrow Sn(white)$, ΔG is negative at T > 13.2°C and ΔG is positive at T < 13.2°C. This is only possible if ΔH is positive and ΔS is positive. Therefore, gray tin has the more ordered structure (has the smaller positional probability).

33. $Sn(s) + 2F_2(g) \rightarrow SnF_4(s)$, tin(IV) fluoride; $Sn(s) + F_2(g) \rightarrow SnF_2(s)$, tin(II) fluoride

35. Pb_3O_4: we assign −2 for the oxidation state of O. The sum of the oxidation states of Pb must be +8. We get this if two of the lead atoms are Pb(II) and one is Pb(IV). Therefore, the mole ratio of lead(II) to lead(IV) is 2:1.

Group 5A Elements

37. NO_4^{3-}

Both NO_4^{3-} and PO_4^{3-} have 32 valence electrons so both have similar Lewis structures. From the Lewis structure for NO_4^{3-}, the central N atom has a tetrahedral arrangement of electron pairs. N is small. There is probably not enough room for all four oxygen atoms around N. P is larger; thus PO_4^{3-} is stable.

PO_3^-

PO_3^- and NO_3^- both have 24 valence electrons, so both have similar Lewis structures. From the Lewis structure, PO_3^- has a trigonal arrangement of electron pairs about the central P atom (two single bonds and one double bond). P=O bonds are not particularly stable, whereas N=O bonds are stable. Thus NO_3^- is stable.

39.   N: $1s^2 2s^2 2p^3$; the extremes of the oxidation states for N can be rationalized by examining the electron configuration of N. Nitrogen is three electrons short of the stable Ne electron configuration of $1s^2 2s^2 2p^6$. Having an oxidation state of -3 makes sense. The $+5$ oxidation state corresponds to N "losing" its five valence electrons. In compounds with oxygen, the N–O bonds are polar covalent, with N having the partial positive end of the bond dipole. In the world of oxidation states, electrons in polar covalent bonds are assigned to the more electronegative atom; this is oxygen in N–O bonds. N can form enough bonds to oxygen to give it a $+5$ oxidation state. This loosely corresponds to losing all the valence electrons.

41.   This is due to nitrogen's ability to form strong π bonds, whereas heavier group 5A elements do not form strong π bonds. Therefore, P_2, As_2, and Sb_2 do not form since two π bonds are required to form these diatomic substances.

43.   a. NO_2, $5 + 2(6) = 17 \ e^-$ $\qquad\qquad\qquad$ N_2O_4, $2(5) + 4(6) = 34 \ e^-$

plus other resonance structures $\qquad\qquad$ plus other resonance structures

b. BF_3, $3 + 3(7) = 24$ e$^-$ NH_3, $5 + 3(1) = 8$ e$^-$

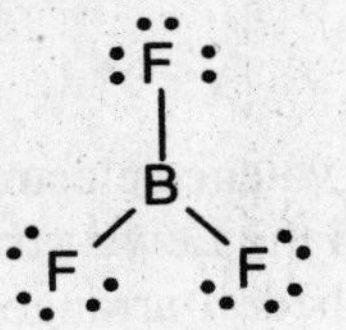

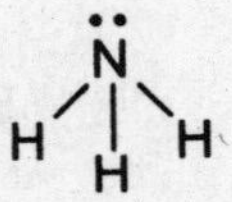

BF_3NH_3, $24 + 8 = 32$ e$^-$

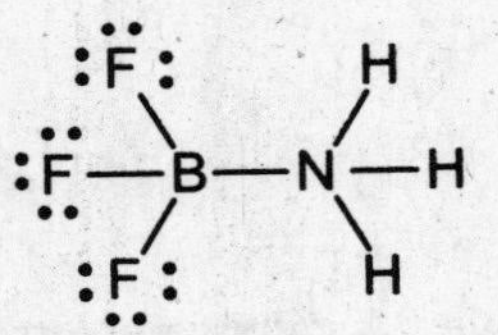

In reaction a, NO_2 has an odd number of electrons, so it is impossible to satisfy the octet rule. By dimerizing to form N_2O_4, the odd electron on two NO_2 molecules can pair up, giving a species whose Lewis structure can satisfy the octet rule. In general, odd-electron species are very reactive. In reaction b, BF_3 can be considered electron-deficient; boron has only six electrons around it. By forming BF_3NH_3, the boron atom satisfies the octet rule by accepting a lone pair of electrons from NH_3 to form a fourth bond.

45.   As temperature increases, the value of K decreases. This is consistent with an exothermic reaction. In an exothermic reaction, heat is a product, and an increase in temperature shifts the equilibrium to the reactant side (as well as lowering the value of K).

47.

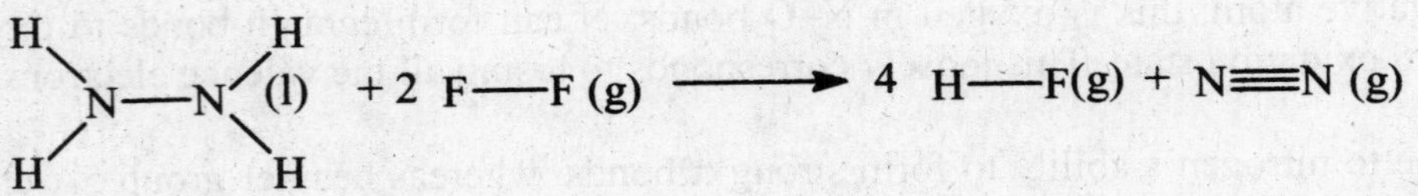

$$\text{H}_2\text{N}-\text{NH}_2 \text{ (l)} + 2 \text{ F}-\text{F (g)} \longrightarrow 4 \text{ H}-\text{F(g)} + \text{N}\equiv\equiv\text{N (g)}$$

Bonds broken: Bonds formed:

     1 N–N (160. kJ/mol)   4 H–F (565 kJ/mol)
     4 N–H (391 kJ/mol)   1 N≡ N (941 kJ/mol)
     2 F–F (154 kJ/mol)

$\Delta H = 160. + 4(391) + 2(154) - [4(565) + 941] = 2032 \text{ kJ} - 3201 \text{ kJ} = -1169 \text{ kJ}$

49.   Production of bismuth:

$2 Bi_2S_3(s) + 9 O_2(g) \rightarrow 2 Bi_2O_3(s) + 6 SO_2(g); \; 2 Bi_2O_3(s) + 3 C(s) \rightarrow 4 Bi(s) + 3 CO_2(g)$

Production of antimony:

$2 Sb_2S_3(s) + 9 O_2(g) \rightarrow 2 Sb_2O_3(s) + 6 SO_2(g); \; 2 Sb_2O_3(s) + 3 C(s) \rightarrow 4 Sb(s) + 3 CO_2(g)$

51.   TSP = Na_3PO_4; PO_4^{3-} is the conjugate base of the weak acid HPO_4^{2-} ($K_a = 4.8 \times 10^{-13}$). All conjugate bases of weak acids are effective bases ($K_b = K_w/K_a = 1.0 \times 10^{-14}/4.8 \times 10^{-13} = 2.1 \times 10^{-2}$). The weak base reaction of PO_4^{3-} with H_2O is $PO_4^{3-}(aq) + H_2O(l) \rightleftharpoons HPO_4^{2-}(aq) + OH^-(aq)$.

53. $4\,As(s) + 3\,O_2(g) \rightarrow As_4O_6(s)$; $4\,As(s) + 5\,O_2(g) \rightarrow As_4O_{10}(s)$

      $As_4O_6(s) + 6\,H_2O(l) \rightarrow 4\,H_3AsO_3(aq)$; $As_4O_{10}(s) + 6\,H_2O(l) \rightarrow 4\,H_3AsO_4(aq)$

55. $1/2\,N_2(g) + 1/2\,O_2(g) \rightarrow NO(g)$ $\Delta G° = \Delta G_{f,\,NO}° = 87$ kJ/mol; by definition, $\Delta G_f°$ for a compound equals the free energy change that would accompany the formation of 1 mole of that compound from its elements in their standard states. NO (and some other oxides of nitrogen) have weaker bonds than the triple bond of N_2 and the double bond of O_2. Because of this, NO (and some other oxides of nitrogen) have higher (positive) standard free energies of formation than the relatively stable N_2 and O_2 molecules.

57.   The pollution provides sources of nitrogen and phosphorus nutrients so that the algae can grow. The algae consume oxygen, which decrease the dissolved oxygen levels below that required for other aquatic life to survive, and fish die.

59. $5\,N_2O_4(l) + 4\,N_2H_3CH_3(l) \rightarrow 12\,H_2O(g) + 9\,N_2(g) + 4\,CO_2(g)$

$$\Delta H° = \left[12\,mol\left(\frac{-242\,kJ}{mol}\right) + 4\,mol\left(\frac{-393.5\,kJ}{mol}\right)\right]$$

$$-\left[5\,mol\left(\frac{-20.\,kJ}{mol}\right) + 4\,mol\left(\frac{54\,kJ}{mol}\right)\right] = -4594\,kJ$$

61.   Hydrazine also can hydrogen bond because it has covalent N–H bonds as well as having a lone pair of electrons on each N. The high boiling point for hydrazine's relatively small size supports this.

Group 6A Elements

63.   The two allotropic forms of oxygen are O_2 and O_3.

      O_2, 2(6) = 12 e⁻ O_3, 3(6) = 18 e⁻

The MO electron configuration of O_2 has two unpaired electrons in the degenerate pi antibonding (π_{2p}^*) orbitals (see Figure 14.41). A substance with unpaired electrons is paramagnetic. Ozone has a V-shape molecular structure with a bond angle of 117°, slightly less than the predicted 120° trigonal planar bond angle.

65.   In the upper atmosphere, O_3 acts as a filter for ultraviolet (UV) radiation:

$$O_3 \xrightarrow{h\nu} O_2 + O \quad \text{(See Exercise 18.64.)}$$

O_3 is also a powerful oxidizing agent. It irritates the lungs and eyes, and at high concentration, it is toxic. The smell of a "spring thunderstorm" is O_3 formed during lightning discharges. Toxic materials don't necessarily smell bad. For example, HCN smells like almonds.

67.   +6 oxidation state: SO_4^{2-}, SO_3, SF_6

      +4 oxidation state: SO_3^{2-}, SO_2, SF_4

      +2 oxidation state: SCl_2

      0 oxidation state: S_8 and all other elemental forms of sulfur

      −2 oxidation state: H_2S, Na_2S

69. $H_2SeO_4(aq) + 3\ SO_2(g) \rightarrow Se(s) + 3\ SO_3(g) + H_2O(l)$

Group 7A Elements

71. O_2F_2 has $2(6) + 2(7) = 26$ valence e^-.

$$:\!\ddot{F}\!—\!\ddot{O}\!—\!\ddot{O}\!—\!\ddot{F}\!:$$

Formal charge   0   0   0   0

Oxidation state   −1   +1   +1   −1

Oxidation states (numbers) are more useful. We are forced to assign +1 as the oxidation state for oxygen. Oxygen is very electronegative, and +1 is not a stable oxidation state for this element.

73.   Fluorine is the most reactive of the halogens because it is the most electronegative atom, and the bond in the F_2 molecule is very weak.

75. $ClO^- + H_2O + 2\ e^- \rightarrow 2\ OH^- + Cl^-$ $E_c^{\circ} = 0.90$ V

       $2\ NH_3 + 2\ OH^- \rightarrow N_2H_4 + 2\ H_2O + 2\ e^-$ $-E_a^{\circ} = 0.10$ V

                                                            ―――――――――――――――――

      $ClO^-(aq) + 2\ NH_3(aq) \rightarrow Cl^-(aq) + N_2H_4(aq) + H_2O(l)$ $E_{cell}^{\circ} = 1.00$ V

Because E_{cell}° is positive for this reaction, ClO^-, at standard conditions, can spontaneously oxidize NH_3 to the somewhat toxic N_2H_4.

77.   a. $AgCl(s) \xrightarrow{h\nu} Ag(s) + Cl$; the reactive chlorine atom is trapped in the crystal. When light is removed, Cl reacts with silver atoms to re-form AgCl; that is, the reverse reaction occurs. In pure AgCl, the Cl atoms escape, making the reverse reaction impossible.

       b.   Over time chlorine is lost, and the presence of the dark silver metal is permanent.

Group 8A Elements

79.   Helium is unreactive and doesn't combine with any other elements. It is a very light gas and would easily escape the earth's gravitational pull as the planet was formed.

81. $10.0 \text{ m} \times 10.0 \text{ m} \times 10.0 \text{ m} = 1.00 \times 10^3 \text{ m}^3$; from Table 18.22, volume percent of Ar = 0.9%.

$$1.00 \times 10^3 \text{ m}^3 \times \left(\frac{10 \text{ dm}}{\text{m}}\right)^3 \times \frac{1 \text{ L}}{\text{dm}^3} \times \frac{0.9 \text{ L Ar}}{100 \text{ L air}} = 9 \times 10^3 \text{ L of Ar in the room}$$

$$PV = nRT, \quad n = \frac{PV}{RT} = \frac{(1.0 \text{ atm})(9 \times 10^3 \text{ L})}{(0.08206 \text{ L atm K}^{-1} \text{ mol}^{-1})(298 \text{ K})} = 4 \times 10^2 \text{ mol Ar}$$

$$4 \times 10^2 \text{ mol Ar} \times \frac{39.95 \text{ g}}{\text{mol}} = 2 \times 10^4 \text{ g Ar in the room}$$

$$4 \times 10^2 \text{ mol Ar} \times \frac{6.022 \times 10^{23} \text{ atoms}}{\text{mol}} = 2 \times 10^{26} \text{ atoms Ar in the room}$$

A 2-L breath contains: $2 \text{ L air} \times \dfrac{0.9 \text{ L Ar}}{100 \text{ L air}} = 2 \times 10^{-2} \text{ L Ar}$

$$n = \frac{PV}{RT} = \frac{(1.0 \text{ atm})(2 \times 10^{-2} \text{ L})}{(0.08206 \text{ L atm K}^{-1} \text{ mol}^{-1})(298 \text{ K})} = 8 \times 10^{-4} \text{ mol Ar}$$

$$8 \times 10^{-4} \text{ mol Ar} \times \frac{6.022 \times 10^{23} \text{ atoms}}{\text{mol}} = 5 \times 10^{20} \text{ atoms of Ar in a 2-L breath}$$

Because Ar and Rn are both noble gases, both species will be relatively unreactive. However, all nuclei of Rn are radioactive, unlike most nuclei of Ar. The radioactive decay products of Rn can cause biological damage when inhaled.

83.   One would expect RnF_2 and RnF_4 to form in fashion similar to XeF_2 and XeF_4. The chemistry of radon is difficult to study because radon isotopes are all radioactive. The hazards of dealing with radioactive materials are immense.

Additional Exercises

85.   Solids have stronger intermolecular forces than liquids. In order to maximize the hydrogen bonding in the solid phase, ice is forced into an open structure. This open structure is why $H_2O(s)$ is less dense than $H_2O(l)$.

87.   Strontium and calcium are both alkaline earth metals, so both have similar chemical properties. Because milk is a good source of calcium, strontium could replace some calcium in milk without much difficulty.

89. $Tl^{3+} + 2\,e^- \rightarrow Tl^+$ $E_c^o = 1.25$ V
         $3\,I^- \rightarrow I_3^- + 2\,e^-$ $-E_a^o = -0.55$ V

         _____

         $Tl^{3+} + 3\,I^- \rightarrow Tl^+ + I_3^-$ $E_{cell}^o = 0.70$ V   (Spontaneous because $E_{cell}^o > 0$.)

         In solution, Tl^{3+} can oxidize I^- to I_3^-. Thus we expect TlI_3 to be thallium(I) triiodide.

91.   The inert pair effect refers to the difficulty of removing the pair of s electrons from some of the elements in the fifth and sixth periods of the periodic table. As a result, multiple oxidation states are exhibited for the heavier elements of Groups 3A and 4A. In^+, In^{3+}, Tl^+, and Tl^{3+} oxidation states are all important to the chemistry of In and Tl.

93. 15 kWh $= \dfrac{15000\,\text{J h}}{\text{s}} \times \dfrac{60\,\text{s}}{\text{min}} \times \dfrac{60\,\text{min}}{\text{h}} = 5.4 \times 10^7$ J or 5.4×10^4 kJ   (Hall process)

         To melt 1.0 kg Al requires: 1.0×10^3 g Al $\times \dfrac{1\,\text{mol Al}}{26.98\,\text{g}} \times \dfrac{10.7\,\text{kJ}}{\text{mol Al}} = 4.0 \times 10^2$ kJ

         It is feasible to recycle Al by melting the metal because, in theory, it takes less than 1% of the energy required to produce the same amount of Al by the Hall process.

95.   The bonds in SnX_4 compounds have a large covalent character. SnX_4 acts as discrete molecules held together by weak London dispersion forces. SnX_2 compounds are ionic and are held in the solid state by strong ionic forces. Because the intermolecular forces are weaker for SnX_4 compounds, they are more volatile (have a lower boiling point).

97.   Carbon cannot form the fifth bond necessary for the transition state because carbon doesn't have low-energy d orbitals available to expand its octet of valence electrons.

99.   a.   The Lewis structures for NNO and NON are (16 valence electrons each):

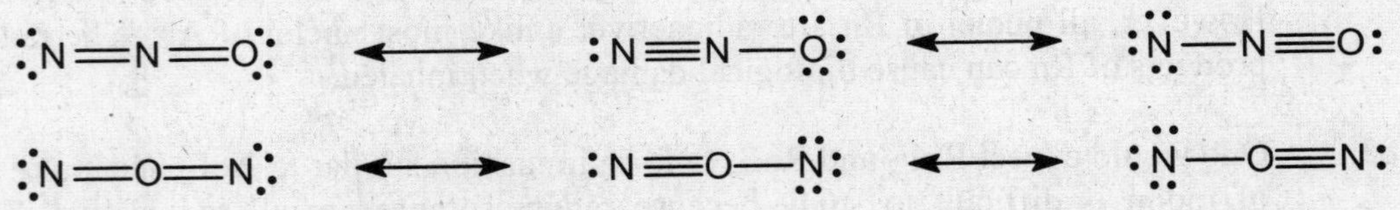

         The NNO structure is correct. From the Lewis structures we would predict both NNO and NON to be linear. However, we would predict NNO to be polar and NON to be nonpolar. Becase experiments show N_2O to be polar, then NNO is the correct structure.

b. Formal charge = number of valence electrons of atoms - [(number of lone pair electrons) + 1/2 (number of shared electrons)].

$$:N\!=\!\!=\!N\!=\!\!=\!O:\quad\longleftrightarrow\quad:N\!\!\equiv\!\!N\!-\!\ddot{O}:\quad\longleftrightarrow\quad:\ddot{N}\!-\!N\!\!\equiv\!\!O:$$

$$\;\;-1\quad\;+1\quad\;\;0\qquad\qquad\qquad 0\quad\;\;+1\quad\;-1\qquad\qquad\qquad -2\quad\;+1\quad\;+1$$

The formal charges for the atoms in the various resonance structures appear below each atom. The central N is sp hybridized in all the resonance structures. We can probably ignore the third resonance structure on the basis of the relatively large formal charges on the various atoms in N_2O as compared with the first two resonance structures.

c. The sp hybrid orbitals on the center N overlap with atomic orbitals (or hybrid orbitals) on the other two atoms to form the two sigma bonds. The remaining two unhybridized p orbitals on the center N overlap with two p orbitals on the peripheral N to form the two π bonds.

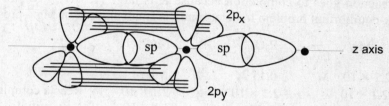

101. For a buffer solution: $pH = pK_a + \log\dfrac{[\text{base}]}{[\text{acid}]}$; $pH = -\log(6.2\times10^{-8}) + \log\dfrac{[HPO_4^{2-}]}{[H_2PO_4^{-}]}$

From the problem, $\dfrac{[H_2PO_4^{-}]}{[HPO_4^{2-}]} = 1.1$, so: $pH = 7.21 + \log\dfrac{1}{1.1}$, $pH = 7.21 - 0.041 = 7.17$

Challenge Problems

103. a. Mol $In(CH_3)_3 = \dfrac{PV}{RT} = \dfrac{2.00\ \text{atm} \times 2.56\ \text{L}}{0.08206\ \text{L atm K}^{-1}\ \text{mol}^{-1} \times 900.\ \text{K}} = 0.0693\ \text{mol}$

Mol $PH_3 = \dfrac{PV}{RT} = \dfrac{3.00\ \text{atm} \times 1.38\ \text{L}}{0.08206\ \text{L atm K}^{-1}\ \text{mol}^{-1} \times 900.\ \text{K}} = 0.0561\ \text{mol}$

Because the reaction requires a 1 : 1 mole ratio between these reactants, the reactant with the small number of moles (PH_3) is limiting.

$0.0561\ \text{mol PH}_3 \times \dfrac{1\ \text{mol InP}}{\text{mol PH}_3} \times \dfrac{145.8\ \text{g InP}}{\text{mol InP}} = 8.18\ \text{g InP}$

The actual yield of InP is $0.87 \times 8.18\ \text{g} = 7.1\ \text{g InP}$.

b. $\lambda = \dfrac{hc}{E} = \dfrac{6.626\times10^{-34}\ \text{J s} \times 2.998\times10^{8}\ \text{m/s}}{2.03\times10^{-19}\ \text{J}} = 9.79\times10^{-7}\ \text{m} = 979\ \text{nm}$

From the Figure 12.3 of the text, visible light has wavelengths between 4×10^{-7} and 7×10^{-7} m. Therefore, this wavelength is not visible to humans; it is in the infrared region of the electromagnetic radiation spectrum.

c. $[Kr]5s^2 4d^{10} 5p^4$ is the electron configuration for tellurium, Te. Because Te has more valence electrons than P, this would form an n-type semiconductor (n-type doping).

105. $Mg^{2+} + P_3O_{10}^{5-} \rightleftharpoons MgP_3O_{10}^{3-}$ pK = −8.60; $[Mg^{2+}]_0 = \dfrac{50. \times 10^{-3} \text{ g}}{L} \times \dfrac{1 \text{ mol}}{24.3 \text{ g}} = 2.1 \times 10^{-3} \, M$

$[P_3O_{10}^{5-}]_0 = \dfrac{40. \text{ g } Na_5P_3O_{10}}{L} \times \dfrac{1 \text{ mol}}{367.9 \text{ g}} = 0.11 \, M$

Assume the reaction goes to completion because K is large (K = $10^{8.60}$ = 4.0×10^8). Then solve the back-equilibrium problem to determine the small amount of Mg^{2+} present.

$$Mg^{2+} \quad + \quad P_3O_{10}^{5-} \quad \rightleftharpoons \quad MgP_3O_{10}^{3-}$$

	Mg^{2+}	$P_3O_{10}^{5-}$	$MgP_3O_{10}^{3-}$	
Before	2.1×10^{-3} M	0.11 M	0	
Change	-2.1×10^{-3}	-2.1×10^{-3} →	$+2.1 \times 10^{-3}$	Reacts completely
After	0	0.11	2.1×10^{-3}	New initial

x mol/L $MgP_3O_{10}^{3-}$ dissociates to reach equilibrium

Change	$+x$	$+x$	$-x$	
Equil.	x	$0.11 + x$	$2.1 \times 10^{-3} - x$	

$K = 4.0 \times 10^8 = \dfrac{[MgP_3O_{10}^{3-}]}{[Mg^{2+}][P_3O_{10}^{5-}]} = \dfrac{2.1 \times 10^{-3} - x}{x(0.11 + x)}$ (assume $x \ll 2.1 \times 10^{-3}$)

$4.0 \times 10^8 \approx \dfrac{2.1 \times 10^{-3}}{x(0.11)}$, $x = [Mg^{2+}] = 4.8 \times 10^{-11} \, M$; assumptions good.

107. $3 O_2(g) \rightleftharpoons 2 O_3(g)$; $\Delta H° = 2(143) = 286$ kJ; $\Delta G° = 2(163) = 326$ kJ

$\ln K_p = \dfrac{-\Delta G°}{RT} = \dfrac{-326 \times 10^3 \text{ J}}{(8.3145 \text{ J K}^{-1} \text{ mol}^{-1})(298 \text{ K})} = -131.573$, $K_p = e^{-131.573} = 7.22 \times 10^{-58}$

Note: We carried extra significant figures for the K_p calculation.

We need the value of K at 230. K. From Section 10.11 of the text: $\ln K = \dfrac{-\Delta H°}{RT} + \dfrac{\Delta S°}{R}$

For two sets of K and T:

$\ln K_1 = \dfrac{-\Delta H°}{R}\left(\dfrac{1}{T_1}\right) + \dfrac{\Delta S°}{R}$; $\ln K_2 = \dfrac{-\Delta H°}{R}\left(\dfrac{1}{T_2}\right) + \dfrac{\Delta S°}{R}$

Subtracting the first expression from the second:

❶

$$\ln K_2 - \ln K_1 = \frac{\Delta H^\circ}{R}\left(\frac{1}{T_1} - \frac{1}{T_2}\right) \quad \text{or} \quad \ln\frac{K_2}{K_1} = \frac{\Delta H^\circ}{R}\left(\frac{1}{T_1} - \frac{1}{T_2}\right)$$

Let $K_2 = 7.22 \times 10^{-58}$, $T_2 = 298$; $K_1 = K_{230}$, $T_1 = 230.$ K; $\Delta H^\circ = 286 \times 10^3$ J

$$\ln\frac{7.22 \times 10^{-58}}{K_{230}} = \frac{286 \times 10^3}{8.3145}\left(\frac{1}{230.} - \frac{1}{298}\right) = 34.13 \quad \text{(Carrying extra sig. figs.)}$$

$$\frac{7.22 \times 10^{-58}}{K_{230}} = e^{34.13} = 6.6 \times 10^{14}, \ K_{230} = 1.1 \times 10^{-72}$$

$$K_{230} = 1.1 \times 10^{-72} = \frac{P_{O_3}^2}{P_{O_2}^3} = \frac{P_{O_3}^2}{(1.0 \times 10^{-3}\ \text{atm})^3}, \ P_{O_3} = 3.3 \times 10^{-41}\ \text{atm}$$

The volume occupied by one molecule of ozone is:

$$V = \frac{nRT}{P} = \frac{(1/6.022 \times 10^{23}\ \text{mol})(0.08206\ \text{L atm K}^{-1}\ \text{mol}^{-1})(230.\,\text{K})}{(3.3 \times 10^{-41}\ \text{atm})}, \ V = 9.5 \times 10^{17}\ \text{L}$$

Equilibrium is probably not maintained under these conditions. When only two ozone molecules are in a volume of 9.5×10^{17} L, the reaction is not at equilibrium. Under these conditions, $Q > K$, and the reaction shifts left. But with only two ozone molecules in this huge volume, it is extremely unlikely that they will collide with each other. At these conditions, the concentration of ozone is not large enough to maintain equilibrium.

109.   a.   The sum of the two steps gives the overall balanced equation.

$$O_3(g) + NO(g) \rightarrow NO_2(g) + O_2(g)$$

$$\underline{NO_2(g) + O(g) \rightarrow NO(g) + O_2(g)}$$

$$O_3(g) + O(g) \rightarrow 2\,O_2(g) \qquad \text{overall equation}$$

b.   NO is the catalyst. NO is present in the first step of the mechanism on the reactant side, but it is not a reactant because it is regenerated in the second step and does not appear in the overall balanced equation.

c. NO_2 is an intermediate. It is produced in the first step, but is consumed in the second step. Intermediates also never appear in the overall balanced equation. In a mechanism, intermediates always appear first on the product side, while catalysts always appear first on the reactant side.

d.   The rate of the slow step in a mechanism gives the rate law for the reaction. From the problem, the rate determining step (the slow step) is step 1. The derived rate law is:

Rate $= k[O_3][NO]$

Because NO is a catalyst and not a proposed intermediate, it can appear in the rate law.

e. The mechanism for the chlorine-catalyzed destruction of ozone is:

$$O_3(g) + Cl(g) \rightarrow O_2(g) + ClO(g) \qquad \text{slow step 1}$$
$$ClO(g) + O(g) \rightarrow O_2(g) + Cl(g) \qquad \text{fast step 2}$$

$$\overline{O_3(g) + O(g) \rightarrow 2\ O_2(g)} \qquad \text{overall equation}$$

111. $$Pb^{2+} + H_2EDTA^{2-} \rightleftharpoons PbEDTA^{2-} + 2\ H^+$$

Before	0.0050 M	0.075 M	0	$1.0 \times 10^{-7}\ M$ (Buffer, $[H^+]$ constant)
Change	-0.0050	-0.0050	$\rightarrow$ $+0.0050$	No change	Reacts completely
After	0	0.070	0.0050	1.0×10^{-7}	New initial conditions

x mol/L PbEDTA^{2-} dissociates to reach equilibrium

Change	$+x$	$+x$	$\leftarrow$ $-x$		
Equil.	x	$0.070 + x$	$0.0050 - x$	1.0×10^{-7}	(Buffer)

$$K = 6.7 \times 10^{21} = \frac{[PbEDTA^{2-}][H^+]^2}{[Pb^{2+}][H_2EDTA^{2-}]} = \frac{(0.0050 - x)(1.0 \times 10^{-7})^2}{(x)(0.070 + x)}$$

$$6.7 \times 10^{21} \approx \frac{(0.0050)(1.0 \times 10^{-14})}{(x)(0.070)}, \quad x = [Pb^{2+}] = 1.1 \times 10^{-37}\ M; \quad \text{assumptions good.}$$

113. a. $K^+(\text{blood}) \rightleftharpoons K^+(\text{muscle}) \quad \Delta G° = 0; \quad \Delta G = RT \ln\left(\dfrac{[K^+]_m}{[K^+]_b}\right); \quad \Delta G = w_{max}$

$$\Delta G = \frac{8.3145\ J}{K\ mol} (310.\ K) \ln\left(\frac{0.15}{0.0050}\right), \quad \Delta G = 8.8 \times 10^3\ J/mol = 8.8\ kJ/mol$$

At least 8.8 kJ of work must be applied to transport 1 mol K$^+$.

b. Other ions will have to be transported in order to maintain electroneutrality. Either anions must be transported into the cells, or cations (Na$^+$) in the cell must be transported to the blood. The latter is what happens: [Na$^+$] in blood is greater than [Na$^+$] in cells as a result of this pumping.

c. $\Delta G° = -RT \ln K = -8.3145\ J\ K^{-1}\ mol^{-1} \times 310.\ K \times \ln(1.7 \times 10^5) = -3.1 \times 10^4\ J/mol$
$$= -31\ kJ/mol$$

The hydrolysis of ATP (at standard conditions) provides 31 kJ/mol of energy to do work. We need 8.8 kJ of work to transport 1.0 mol of K$^+$.

$$8.8\ kJ \times \frac{1\ mol\ ATP}{31\ kJ} = 0.28\ mol\ ATP\ \text{must be hydrolyzed}$$

CHAPTER 19

TRANSITION METALS AND COORDINATION CHEMISTRY

Transition Metals

7. Cr and Cu are exceptions to the normal filling order of electrons.

   a. Cr: $[Ar]4s^1 3d^5$ b. Cu: $[Ar]4s^1 3d^{10}$ c. V: $[Ar]4s^2 3d^3$

   Cr^{2+}: $[Ar]3d^4$ Cu^+: $[Ar]3d^{10}$ V^{2+}: $[Ar]3d^3$

   Cr^{3+}: $[Ar]3d^3$ Cu^{2+}: $[Ar]3d^9$ V^{3+}: $[Ar]3d^2$

9. The lanthanide elements are located just before the 5d transition metals. The lanthanide contraction is the steady decrease in the atomic radii of the lanthanide elements when going from left to right across the periodic table. As a result of the lanthanide contraction, the sizes of the 4d and 5d elements are very similar (see the following exercise). This leads to a greater similarity in the chemistry of the 4d and 5d elements in a given vertical group.

11. a. molybdenum(IV) sulfide; molybdenum(VI) oxide

    b. MoS_2, +4; MoO_3, +6; $(NH_4)_2Mo_2O_7$, +6; $(NH_4)_6Mo_7O_{24} \bullet 4\ H_2O$, +6

13. TiF_4: ionic compound containing Ti^{4+} ions and F^- ions. $TiCl_4$, $TiBr_4$, and TiI_4: covalent compounds containing discrete, tetrahedral TiX_4 molecules. As these covalent molecules get larger, the boiling points and melting points increase because the London dispersion forces increase. TiF_4 has the highest boiling point because the interparticle forces are stronger in ionic compounds than in covalent compounds.

15. $H^+ + OH^- \rightarrow H_2O$; sodium hydroxide (NaOH) will react with the H^+ on the product side of the reaction. This effectively removes H^+ from the equilibrium, which will shift the reaction to the right to produce more H^+ and CrO_4^{2-}. As more CrO_4^{2-} is produced, the solution turns yellow.

17. Because transition metals form bonds to species that donate lone pairs of electrons, transition metals are Lewis acids (electron pair acceptors). The Lewis bases in coordination compounds are the ligands, all of which have an unshared pair of electrons to donate. The coordinate covalent bond between the ligand and the transition metal just indicates that both electrons in the bond originally came from one of the atoms in the bond. Here, the electrons in the bond come from the ligand.

19.   The complex ion is $PtCl_4^{2-}$, which is composed of Pt^{2+} and four Cl^- ligands. Pt^{2+}: $[Xe]4f^{14}5d^8$. With square planar geometry, geometric (cis-trans) isomerism is possible. Cisplatin is the cis isomer of the compound and has the following structural formula.

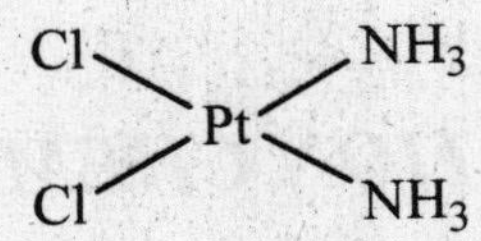

21.   To determine the oxidation state of the metal, you must know the charges of the various common ligands (see Table 19.13 of the text).

   a. hexacyanomanganate(II) ion

   b. cis-tetraamminedichlorocobalt(III) ion

   c. pentaamminechlorocobalt(II) ion

23.   To determine the oxidation state of the metal, you must know the charges of the various common ligands (see Tables 19.13 and 19.14 of the text).

   a. pentaamminechlororuthenium(III) ion   b. hexacyanoferrate(II) ion

   c. tris(ethylenediamine)manganese(II) ion   d. pentaamminenitrocobalt(III) ion

25.   a. $K_2[CoCl_4]$ b. $[Pt(H_2O)(CO)_3]Br_2$

   c. $Na_3[Fe(CN)_2(C_2O_4)_2]$ d. $[Cr(NH_3)_3Cl(NH_2CH_2CH_2NH_2)]I_2$

27. $BaCl_2$ gives no precipitate, so SO_4^{2-} must be in the coordination sphere ($BaSO_4$ is insoluble). A precipitate with $AgNO_3$ means the Cl^- is not in the coordination sphere, that is, Cl^- is a counter ion. Because there are only four ammonia molecules in the coordination sphere, SO_4^{2-} must be acting as a bidentate ligand. The structure is:

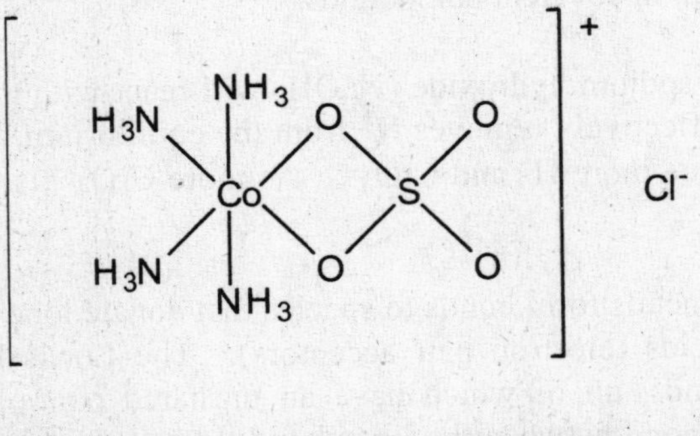

29.   Because each compound contains an octahedral complex ion, the formulas for the compounds are $[Co(NH_3)_6]I_3$, $[Pt(NH_3)_4I_2]I_2$, $Na_2[PtI_6]$, and $[Cr(NH_3)_4I_2]I$. Note that in some cases the I^- ions are ligands bound to the transition metal ion as required for a coordination number of 6, whereas in other cases the I^- ions are counter ions required to balance the charge of the complex ion. The $AgNO_3$ solution will only precipitate the I^- counter ions and will not

precipitate the I⁻ ligands. Therefore, 3 mol AgI will precipitate per mole of $[Co(NH_3)_6]I_3$, 2 mol AgI will precipitate per mole of $[Pt(NH_3)_4I_2]I_2$, 0 mol AgI will precipitate per mole of $Na_2[PtI_6]$, and 1 mol AgI will precipitate per mole of $[Cr(NH_3)_4I_2]I$.

31.   a.   Isomers: species with the same formulas but different properties; they are different compounds. See the text for examples of the following types of isomers.

       b.   Structural isomers: isomers that have one or more bonds that are different.

       c.   Stereoisomers: isomers that contain the same bonds but differ in how the atoms are arranged in space.

       d.   Coordination isomers: structural isomers that differ in the atoms that make up the complex ion.

       e.   Linkage isomers: structural isomers that differ in how one or more ligands are attached to the transition metal.

       f.   Geometric isomers: (cis-trans isomerism); stereoisomers that differ in the positions of atoms with respect to a rigid ring, bond, or each other.

       g.   Optical isomers: stereoisomers that are nonsuperimposable mirror images of each other; that is, they are different in the same way that our left and right hands are different.

33.   a.

cis   trans

Note: $C_2O_4^{2-}$ is a bidentate ligand. Bidentate ligands bond to the metal at two positions that are 90° apart from each other in octahedral complexes. Bidentate ligands do not bond to the metal at positions 180° apart from each other.

b.

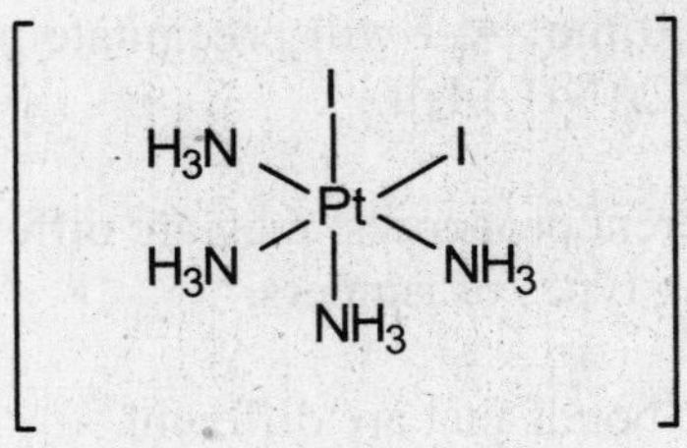

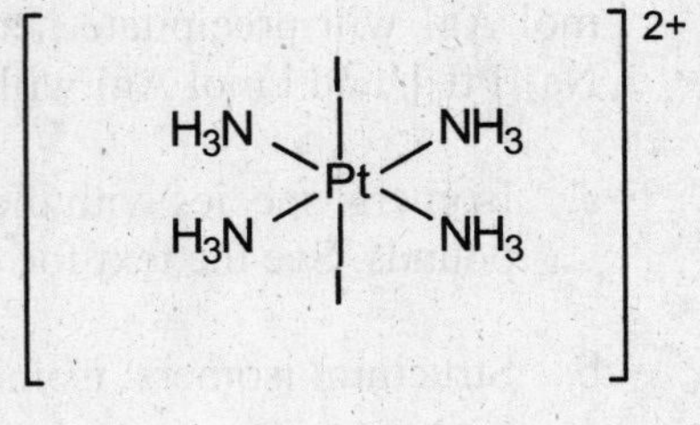

cis

trans

c.

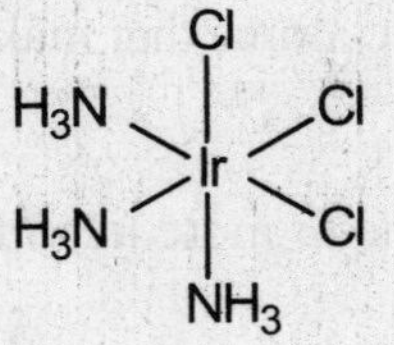

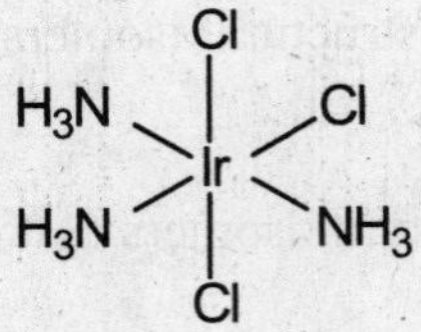

cis

trans

d.

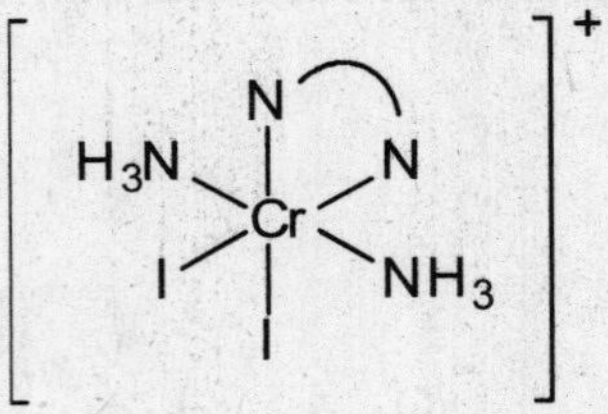

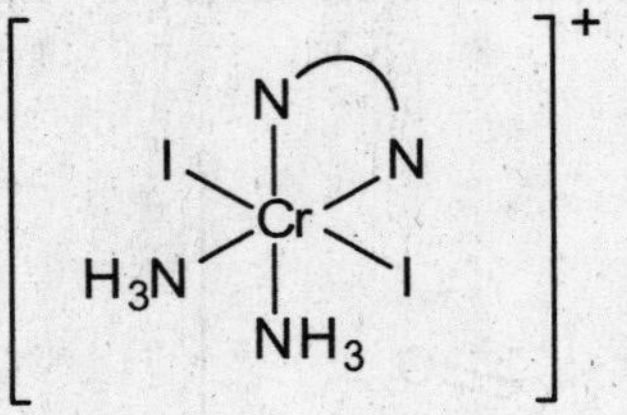

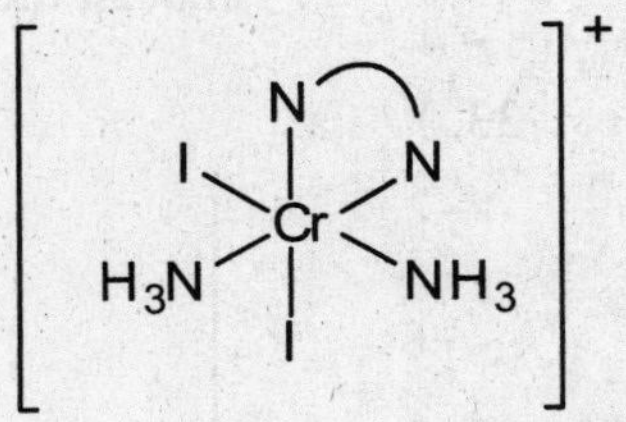

Note: N⌣N is an abbreviation for the bidentate en ligand (ethylenediamine, NH₂CH₂CH₂NH₂).

35.

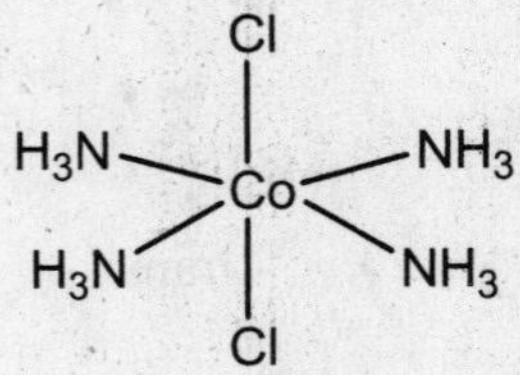

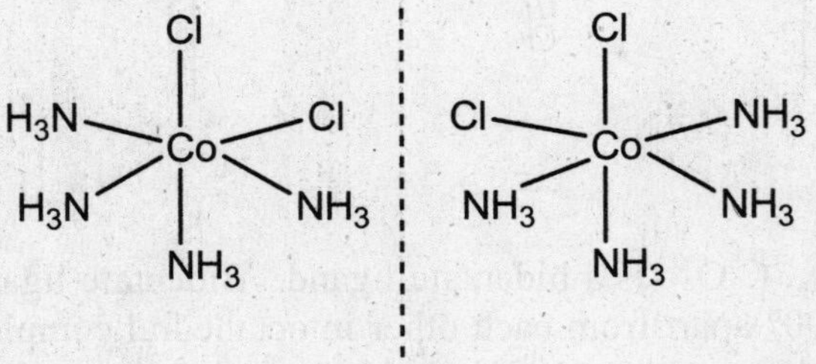

mirror

trans
(mirror image is
superimposable)

cis

The mirror image of the cis
isomer is also superimposable.

No; both the trans and the cis forms of $Co(NH_3)_4Cl_2^+$ have mirror images that are superimposable. For the cis form, the mirror image only needs a $90°$ rotation to produce the original structure. Hence neither the trans nor cis form is optically active.

37.

M = transition metal ion

39.   Linkage isomers differ in the way that the ligand bonds to the metal. SCN^- can bond through the sulfur or through the nitrogen atom. NO_2^- can bond through the nitrogen or through the oxygen atom. OCN^- can bond through the oxygen or through the nitrogen atom. N_3^-, $NH_2CH_2CH_2NH_2$ and I^- are not capable of linkage isomerism.

41.   Similar to the molecules discussed in Figures 19.17 and 19.18 of the text, $Cr(acac)_3$ and cis-$Cr(acac)_2(H_2O)_2$ are optically active.   The mirror images of these two complexes are nonsuperimposable. There is a plane of symmetry in trans-$Cr(acac)_2(H_2O)_2$, so it is not optically active. A molecule with a plane of symmetry is never optically active because the mirror images are always superimposable.   A plane of symmetry is a plane through a molecule where one side exactly reflects the other side of the molecule.

43.   There are four geometrical isomers (labeled i-iv). Isomers iii and iv are optically active, and the nonsuperimposable mirror images are shown.

i.

ii.

iii.

optically active   mirror   mirror image of iii
                                           (nonsuperimposable)

iv.

$$\left[\begin{array}{c} \text{optically active} \end{array}\right]^+$$

optically active   mirror   mirror image of iv
                                                              (nonsuperimposable)

Bonding, Color, and Magnetism in Coordination Compounds

45.   a.   Ligand that will give complex ions with the maximum number of unpaired electrons.

      b.   Ligand that will give complex ions with the minimum number of unpaired electrons.

      c.   Complex with a minimum number of unpaired electrons (low spin = strong field).

      d.   Complex with a maximum number of unpaired electrons (high spin = weak field).

47. Sc^{3+} has no electrons in d orbitals. Ti^{3+} and V^{3+} have d electrons present.   The color of transition metal complexes results from electron transfer between split d orbitals.   If no d electrons are present, no electron transfer can occur, and the compounds are not colored.

49.   The metal ions are both d^5 (Fe^{3+}: $[Ar]d^5$ and Mn^{2+}: $[Ar]d^5$).   One of the diagrams (a) is for a weak-field (high-spin) d^5 complex ion while the other diagram (b) is for a strong-field (low-spin) d^5 complex ion.   From the spectrochemical series, CN^- is a strong-field ligand while H_2O is in the middle of the series.   Because the iron complex ion has CN^- for the ligands as well as having a higher metal ion charge (3+ vs. 2+), one would expect $[Fe(CN)_6]^{3-}$ to have the strong-field diagram in b, while $[Mn(H_2O)_6]^{2+}$ would have the weak-field diagram in a.

51.   a. Fe^{2+}: $[Ar]3d^6$

High spin, small Δ Low spin, large Δ

b. Fe^{3+}: $[Ar]3d^5$

$\underline{\uparrow}\qquad\underline{\uparrow}$

$\underline{\uparrow}\qquad\underline{\uparrow}\qquad\underline{\uparrow}$

High spin, small Δ

c. Ni^{2+}: $[Ar]3d^8$

$\underline{\uparrow}\qquad\underline{\uparrow}$

$\underline{\uparrow\downarrow}\qquad\underline{\uparrow\downarrow}\qquad\underline{\uparrow\downarrow}$

d. Zn^{2+}: $[Ar]3d^{10}$

$\underline{\uparrow\downarrow}\qquad\underline{\uparrow\downarrow}$

$\underline{\uparrow\downarrow}\qquad\underline{\uparrow\downarrow}\qquad\underline{\uparrow\downarrow}$

e. Co^{2+}: $[Ar]3d^7$

$\underline{\uparrow}\qquad\underline{\uparrow}$

$\underline{\uparrow\downarrow}\qquad\underline{\uparrow\downarrow}\qquad\underline{\uparrow}$

High spin, small Δ

$\underline{\uparrow}\qquad\underline{\quad}$

$\underline{\uparrow\downarrow}\qquad\underline{\uparrow\downarrow}\qquad\underline{\uparrow\downarrow}$

Low spin, large Δ

53. To determine the crystal field diagrams, you need to determine the oxidation state of the transition metal, which can only be determined if you know the charges of the ligands (see Table 19.13). The electron configurations and the crystal field diagrams follow.

a. Ru^{2+}: $[Kr]4d^6$, no unpaired e^-

$\underline{\quad}\qquad\underline{\quad}$

$\underline{\uparrow\downarrow}\qquad\underline{\uparrow\downarrow}\qquad\underline{\uparrow\downarrow}$

Low spin, large Δ

b. Ni^{2+}: $[Ar]3d^8$, 2 unpaired e^-

$\underline{\uparrow}\qquad\underline{\uparrow}$

$\underline{\uparrow\downarrow}\qquad\underline{\uparrow\downarrow}\qquad\underline{\uparrow\downarrow}$

c. V^{3+}: $[Ar]3d^2$, 2 unpaired e^-

$\underline{\quad}\qquad\underline{\quad}$

$\underline{\uparrow}\qquad\underline{\uparrow}\qquad\underline{\quad}$

Note: Ni^{2+} must have two unpaired electrons, whether high spin or low spin, and V^{3+} must have two unpaired electrons, whether high spin or low spin.

55.   Replacement of water ligands by ammonia ligands resulted in shorter wavelengths of light being absorbed. Energy and wavelength are inversely related, so the presence of the NH_3 ligands resulted in a larger d-orbital splitting (larger Δ). Therefore, NH_3 is a stronger-field ligand than H_2O.

57.   From Table 19.16 of the text, the violet complex ion absorbs yellow-green light ($\lambda \approx 570$ nm), the yellow complex ion absorbs blue light ($\lambda \approx 450$ nm), and the green complex ion absorbs red light ($\lambda \approx 650$ nm). The spectrochemical series shows that NH_3 is a stronger-field ligand than H_2O, which is a stronger-field ligand than Cl^-. Therefore, $Cr(NH_3)_6^{3+}$ will have the largest d-orbital splitting and will absorb the lowest wavelength electromagnetic radiation ($\lambda \approx 450$ nm) because energy and wavelength are inversely related ($\lambda = hc/E$).   Thus the yellow solution contains the $Cr(NH_3)_6^{3+}$ complex ion.   Similarly, we would expect the $Cr(H_2O)_4Cl_2^+$ complex ion to have the smallest d-orbital splitting because it contains the weakest-field ligands.   The green solution with the longest wavelength of absorbed light contains the $Cr(H_2O)_4Cl_2^+$ complex ion.   This leaves the violet solution, which contains the $Cr(H_2O)_6^{3+}$ complex ion.   This makes sense because we would expect $Cr(H_2O)_6^{3+}$ to absorb light of a wavelength between that of $Cr(NH_3)_6^{3+}$ and $Cr(H_2O)_4Cl_2^+$.

59.   a. $Ru(phen)_3^{2+}$ exhibits optical isomerism [similar to $Co(en)_3^{3+}$ in Figure 19.17 of the text].

    b. Ru^{2+}: $[Kr]4d^6$; because there are no unpaired electrons, Ru^{2+} is a strong-field (low-spin) case.

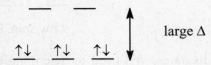

61.   The crystal field diagrams are different because the geometries of where the ligands point are different. The tetrahedrally oriented ligands point differently in relationship to the d orbitals than do the octahedrally oriented ligands. Also, we have more ligands in an octahedral complex.

    See Figure 19.28 for the tetrahedral crystal field diagram. Notice that the orbitals are reverse of that in the octahedral crystal field diagram. The degenerate d_{z^2} and $d_{x^2-y^2}$ are at a lower energy than the degenerate d_{xy}, d_{xz}, and d_{yz} orbitals. Again, the reason for this is that tetrahedral ligands are oriented differently than octahedral field ligands, so the interactions with specifically oriented d orbitals are different. Also notice that the difference in magnitude of the d-orbital splitting for the two geometries. The d-orbital splitting in tetrahedral complexes is less than one-half the d-orbital splitting in octahedral complexes. There are no known ligands powerful enough to produce the strong-field case; hence all tetrahedral complexes are weak field or high spin.

63. $CoBr_6^{4-}$ has an octahedral structure, and $CoBr_4^{2-}$ has a tetrahedral structure (as do most Co^{2+} complexes with four ligands).   Coordination complexes absorb electromagnetic radiation (EMR) of energy equal to the energy difference between the split d orbitals.   Because the tetrahedral d-orbital splitting is less than one-half the octahedral d-orbital splitting, tetrahedral complexes will absorb lower-energy EMR, which corresponds to longer-wavelength EMR ($E = hc/\lambda$). Therefore, $CoBr_6^{4-}$ will absorb EMR having a wavelength shorter than 3.4×10^{-6} m.

Additional Exercises

65. $Hg^{2+}(aq) + 2\ I^-(aq) \rightarrow HgI_2(s)$, orange precipitate

$HgI_2(s) + 2\ I^-(aq) \rightarrow HgI_4^{2-}(aq)$, soluble complex ion

Hg^{2+} is a d^{10} ion. Color is the result of electron transfer between split d orbitals. This cannot occur for the filled d orbitals in Hg^{2+}. Therefore, we would not expect Hg^{2+} complex ions to form colored solutions.

67. $Ni(CO)_4$ is composed of 4 CO molecules and Ni. Thus nickel has an oxidation state of zero.

69.   At high altitudes, the oxygen content of air is lower, so less oxyhemoglobin is formed, which diminishes the transport of oxygen in the blood. A serious illness called high-altitude sickness can result from the decrease of O_2 in the blood. High-altitude acclimatization is the phenomenon that occurs with time in the human body in response to the lower amounts of oxyhemoglobin in the blood. This response is to produce more hemoglobin and hence, increase the oxyhemoglobin in the blood. High-altitude acclimatization takes several weeks to take hold for people moving from lower altitudes to higher altitudes.

71.   i. $0.0203 \text{ g CrO}_3 \times \dfrac{52.00 \text{ g Cr}}{100.0 \text{ g CrO}_3} = 0.0106 \text{ g Cr};\quad \% \text{ Cr} = \dfrac{0.0106 \text{ g}}{0.105 \text{ g}} \times 100 = 10.1\% \text{ Cr}$

ii. $32.93 \text{ mL HCl} \times \dfrac{0.100 \text{ mmol HCl}}{\text{mL}} \times \dfrac{1 \text{ mmol NH}_3}{\text{mmol HCl}} \times \dfrac{17.03 \text{ mg NH}_3}{\text{mmol}} = 56.1 \text{ mg NH}_3$

$\% \text{ NH}_3 = \dfrac{56.1 \text{ mg}}{341 \text{ mg}} \times 100 = 16.5\% \text{ NH}_3$

iii. $73.53\% + 16.5\% + 10.1\% = 100.1\%$; the compound must be composed of only Cr, NH_3, and I.

Out of 100.00 g of compound:

$10.1 \text{ g Cr} \times \dfrac{1 \text{ mol}}{52.00 \text{ g}} = 0.194 \text{ mol};\qquad \dfrac{0.194}{0.194} = 1.00$

$16.5 \text{ g NH}_3 \times \dfrac{1 \text{ mol}}{17.03 \text{ g}} = 0.969 \text{ mol};\qquad \dfrac{0.969}{0.194} = 4.99$

$73.53 \text{ g I} \times \dfrac{1 \text{ mol}}{126.9 \text{ g}} = 0.5794 \text{ mol};\qquad \dfrac{0.5794}{0.194} = 2.99$

$Cr(NH_3)_5I_3$ is the empirical formula. Cr^{3+} forms octahedral complexes. So compound A is made of the octahedral $[Cr(NH_3)_5I]^{2+}$ complex ion and two I^- ions as counter ions; the formula is $[Cr(NH_3)_5I]I_2$. Lets check this proposed formula using the freezing-point data.

iv. $\Delta T_f = iK_f m$; for $[Cr(NH_3)_5I]I_2$, $i = 3.0$ (assuming complete dissociation).

$$\text{Molality} = m = \frac{0.601\,\text{g complex}}{1.000 \times 10^{-2}\,\text{kg H}_2\text{O}} \times \frac{1\,\text{mol complex}}{517.9\,\text{g complex}} = 0.116\,\text{mol/kg}$$

$$\Delta T_f = 3.0 \times 1.86\,°\text{C kg/mol} \times 0.116\,\text{mol/kg} = 0.65°\text{C}$$

Because ΔT_f is close to the measured value, this is consistent with the formula $[Cr(NH_3)_5I]I_2$.

73.   M = metal ion

75.   No; in all three cases, six bonds are formed between Ni^{2+} and nitrogen, so ΔH values should be similar. $\Delta S°$ for formation of the complex ion is most negative for 6 NH_3 molecules reacting with a metal ion (seven independent species become one). For penten reacting with a metal ion, two independent species become one, so $\Delta S°$ is least negative for this reaction compared to the other reactions. Thus the chelate effect occurs because the more bonds a chelating agent can form to the metal, the less unfavorable $\Delta S°$ is for the formation of the complex ion, and the larger the formation constant.

77.
$$\overset{\text{II}\qquad\text{III}}{(H_2O)_5Cr-Cl-Co(NH_3)_5} \rightarrow \overset{\text{III}\qquad\text{II}}{(H_2O)_5Cr-Cl-Co(NH_3)_5} \rightarrow Cr(H_2O)_5Cl^{2+} + Co(II)\ \text{complex}$$

Yes; this is consistent. After the oxidation, the ligands on Cr(III) won't exchange. Because Cl^- is in the coordination sphere, it must have formed a bond to Cr(II) before the electron transfer occurred (as proposed through the formation of the intermediate).

79.   a.
$$Fe(H_2O)_6^{3+} + H_2O \rightleftharpoons Fe(H_2O)_5(OH)^{2+} + H_3O^+$$

Initial	0.10 M	0	~0
Equil.	$0.10 - x$	x	x

$$K_a = \frac{[Fe(H_2O)_5(OH)^{2+}][H_3O^+]}{[Fe(H_2O)_6^{3+}]} = 6.0 \times 10^{-3} = \frac{x^2}{0.10 - x} \approx \frac{x^2}{0.10}$$

$x = 2.4 \times 10^{-2}$; assumption is poor (x is 24% of 0.10). Using successive approximations:

$$\frac{x^2}{0.10 - 0.024} = 6.0 \times 10^{-3},\ x = 0.021$$

$$\frac{x^2}{0.10 - 0.021} = 6.0 \times 10^{-3}, \ x = 0.022; \quad \frac{x^2}{0.10 - 0.022} = 6.0 \times 10^{-3}, \ x = 0.022$$

$$x = [H^+] = 0.022 \ M; \ pH = 1.66$$

b. Because of the lower charge, $Fe^{2+}(aq)$ will not be as strong an acid as $Fe^{3+}(aq)$. A solution of iron(II) nitrate will be less acidic (have a higher pH) than a solution with the same concentration of iron(III) nitrate.

81. a. In the lungs, there is a lot of O_2, and the equilibrium favors $Hb(O_2)_4$. In the cells, there is a lower concentration of O_2, and the equilibrium favors HbH_4^{4+}.

b. CO_2 is a weak acid in water; $CO_2 + H_2O \rightleftharpoons HCO_3^- + H^+$. Removing CO_2 essentially decreases H^+, which causes the hemoglobin reaction to shift right. $Hb(O_2)_4$ is then favored, and O_2 is not released by hemoglobin in the cells. Breathing into a paper bag increases $[CO_2]$ in the blood, thus increasing $[H^+]$, which shifts the hemoglobin reaction left.

c. CO_2 builds up in the blood, and it becomes too acidic, driving the equilibrium to the left. Hemoglobin can't bind O_2 as strongly in the lungs. Bicarbonate ion acts as a base in water and neutralizes the excess acidity.

Challenge Problems

83.

The $d_{x^2-y^2}$ and d_{xy} orbitals are in the plane of the three ligands and should be destabilized the most. The amount of destabilization should be about equal when all the possible interactions are considered. The d_{z^2} orbital has some electron density in the xy plane (the doughnut) and should be destabilized a lesser amount than the $d_{x^2-y^2}$ and d_{xy} orbitals. The d_{xz} and d_{yz} orbitals have no electron density in the plane and should be lowest in energy.

85.

The d_{z^2} orbital will be destabilized much more than in the trigonal planar case (see Exercise 19.83). The d_{z^2} orbital has electron density on the z axis directed at the two axial ligands. The $d_{x^2-y^2}$ and d_{xy} orbitals are in the plane of the three trigonal planar ligands and should be destabilized a lesser amount than the d_{z^2} orbital; only a portion of the electron density in the $d_{x^2-y^2}$ and d_{xy} orbitals is directed at the ligands. The d_{xz} and d_{yz} orbitals will be destabilized the least since the electron density is directed between the ligands.

87.   a.   Consider the following electrochemical cell:

$$Co^{3+} + e^- \rightarrow Co^{2+} \qquad\qquad E_c^\circ = 1.82\ V$$
$$Co(en)_3^{2+} \rightarrow Co(en)_3^{3+} + e^- \qquad\qquad -E_a^\circ = ?$$

$$\overline{Co^{3+} + Co(en)_3^{2+} \rightarrow Co^{2+} + Co(en)_3^{3+} \qquad\qquad E_{cell}^\circ = 1.82 - E_a^\circ}$$

The equilibrium constant for this overall reaction is:

$$Co^{3+} + 3\ en \rightarrow Co(en)_3^{3+} \qquad\qquad K_1 = 2.0 \times 10^{47}$$
$$Co(en)_3^{2+} \rightarrow Co^{2+} + 3\ en \qquad\qquad K_2 = 1/1.5 \times 10^{12}$$

$$\overline{Co^{3+} + Co(en)_3^{2+} \rightarrow Co(en)_3^{3+} + Co^{2+} \qquad K = K_1 K_2 = \frac{2.0 \times 10^{47}}{1.5 \times 10^{12}} = 1.3 \times 10^{35}}$$

From the Nernst equation for the overall reaction:

$$E_{cell}^\circ = \frac{0.0591}{n} \log K = \frac{0.0591}{1} \log(1.3 \times 10^{35}), \quad E_{cell}^\circ = 2.08\ V$$

$$E_{cell}^\circ = 1.82 - E_a^\circ = 2.08\ V, \quad -E_a^\circ = 2.08\ V - 1.82\ V = 0.26\ V, \text{ so } E_c^\circ = -0.26\ V$$

b.   The stronger oxidizing agent will be the more easily reduced species and will have the more positive standard reduction potential. From the reduction potentials, Co^{3+} ($E^\circ = 1.82\ V$) is a much stronger oxidizing agent than $Co(en)_3^{3+}$ ($E^\circ = -0.26\ V$).

c.   In aqueous solution, Co^{3+} forms the hydrated transition metal complex $Co(H_2O)_6^{3+}$. In both complexes, $Co(H_2O)_6^{3+}$ and $Co(en)_3^{3+}$, cobalt exists as Co^{3+}, which has six d electrons. Assuming a strong-field case for each complex ion, the d-orbital splitting diagram for each is:

$$\underline{\quad}\quad\underline{\quad}\qquad e_g$$

$$\underline{\uparrow\downarrow}\quad\quad\underline{\uparrow\downarrow}\quad\quad\underline{\uparrow\downarrow}\qquad t_{2g}$$

When each complex gains an electron, the electron enters a higher-energy e_g orbital. Because en is a stronger-field ligand than H_2O, the d-orbital splitting is larger for $Co(en)_3^{3+}$, and it takes more energy to add an electron to $Co(en)_3^{3+}$ than to $Co(H_2O)_6^{3+}$. Therefore, it is more favorable for $Co(H_2O)_6^{3+}$ to gain an electron than for $Co(en)_3^{3+}$ to gain an electron.

89. a.

$$AgBr(s) \rightleftharpoons Ag^+ + Br^- \quad K_{sp} = [Ag^+][Br^-] = 5.0 \times 10^{-13}$$

Initial	s = solubility (mol/L)	0	0
Equil.		s	s

$$K_{sp} = 5.0 \times 10^{-13} = s^2, \quad s = 7.1 \times 10^{-7} \text{ mol/L}$$

b.

$$AgBr(s) \rightleftharpoons Ag^+ + Br^- \qquad K_{sp} = 5.0 \times 10^{-13}$$
$$Ag^+ + 2 NH_3 \rightleftharpoons Ag(NH_3)_2^+ \qquad K_f = 1.7 \times 10^7$$

$$AgBr(s) + 2 NH_3(aq) \rightleftharpoons Ag(NH_3)_2^+(aq) + Br^-(aq) \quad K = K_{sp} \times K_f = 8.5 \times 10^{-6}$$

$$AgBr(s) + 2 NH_3 \rightleftharpoons Ag(NH_3)_2^+ + Br^-$$

Initial	3.0 M	0	0
	s mol/L of AgBr(s) dissolves to reach equilibrium = molar solubility		
Equil.	$3.0 - 2s$	s	s

$$K = \frac{[Ag(NH_3)_2^+][Br^-]}{[NH_3]^2} = \frac{s^2}{(3.0 - 2s)^2}, \quad 8.5 \times 10^{-6} \approx \frac{s^2}{(3.0)^2}, \quad s = 8.7 \times 10^{-3} \text{ mol/L}$$

Assumption good.

c. The presence of NH_3 increases the solubility of AgBr. Added NH_3 removes Ag^+ from solution by forming the complex ion $Ag(NH_3)_2^+$. As Ag^+ is removed, more AgBr(s) will dissolve to replenish the Ag^+ concentration.

d. Mass AgBr = $0.2500 \text{ L} \times \dfrac{8.7 \times 10^{-3} \text{ mol AgBr}}{\text{L}} \times \dfrac{187.8 \text{ g AgBr}}{\text{mol AgBr}} = 0.41 \text{ g AgBr}$

e. Added HNO_3 will have no effect on the AgBr(s) solubility in pure water. Neither H^+ nor NO_3^- reacts with Ag^+ or Br^- ions. Br^- is the conjugate base of the strong acid HBr, so it is a terrible base. Added H^+ will not react with Br^- to any great extent. However, added HNO_3 will reduce the solubility of AgBr(s) in the ammonia solution. NH_3 is a weak base ($K_b = 1.8 \times 10^{-5}$). Added H^+ will react with NH_3 to form NH_4^+. As NH_3 is removed, a smaller amount of the $Ag(NH_3)_2^+$ complex ion will form, resulting in a smaller amount of AgBr(s) that will dissolve.

CHAPTER 20

THE NUCLEUS: A CHEMIST'S VIEW

Radioactive Decay and Nuclear Transformations

1. a. Thermodynamic stability: the potential energy of a particular nucleus compared to the sum of the potential energies of its component protons and neutrons.

    b. Kinetic stability: the probability that a nucleus will undergo decomposition to form a different nucleus.

    c. Radioactive decay: a spontaneous decomposition of a nucleus to form a different nucleus.

    d. Beta-particle production: a decay process for radioactive nuclides where an electron is produced; the mass number remains constant and the atomic number changes.

    e. Alpha-particle production: a common mode of decay for heavy radioactive nuclides where a helium nucleus is produced, causing the atomic number and the mass number to change.

    f. Positron production: a mode of nuclear decay in which a particle is formed having the same mass as an electron but opposite in charge.

    g. Electron capture: a process in which one of the inner-orbital electrons in an atom is captured by the nucleus.

    h. Gamma-ray emissions; the production of high-energy photons (gamma rays) that frequently accompany nuclear decays and particle reactions.

3. All nuclear reactions must be charge balanced and mass balanced. To charge balance, balance the sum of the atomic numbers on each side of the reaction, and to mass balance, balance the sum of the mass numbers on each side of the reaction.

    a. $^{238}_{92}U \rightarrow {}^{4}_{2}He + {}^{234}_{90}Th$; this is alpha-particle production.

    b. $^{234}_{90}Th \rightarrow {}^{234}_{91}Pa + {}^{0}_{-1}e$; this is β-particle production.

5. All nuclear reactions must be charge-balanced and mass-balanced. To charge-balance, balance the sum of the atomic numbers on each side of the reaction, and to mass-balance, balance the sum of the mass numbers on each side of the reaction.

a. $^{51}_{24}\text{Cr} + ^{0}_{-1}\text{e} \rightarrow ^{51}_{23}\text{V}$ b. $^{131}_{53}\text{I} \rightarrow ^{0}_{-1}\text{e} + ^{131}_{54}\text{Xe}$

c. $^{32}_{15}\text{P} \rightarrow ^{0}_{-1}\text{e} + ^{32}_{16}\text{S}$

7. a. $^{68}_{31}\text{Ga} + ^{0}_{-1}\text{e} \rightarrow ^{68}_{30}\text{Zn}$ b. $^{62}_{29}\text{Cu} \rightarrow ^{0}_{+1}\text{e} + ^{62}_{28}\text{Ni}$

c. $^{212}_{87}\text{Fr} \rightarrow ^{4}_{2}\text{He} + ^{208}_{85}\text{At}$ d. $^{129}_{51}\text{Sb} \rightarrow ^{0}_{-1}\text{e} + ^{129}_{52}\text{Te}$

9. $^{247}_{97}\text{Bk} \rightarrow ^{207}_{82}\text{Pb} + ?\, ^{4}_{2}\text{He} + ?\, ^{0}_{-1}\text{e}$; The change in mass number (247 - 207 = 40) is due exclusively to the alpha particles. A change in mass number of 40 requires 10 $^{4}_{2}\text{He}$ particles to be produced. The atomic number only changes by 97 − 82 = 15. The 10 alpha particles change the atomic number by 20, so 5 $^{0}_{-1}\text{e}$ (five beta particles) are produced in the decay series of ^{247}Bk to ^{207}Pb.

11. Reference Table 20.2 of the text for potential radioactive decay processes. ^{17}F and ^{18}F contain too many protons or too few neutrons. Electron capture and positron production are both possible decay mechanisms that increase the neutron-to-proton ratio. Alpha-particle production also increases the neutron-to-proton ratio, but it is not likely for these light nuclei. ^{21}F contains too many neutrons or too few protons. Beta-particle production lowers the neutron-to-proton ratio, so we expect ^{21}F to be a β-emitter.

13. a. $^{249}_{98}\text{Cf} + ^{18}_{8}\text{O} \rightarrow ^{263}_{106}\text{Sg} + 4\,^{1}_{0}\text{n}$ b. $^{259}_{104}\text{Rf}$; $^{263}_{106}\text{Sg} \rightarrow ^{4}_{2}\text{He} + ^{259}_{104}\text{Rf}$

Kinetics of Radioactive Decay

15. $k = \dfrac{\ln 2}{t_{1/2}} = \dfrac{0.69315}{433\,\text{yr}} \times \dfrac{1\,\text{yr}}{365\,\text{d}} \times \dfrac{1\,\text{d}}{24\,\text{h}} \times \dfrac{1\,\text{h}}{3600\,\text{s}} = 5.08 \times 10^{-11}\,\text{s}^{-1}$

Rate $= kN = 5.08 \times 10^{-11}\,\text{s}^{-1} \times 5.00\,\text{g} \times \dfrac{1\,\text{mol}}{241\,\text{g}} \times \dfrac{6.022 \times 10^{23}\,\text{nuclei}}{\text{mol}} = 6.35 \times 10^{11}\,\text{decays/s}$

6.35×10^{11} alpha particles are emitted each second from a 5.00-g ^{241}Am sample.

17. a. $k = \dfrac{\ln 2}{t_{1/2}} = \dfrac{0.6931}{12.8\,\text{d}} \times \dfrac{1\,\text{d}}{24\,\text{h}} \times \dfrac{1\,\text{h}}{3600\,\text{s}} = 6.27 \times 10^{-7}\,\text{s}^{-1}$

b. Rate $= kN = 6.27 \times 10^{-7}\,\text{s}^{-1} \left(28.0 \times 10^{-3}\,\text{g} \times \dfrac{1\,\text{mol}}{64.0\,\text{g}} \times \dfrac{6.022 \times 10^{23}\,\text{nuclei}}{\text{mol}} \right)$

Rate $= 1.65 \times 10^{14}\,\text{decays/s}$

c. 25% of the ^{64}Cu will remain after 2 half-lives (100% decays to 50% after one half-life which decays to 25% after a second half-life). Hence 2(12.8 days) = 25.6 days is the time frame for the experiment.

19. $t = 67.0$ yr; $k = \dfrac{\ln 2}{t_{1/2}}$; $\ln\left(\dfrac{N}{N_0}\right) = -kt = \dfrac{-(0.6931)67.0 \text{ yr}}{28.9 \text{ yr}} = -1.61$, $\left(\dfrac{N}{N_0}\right) = e^{-1.61} = 0.200$

20.0% of the ^{90}Sr remains as of July 16, 2012.

21. 175 mg $Na_3{}^{32}PO_4 \times \dfrac{32.0 \text{ mg } {}^{32}P}{165.0 \text{ mg } Na_3{}^{32}PO_4} = 33.9$ mg ^{32}P; $k = \dfrac{\ln 2}{t_{1/2}}$

$\ln\left(\dfrac{N}{N_0}\right) = -kt = \dfrac{-(0.6931)t}{t_{1/2}}$, $\ln\left(\dfrac{m}{33.9 \text{ mg}}\right) = \dfrac{-0.6931(35.0 \text{ d})}{14.3 \text{ d}}$; carrying extra sig. figs.:

$\ln(m) = -1.696 + 3.523 = 1.827$, $m = e^{1.827} = 6.22$ mg ^{32}P remains

23.   Plants take in CO_2 during the photosynthesis process, which incorporates carbon, including ^{14}C, into its molecules. As long as the plant is alive, the $^{14}C/^{12}C$ ratio in the plant will equal the ratio in the atmosphere. When the plant dies, ^{14}C is not replenished because ^{14}C decays by beta-particle production. By measuring the ^{14}C activity today in the artifact and comparing this to the assumed ^{14}C activity when the plant died to make the artifact, an age can be determined for the artifact. The assumptions are that the ^{14}C level in the atmosphere is constant or that the ^{14}C level at the time the plant died can be calculated. A constant ^{14}C level is a pure assumption, and accounting for variation is complicated. Another problem is that some of the material must be destroyed to determine the ^{14}C level.

25. $t_{1/2} = 5730$ y; $k = (\ln 2)/t_{1/2}$; $\ln (N/N_0) = -kt$; $\ln \dfrac{15.1}{15.3} = \dfrac{-(\ln 2)t}{5730 \text{ yr}}$, $t = 109$ years

No; from ^{14}C dating, the painting was produced during the early 1900s.

27. $\ln\left(\dfrac{N}{N_0}\right) = -kt = \dfrac{-(\ln 2)t}{12.3 \text{ yr}}$, $\ln\left(\dfrac{0.17 \times N_0}{N_0}\right) = -(5.64 \times 10^{-2})t$, $t = 31.4$ years

It takes 31.4 years for the tritium to decay to 17% of the original amount. Hence the watch stopped fluorescing enough to be read in 1975 $(1944 + 31.4)$.

29.   Assuming 1.000 g ^{238}U present in a sample, then 0.688 g ^{206}Pb is present. Because 1 mol ^{206}Pb is produced per mol ^{238}U decayed:

^{238}U decayed $= 0.688$ g Pb $\times \dfrac{1 \text{ mol Pb}}{206 \text{ g Pb}} \times \dfrac{1 \text{ mol U}}{\text{mol Pb}} \times \dfrac{238 \text{ g U}}{\text{mol U}} = 0.795$ g ^{238}U

Original mass ^{238}U present $= 1.000$ g $+ 0.795$ g $= 1.795$ g ^{238}U

$\ln\left(\dfrac{N}{N_0}\right) = -kt = \dfrac{-(\ln 2)t}{t_{1/2}}$, $\ln\left(\dfrac{1.000 \text{ g}}{1.795 \text{ g}}\right) = \dfrac{-(0.693)t}{4.5 \times 10^9 \text{ yr}}$, $t = 3.8 \times 10^9$ years

Energy Changes in Nuclear Reactions

31. $\Delta E = \Delta mc^2, \ \Delta m = \dfrac{\Delta E}{c^2} = \dfrac{3.9 \times 10^{23} \text{ kg m}^2/\text{s}^2}{(3.00 \times 10^8 \text{ m/s})^2} = 4.3 \times 10^6 \text{ kg}$

The sun loses 4.3×10^6 kg of mass each second. *Note*: 1 J = 1 kg m^2/s^2.

33.   From the text, the mass of a proton = 1.00728 amu, the mass of a neutron = 1.00866 amu, and the mass of an electron = 5.486×10^{-4} amu.

Mass of $_{26}^{56}$Fe nucleus = mass of atom − mass of electrons = 55.9349 − 26(0.0005486)

$= 55.9206$ amu

$26\ _1^1\text{H} + 30\ _1^1\text{n} \rightarrow\ _{26}^{56}\text{Fe}; \ \Delta m = 55.9206 \text{ amu} - [26(1.00728) + 30(1.00866)] \text{ amu}$

$= -0.5285$ amu

$\Delta E = \Delta mc^2 = -0.5285 \text{ amu} \times \dfrac{1.6605 \times 10^{-27} \text{ kg}}{\text{amu}} \times (2.9979 \times 10^8 \text{ m/s})^2 = -7.887 \times 10^{-11} \text{ J}$

$\dfrac{\text{Binding energy}}{\text{Nucleon}} = \dfrac{7.887 \times 10^{-11} \text{ J}}{56 \text{ nucleons}} = 1.408 \times 10^{-12} \text{ J/nucleon}$

35.   Let m_e = mass of electron; for ^{12}C (6e, 6p, 6n): mass defect = Δm = [mass of ^{12}C nucleus] − [mass of 6 protons + mass of 6 neutrons]. *Note*: Atomic masses given include the mass of the electrons.

$\Delta m = 12.00000 \text{ amu} - 6m_e - [6(1.00782 - m_e) + 6(1.00866)];$ mass of electrons cancel.

$\Delta m = 12.00000 - [6(1.00782) + 6(1.00866)] = -0.09888$ amu

$\Delta E = \Delta mc^2 = -0.09888 \text{ amu} \times \dfrac{1.6605 \times 10^{-27} \text{ kg}}{\text{amu}} \times (2.9979 \times 10^8 \text{ m/s})^2$

$= -1.476 \times 10^{-11} \text{ J}$

$\dfrac{\text{Binding energy}}{\text{Nucleon}} = \dfrac{1.476 \times 10^{-11} \text{ J}}{12 \text{ nucleons}} = 1.230 \times 10^{-12} \text{ J/nucleon}$

For ^{235}U (92e, 92p, 143n):

$\Delta m = 235.0439 - 92m_e - [92(1.00782 - m_e) + 143(1.00866)] = -1.9139$ amu

$\Delta E = \Delta mc^2 = -1.9139 \times \dfrac{1.66054 \times 10^{-27} \text{ kg}}{\text{amu}} \times (2.99792 \times 10^8 \text{ m/s})^2 = -2.8563 \times 10^{-10} \text{ J}$

$\dfrac{\text{Binding energy}}{\text{Nucleon}} = \dfrac{2.8563 \times 10^{-10} \text{ J}}{235 \text{ nucleons}} = 1.2154 \times 10^{-12} \text{ J/nucleon}$

Because ^{56}Fe is the most stable known nucleus, the binding energy per nucleon for ^{56}Fe (1.408×10^{-12} J/nucleon) will be larger than that of ^{12}C or ^{235}U (see Figure 20.9 of the text).

37. Binding energy $= \dfrac{1.326 \times 10^{-12} \text{ J}}{\text{nucleon}} \times 27 \text{ nucleons} = 3.580 \times 10^{-11}$ J for each ^{27}Mg nucleus

$\Delta E = \Delta mc^2$, $\Delta m = \dfrac{\Delta E}{c^2} = \dfrac{-3.580 \times 10^{-11} \text{ J}}{(2.9979 \times 10^8 \text{ m/s})^2} = -3.983 \ 10^{-28}$ kg

$\Delta m = -3.983 \ 10^{-28} \text{ kg} \times \dfrac{1 \text{ amu}}{1.6605 \times 10^{-27} \text{ kg}} = -0.2399 \text{ amu} = \text{mass defect}$

Let m_{Mg} = mass of ^{27}Mg nucleus; an ^{27}Mg nucleus has 12 p and 15 n.

$-0.2399 \text{ amu} = m_{Mg} - (12m_p + 15m_n) = m_{Mg} - [12(1.00728 \text{ amu}) + 15(1.00866 \text{ amu})]$

$m_{Mg} = 26.9764$ amu

Mass of ^{27}Mg atom $= 26.9764 \text{ amu} + 12m_e$, $26.9764 + 12(5.49 \times 10^{-4} \text{ amu}) = 26.9830$ amu (includes mass of 12 e$^-$)

39. $^2_1\text{H} + ^3_1\text{H} \rightarrow ^4_2\text{He} + ^1_0\text{n}$; mass of electrons cancel when determining Δm for this nuclear reaction.

$\Delta m = [4.00260 + 1.00866 - (2.01410 + 3.01605)] \text{ amu} = -1.889 \times 10^{-2}$ amu

For the production of 1 mol of ^4_2He: $\Delta m = -1.889 \times 10^{-2} \text{ g} = -1.889 \times 10^{-5}$ kg

$\Delta E = \Delta mc^2 = -1.889 \times 10^{-5} \text{ kg} \times (2.9979 \times 10^8 \text{ m/s})^2 = -1.698 \times 10^{12}$ J/mol

For one nucleus of ^4_2He:

$$\dfrac{-1.698 \times 10^{12} \text{ J}}{\text{mol}} \times \dfrac{1 \text{ mol}}{6.0221 \times 10^{23} \text{ nuclei}} = -2.820 \times 10^{-12} \text{ J/nucleus}$$

Detection, Uses, and Health Effects of Radiation

41. The Geiger-Müller tube has a certain response time. After the gas in the tube ionizes to produce a "count," some time must elapse for the gas to return to an electrically neutral state. The response of the tube levels out because at high activities, radioactive particles are entering the tube faster than the tube can respond to them.

43. Fission: Splitting of a heavy nucleus into two (or more) lighter nuclei.

Fusion: Combining two light nuclei to form a heavier nucleus.

The maximum binding energy per nucleon occurs at Fe. Nuclei smaller than Fe become more stable by fusing to form heavier nuclei closer in mass to Fe. Nuclei larger than Fe form more stable nuclei by splitting to form lighter nuclei closer in mass to Fe.

45.   Moderator: Slows the neutrons to increase the efficiency of the fission reaction.

Control rods: Absorbs neutrons to slow or halt the fission reaction.

47.   In order to sustain a nuclear chain reaction, the neutrons produced by the fission process must be contained within the fissionable material so that they can go on to cause other fissions. The fissionable material must be closely packed together to ensure that neutrons are not lost to the outside. The critical mass is the mass of material in which exactly one neutron from each fission event causes another fission event so that the process sustains itself. A supercritical situation occurs when more than one neutron from each fission event causes another fission event. In this case the process rapidly escalates, and the heat buildup causes a violent explosion.

49.   A nonradioactive substance can be put in equilibrium with a radioactive substance. The two materials can then be checked to see whether all the radioactivity remains in the original material or if it has been scrambled by the equilibrium.

51.   All evolved oxygen in O_2 comes from water and not from carbon dioxide.

53.   Some factors for the biological effects of radiation exposure are:

a.   The energy of the radiation. The higher the energy, the more damage it can cause.

b.   The penetrating ability of radiation. The ability of specific radiation to penetrate human tissue where it can do damage must be considered.

c.   The ionizing ability of the radiation. When biomolecules are ionized, their function is usually disturbed.

d.   The chemical properties of the radiation source. Specifically, can the radioactive substance be readily incorporated into the body, or is the radiation source inert chemically so that it passes through the body relatively quickly.

^{90}Sr will be incorporated into the body by replacing calcium in the bones. Once incorporated, ^{90}Sr can cause leukemia and bone cancer. Krypton is chemically inert, so it will not be incorporated into the body.

55.   (i) and (ii) mean that Pu is not a significant threat outside the body. Our skin is sufficient to keep out the α particles. If Pu gets inside the body, it is easily oxidized to Pu^{4+} (iv), which is chemically similar to Fe^{3+} (iii). Thus Pu^{4+} will concentrate in tissues where Fe^{3+} is found, including the bone marrow, where red blood cells are produced. Once inside the body, α particles can cause considerable damage.

Additional Exercises

57. $$N = 180 \text{ lb} \times \frac{453.6 \text{ g}}{\text{lb}} \times \frac{18 \text{ g C}}{100 \text{ g body}} \times \frac{1.6 \times 10^{-10} \text{ g }^{14}C}{100 \text{ g C}} \times \frac{1 \text{ mol }^{14}C}{14 \text{ g }^{14}C}$$

$$\times \frac{6.022 \times 10^{23} \text{ nuclei }^{14}C}{\text{mol }^{14}C} = 1.0 \times 10^{15} \text{ nuclei }^{14}C$$

$$\text{Rate} = kN; \quad k = \frac{\ln 2}{t_{1/2}} = \frac{0.693}{5730 \text{ yr}} \times \frac{1 \text{ yr}}{365 \text{ d}} \times \frac{1 \text{ d}}{24 \text{ h}} \times \frac{1 \text{ h}}{3600 \text{ s}} = 3.8 \times 10^{-12} \text{ s}^{-1}$$

$$\text{Rate} = kN; \quad k = 3.8 \times 10^{-12} \text{ s}^{-1}(1.0 \times 10^{15} \text{ nuclei }^{14}C) = 3800 \text{ decays/s}$$

A typical 180 lb person produces 3800 beta particles each second.

59. $$20,000 \text{ ton TNT} \times \frac{4 \times 10^9 \text{ J}}{\text{ton TNT}} \times \frac{1 \text{ mol }^{235}U}{2 \times 10^{13} \text{ J}} \times \frac{235 \text{ g }^{235}U}{\text{mol }^{235}U} = 940 \text{ g }^{235}U \approx 900 \text{ g }^{235}U$$

This assumes that all of the ^{235}U undergoes fission.

61. The only product in the fast-equilibrium step is assumed to be $N^{16}O^{18}O_2$, where N is the central atom. However, this is a reversible reaction where $N^{16}O^{18}O_2$ will decompose to NO and O_2. Because any two oxygen atoms can leave $N^{16}O^{18}O_2$ to form O_2, we would expect (at equilibrium) one-third of the NO present in this fast equilibrium step to be $N^{16}O$ and two-thirds to be $N^{18}O$. In the second step (the slow step), the intermediate $N^{16}O^{18}O_2$ reacts with the scrambled NO to form the NO_2 product, where N is the central atom in NO_2. Any one of the three oxygen atoms can be transferred from $N^{16}O^{18}O_2$ to NO when the NO_2 product is formed. The distribution of ^{18}O in the product can best be determined by forming a probability table.

	$N^{16}O$ (1/3)	$N^{18}O$ (2/3)
^{16}O (1/3) from $N^{16}O^{18}O_2$	$N^{16}O_2$ (1/9)	$N^{18}O^{16}O$ (2/9)
^{18}O (2/3) from $N^{16}O^{18}O_2$	$N^{16}O^{18}O$ (2/9)	$N^{18}O_2$ (4/9)

From the probability table, 1/9 of the NO_2 is $N^{16}O_2$, 4/9 of the NO_2 is $N^{18}O_2$, and 4/9 of the NO_2 is $N^{16}O^{18}O$ (2/9 + 2/9 = 4/9). *Note*: $N^{16}O^{18}O$ is the same as $N^{18}O^{16}O$. In addition, $N^{16}O^{18}O_2$ is not the only NO_3 intermediate formed; $N^{16}O_2^{18}O$ and $N^{18}O_3$ can also form in the fast-equilibrium first step. However, the distribution of ^{18}O in the NO_2 product is the same as calculated above, even when these other NO_3 intermediates are considered.

63. Assuming that the radionuclide is long lived enough that no significant decay occurs during the time of the experiment, the total counts of radioactivity injected are:

$$0.10 \text{ mL} \times \frac{5.0 \times 10^3 \text{ cpm}}{\text{mL}} = 5.0 \times 10^2 \text{ cpm}$$

Assuming that the total activity is uniformly distributed only in the rat's blood, the blood volume is:

$$V \times \frac{48 \text{ cpm}}{\text{mL}} = 5.0 \times 10^2 \text{ cpm}, \; V = 10.4 \text{ mL} = 10. \text{ mL}$$

Challenge Problems

65. $$k = \frac{\ln 2}{t_{1/2}}; \quad \ln\left(\frac{N}{N_0}\right) = -kt = \frac{-(0.693)t}{t_{1/2}}$$

For ^{238}U: $$\ln\left(\frac{N}{N_0}\right) = \frac{-(0.693)(4.5 \times 10^9 \text{ yr})}{4.5 \times 10^9 \text{ yr}} = -0.693, \; \frac{N}{N_0} = e^{-0.693} = 0.50$$

For ^{235}U: $$\ln\left(\frac{N}{N_0}\right) = \frac{-(0.693)(4.5 \times 10^9 \text{ yr})}{7.1 \times 10^8 \text{ yr}} = -4.39, \; \frac{N}{N_0} = e^{-4.39} = 0.012$$

If we have a current sample of 10,000 uranium nuclei, 9928 nuclei of ^{238}U and 72 nuclei of ^{235}U are present. Now let's calculate the initial number of nuclei that must have been present 4.5×10^9 years ago to produce these 10,000 uranium nuclei.

For ^{238}U: $$\frac{N}{N_0} = 0.50, \; N_0 = \frac{N}{0.50} = \frac{9928 \text{ nuclei}}{0.50} = 2.0 \times 10^4 \; ^{238}\text{U nuclei}$$

For ^{235}U: $$N_0 = \frac{N}{0.012} = \frac{72 \text{ nuclei}}{0.012} = 6.0 \times 10^3 \; ^{235}\text{U nuclei}$$

So 4.5 billion years ago, the 10,000-nuclei sample of uranium was composed of 2.0×10^4 ^{238}U nuclei and 6.0×10^3 ^{235}U nuclei. The percent composition 4.5 billion years ago would have been:

$$\frac{2.0 \times 10^4 \; ^{238}\text{U nuclei}}{(6.0 \times 10^3 + 2.0 \times 10^4) \text{ total nuclei}} \times 100 = 77\% \; ^{238}\text{U and } 23\% \; ^{235}\text{U}$$

67. $^2_1H + {}^2_1H \rightarrow {}^4_2He$; Q for $^2_1H = 1.6 \times 10^{-19}$ C; mass of deuterium = 2 amu.

$$E = \frac{9.0 \times 10^9 \text{ J} \bullet \text{m/C}^2 (Q_1 Q_2)}{r} = \frac{9.0 \times 10^9 \text{ J} \bullet \text{m/C}^2 (1.6 \times 10^{-19} \text{ C})^2}{2 \times 10^{-15} \text{ m}}$$

$$= 1 \times 10^{-13} \text{ J per alpha particle}$$

$KE = 1/2 \, mv^2$; 1×10^{-13} J $= 1/2 \, (2 \text{ amu} \times 1.66 \times 10^{-27} \text{ kg/amu})v^2$, $v = 8 \times 10^6$ m/s

From the kinetic molecular theory discussed in Chapter 5:

$$u_{rms} = \left(\frac{3RT}{M}\right)^{1/2} \text{ where M = molar mass in kilograms} = 2 \times 10^{-3} \text{ kg/mol for deuterium}$$

$$8 \times 10^6 \text{ m/s} = \left[\frac{3(8.3145 \text{ J K}^{-1} \text{ mol}^{-1})(T)}{2 \times 10^{-3} \text{ kg}}\right]^{1/2}, \quad T = 5 \times 10^9 \text{ K}$$

69. $\text{mol I} = \dfrac{33 \text{ counts}}{\text{min}} \times \dfrac{1 \text{ mol I} \bullet \text{min}}{5.0 \times 10^{11} \text{ counts}} = 6.6 \times 10^{-11} \text{ mol I}$

$$[\text{I}^-] = \frac{6.6 \times 10^{-11} \text{ mol I}^-}{0.150 \text{ L}} = 4.4 \times 10^{-10} \text{ mol/L}$$

$$Hg_2I_2(s) \rightarrow Hg_2^{2+}(aq) + 2 \, I^-(aq) \qquad K_{sp} = [Hg_2^{2+}][I^-]^2$$

Initial s = solubility (mol/L)   0   0
Equil. s $2s$

From the problem, $2s = 4.4 \times 10^{-10}$ mol/L, $s = 2.2 \times 10^{-10}$ mol/L.

$K_{sp} = (s)(2s)^2 = (2.2 \times 10^{-10})(4.4 \times 10^{-10})^2 = 4.3 \times 10^{-29}$

71.   a. For a gas, $u_{avg} = \sqrt{8RT/\pi R}$ where M is the molar mass in kg. From the equation, the lighter the gas molecule, the faster is the average velocity. Therefore, $^{235}UF_6$ will have the greater average velocity at a certain temperature because it is the lighter molecule.

b. From Graham's law (see Section 5.7 of the text):

$$\frac{\text{diffusion rate for } ^{235}UF_6}{\text{diffusion rate for } ^{238}UF_6} = \sqrt{\frac{M(^{238}UF_6)}{M(^{235}UF_6)}} = \sqrt{\frac{352.05 \text{ g/mol}}{349.03 \text{ g/mol}}} = 1.0043$$

Each diffusion step increases the $^{235}UF_6$ concentration by a factor of 1.0043. To determine the number of steps to get to the desired 3.00% ^{235}U, we use the following formula:

$$\frac{0.700 \ ^{235}UF_6}{99.3 \ ^{238}UF_6} \times (1.0043)^N = \frac{3.00 \ ^{235}UF_6}{97.0 \ ^{238}UF_6}$$

original ratio   final ratio

where N represents the number of steps required.

Solving (and carrying extra sig. figs.):

$$(1.0043)^N = \frac{297.9}{67.9} = 4.387, \ N \log(1.0043) = \log(4.387)$$

$$N = \frac{0.6422}{1.863 \times 10^{-3}} = 345 \text{ steps}$$

Thus 345 steps are required to obtain the desired enrichment.

c. $\dfrac{^{235}UF_6}{^{238}UF_6} \times (1.0043)^{100} = \dfrac{1526}{1.000 \times 10^5 - 1526}, \ \dfrac{^{235}UF_6}{^{238}UF_6} \times 1.5358 = \dfrac{1526}{98500}$

original ratio   final ratio

$$\frac{^{235}UF_6}{^{238}UF_6} = 1.01 \times 10^{-2} = \text{initial } ^{235}U \text{ to } ^{238}U \text{ atom ratio}$$

CHAPTER 21

ORGANIC CHEMISTRY

Hydrocarbons

1. A hydrocarbon is a compound composed of only carbon and hydrogen. A saturated hydrocarbon has only carbon-carbon single bonds in the molecule. An unsaturated hydrocarbon has one or more carbon-carbon multiple bonds but may also contain carbon-carbon single bonds. A normal hydrocarbon has one chain of consecutively bonded carbon atoms, with each carbon atom in the chain bonded to one or two other carbon atoms. A branched hydrocarbon has at least one carbon atom in the structure that forms bonds to three or four other carbon atoms; the structure is not one continuous chain of carbon atoms.

3. In order to form, cyclopropane and cyclobutane are forced to form bond angles much smaller than the preferred 109.5° bond angles. Cyclopropane and cyclobutane easily react in order to obtain the preferred 109.5° bond angles.

5. A difficult task in this problem is recognizing different compounds from compounds that differ by rotations about one or more C–C bonds (called conformations). The best way to distinguish different compounds from conformations is to name them. Different name = different compound; same name = same compound, so it is not an isomer but instead is a conformation.

a.

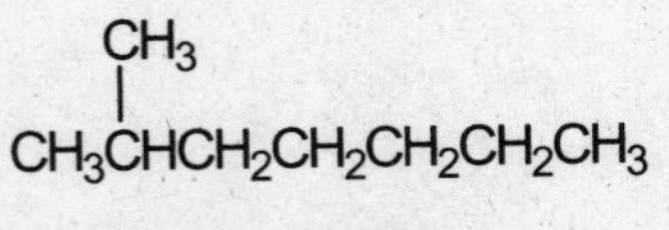

$$CH_3$$
$$|$$
$$CH_3CHCH_2CH_2CH_2CH_2CH_3$$

2-methylheptane

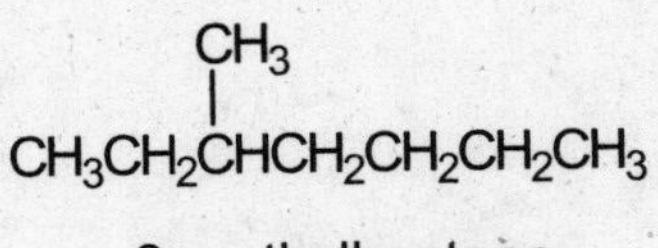

$$CH_3$$
$$|$$
$$CH_3CH_2CHCH_2CH_2CH_2CH_3$$

3-methylheptane

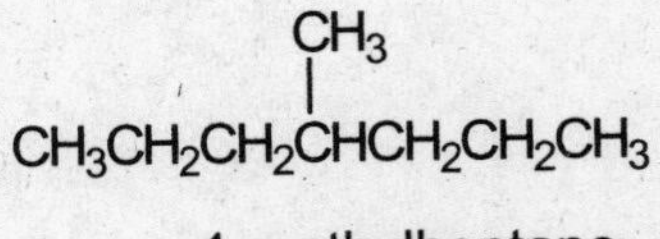

$$CH_3$$
$$|$$
$$CH_3CH_2CH_2CHCH_2CH_2CH_3$$

4-methylheptane

b.

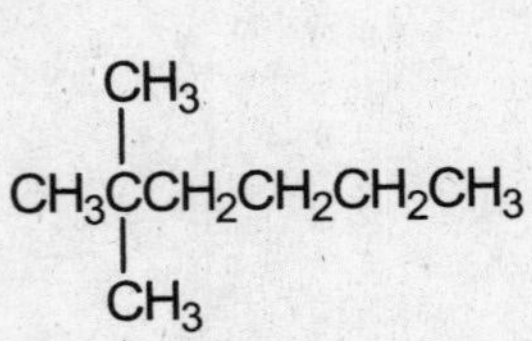

$$CH_3$$
$$|$$
$$CH_3CCH_2CH_2CH_2CH_3$$
$$|$$
$$CH_3$$

2,2-dimethylhexane

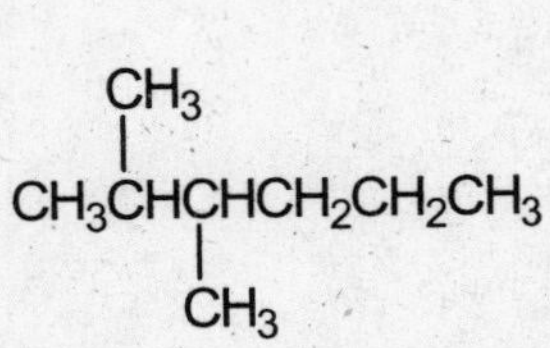

$$CH_3$$
$$|$$
$$CH_3CHCHCH_2CH_2CH_3$$
$$|$$
$$CH_3$$

2,3-dimethylhexane

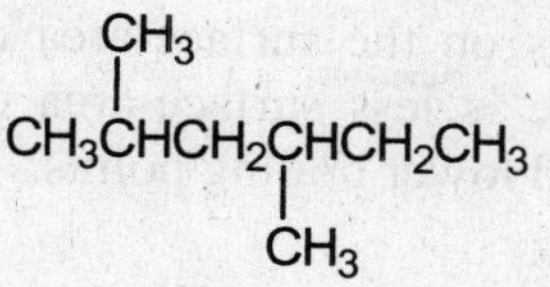

$$CH_3CHCH_2CHCH_2CH_3$$

2,4-dimethylhexane

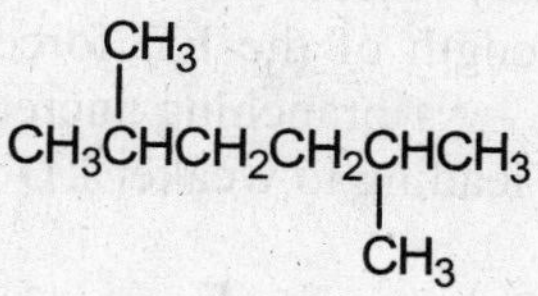

$$CH_3CHCH_2CH_2CHCH_3$$

2,5-dimethylhexane

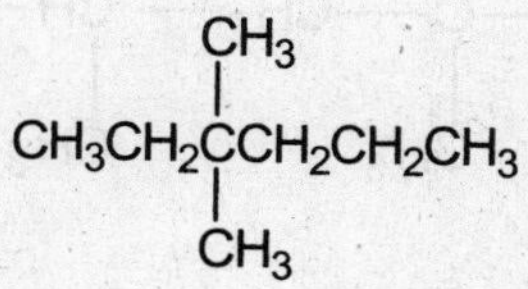

$$CH_3CH_2CCH_2CH_2CH_3$$

3,3-dimethylhexane

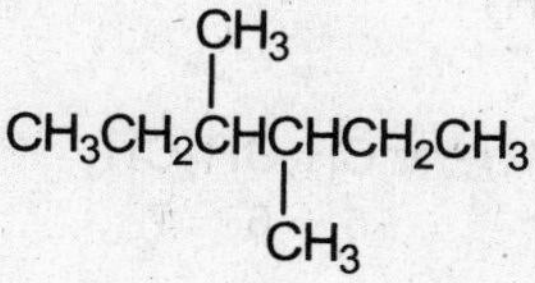

$$CH_3CH_2CHCHCH_2CH_3$$

3,4-dimethylhexane

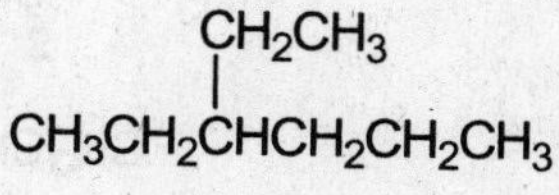

$$CH_3CH_2CHCH_2CH_2CH_3$$

3-ethylhexane

c.

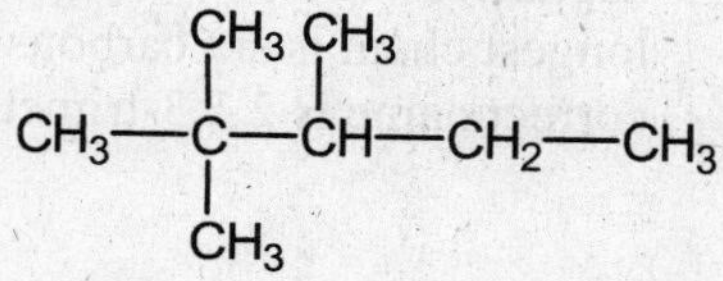

$$CH_3-C-CH-CH_2-CH_3$$

2,2,3-trimethylpentane

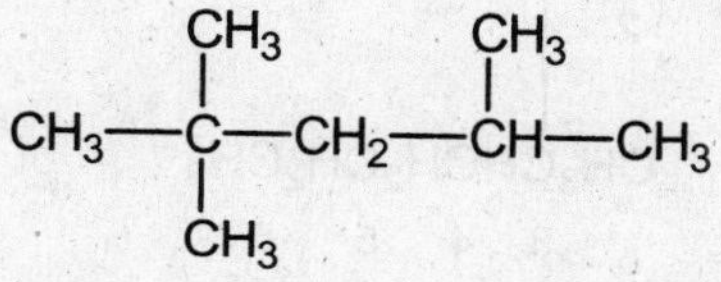

$$CH_3-C-CH_2-CH-CH_3$$

2,2,4-trimethylpentane

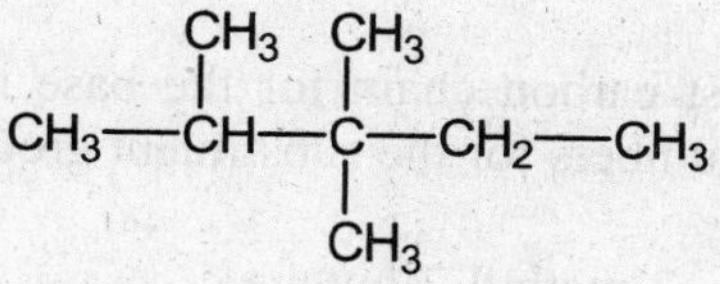

$$CH_3-CH-C-CH_2-CH_3$$

2,3,3-trimethylpentane

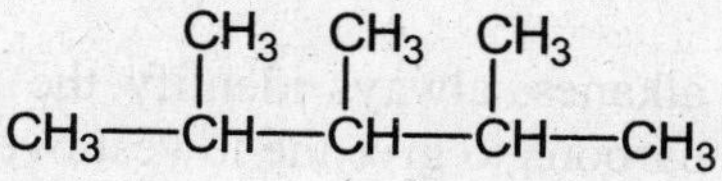

$$CH_3-CH-CH-CH-CH_3$$

2,3,4-trimethylpentane

$$CH_3-CH-CH-CH_2-CH_3$$

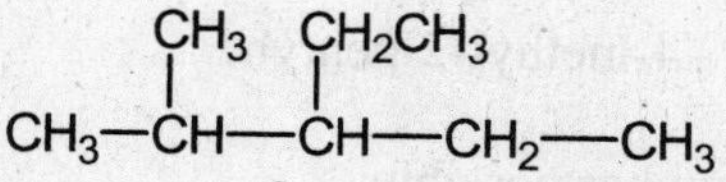

3-ethyl-2-methylpentane

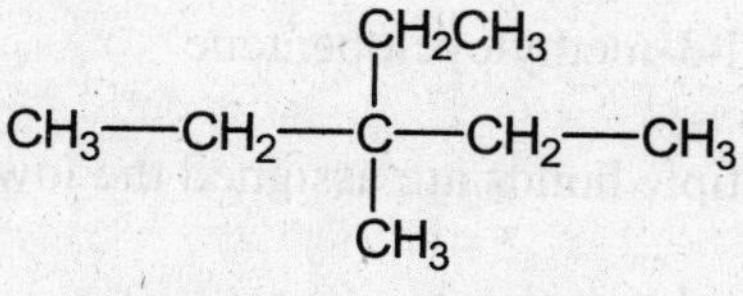

$$CH_3-CH_2-C-CH_2-CH_3$$

3-ethyl-3-methylpentane

d.

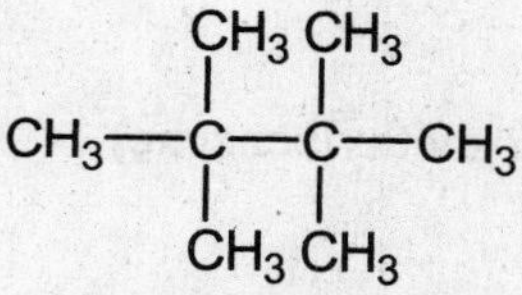

$$CH_3-C-C-CH_3$$

2,2,3,3-tetramethylbutane

7.　London dispersion (LD) forces are the primary intermolecular forces exhibited by hydrocarbons. The strength of the LD forces depends on the surface-area contact among neighboring molecules. As branching increases, there is less surface-area contact among neighboring molecules, leading to weaker LD forces and lower boiling points.

9.　a.

$$CH_3-\underset{2}{\overset{\overset{\displaystyle CH_3}{|}}{CH}}-\underset{3}{CH_2}$$

$$CH_3CH_2-\underset{4}{CH}-\underset{5\ \ 6\ \ 7}{CH_2CH_2CH_3}$$

b.

$$CH_3-\underset{\underset{\displaystyle CH_3}{|}}{\overset{\overset{\displaystyle CH_3}{|}}{C}}-CH_2-\underset{\underset{\displaystyle CH_3}{|}}{CH}-CH_3$$

c.

$$CH_3-\underset{2}{\overset{\overset{\displaystyle CH_3}{|}}{C}}-CH_3$$

$$\underset{3\ \ 4\ \ 5\ \ 6}{CH_3CHCH_2CH_2CH_3}$$

d.　For 3-isobutylhexane, the longest chain is seven carbons long. The correct name is 4-ethyl-2-methyl-heptane. For 2-tert-butylpentane, the longest chain is six carbons long. The correct name is 2,2,3-trimethylhexane.

11.　a.　2,2,4-trimethylhexane　　b.　5-methylnonane　　c.　2,2,4,4-tetramethylpentane

d.　3-ethyl-3-methyloctane

Note: For alkanes, always identify the longest carbon chain for the base name first, then number the carbons to give the lowest overall numbers for the substituent groups.

13.　a.　1-butene　　　　　　　　　　b.　2-methyl-2-butene

c.　2,5-dimethyl-3-heptene　　　　d.　2,3-dimethyl-1-pentene

e.　1-ethyl-3-methylcyclopentene (double bond assumed between C_1 and C_2)

f.　4-ethyl-3-methylcyclopentene　　g.　4-methyl-2-pentyne

Note: Multiple bonds are assigned the lowest number possible.

15.　a.　1,3-dichlorobutane　　　　　b.　1,1,1-trichlorobutane

c.　2,3-dichloro-2,4-dimethylhexane　　d.　1,2-difluoroethane

e.　3-iodo-1-butene

f.　2-bromotoluene (or o-bromotoluene or 1-bromo-2-methylbenzene)

g.　1-bromo-2-methylcyclohexane

h.　4-bromo-3-methylcyclohexene (double bond assumed between C_1 and C_2)

17.   isopropylbenzene or 2-phenylpropane

Isomerism

19.   a.   1-sec-butylpropane

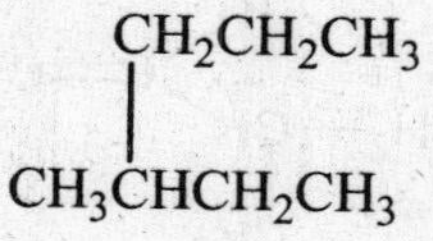

$$CH_2CH_2CH_3$$
$$|$$
$$CH_3CHCH_2CH_3$$

3-Methylhexane is correct.

b.   4-methylhexane

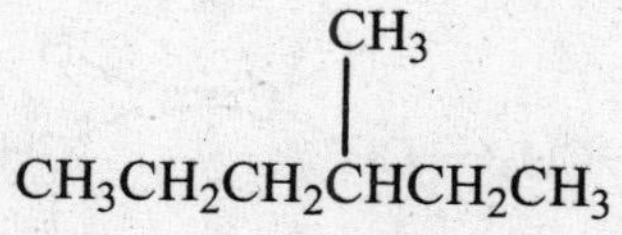

$$CH_3$$
$$|$$
$$CH_3CH_2CH_2CHCH_2CH_3$$

3-Methylhexane is correct.

c.   2-ethylpentane

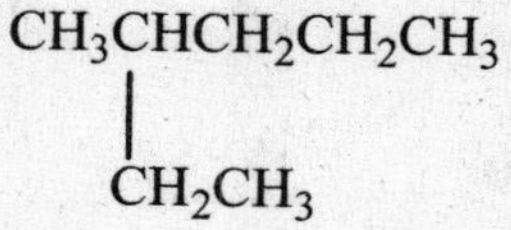

$$CH_3CHCH_2CH_2CH_3$$
$$|$$
$$CH_2CH_3$$

3-Methylhexane is correct.

d.   1-ethyl-1-methylbutane

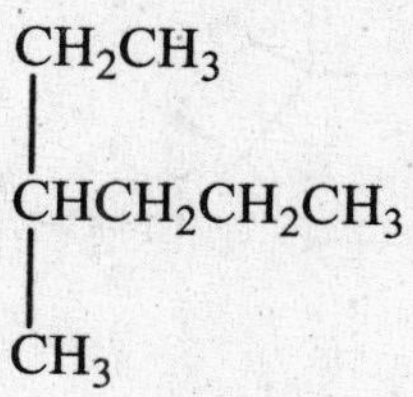

$$CH_2CH_3$$
$$|$$
$$CHCH_2CH_2CH_3$$
$$|$$
$$CH_3$$

3-Methylhexane is correct.

e.   3-methylhexane

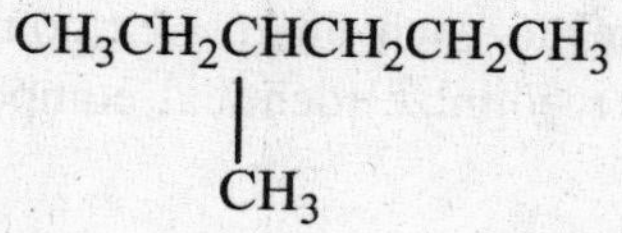

$$CH_3CH_2CHCH_2CH_2CH_3$$
$$|$$
$$CH_3$$

This is a correct name.

f.   4-ethylpentane

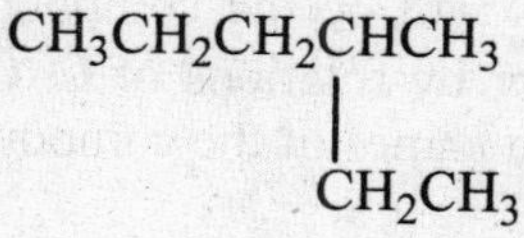

$$CH_3CH_2CH_2CHCH_3$$
$$|$$
$$CH_2CH_3$$

3-Methylhexane is correct.

All six of these compounds are the same. They only differ from each other by rotations about one or more carbon-carbon single bonds. Only one isomer of C_7H_{16} is present in all of these names, 3-methylhexane.

21.   To exhibit cis-trans isomerism, a compound must first have restricted rotation about a carbon-carbon bond. This occurs in compounds with double bonds and ring compounds. Second, the compound must have two carbons in the restricted rotation environment that each have two different groups bonded. For example, the compound in Exercise 21.13a has a double bond, but the first carbon in the double bond has two H atoms attached. This compound does not exhibit cis-trans isomerism. To see this, let's draw the potential cis-trans isomers:

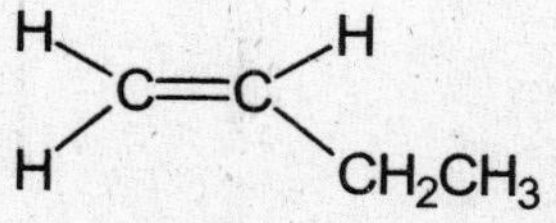

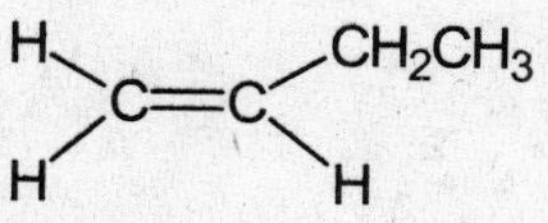

These are the same compounds; they only differ by a simply rotation of the molecule. Therefore, they are not isomers of each other, but instead, they are the same compound. The only compounds that fulfill the restricted rotation requirement *and* have two different groups attached to carbons in the restricted rotation are compounds c and f. The cis-trans isomerism for these follows.

c.

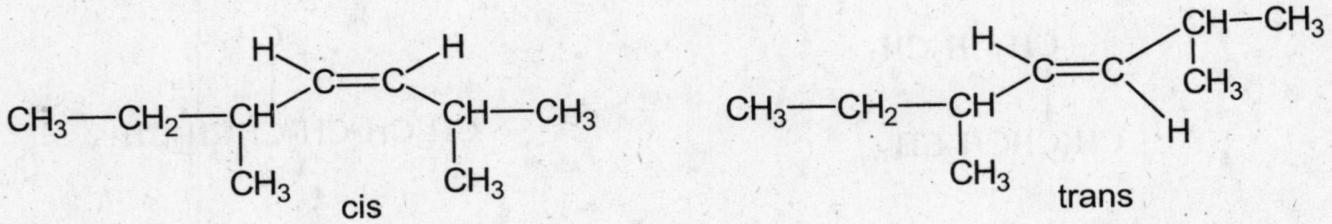
H
H
C=C
CH3—CH2—CH
CH—CH3
CH3
CH3
cis
CH—CH3
H
C=C
CH3
CH3—CH2—CH
H
CH3
trans

f.

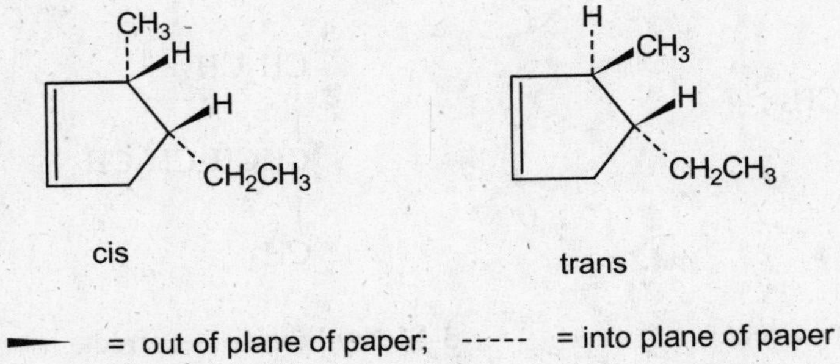
CH3
H
H
CH2CH3
cis
H
CH3
H
CH2CH3
trans
= out of plane of paper; ----- = into plane of paper

23. a. All these structures have the formula C_5H_8. The compounds with the same physical properties will be the compounds that are identical to each other, i.e., compounds that only differ by rotations of C–C single bonds. To recognize identical compounds, name them. The names of the compounds are:

i. trans-1,3-pentadiene ii. cis-1,3-pentadiene

iii. cis-1,3-pentadiene iv. 2-methyl-1,3-butadiene

Compounds ii and iii are identical compounds, so they would have the same physical properties.

b. Compound i is a trans isomer because the bulkiest groups bonded to the carbon atoms in the C_3=C_4 double bond are as far apart as possible.

c. Compound iv does not have carbon atoms in a double bond that each have two different groups attached. Compound iv does not exhibit cis-trans isomerism.

25.   The cis isomer has the CH_3 groups on the same side of the ring. The trans isomer has the
       CH_3 groups on opposite sides of the ring.

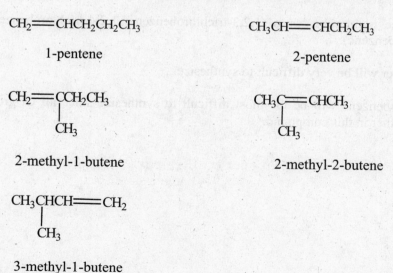

                        cis   trans

27.   To help distinguish the different isomers, we will name them.

       cis-1-chloro-1-propene   trans-1-chloro-1-propene

       2-chloro-1-propene   3-chloro-1-propene   chlorocyclopropane

29. C_5H_{10} has the general formula for alkenes, C_nH_{2n}. To distinguish the different isomers from
       each other, we will name them. Each isomer must have a different name.

       $CH_2{=\!\!=}CHCH_2CH_2CH_3$ $CH_3CH{=\!\!=}CHCH_2CH_3$

              1-pentene   2-pentene

       $CH_2{=\!\!=}CCH_2CH_3$ $CH_3C{=\!\!=}CHCH_3$
                    | |
                   CH_3 CH_3

       2-methyl-1-butene   2-methyl-2-butene

       $CH_3CHCH{=\!\!=}CH_2$
             |
            CH_3

       3-methyl-1-butene

Only 2-pentene exhibits cis-trans isomerism. The isomers are:

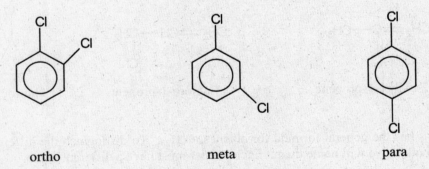

cis   trans

The other isomers of C_5H_{10} do not contain carbons in the double bond to which two different groups are attached.

31.   a. cis-1-bromo-1-propene   b. cis-4-ethyl-3-methyl-3-heptene

       c. trans-1,4-diiodo-2-propyl-1-pentene

Note: In general, cis-trans designations refer to the relative positions of the largest groups. In compound b, the largest group off the first carbon in the double bond is CH_2CH_3, and the largest group off the second carbon in the double bond is $CH_2CH_2CH_3$. Since their relative placement is on the same side of the double bond, this is the cis isomer.

33.   a.

Cl   Cl   Cl

Cl   Cl   Cl

ortho   meta   para

       b. There are three trichlorobenzenes (1,2,3-trichlorobenzene, 1,2,4-trichlorobenzene, and 1,3,5-trichlorobenzene).

       c. The meta isomer will be very difficult to synthesize.

       d. 1,3,5-Trichlorobenzene will be the most difficult to synthesize since all Cl groups are meta to each other in this compound.

Functional Groups

35.

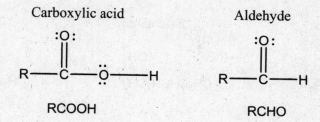

The R designation refers to the rest of the organic molecule beyond the specific functional group indicated in the formula. The R group may be a hydrogen but is usually a hydrocarbon fragment. The major point in the R group designation is that if the R group is an organic fragment, then the first atom in the R group is a carbon atom. What the R group has after the first carbon is not important to the functional group designation.

37.   Reference Table 21.4 for the common functional groups.

   a. ketone   b. aldehyde   c. carboxylic acid   d. amine

39.   a.

```
      H           H    H     H
       \         /      \   /
        C=======C        N  amine
       /         \      /
  O===C           N----C----C----C----O---H
      ketone       |    |    |    ‖
       \           H    H    H    O   carboxylic acid
        C=======C           amine
       /         \
  H---O           H
      alcohol
```

   b. 5 carbons in ring and the carbon in $-CO_2H$: sp^2; the other two carbons: sp^3

   c. 24 sigma bonds; 4 pi bonds

41.   a. 3-chloro-1-butanol; since the carbon containing the OH group is bonded to just one other carbon (one R group), this is a primary alcohol.

   b. 3-methyl-3-hexanol; since the carbon containing the OH group is bonded to three other carbons (three R groups), this is a tertirary alcohol

   c. 2-methylcyclopentanol; secondary alcohol (two R groups bonded to carbon containing the OH group). *Note*: In ring compounds, the alcohol group is assumed to be bonded to C_1, so the number designation is commonly omitted for the alcohol group.

43.

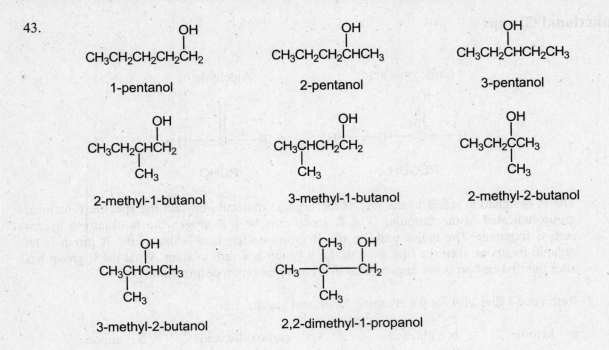

1-pentanol   2-pentanol   3-pentanol

2-methyl-1-butanol   3-methyl-1-butanol   2-methyl-2-butanol

3-methyl-2-butanol   2,2-dimethyl-1-propanol

There are six isomeric ethers with formula $C_5H_{12}O$. The structures follow.

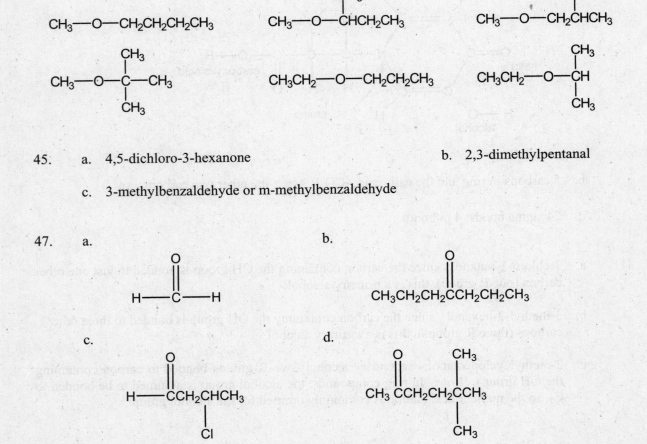

CH_3—O—$CH_2CH_2CH_2CH_3$

CH_3—O—$CHCH_2CH_3$

CH_3—O—CH_2CHCH_3

CH_3—O—C—CH_3

CH_3CH_2—O—$CH_2CH_2CH_3$

CH_3CH_2—O—CH

45.   a.   4,5-dichloro-3-hexanone   b.   2,3-dimethylpentanal

c.   3-methylbenzaldehyde or m-methylbenzaldehyde

47.   a.   b.

H—C—H

$CH_3CH_2CH_2CCH_2CH_2CH_3$

c.   d.

H—CCH$_2$CHCH$_3$

CH_3 $CCH_2CH_2CCH_3$

Cl

CH_3

49.   a.   trans-2-butene: $CH_3\!\!-\!\!C\!\!=\!\!C\!\!-\!\!H$ with H and CH_3 , formula = C_4H_8

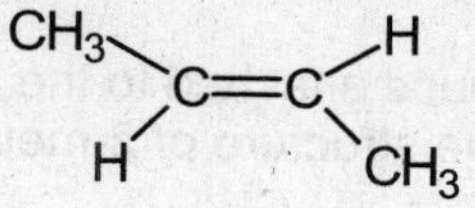

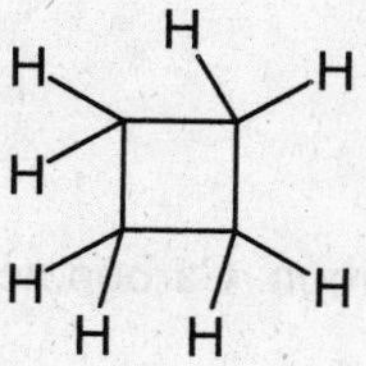

or

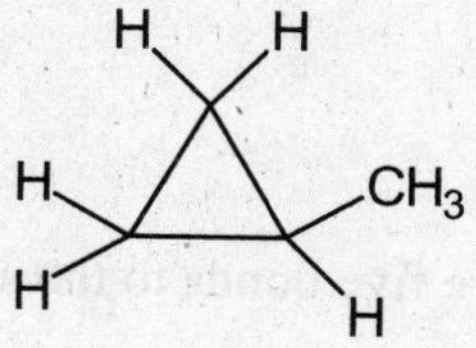

b.   propanoic acid: $CH_3CH_2C\!-\!OH$ (with $=O$),   formula = $C_3H_6O_2$

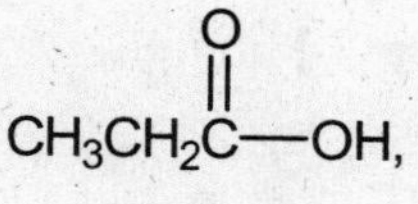

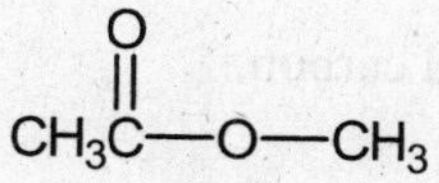

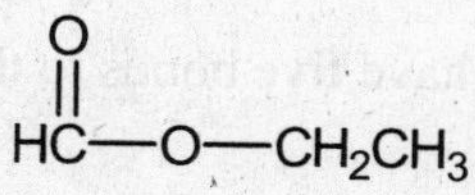

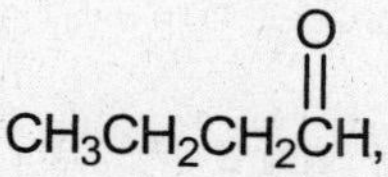

$CH_3C\!-\!O\!-\!CH_3$ (with $=O$)   or $HC\!-\!O\!-\!CH_2CH_3$ (with $=O$)

c.   butanal: $CH_3CH_2CH_2CH$ (with $=O$),   formula = C_4H_8O

$CH_3CH_2CCH_3$ (with $=O$)

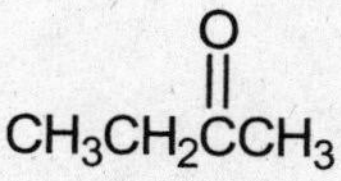

d.   butylamine: $CH_3CH_2CH_2CH_2NH_2$,   formula = $C_4H_{11}N$:

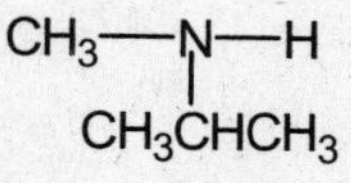

A secondary amine has two R groups bonded to N.

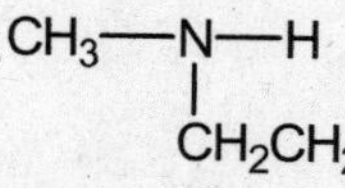

$CH_3\!-\!N\!-\!H$ with $CH_2CH_2CH_3$ or $CH_3\!-\!N\!-\!H$ with CH_3CHCH_3 or $CH_3CH_2\!-\!N\!-\!H$ with CH_2CH_3

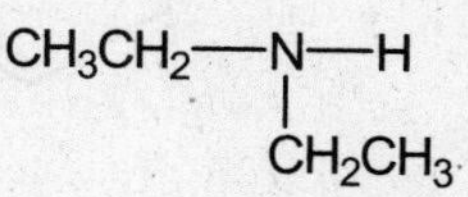

e.   A tertiary amine has two R groups bonded to N.
     (See answer d for the structure of butylamine.)

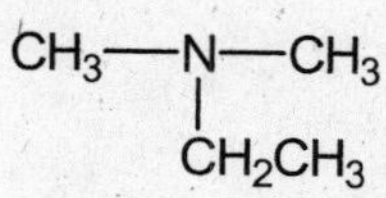

$CH_3\!-\!N\!-\!CH_3$ with CH_2CH_3

f.   2-methyl-2-propanol: CH_3CCH_3 with CH_3 above and OH below ,   formula = $C_4H_{10}O$

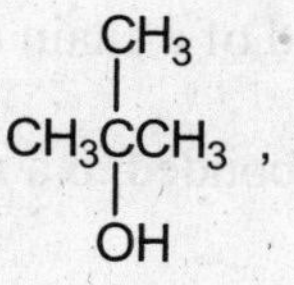

$CH_3\!-\!O\!-\!CH_2CH_2CH_3$ or $CH_3\!-\!O\!-\!CH$ with CH_3 and CH_3 or $CH_3CH_2\!-\!O\!-\!CH_2CH_3$

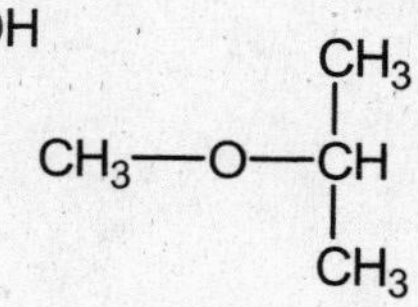

g.   A secondary alcohol has two R groups attached to the carbon bonded to
the OH group. (See answer f for the structure of 2-methyl-2-propanol.)

$$\underset{CH_3CHCH_2CH_3}{\overset{\overset{\textstyle OH}{|}}{}}$$

51.   a.   2-Chloro-2-butyne would have five bonds to the second carbon. Carbon never expands
its octet.

$$\underset{CH_3-C\equiv CCH_3}{\overset{\overset{\textstyle Cl}{|}}{}}$$

b.   2-Methyl-2-propanone would have five bonds to the second carbon.

$$CH_3-\overset{\overset{\textstyle O}{\|}}{\underset{\underset{\textstyle CH_3}{|}}{C}}-CH_3$$

c.   Carbon-1 in 1,1-dimethylbenzene would have five bonds.

d.   You cannot have an aldehyde functional group bonded to a middle carbon in a chain.
Aldehyde groups, i.e.,

$$-\overset{\overset{\textstyle O}{\|}}{C}-H$$

can only be at the beginning and/or the end of a chain of carbon atoms.

e.   You cannot have a carboxylic acid group bonded to a middle carbon in a chain.
Carboxylic groups, i.e.,

$$-\overset{\overset{\textstyle O}{\|}}{C}-OH$$

must be at the beginning and/or the end of a chain of carbon atoms.

f. In cyclobutanol, the 1 and 5 positions refer to the same carbon atom. 5,5-Dibromo-1-cyclobutanol would have five bonds to carbon-1. This is impossible; carbon never expands its octet.

Reactions of Organic Compounds

53.   Substitution: An atom or group is replaced by another atom or group.

For example: H in benzene is replaced by Cl.

$$C_6H_6 + Cl_2 \xrightarrow{\text{catalyst}} C_6H_5Cl + HCl$$

Addition: Atoms or groups are added to a molecule.

For example: Cl_2 adds to ethene. $CH_2 = CH_2 + Cl_2 \rightarrow CH_2Cl–CH_2Cl$

55.   a.

b.

c.

d.

57.

a. $\underset{\text{H}}{\text{CH}_3\text{CH}}-\underset{\text{H}}{\text{CHCH}_3}$

b. $\underset{\text{Cl}}{\text{CH}_2}-\underset{\text{Cl}}{\text{CHCHCH}}-\underset{\text{Cl}}{\text{CH}}$ with CH_3 and CH_3 groups

c.   (benzene ring)—Cl + HCl

d. $C_4H_8(g) + 6 O_2(g) \rightarrow 4 CO_2(g) + 4 H_2O(g)$

59.   When $CH_2{=}CH_2$ reacts with HCl, there is only one possible product, chloroethane. When Cl_2 is reacted with CH_3CH_3 (in the presence of light), there are six possible products because any number of the six hydrogens in ethane can be substituted for by Cl. The light-catalyzed substitution reaction is very difficult to control; hence it is not a very efficient method of producing monochlorinated alkanes.

61.   Primary alcohols (a, d, and f) are oxidized to aldehydes, which can be oxidized further to carboxylic acids. Secondary alcohols (b, e, and f) are oxidized to ketones, and tertiary alcohols (c and f) do not undergo this type of oxidation reaction. Note that compound f contains a primary, a secondary, and a tertiary alcohol. For the primary alcohols (a, d, and f), we listed both the aldehyde and the carboxylic acid as possible products.

a. $\underset{\quad}{\text{H}}-\overset{\text{O}}{\underset{\quad}{\text{C}}}-\text{CH}_2\text{CHCH}_3$ (with CH_3) + $\text{HO}-\overset{\text{O}}{\underset{\quad}{\text{C}}}-\text{CH}_2\text{CHCH}_3$ (with CH_3)

b. $\text{CH}_3-\overset{\text{O}}{\underset{\quad}{\text{C}}}-\text{CHCH}_3$ (with CH_3)   c.   No reaction

d.   (benzene ring)$-\overset{\text{O}}{\underset{\quad}{\text{C}}}-\text{H}$ + (benzene ring)$-\overset{\text{O}}{\underset{\quad}{\text{C}}}-\text{OH}$

e.   (cyclohexanone ring with $-CH_3$ group)

f.   (cyclohexanone ring with OH, CH_3, and $-C(=O)-H$ groups) + (cyclohexanone ring with OH, CH_3, and $-C(=O)-OH$ groups)

63.   KMnO$_4$ will oxidize primary alcohols to aldehydes and then to carboxylic acids. Secondary alcohols are oxidized to ketones by KMnO$_4$. Tertiary alcohols and ethers are not oxidized by KMnO$_4$.

The three isomers and their reactions with KMnO$_4$ are:

$$CH_3—O—CH_2CH_3 \xrightarrow{KMnO_4} \text{no reaction}$$

ether

$$CH_3—\overset{\overset{OH}{|}}{CH}—CH_3 \xrightarrow{KMnO_4} CH_3—\overset{\overset{O}{\parallel}}{C}—CH_3$$

2° alcohol   2-propanone (acetone)

$$CH_3CH_2CH_2 \xrightarrow{KMnO_4} CH_3CH_2\overset{\overset{O}{\parallel}}{C}H \xrightarrow{KMnO_4} CH_3CH_2\overset{\overset{O}{\parallel}}{C}—OH$$

1° alcohol   propanal   propanoic acid

The products of the reactions with excess KMnO$_4$ are 2-propanone and propanoic acid.

65.   a. $CH_3CH = CH_2 + Br_2 \rightarrow CH_3CHBrCH_2Br$ (Addition reaction of Br$_2$ with propene)

       b. $$CH_3—\overset{\overset{OH}{|}}{CH}—CH_3 \xrightarrow{oxidation} CH_3—\overset{\overset{O}{\parallel}}{C}—CH_3$$

Oxidation of 2-propanol yields acetone (2-propanone).

       c. $$CH_2 =\overset{\overset{CH_3}{|}}{C}—CH_3 + H_2O \xrightarrow{H^+} CH_2—\overset{\overset{CH_3}{|}}{\underset{\underset{OH}{|}}{C}}—CH_3$$

Addition of H$_2$O to 2-methylpropene would yield tert-butyl alcohol (2-methyl-2-propanol) as the major product.

       d. $$CH_3CH_2CH_2OH \xrightarrow{KMnO_4} CH_3CH_2\overset{\overset{O}{\parallel}}{C}—OH$$

Oxidation of 1-propanol would eventually yield propanoic acid. Propanal is produced first in this reaction and is then oxidized to propanoic acid.

67.   When an alcohol is reacted with a carboxylic acid, an ester is produced.

a.

$$CH_3\overset{\displaystyle O}{\overset{\|}{C}}\text{---}OH + HO\text{---}CH_3 \longrightarrow CH_3\overset{\displaystyle O}{\overset{\|}{C}}\text{---}O\text{---}CH_3 + H_2O$$

b.

$$H\text{---}\overset{\displaystyle O}{\overset{\|}{C}}\text{---}OH + HO\text{---}CH_2CH_2CH_3 \longrightarrow H\text{---}\overset{\displaystyle O}{\overset{\|}{C}}\text{---}O\text{---}CH_2CH_2CH_3 + H_2O$$

Polymers

69.   a.   Addition polymer: a polymer that forms by adding monomer units together (usually by reacting double bonds). Teflon, polyvinyl chloride, and polyethylene are examples of addition polymers.

b.   Condensation polymer: a polymer that forms when two monomers combine by eliminating a small molecule (usually H_2O or HCl). Nylon and Dacron are examples of condensation polymers.

c.   Copolymer: a polymer formed from more than one type of monomer. Nylon and Dacron are copolymers.

d.   Homopolymer: a polymer formed from the polymerization of only one type of monomer. Polyethylene, Teflon, and polystyrene are examples of homopolymers.

e.   Polyester: a condensation polymer whose monomers link together by formation of the ester functional group. Dacron is a polyester.

$$\text{Ester} = \quad R\text{---}\overset{\displaystyle O}{\overset{\|}{C}}\text{---}O\text{---}R'$$

f.   Polyamide: a condensation polymer whose monomers link together by formation of the amide functional group. Nylon is a polyamide as are proteins in the human body.

$$\text{Amide} = \quad R\text{---}\overset{\displaystyle O}{\overset{\|}{C}}\text{---}\overset{\displaystyle H}{\overset{|}{N}}\text{---}R$$

71.   a.   A polyester forms when an alcohol functional group reacts with a carboxylic acid functional group. The monomer for a homopolymer polyester must have an alcohol functional group and a carboxylic acid functional group present within the structure of the monomer.

      b.   A polyamide forms when an amine functional group reacts with a carboxylic acid functional group. For a copolymer polyamide, one monomer would have at least two amine functional groups present, and the other monomer would have at least two carboxylic acid functional groups present. For polymerization to occur, each monomer must have two reactive functional groups present.

      c.   To form an addition polymer, a carbon-carbon double bond must be present. Polyesters and polyamides are condensation polymers. To form a polyester, the monomer would need the alcohol and carboxylic acid functional groups present. To form a polyamide, the monomer would need the amine and carboxylic acid functional groups present. The two possibilities are for the monomer to have a carbon-carbon double bond, an alcohol functional group, and a carboxylic acid functional group present or to have a carbon-carbon double bond, an amine functional group, and a carboxylic acid functional group present.

73.   The backbone of the polymer contains only carbon atoms, which indicates that Kel-F is an addition polymer. The smallest repeating unit of the polymer and the monomer used to produce this polymer are:

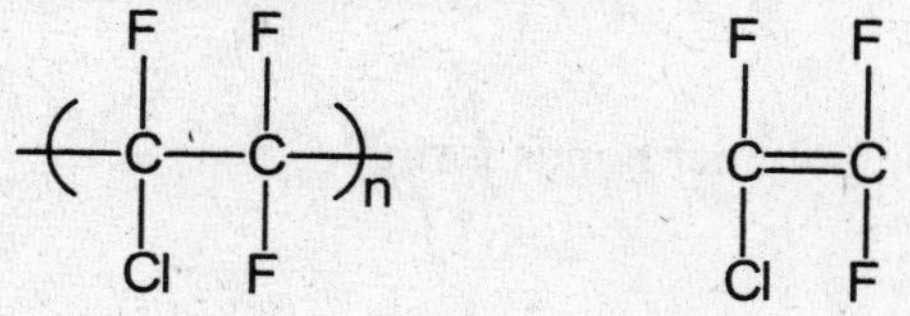

75.   a.

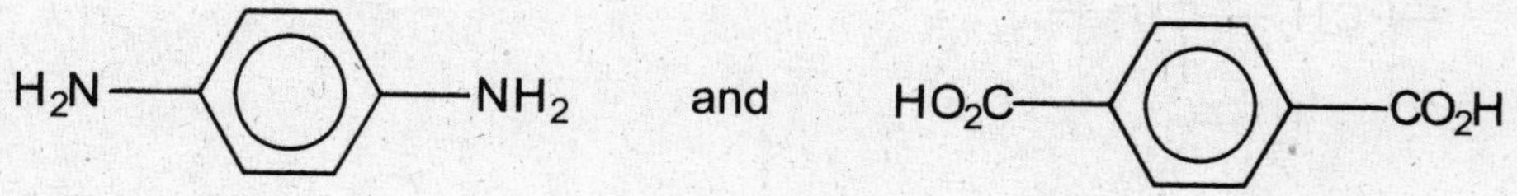

      b.   Repeating unit:

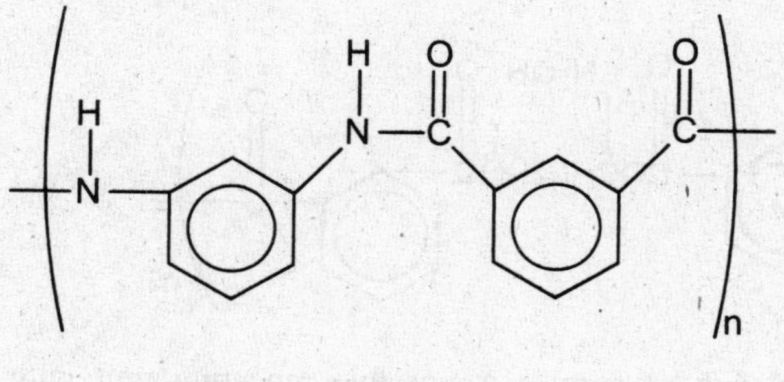

      The two polymers differ in the substitution pattern on the benzene rings. The Kevlar chain is straighter, and there is more efficient hydrogen bonding between Kevlar chains than between Nomex chains.

77.

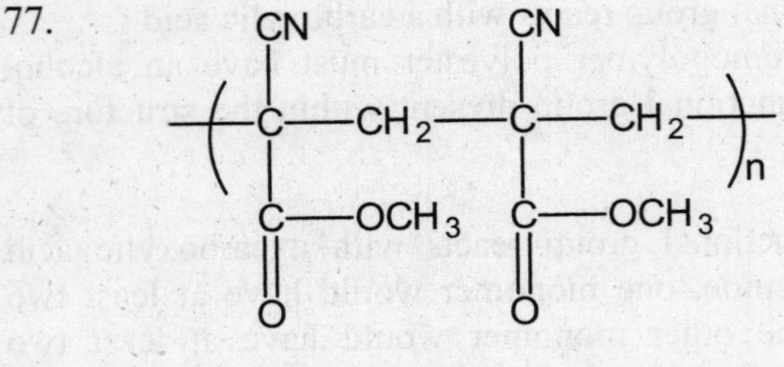

"Super glue" is an addition polymer formed by reaction of the C=C bond in methyl cyanoacrylate.

79. Divinylbenzene is a crosslinking agent. Divinylbenzene has two reactive double bonds that are both reacted when divinylbenzene inserts itself into two adjacent polymer chains during the polymerization process. The chains cannot move past each other because the crosslinks bond adjacent polymer chains together, making the polymer more rigid.

81. a.

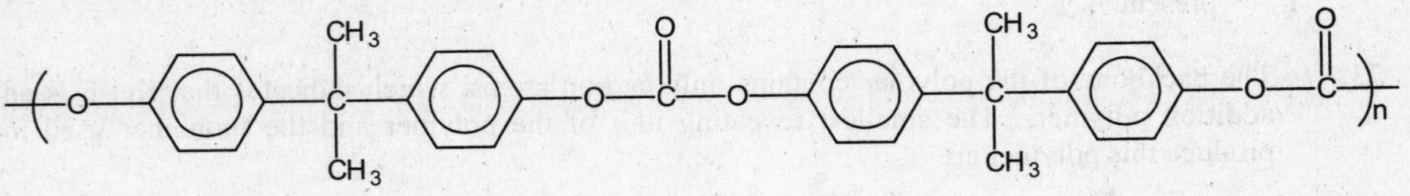

b. Condensation; HCl is eliminated when the polymer bonds form.

83. Polyacrylonitrile:

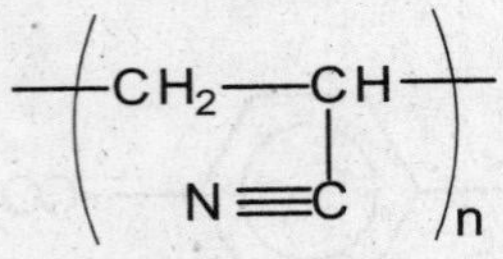

The CN triple bond is very strong and will not easily break in the combustion process. A likely combustion product is the toxic gas hydrogen cyanide, HCN(g).

85.

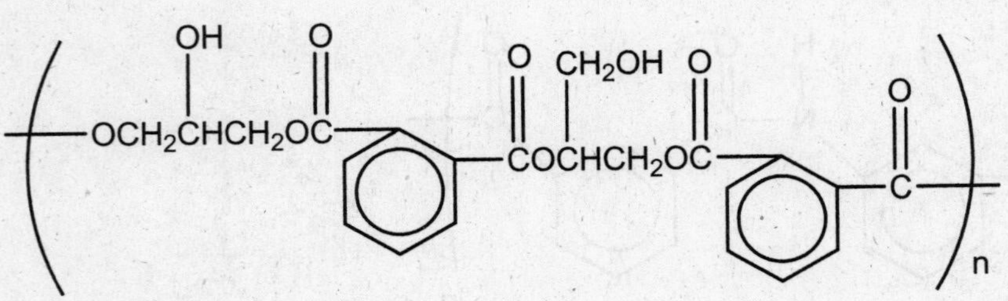

Two linkages are possible with glycerol. A possible repeating unit with both types of linkages is shown above. With either linkage, there are unreacted OH groups on the polymer chains. These unreacted OH groups on adjacent polymer chains can react with the acid groups of phthalic acid to form crosslinks (bonds) between various polymer chains.

Natural Polymers

87.

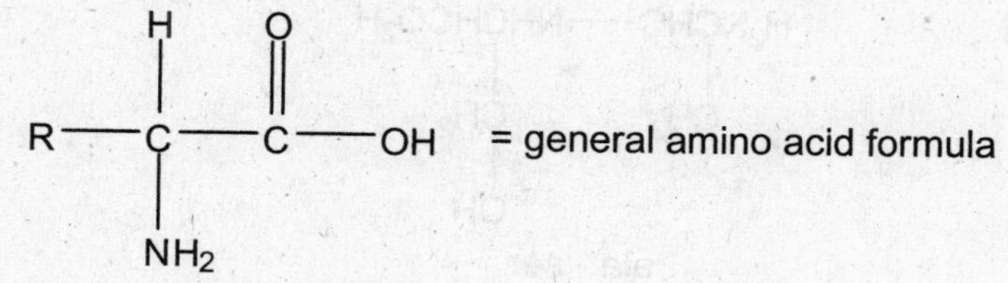

R—C—C—OH   = general amino acid formula

Hydrophilic ("water-loving") and hydrophobic ("water-fearing") refer to the polarity of the R group. When the R group consists of a polar group, then the amino acid is hydrophilic. When the R group consists of a nonpolar group, then the amino acid is hydrophobic.

89. Denaturation changes the three-dimensional structure of a protein. Once the structure is affected, the function of the protein will also be affected.

91. a. Serine, tyrosine, and threonine contain the –OH functional group in the R group.

b. Aspartic acid and glutamic acid contain the –COOH functional group in the R group.

c. An amine group has a nitrogen bonded to other carbon and/or hydrogen atoms. Histidine, lysine, arginine, and tryptophan contain the amine functional group in the R group.

d. The amide functional group is:

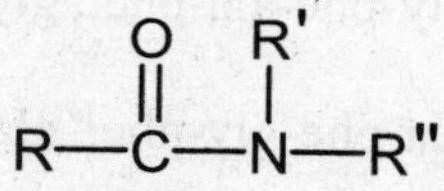

R—C—N—R"

This functional group is formed when individual amino acids bond together to form the peptide linkage. Glutamine and asparagine have the amide functional group in the R group.

93. a. Aspartic acid and phenylalanine make up aspartame.

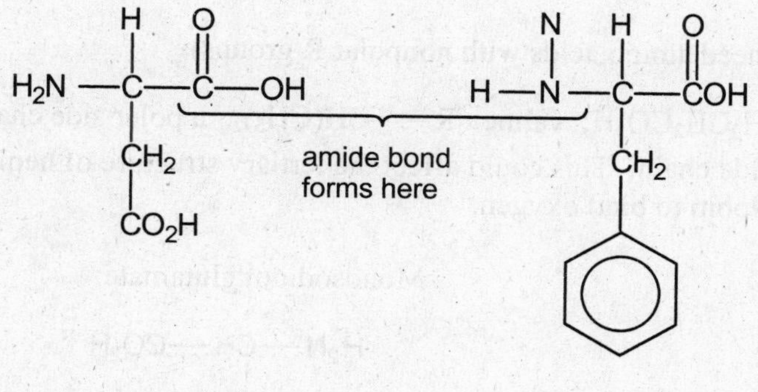

amide bond
forms here

b. Aspartame contains the methyl ester of phenylalanine. This ester can hydrolyze to form methanol:

$$R—CO_2CH_3 + H_2O \rightleftharpoons RCO_2H + HOCH_3$$

95.

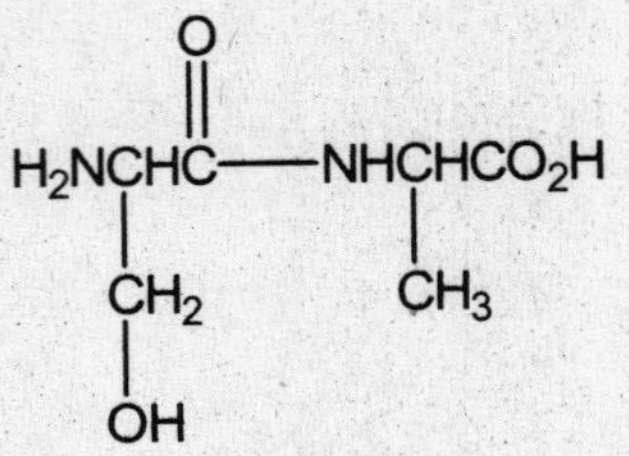

ser - ala 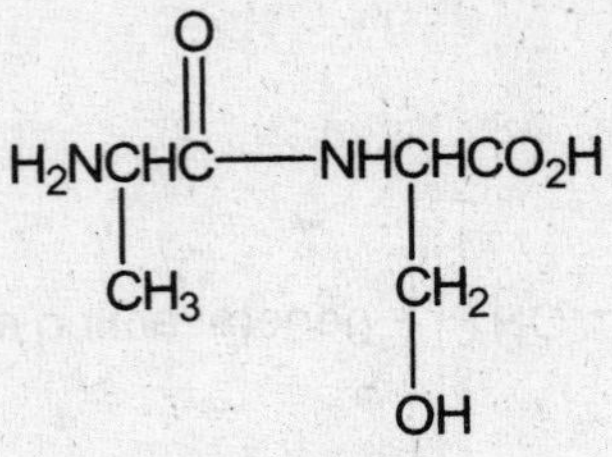   ala - ser

97.   a.   Six tetrapeptides are possible. From NH_2 to CO_2H end:

phe-phe-gly-gly, gly-gly-phe-phe, gly-phe-phe-gly,

phe-gly-gly-phe, phe-gly-phe-gly, gly-phe-gly-phe

   b.   Twelve tetrapeptides are possible. From NH_2 to CO_2H end:

phe-phe-gly-ala, phe-phe-ala-gly, phe-gly-phe-ala,

phe-gly-ala-phe, phe-ala-phe-gly, phe-ala-gly-phe,

gly-phe-phe-ala, gly-phe-ala-phe, gly-ala-phe-phe,

ala-phe-phe-gly, ala-phe-gly-phe, ala-gly-phe-phe

99.   a.   Covalent (forms a disulfide linkage)

   b.   Hydrogen bonding (need N-H or O-H bond in side chain)

   c.   Ionic (need NH_2 group on side chain of one amino acid with CO_2H group on side chain of the other amino acid)

   d.   London dispersion (need amino acids with nonpolar R groups)

101.   Glutamic acid: $R = -CH_2CH_2CO_2H$; valine: $R = -CH(CH_3)_2$; a polar side chain is replaced by a nonpolar side chain. This could affect the tertiary structure of hemoglobin and the ability of hemoglobin to bind oxygen.

103.   Glutamic acid: Monosodium glutamate:

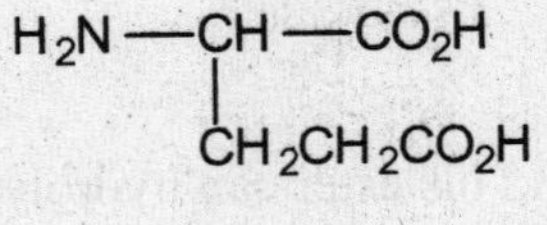

One of the two acidic protons in the carboxylic acid groups is lost to form MSG. Which proton is lost is impossible for you to predict.

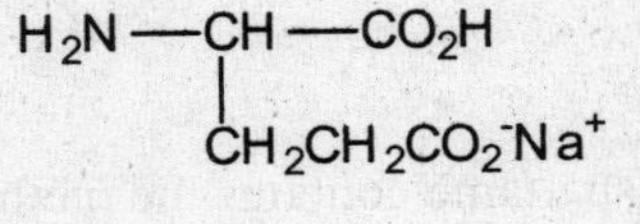

In MSG, the acidic proton from the carboxylic acid in the R group is lost, allowing formation of the ionic compound.

105.   See Figures 21.30 and 21.31 of the text for examples of the cyclization process.

D-Ribose   D-Mannose

107.   The aldohexoses contain six carbons and the aldehyde functional group. Glucose, mannose, and galactose are aldohexoses. Ribose and arabinose are aldopentoses since they contain five carbons with the aldehyde functional group. The ketohexose (six carbons + ketone functional group) is fructose, and the ketopentose (five carbons + ketone functional group) is ribulose.

109.   A disaccharide is a carbohydrate formed by bonding two monosaccharides (simple sugars) together. In sucrose, the simple sugars are glucose and fructose, and the bond formed between these two monosaccharides is called a glycoside linkage.

111.   The α and β forms of glucose differ in the orientation of a hydroxy group on one specific carbon in the cyclic forms (see Figure 21.31 of the text). Starch is a polymer composed of only α-D-glucose, and cellulose is a polymer composed of only β-D-glucose.

113.   A chiral carbon has four different groups attached to it. A compound with a chiral carbon is optically active. Isoleucine and threonine contain more than the one chiral carbon atom (see asterisks).

isoleucine   threonine

115.   Only one of the isomers is optically active. The chiral carbon in this optically active isomer is marked with an asterisk.

117.   They all contain nitrogen atoms with lone pairs of electrons.

119.   The complementary base pairs in DNA are cytosine (C) and guanine (G), and thymine (T) and adenine (A). The complementary sequence is C-C-A-G-A-T-A-T-G.

121.   Uracil will hydrogen bond to adenine.

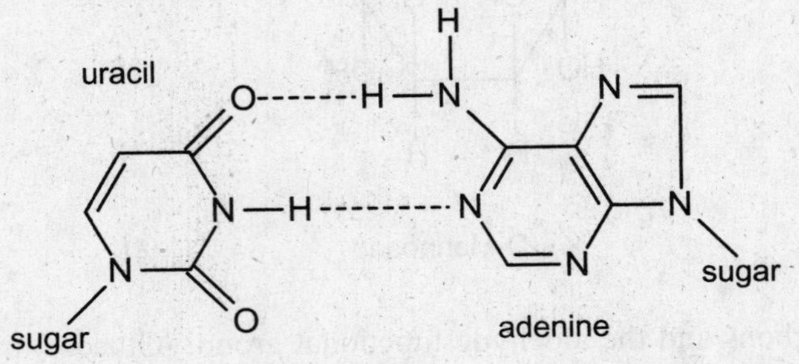

123.   Base pair:

RNA   DNA

A . . . . T

G C

C . . . . G

U . . . . A

a.   Glu: CTT, CTC   Val: CAA, CAG, CAT, CAC

Met: TAC   Trp: ACC

Phe: AAA, AAG   Asp: CTA, CTG

b.   DNA sequence for trp-glu-phe-met:

ACC – CTT – AAA – TAC
       or   or
     CTC   AAG

c.   Due to glu and phe, there is a possibility of four different DNA sequences. They are:

ACC–CTT–AAA–TAC or ACC–CTC–AAA–TAC or

ACC–CTT–AAG –TAC or ACC–CTC–AAG–TAC

d.

$$T-A-C-C-T-G-A-A-G$$
$$\underbrace{\qquad}_{met}\;\underbrace{\qquad}_{asp}\;\underbrace{\qquad}_{phe}$$

e. TAC–CTA–AAG; TAC–CTA–AAA; TAC–CTG–AAA

125. A deletion may change the entire code for a protein, thus giving an entirely different sequence of amino acids. A substitution will change only one single amino acid in a protein.

Additional Exercises

127. $CH_3CH_2CH_2CH_2CH_2CH_2CH_2COOH + OH^- \rightarrow CH_3-(CH_2)_6-COO^- + H_2O$; octanoic acid is more soluble in 1 M NaOH. Added OH^- will remove the acidic proton from octanoic acid, creating a charged species. As is the case with any substance with an overall charge, solubility in water increases. When morphine is reacted with H^+, the amine group is protonated, creating a positive charge on morphine ($R_3N + H^+ \rightarrow R_3NH^+$). By treating morphine with HCl, an ionic compound results that is more soluble in water and in the bloodstream than is the neutral covalent form of morphine.

129.

To substitute for the benzene ring hydrogens, an iron(III) catalyst must be present. Without this special iron catalyst, the benzene ring hydrogens are unreactive. To substitute for an alkane hydrogen, specific wavelengths of light must be present. For toluene, the light-catalyzed reaction substitutes a chlorine for a hydrogen in the methyl group attached to the benzene ring.

131. $85.63 \text{ g C} \times \dfrac{1 \text{ mol C}}{12.011 \text{ g C}} = 7.129 \text{ mol C}; \quad 14.37 \text{ g H} \times \dfrac{1 \text{ mol H}}{1.0079 \text{ g H}} = 14.26 \text{ mol H}$

Because the moles of H to moles of C ratio is 2:1 (14.26/7.129 = 2.000), the empirical formula is CH_2. The empirical formula mass $\approx 12 + 2(1) = 14$. Because $4 \times 14 = 56$ puts the molar mass between 50 and 60, the molecular formula is C_4H_8.

The isomers of C_4H_8 are:

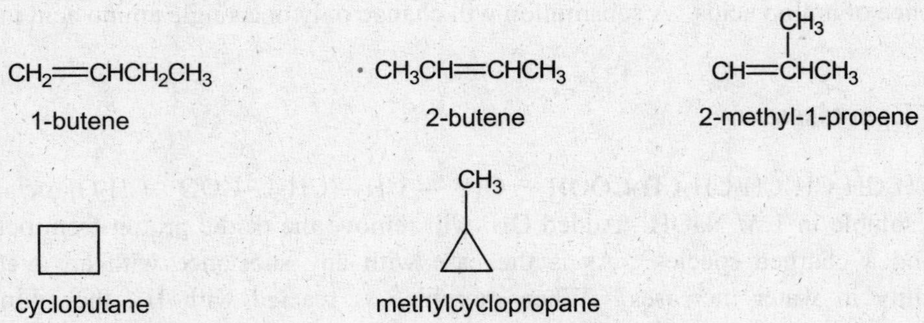

$CH_2{=\!=}CHCH_2CH_3$ $CH_3CH{=\!=}CHCH_3$ $\overset{\displaystyle CH_3}{\underset{}{CH{=\!=}CHCH_3}}$

1-butene   2-butene   2-methyl-1-propene

cyclobutane   methylcyclopropane

Only the alkenes will react with H_2O to produce alcohols, and only 1-butene will produce a secondary alcohol for the major product and a primary alcohol for the minor product.

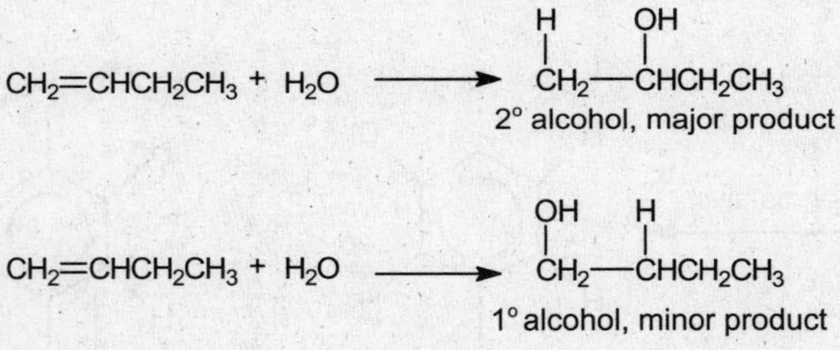

$CH_2{=\!=}CHCH_2CH_3 + H_2O \longrightarrow \overset{\text{H}\quad\text{OH}}{CH_2{-}CHCH_2CH_3}$

2° alcohol, major product

$CH_2{=\!=}CHCH_2CH_3 + H_2O \longrightarrow \overset{\text{OH}\quad\text{H}}{CH_2{-}CHCH_2CH_3}$

1° alcohol, minor product

2-Butene will produce only a secondary alcohol when reacted with H_2O, and 2-methyl-1-propene will produce a tertiary alcohol as the major product and a primary alcohol as the minor product.

133.   a.   The bond angles in the ring are about 60°. VSEPR predicts bond angles close to 109°. The bonding electrons are closer together than they prefer, resulting is strong electron-electron repulsions. Thus ethylene oxide is unstable (reactive).

b.   The ring opens up during polymerization; the monomers link together through the formation of O—C bonds.

$\left(\text{O}{-}CH_2CH_2{-}\text{O}{-}CH_2CH_2{-}\text{O}{-}CH_2CH_2\right)_n$

135.

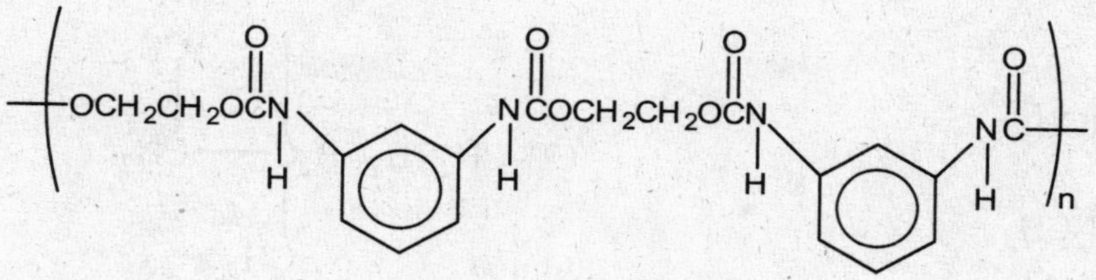

137.   The structures, the types of intermolecular forces exerted, and the boiling points for the compounds are:

butanoic acid, 164°C
LD + dipole + H-bonding

1-pentanol, 137°C
LD + H-bonding

pentanal, 103°C
LD + dipole

n-hexane, 69°C
LD only

All these compounds have about the same molar mass. Therefore, the London dispersion (LD) forces in each are about the same. The other types of forces determine the boiling-point order. Because butanoic acid and 1-pentanol both exhibit hydrogen-bonding (H–bonding) interactions, these two compounds will have the two highest boiling points. Butanoic acid has the highest boiling point since it exhibits H-bonding along with dipole-dipole forces due to the polar C=O bond.

139.   When addition polymerization of monomers with C=C bonds occurs, the backbone of the polymer chain consists of only carbon atoms. Because the backbone contains oxygen atoms, this is not an addition polymer; it is a condensation polymer. Because the ester functional group is present, we have a polyester condensation polymer. To form an ester functional group, we need the carboxylic acid and alcohol functional groups present in the monomers. From the structure of the polymer, we have a copolymer formed by the following monomers.

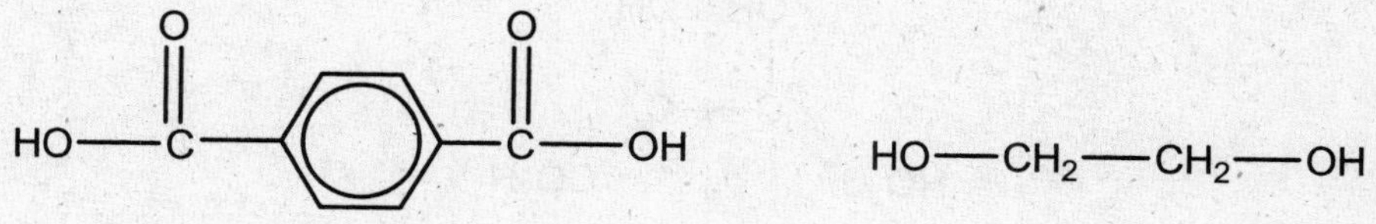

141.   a.

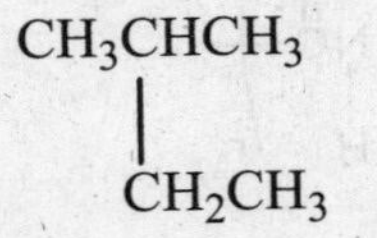

CH_3CHCH_3
|
CH_2CH_3

The longest chain is four carbons long. The correct name is 2-methylbutane.

b.

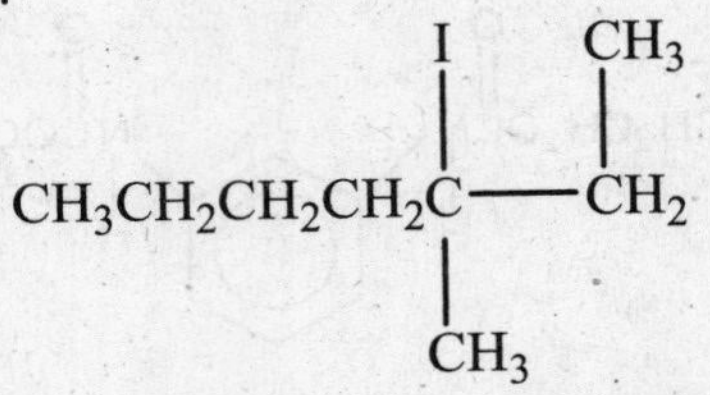

$CH_3CH_2CH_2CH_2\overset{\displaystyle I}{\underset{\displaystyle CH_3}{C}}\!\!-\!\!\overset{\displaystyle CH_3}{CH_2}$

The longest chain is seven carbons long and we would start the numbering system at the other end for lowest possible numbers. The correct name is 3-iodo-3-methylheptane.

c.

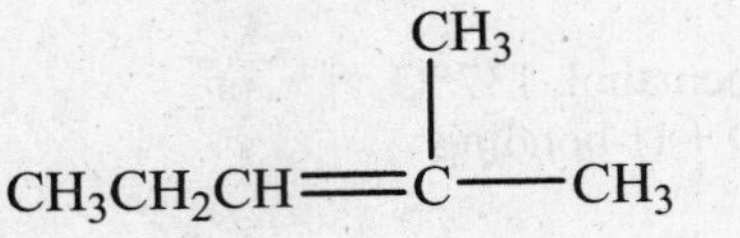

$CH_3CH_2CH\!\!=\!\!\overset{\displaystyle CH_3}{C}\!\!-\!\!CH_3$

This compound cannot exhibit cis–trans isomerism since one of the double-bonded carbons has the same two groups (CH_3) attached. The numbering system should also start at the other end to give the double bond the lowest possible number. 2-Methyl-2-pentene is correct.

d.

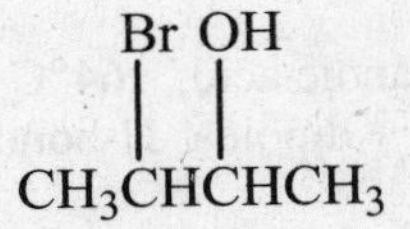

$CH_3\overset{\displaystyle Br}{C}\!H\overset{\displaystyle OH}{C}\!HCH_3$

The OH functional group gets the lowest number. 3-Bromo-2-butanol is correct.

143. $\Delta G = \Delta H - T\Delta S$; for the reaction, we break a P–O and O–H bond and form a P–O and O–H bond, so $\Delta H \approx 0$ based on bond dissociation energies. ΔS for this process is negative (unfavorable) because positional probability decreases. $\Delta G = \Delta H - T\Delta S$; thus, $\Delta G > 0$ due to the unfavorable ΔS term, and the reaction is not expected to be spontaneous.

145.   a.   Even though this form of tartaric acid contains two chiral carbon atoms (see asterisks in the following structure), the mirror image of this form of tartaric acid is superimposable. Therefore, it is not optically active. One way to identify optical activity in molecules with two or more chiral carbon atoms is to look for a plane of symmetry in the molecule. If a molecule has a plane of symmetry, then it is never optically active. A plane of symmetry is a plane that bisects the molecule where one side exactly reflects the other side.

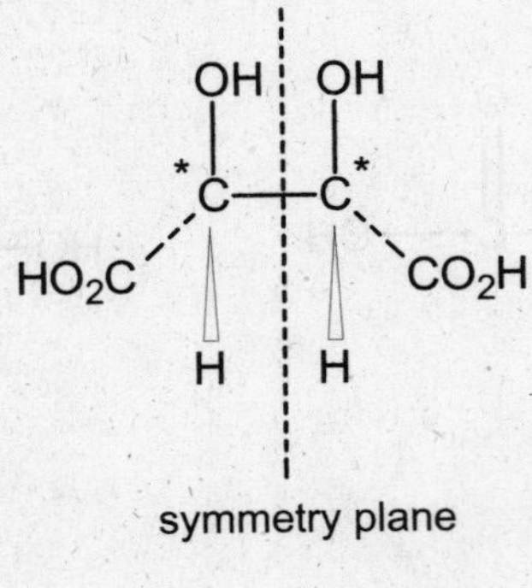

b. The optically active forms of tartaric acid have no plane of symmetry. The structures of the optically active forms of tartaric acid are:

mirror

These two forms of tartaric acid are nonsuperimposable.

147.   For the reaction:

$$^+H_3NCH_2CO_2H \rightleftharpoons 2\,H^+ + H_2NCH_2CO_2^- \qquad K_{eq} = 7.3 \times 10^{-13} = K_a(-CO_2H) \times K_a(-NH_3^+)$$

$$7.3 \times 10^{-13} = \frac{[H^+]^2[H_2NCH_2CO_2^-]}{[^+H_3NCH_2CO_2H]} = [H^+]^2, \;\; [H^+] = (7.3 \times 10^{-13})^{1/2}$$

$$[H^+] = 8.5 \times 10^{-7}\,M; \;\; pH = -\log\,[H^+] = 6.07 = \text{isoelectric point}$$

149. $4.2 \times 10^{-3}\,g\,K_2CrO_7 \times \dfrac{1\,mol\,K_2Cr_2O_7}{294.20\,g} \times \dfrac{1\,mol\,Cr_2O_7^{2-}}{mol\,K_2Cr_2O_7} \times \dfrac{3\,mol\,C_2H_5OH}{2\,mol\,Cr_2O_7^{2-}}$

$$= 2.1 \times 10^{-5}\,mol\,C_2H_5OH$$

$$n_{breath} = \frac{PV}{RT} = \frac{\left(750.\,mm\,Hg \times \dfrac{1\,atm}{760\,mm\,Hg}\right) \times 0.500\,L}{\dfrac{0.08206\,L\,atm}{K\,mol} \times 303\,K} = 0.0198\,mol\,breath$$

$$\text{Mol \% } C_2H_5OH = \frac{2.1 \times 10^{-5}\,mol\,C_2H_5OH}{0.0198\,mol\,total} \times 100 = 0.11\% \text{ alcohol}$$

Challenge Problems

151.   Assuming 1.000 L of the hydrocarbon (C_xH_y), then the volume of products will be 4.000 L and the mass of products ($H_2O + CO_2$) will be:

$$1.391\,g/L \times 4.000\,L = 5.564\,g\,products$$

$$\text{Mol } C_xH_y = n_{C_xH_y} = \frac{PV}{RT} = \frac{0.959 \text{ atm} \times 1.000 \text{ L}}{\dfrac{0.08206 \text{ L atm}}{K \text{ mol}} \times 298 \text{ K}} = 0.0392 \text{ mol}$$

$$\text{Mol products} = n_p = \frac{PV}{RT} = \frac{1.51 \text{ atm} \times 4.000 \text{ L}}{\dfrac{0.08206 \text{ L atm}}{K \text{ mol}} \times 375 \text{ K}} = 0.196 \text{ mol}$$

$C_xH_y + \text{oxygen} \rightarrow x\ CO_2 + y/2\ H_2O;$ setting up two equations:

$(0.0392)x + 0.0392(y/2) = 0.196$ (moles of products)

$(0.0392)x(44.01 \text{ g/mol}) + 0.0392(y/2)(18.02 \text{ g/mol}) = 5.564 \text{ g}$ (mass of products)

Solving: $x = 2$ and $y = 6$, so the formula of the hydrocarbon is C_2H_6 which is ethane.

153.   a.   The three structural isomers of C_5H_{12} are:

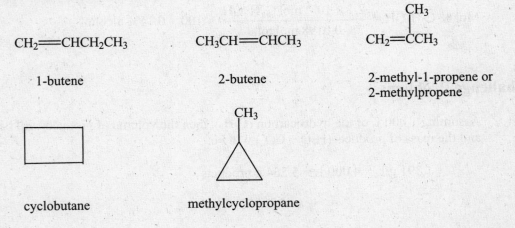

n-Pentane will form three different monochlorination products: 1-chloropentane, 2-chloropentane, and 3-chloropentane (the other possible monochlorination products differ by a simple rotation of the molecule; they are not different products from the ones listed). 2,2-Dimethylpropane will only form one monochlorination product: 1-chloro-2,2-dimethylpropane. 2-Methylbutane is the isomer of C_5H_{12} that forms four different monochlorination products: 1-chloro-2-methylbutane, 2-chloro-2-methylbutane, 3-chloro-2-methylbutane (or we could name this compound as 2-chloro-3-methylbutane), and 1-chloro-3-methylbutane.

b.   The isomers of C_4H_8 are:

The cyclic structures will not react with H_2O; only the alkenes will add H_2O to the double bond. From Exercise 21.58, the major product of the reaction of 1-butene and H_2O is 2-butanol (a 2° alcohol). 2-Butanol is also the major (and only) product when 2-butene and H_2O react. 2-Methylpropene forms 2-methyl-2-propanol as the major product when reacted with H_2O; this product is a tertiary alcohol. Therefore, the C_4H_8 isomer is 2-methylpropene.

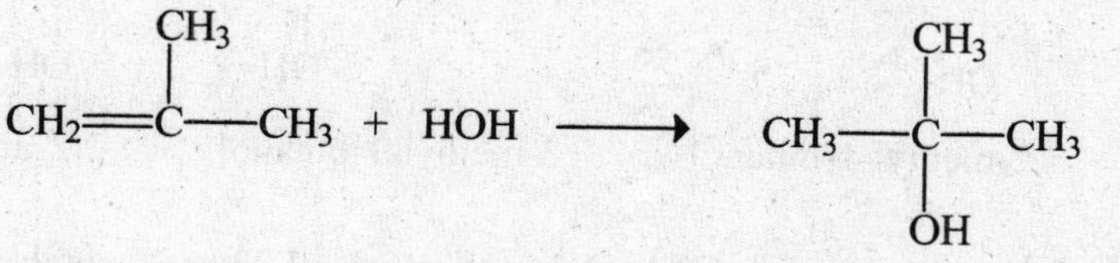

$$CH_2{=}\overset{\overset{\displaystyle CH_3}{|}}{C}{-}CH_3 \; + \; HOH \longrightarrow CH_3{-}\overset{\overset{\displaystyle CH_3}{|}}{\underset{\underset{\displaystyle OH}{|}}{C}}{-}CH_3$$

2-methyl-2-propanol (a 3° alcohol, 3 R groups)

c. The structure of 1-chloro-1-methylcyclohexane is:

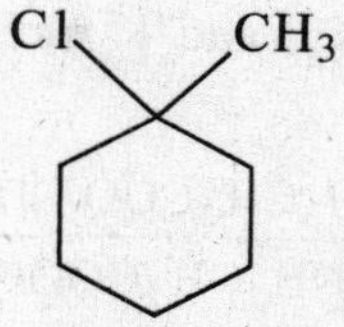

The addition reaction of HCl with an alkene is a likely choice for this reaction (see Exercise 21.58). The two isomers of C_7H_{12} that produce 1-chloro-1-methylcyclohexane as the major product are:

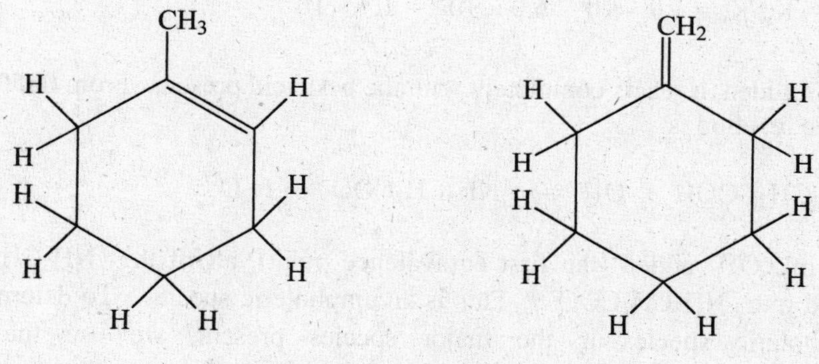

d. Working backwards, 2° alcohols produce ketones when they are oxidized (1° alcohols produce aldehydes, and then carboxylic acids). The easiest way to produce the 2° alcohol from a hydrocarbon is to add H_2O to an alkene. The alkene reacted is 1-propene (or propene).

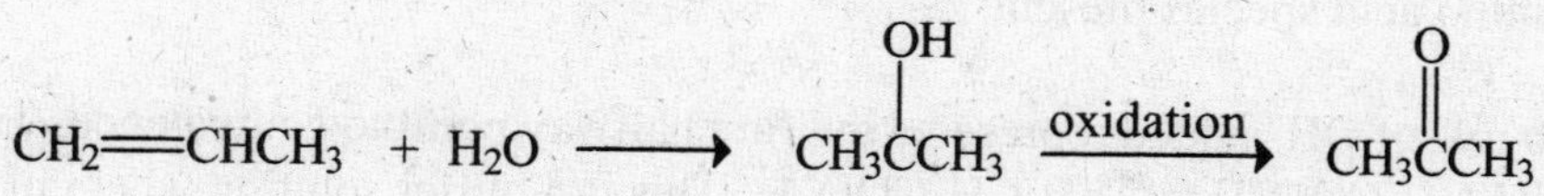

$$CH_2{=}CHCH_3 \; + \; H_2O \longrightarrow CH_3\overset{\overset{\displaystyle OH}{|}}{C}CH_3 \xrightarrow{\text{oxidation}} CH_3\overset{\overset{\displaystyle O}{\|}}{C}CH_3$$

propene   acetone

e. The $C_5H_{12}O$ formula has too many hydrogens to be anything other than an alcohol (or an unreactive ether). 1° alcohols are first oxidized to aldehydes, and then to carboxylic acids. Therefore, we want a 1° alcohol. The 1° alcohols with formula $C_5H_{12}O$ are:

1-pentanol   2-methyl-1-butanol   3-methyl-1-butanol   2,2-dimethyl-1-propanol

There are other alcohols with formula $C_5H_{12}O$, but they are all 2° or 3° alcohols, which do not produce carboxylic acids when oxidized.

155.   Treat this problem like a diprotic acid (H_2A) titration. The K_{a_1} and K_{a_2} reactions are:

$$^+NH_3CH_2COOH \rightleftharpoons {}^+NH_3CH_2COO^- + H^+ \quad K_{a_1} = \frac{[^+NH_3CH_2COO^-][H^+]}{[^+NH_3CH_2COOH]} = 4.3 \times 10^{-3}$$

$$^+NH_3CH_2COO^- \rightleftharpoons NH_2CH_2COO^- + H^+ \quad K_{a_2} = \frac{[NH_2CH_2COO^-][H^+]}{[^+NH_3CH_2COO^-]}$$

$$K_{a_2} = K_w/K_b = 1.0 \times 10^{-14}/6.0 \times 10^{-5} = 1.7 \times 10^{-10}$$

As OH^- is added, it reacts completely with the best acid present. From 0–50.0 mL of OH^- added, the reaction is:

$$^+NH_3CH_2COOH + OH^- \rightarrow {}^+NH_3CH_2COO^- + H_2O$$

At 50.0 mL OH^- added (the first equivalence point), all of the $^+NH_3CH_2COOH$ has been converted into $^+NH_3CH_2COO^-$. This is an amphoteric species. To determine the pH when an amphoteric species is the major species present, we use the formula $pH = (pK_{a_1} + pK_{a_2})/2$. From 50.1–100.0 mL of OH^- added, the reaction that occurs is:

$$^+NH_3CH_2COO^- + OH^- \rightarrow NH_2CH_2COO^- + H_2O$$

100.0 mL of OH^- added represents the second equivalence point where $H_2NCH_2COO^-$ is the major amino acid species present.

a. 25.0 mL of OH^- added represents the first halfway point to equivalence. Here, $[^+NH_3CH_2COOH] = [^+NH_3CH_2COO^-]$. This is a buffer solution where $pH = pK_{a_1}$.

    At 25.0 mL OH^- added:

$$pH = pK_{a_1} = -\log(4.3 \times 10^{-3}) = 2.37$$

50.0 mL of OH⁻ added represents the first equivalence point. Here, $^+NH_3CH_2COO^-$ is the major amino acid species present. This is amphoteric species. At 50.0 mL OH⁻ added:

$$pH = \frac{pK_{a_1} + pK_{a_2}}{2} = \frac{2.37 - \log(1.7 \times 10^{-10})}{2} = \frac{2.37 + 9.77}{2} = 6.07$$

75.0 mL of OH⁻ added represents the second halfway point to equivalence. Here, $[^+NH_3CH_2COO^-] = [NH_2CH_2COO^-]$ and $pH = pK_{a_2}$. At 75.0 mL OH⁻ added:

$$pH = pK_{a_2} = -\log(1.7 \times 10^{-10}) = 9.77$$

b.

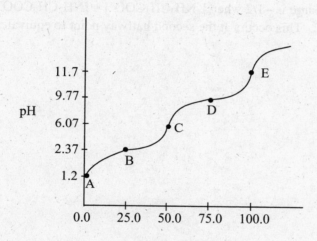

Volume of OH⁻ added

The major amino acid species present are:

point A (0.0 mL OH⁻): $^+NH_3CH_2COOH$

point B (25.0 mL OH⁻): $^+NH_3CH_2COOH$ and $^+NH_3CH_2COO^-$

point C (50.0 mL OH⁻): $^+NH_3CH_2COO^-$

point D (75.0 mL OH⁻): $^+NH_3CH_2COO^-$ and $NH_2CH_2COO^-$

point E (100.0 mL OH⁻): $NH_2CH_2COO^-$

c. The various charged amino acid species are:

$^+NH_3CH_2COOH$ net charge = +1

$^+NH_3CH_2COO^-$ net charge = 0

$NH_2CH_2COO^-$ net charge = −1

The net charge is zero at the pH when the major amino acid species present is the $^+NH_3CH_2COO^-$ form; this happens at the first equivalence point in our titration problem. From part a, this occurs at pH = 6.07 (the isoelectric point).

d. The net charge is +1/2 when = $[^+NH_3CH_2COOH] = [^+NH_3CH_2COO^-]$; net charge = (+1 + 0)/2 = +1/2. This occurs at the first halfway point to equivalence, where pH = pK_{a_1} = 2.37.   The net charge is −1/2 when $[^+NH_3CH_2COO^-] = [NH_2CH_2COO^-]$; net charge = (0 − 1)/2 = −1/2. This occurs at the second halfway point to equivalence, where pH = pK_{a_2} = 9.77.